The Elusive West

AND THE CONTEST FOR

EMPIRE,

1713–1763

D1483443

MER DE L'OU

C. Blanc

C. Mendocin

CALIFORNIE

Cibola

N

M

Mer

MER

The Elusive West

AND THE CONTEST FOR

EMPIRE,

1713–1763

Paul W. Mapp

Published for the
Omohundro Institute of Early American History
and Culture, Williamsburg, Virginia,
by the University of North Carolina Press, Chapel Hill

THE OMOHUNDRO INSTITUTE OF EARLY AMERICAN HISTORY AND CULTURE *is sponsored jointly by the College of William and Mary and the Colonial Williamsburg Foundation. On November 15, 1996, the Institute adopted the present name in honor of a bequest from Malvern H. Omohundro, Jr.*

Library of Congress Cataloging-in-Publication Data
Mapp, Paul W.
The elusive West and the contest for empire, 1713–1763 / Paul W. Mapp.
p. cm.
Includes bibliographical references and index.
ISBN 978-0-8078-3395-7 (cloth : alk. paper)
ISBN 978-1-4696-0086-4 (pbk. : alk. paper)
1. North America—Colonization. 2. West (U.S.)—Colonization. 3. North America—Discovery
and exploration—Spanish. 4. North America—Discovery and exploration—British. 5. North
America—Discovery and exploration—French. 6. North America—History—Colonial period,
ca. 1600–1775. 7. Seven Years' War, 1756-1763—Causes. 8. Imperialism—History—18th century.
I. Omohundro Institute of Early American History & Culture. II. Title.
E46.M37 2011
970.01—dc22
2010027108

colth 15 14 13 12 11 5 4 3 2 1
paper 17 16 15 14 13 5 4 3 2 1

THIS BOOK WAS DIGITALLY PRINTED

To Anne Florence, Anouk,
and whatever the future holds

ACKNOWLEDGMENTS

Though writing a long and complicated book sometimes seems a solitary pursuit, I've found, like all the authors I know, that I've relied on a great deal of help from a great many people. I owe many thanks.

Princeton's Peter Brown, William Jordan, and Anthony Grafton showed an undergraduate from Oregon what historians could do and be. They inspired me to become a historian myself. The way they talked about early European history in class shaped the way I've thought about early American history in *The Elusive West*.

Harvard's Bernard Bailyn, David Hancock, Laurel Ulrich, John Coatsworth, Patrice Higonnet, Tom Bisson, Michael McCormick, and my late and much-missed adviser Ernest May provided examples of intellectual rigor and excellence, encouraged and supported my research in its fragile early phases, and generally whipped me into shape. The project simply could never have got started without their finely judged combination of guidance and latitude, aid and exhortation. The Andrew W. Mellon Foundation and the Jacob K. Javits Fellowship Program paid for my graduate studies with these scholars.

The Fulbright and Krupp Foundations and the Harvard University Graduate Society, Committee on General Scholarships, and Charles Warren Center for Studies in American History financed doctoral research in France and Spain; the Omohundro Institute of Early American History and Culture and the National Endowment for the Humanities did the same for postdoctoral investigations in England. The amiable and patient staffs of the Archives du ministère des affaires étrangères and Archives nationales in Paris, Archivo General de Simancas, Archivo Histórico Nacional in Madrid, Archivo General de Indias in Seville, British Library in London, and National Archives in Kew made historical research productive and pleasurable. François Weil, Hamish Scott, Glyndwr Williams, Gary McDowell, and Pierre Boulle provided invaluable counsel and support while I was conducting archival research.

David Buisseret, the Society for the History of Discoveries, *Terrae Incognitae*, Bradley Bond, and the Louisiana State University Press published essays based on dissertation research and have kindly allowed me to draw on material from those works for *Elusive West*.

A National Endowment for the Humanities postdoctoral fellowship at the Omohundro Institute of Early American History and Culture in Williamsburg, Virginia, made it possible to reconceptualize my dissertation. At the Institute, Ron Hoffman, Fredrika Teute, Chris Grasso, and Bob Gross created a supportive and stimulating environment for scholarship. Fred Anderson, Dorinda Outram, John TePaske, and fellow Institute fellow Ben Mutschler read the dissertation from which *Elusive West* derives and offered important advice about how to move forward.

The College of William and Mary and its Department of History fostered the book and were remarkably supportive of and patient with efforts to make it as strong as possible. The college's enthusiastic, open-minded, and acute students, particularly in seminars on North American exploration, the Lewis and Clark Expedition, and the Seven Years' War, taught me much about the topics I was ostensibly teaching them.

Roy Ritchie and the staff of the Huntington Library and Ted Widmer and the staff of the John Carter Brown Library made possible extensive and essential research and writing. They've created model academic communities where scholars of a range of subjects can come together and teach and learn from one another. I could never have pushed the investigations of *Elusive West* where they needed to go without the opportunity to ponder in Pasadena and Providence.

Library and Archives Canada, Special Collections at the University of Virginia Library, the Library of Congress Geography and Map Division, the John Carter Brown and Bancroft Libraries, the Provincial Archives of Manitoba, the Bridgeman Art Library, and the North Carolina Museum of History made images available for use as illustrations. Jim DeGrand drew maps for the book with skill, speed, and elegance.

Kris Lane, Brett Rushforth, Jim Drake, Max Edelson, Peter Wood, and an anonymous reviewer examined a later version of the manuscript and made crucial suggestions about how to bring it to completion. Fredrika Teute, Gil Kelly, Mendy Gladden, Justin Clement, and especially the indispensably instructive and alert Kathy Burdette of the Institute's book program guided the manuscript through its challenging final steps.

My indulgent and understanding family and friends sustained me in this long endeavor. Anne Florence and Anouk, reminding me always of what's really important, became its purpose.

The book's shortcomings are my own, its virtues the product of collective effort. I thank all for all their help.

CONTENTS

ILLUSTRATIONS

Maps

Plates

ABBREVIATIONS

Journals

AHR
American Historical Review

CLAHR
Colonial Latin American Historical Review

EHR
English Historical Review

FHS
French Historical Studies

HAHR
Hispanic American Historical Review

IHR
International History Review

JAH
Journal of American History

JHG
Journal of Historical Geography

JMH
Journal of Modern History

JWH
Journal of World History

MVHR
Mississippi Valley Historical Review

NMHR
New Mexico Historical Review

PHR
Pacific Historical Review

PSQ
Political Science Quarterly

SWHQ
Southwestern Historical Quarterly

WHQ
Western Historical Quarterly

WMQ
William and Mary Quarterly

Archives

AC
Archives coloniales, Archives nationales, Paris

Add. Mss.
Additional Manuscripts, British Library, London

AGI
Archivo General de Indias, Seville

AGS
Archivo General de Simancas, Valladolid

AHN
Archivo Histórico Nacional, Madrid

CO
Colonial Office Records, National Archives, Kew, Surrey, U.K.

CP
Correspondance politique, Archives des affaires étrangères, Paris

MD
Mémoires et documents, Archives des affaires étrangères, Paris

PRO
Domestic Records of the Public Record Office, Gifts, Deposits, Notes and
Transcripts, National Archives, Kew, Surrey, U.K.

SP
State Papers, National Archives, Kew, Surrey, U.K.

The Elusive West

AND THE CONTEST FOR

EMPIRE,

1713–1763

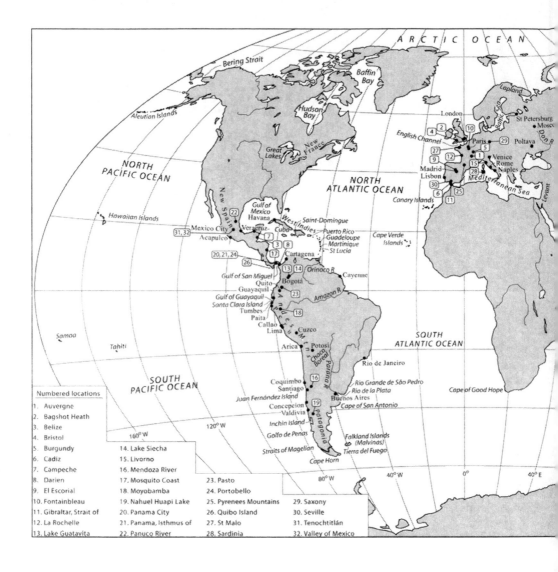

ARCTIC OCEAN

Bering Strait

Baffin Bay

Aleutian Islands

Hudson Bay

Lapland

Baltic Sea

London

St Petersburg
Moscow

NORTH PACIFIC OCEAN

Great Lakes

New France

English Channel

Paris

Poltava
Don R

29

Saxony

4 2 10

27 9 12

1

15

Venice
Rome
Naples

Madrid
Lisbon

5

30 6 25

28

NORTH ATLANTIC OCEAN

Mediterranean Sea

Levant

Hawaiian Islands

22

Gulf of Mexico
Havana
Veracruz

Saint-Domingue

Canary Islands

11

Cape Verde Islands

Mexico City

31, 32

Acapulco

7

West Indies
Cuba

Puerto Rico
Guadeloupe
Martinique
St Lucia

8

20, 21, 24

17

Cartagena

26

13 14

Orinoco R

Cayenne

New Spain

Gulf of San Miguel
Quito
Guayaquil
Gulf of Guayaquil
Santa Clara Island
Tumbes
Paita
Callao
Lima

Bogotá

23

Amazon R

18

Cuzco

Arica

Potosí
Choco Boreal

Rio de Janeiro

SOUTH ATLANTIC OCEAN

Samoa

Tahiti

SOUTH PACIFIC OCEAN

Coquimbo
Santiago

16

Rio Grande de São Pedro
Rio de la Plata

Cape of Good Hope

Juan Fernández Island

Concepcion
Valdivia

19

Buenos Aires
Cape of San Antonio

160° W

120° W

Inchin Island

Golfo de Penas

Straits of Magellan

Falkland Islands
(Malvinas)
Tierra del Fuego

Cape Horn

80° W

40° W

0°

40° E

Andes

Parana

Patagonia

Numbered locations

1. Auvergne
2. Bagshot Heath
3. Belize
4. Bristol
5. Burgundy
6. Cadiz
7. Campeche
8. Darien
9. El Escorial
10. Fontainbleau
11. Gibraltar, Strait of
12. La Rochelle
13. Lake Guatavita

14. Lake Siecha
15. Livorno
16. Mendoza River
17. Mosquito Coast
18. Moyobamba
19. Nahuel Huapi Lake
20. Panama City
21. Panama, Isthmus of
22. Panuco River

23. Pasto
24. Portobello
25. Pyrenees Mountains
26. Quibo Island
27. St Malo
28. Sardinia

29. Saxony
30. Seville
31. Tenochtitlán
32. Valley of Mexico

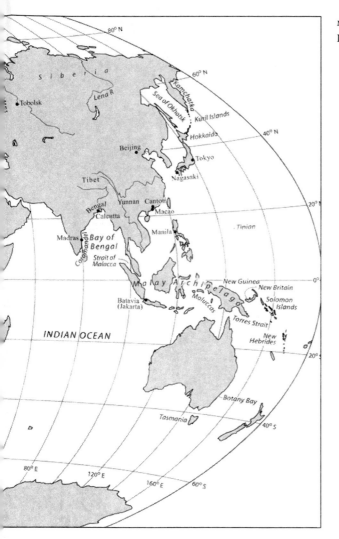

MAP 1. World map.
Drawn by Jim DeGrand

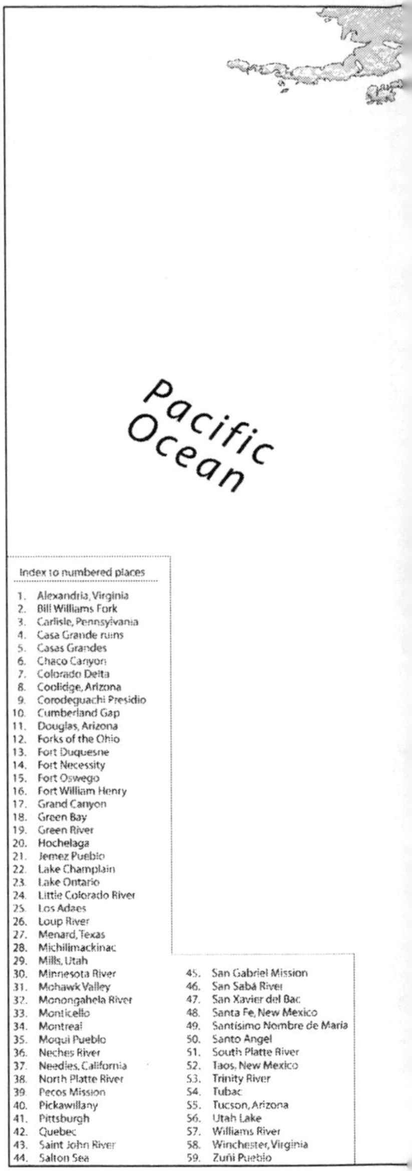

Pacific Ocean

MAP 2. North America.
Drawn by Jim DeGrand

Alaska
Cooper R
Anchorage

Lituya Bay
des Français
Juneau

Inside Passage
Nootka Sound
Vancouver Island
Strait of Juan de Fuca
Puget Sound

Tillamook

Cape Blanco

Trinidad Harbor
Cape Mendocino

n Francisco Bay
Golden Gate
San Francisco
Monterey Bay
Monterey

n Barbara Channel
nta Catalina Island

San Diego Bay
San Diego

Cape Disappointment
(Punta Baja)

Cape San Lucas

Tres Marias Islands

Boothia Peninsula

Great Bear
Lake

Coppermine
River

Melville
Peninsula

Repulse
Bay

Great Slave
Lake

Chesterfield
Inlet

Hudson
Bay

Labrador

Mackenzie River

Lake
Athabasca

Fort
Chipewyan

Churchill

York

Fort
Bourbon

Newfoundland

Fraser River

Saskatchewan R

Fort
Le Pas

Nelson R

Forks of the
Saskatchewan

Lake
Winnipeg

Lake
Manitoba

Winnipeg R

Lake
of the
Woods

New France

Acadia

Louisbourg

Columbia R

Bitterroot Mtns

Oregon

The Dalles

Idaho

Rogue R

Snake R

Rocky Mtns

Knife R

N Dakota
Bismarck

Red River

S Dakota

Black
Hills

Lake Superior

Wisconsin

43
42
20
14

St Lawrence R

Lake Huron

23

81

16

Boston

Massachusetts

Nevada

Great
Basin

Great
Salt
Lake

56

Wyoming

Utah

29

19

33

Nebraska

26

Missouri River

30

18

28

Lake Michigan

New York

12

32

13

41

40

14

3

New York City

Pennsylvania

Maryland

California

Colorado

Platte R

Illinois

Fort
Chartres

Kentucky

Ohio River

10

Appalachian Mtns

1

73

58

N Carolina

Colorado R

35

6

52

Arkansas R

Kansas

Los Angeles

37

12

57

24

21

48

39

New
Mexico

Ozark
Mtns

Arkansas

S Carolina

45

50

2

Arizona

59

Oklahoma

Red River

Charleston

Atlantic
Ocean

44

4

47

35

Gila River

7

8

54

11

9

5

Pecos River

Texas

53

36

Louisiana

Georgia

Florida

Saint Augustine

Baja California

Rio Grande

27

46

49

25

San
Antonio

San José
de Cabo

Sonora

Nueva Vizcaya

Sinaloa

Chihuahua

Zacatecas

Gulf of California

Mississippi
Delta

Mobile
Pensacola

Gulf of
Mexico

Straits of Florida

Havana

Cuba

Yucatan
Peninsula

Santiago de Cuba
St. Domingo

INTRODUCTION

Histories of the Seven Years' War, especially those written in the United States, often begin with George Washington's blunderings in the Ohio Valley in 1754. It's a good place to start. Competing British, French, and Indian claims to lands west of the Appalachians formed one of the principal sources of international tension in the early 1750s, and when Washington's Virginia Regiment and Indian allies made contact with a larger French and Indian force east of the Forks of the Ohio, the sparks thrown off by the collision helped ignite a global conflagration. Later in life, when immersed in adversity, Washington enjoyed the inestimable advantage of having made and recovered from serious mistakes before. Examination of 1754 Ohio Valley events clarifies the causes of a major war and the career of a prominent figure.[1]

Beginning with Washington's march toward the Ohio possesses other virtues as well. One of these concerns his direction of travel and disposition of mind. When not looking precariously down into Jumonville's Glen or nervously up at the wooded slopes around Fort Necessity, Washington was one of those eighteenth-century Anglo-Americans most notably turning his eyes to the west. Informed in part by what he saw on his 1750s treks in the Ohio country, Washington then and later intuited the significance of the lands stretching boundlessly away from the Atlantic and Appalachians. He saw a site of youthful adventure and precocious recognition, a source of lands from which speculation could wring a coveted fortune, and a promising field for Anglo-American expansion. Farsighted as he was, for the purposes of eighteenth-century American and imperial history, Washington didn't see the half of it. His youthful journeys took him in the right direction but covered insufficient distance. The case of another celebrated figure of Seven Years' War history suggests why.[2]

Accounts of the Seven Years' War in North America sometimes finish with Robert Rogers and his frustrated dream of finding a Northwest Passage. This is

---

1. The outstanding modern example of a Seven Years' War history beginning in the Ohio Valley is Fred Anderson, *Crucible of War: The Seven Years' War and the Fate of Empire in British North America, 1754–1766* (New York, 2001).

2. For a modern discussion of Washington's western inclinations, see Joseph J. Ellis, *His Excellency: George Washington* (New York, 2004), 154–156, 209–210.

not a bad place to end. The westward extension of British imperial claims and projects was one of the most significant results of the war, and the quest for a practicable connection between eastern North America and the Pacific would impel American history until the driving of the golden spike in 1869 made an iron reality of a cherished fancy. Rogers presented a proposal to the British government for an expedition in search of a Northwest Passage in 1765. In 1766, in his new position as commander of the former French post at Michilimackinac (at the junction of lakes Michigan and Huron), he dispatched a party in search of it. Rogers himself never made it beyond the Great Lakes, but he did anticipate the course more fortunate explorers like Lewis and Clark would follow. His projects demonstrate, moreover, that at least some participants in the Great War for Empire perceived its stakes in the grandest geographical terms. Their ideas reached beyond the eastern third of North America, running all the way through the continent to the great ocean to the west.[3]

With a century and a half of strong work on the Seven Years' War in North America by Francis Parkman, Lawrence Henry Gipson, Fred Anderson, and many others standing on our shelves, we know a great deal about figures like Washington and Rogers and the conflict in which they fought. Despite the quality, capaciousness, and geographic reach of volumes written by these and other scholars, important aspects of the war remain incompletely understood. The French officials assessing Washington's amateurish mid-1750s forays dearly wanted to avoid war with Britain and seriously questioned the larger significance of the Ohio Valley region where Anglo-French skirmishes were taking place. Yet they responded aggressively to these awkward British incursions, adding to the series of increasingly provocative diplomatic and military actions that would lead to a global struggle. In the years that followed, Rogers was not alone in wondering about the existence of a passage running through the West to the Pacific. If such a passage existed, it could make the area west of the Mississippi and Great Lakes immeasurably valuable. French explorers and officials had long been interested in this possibility. Nevertheless, French statesmen at the end of the Seven Years' War decided to cede France's unconquered trans-Mississippi

3. The literary example of an account ending with Rogers and the Northwest Passage is Kenneth Roberts's 1937 novel, *Northwest Passage* (New York). On Rogers at Michilimackinac, see John R. Cuneo, *Robert Rogers of the Rangers* ([New York], 1987), 177–180, 185–186, 191–193; David A. Armour, ed., *Treason? At Michilimackinac: The Proceedings of a General Court Martial held at Montreal in October 1768 for the Trial of Major Robert Rogers* (Mackinac Island, Mich., 1967), 47–56; Norman Gelb, ed., *Jonathan Carver's Travels through America, 1766–1768: An Eighteenth-Century Explorer's Account of Unchartered Territory* (New York, 1993), 15–24; John Parker, ed., *The Journals of Jonathan Carver and Related Documents, 1766–1770* ([Saint Paul, Minn.], 1976), 12–16.

territories to Spain. Having just fought a desperate war for American empire, France rather oddly gave up the last and potentially most precious continental piece of it. And Spain — despite acute awareness of the dangers of having neighbors like Rogers and Washington — not only refrained from aiding its traditional French ally in the war against British imperial expansion until it was too late to prevent a British victory but also chose to accept France's offer of western Louisiana, thereby cooperating in the removal of the French barrier protecting New Spain from Anglo-American westward expansion.

To address the issues raised by these imperial actions, and to comprehend more fully the war Washington helped start and Rogers helped wage, we need to look more closely at the ideas about American geography people of Washington and Rogers's century held. We need to look farther west than Washington and Rogers could and bring into the story characters far less familiar. As many of the great historians of the Seven Years' War have recognized, significant as eastern North America and the Ohio Valley in particular were in shaping the Seven Years' War, the extent of the war's effects and the scale and dynamics of the imperial rivalry that generated the conflict beckon scholars to a perspective comprehending more than the Great Lakes, the Belle Rivière, and the Atlantic's shores. We need to direct our gaze not just into the trans-Appalachian lands Washington saw from the ridges around the Forks of the Ohio but into the far western regions Rogers wished he could discern from Michilimackinac. For the Seven Years' War in the Americas arose, proceeded, and expired not only in response to events in Atlantic America's Ohio Valley backyard but also as a result of imperial perceptions of present happenings and future possibilities in far-off places like the western shores of Hudson Bay, the forbidding lands around New Mexico, the enticing reaches of trans-Mississippi Louisiana, and the Pacific littoral of North America. To most mid-eighteenth-century Europeans, these areas were entirely unknown. This ignorance of western American geography influenced in unfamiliar and surprising ways the contest for America and empire, and it is to those recondite portions of eighteenth-century North America far to the west of the Ohio Valley that we must now turn.

The Elusive Eighteenth-Century West

Nothing now remained save to investigate the unexplored coast between the Pánuco River and the coast of Florida. . . . For it is believed that there is on that coast a strait leading to the Southern Sea. . . . And if Our Lord God be pleased that we find this strait, it will prove a very good and very short route from the Spice Islands to Your Majesty's realms. . . .

... Even if it is not, many great and rich lands must surely be discovered, where your Caesarean Majesty may be served and the realms and dominions of Your Royal Crown much increased.

—HERNÁN CORTÉS to CHARLES V, 1524

This Indian said that he was the son of a merchant. . . . When he was young, . . . his father used to go into the interior of the land as a merchant, with sumptuous feathers to use as ornaments. In exchange, he brought back a great amount of gold and silver, of which there is much in that land. [Tejo also said] that once or twice he went with [his father] and saw *pueblos* so grand that he liked to compare them with [the Ciudad de] México and its environs. [He said] he had seen seven very grand *pueblos* where there were streets of silver workshops. . . .

. . . The name of the Seven Ciudades and the search for them have endured. [Still] today they have not been reconnoitered.

—PEDRO DE CASTAÑEDA DE NÁJERA, 1560s

It may be asked, why I insert the Mammoth, as if it still existed? I ask in return, why I should omit it, as if it did not exist? . . . Those parts [the northern and western parts of America] still remain in their aboriginal state, unexplored and undisturbed by us, or by others for us. He may as well exist there now, as he did formerly where we find his bones.

—THOMAS JEFFERSON, *Notes on the State of Virginia*, 1787

The object of your mission is to explore the Missouri river, and such principal stream of it, as, by it's [sic] course and communication with the waters of the Pacific Ocean, may offer the most direct and practicable water communication across this continent, for the purposes of commerce.

—THOMAS JEFFERSON to MERIWETHER LEWIS, 1803

The quotations above serve as a reminder of the uncertainty long character-izing European visions of the North American West and a testimonial of the products of desire and imagination with which European dreamers filled these mysterious regions. In the decades after 1492, intoxicated by New World won-ders glimpsed farther south, Spanish visionaries like Cortés were already specu-lating about a watery passage running through the North American continent and toward the Asia of Columbus's reveries. Aspiring conquerors like Coronado were already seeking the fabulous riches rumored to exist beyond the North American horizon. Some three hundred years later, after more than two cen-turies of French and Anglo-American experience exploring, settling, and trad-ing in North America had been added to that of Spain, the United States' lead-ing expert on the West could still imagine mammoths roaming the continent's unknown expanses and a practicable water route extending through western territories to the South Sea. Well into the nineteenth century, North American regions familiar for millennia to western Indians remained unexplored by and unknown to Europeans and Euro-Americans.[4]

During the years after the first Spanish wanderings and conquests, and before the Louisiana Purchase and the Lewis and Clark Expedition, European imperial governments vying for mastery in North America possessed their own visions of western territories and responded in their own fashion to the challenges posed by western geographic obscurities. Competing for empire and coping with geographic ignorance went hand in hand in the early modern Americas, the demands of each shaping the conduct of the other. The potential value of North American resources and routes heightened European interest in locating such treasures and keeping them from imperial rivals. At the same time, lingering uncertainties about western American geography made it difficult for European officials to judge the likelihood that the passages and palaces of explorers' fancies actually existed and therefore to assess the threat posed by competing empires' quest for them.

The decades leading to the great imperial reconfiguration at the end of the Seven Years' War formed an especially consequential phase of this ongoing relationship between European geographic uncertainty and North American imperial rivalry. In the years between the 1713 Treaty of Utrecht and the 1763 Peace of Paris, large parts of the western two-thirds of North America were still untrodden by European scouts, traders, or settlers and only unreliably and incompletely described in the reports trickling back across the Atlantic. As a result, for officials in London, Paris, or Madrid, the Pacific coast north of Baja California, the mountain West north of New Mexico and Sonora, large parts of the western plains, and nearly all the lands west of Hudson Bay remained realms of rumor and imagination rather than of reliable information. The geographers, explorers, and promoters to whom officials turned for counsel speculated that a Northwest Passage might extend from Hudson Bay to the Pacific. A great navigable river might run from a low plateau at the continent's center to a Mediterranean-like extension of the Pacific. Silver deposits rivaling those of Peru and Mexico might lie under western mountains. Chinese or Japanese trading outposts or rich civili-

4. See Hernán Cortés, *Letters from Mexico,* ed. and trans. Anthony Pagden (New Haven, Conn., 1986), 326–327 (for the Spanish text, see Cortés, *Cartas de relación,* ed. Manuel Alcalá [Mexico City, 1988], 199–200); "The Relación de la Jornada de Cíbola, Pedro de Castañeda de Nájera's Narrative, 1560s," in Richard Flint and Shirley Cushing Flint, eds. and trans., *Documents of the Coronado Expedition, 1539–1542* (Dallas, Tex., 2005), 386–387, 437; Thomas Jefferson, *Notes on the State of Virginia,* ed. William Peden (Chapel Hill, N.C., 1954), 53–54; Frank Bergon, ed., *The Journals of Lewis and Clark* (New York, 1989), xxiv. See also Jean-Bernard Bossu, *Travels in the Interior of North America, 1751–1762,* ed. and trans. Seymour Feiler (Norman, Okla., 1962), 104.

zations comparable to those of the Aztecs or Incas might exist somewhere on the northwest coast.

Although it's easy enough to mention such geographic conceptions and misconceptions and to note the continental inexperience giving rise to them, appreciating the historical implications of such notion and nescience is a trickier and rarer matter. Good examples of skirting the issue can be found in many textbook maps displaying European territorial pretensions in North America. Such maps generally use the outline of North America familiar to anyone who has glanced at a modern globe and then mark the territories claimed by the respective European empires in different colors. Though useful as an indication of the continental implications of European ambitions, such maps can give an exaggerated sense of the precision and completeness of European geographic thought. Early modern Europeans generally knew little about much of the North American territory they ostensibly possessed. They could not have produced a sketch of the continent's shape meeting the standards of a modern issue of *National Geographic* and could rarely state with confidence the longitude at which North American land ended and Pacific water began.

In his classic 1952 history of North American exploration, *The Course of Empire*, Bernard DeVoto portrayed the continent in a fashion more clearly indicating the limited extent of eighteenth-century European geographic understanding. In this example, the shaded areas represent North American territories unfamiliar to Europeans from roughly 1728 to 1763. The map conveys a general sense of Europeans' circumscribed geographic horizons. Here again, however, by including a sketch of North American regions with which Europeans were as yet unacquainted, DeVoto filled the geographic unknown with the fruits of modern cartographic comprehension rather than the products of eighteenth-century geographic speculation. He left the viewer without a full appreciation of the doubts unreliable, incomprehensible, and often nonexistent information imposed upon European statesmen.

Not surprisingly, old European maps represent North America differently. Consider this 1752 reproduction of an early-eighteenth-century chart of North America by Guillaume Delisle (1675–1726), Europe's leading geographer for much of his lifetime. The image shows Delisle's attempt to construct a useful representation of western geography on the basis of reports from a variety of Amerindian and European sources. The map is elegant in its way, but eighteenth-century explorers looked in vain for the inland sea dominating Delisle's American West. A related vision appears in a 1708 map by J. B. Nolin. It displays the recurring hope that some kind of Northwest Passage extended from Hudson Bay to a western sea or to the Pacific itself. The 1720 Herman Moll and Guillaume

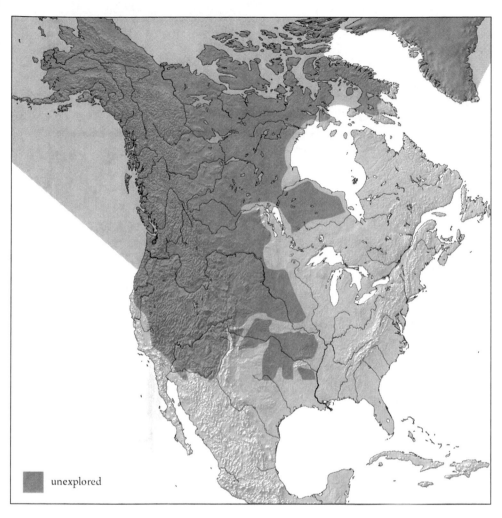

unexplored

MAP 3. European Knowledge of North America, c. 1728–1763.
Drawn by Jim DeGrand, after "Map 13: The World Turned Upside Down,"
in Bernard DeVoto, *The Course of Empire* (1952; rpt. Boston, 1998), 193

Delisle maps following Nolin's are less imaginative, but the large northwest American voids reveal the difficulty Europeans confronted in determining the boundary between American land and Pacific water, and Moll's representation of an insular California shows a misconception that lingered even after Father Francisco Eusebio Kino's late-seventeenth- and early-eighteenth-century explorations had again indicated its falsehood. Perhaps most famously for readers of literature as well as for students of history, Jonathan Swift attached Gulliver's Brobdingnag to the northwest coast of North America in 1726, demonstrating

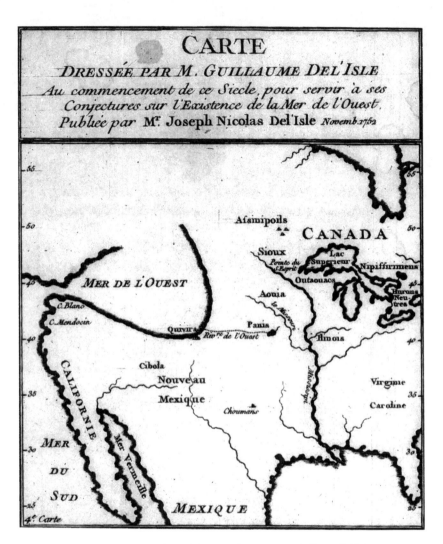

MAP 4. Joseph-Nicolas Delisle, *Carte dressée par M. Guillaume Del'Isle:*
Au commencement de ce siecle, pour servir à ses conjectures sur l'existence de la mer
de l'Ouest. 1752. Library and Archives Canada, NMC 38506

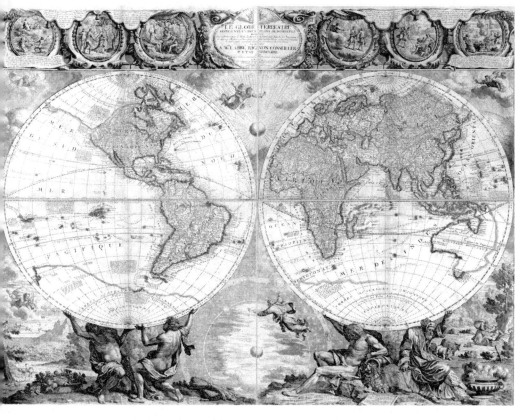

MAP 5. Jean Baptiste Nolin, *Le globe terrestré représenté en deux plans-hemispheres*. 1708. Geography and Map Division, Library of Congress, g3200 ct001475

that the region remained for Europeans a realm of literary fantasy rather than geographic familiarity. The incompleteness, whimsy, and, by conventional standards, inaccuracy of such images offer one indication that the acquisition of reliable and useable information about western American geography was a slow and frustrating process for Europe's early modern empires.[5]

Although European comprehension of the West came slowly, North American territorial conflicts arose quickly. Between 1713 and 1763, when cartogra-

5. On an insular California, see William H. Goetzmann and Glyndwr Williams, *The Atlas of North American Exploration: From the Norse Voyages to the Race to the Pole* (New York, 1992), 38–39, 120–121; Dora Beale Polk, *The Island of California: A History of the Myth* (Spokane, Wash., 1991), 297–307. On Swift and the northwest coast, see James R. Gibson, "The Exploration of the Pacific Coast," in John Logan Allen, ed., *North American Exploration*, II, *A Continent Defined* (Lincoln, Neb., 1997), 328–396, esp. 328; Glyndwr Williams, *The British Search for the Northwest Passage in the Eighteenth Century* (London, 1962), 138.

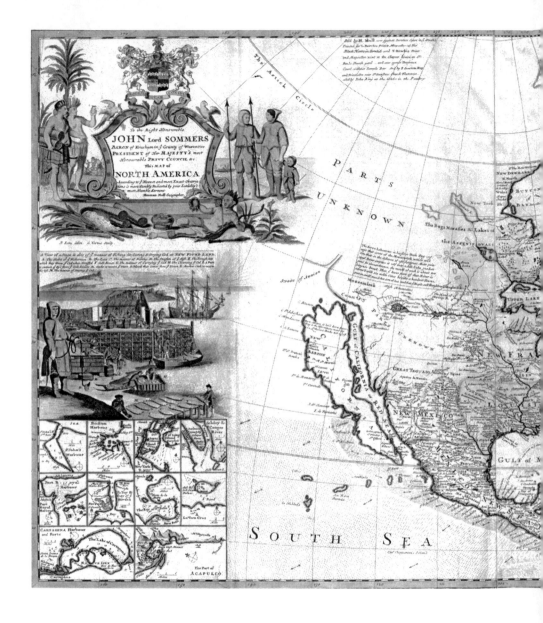

MAP 6. Herman Moll, *Map of North America According to the Newest and Most Exact Observations.* c. 1715. Courtesy, Colonial Williamsburg Foundation

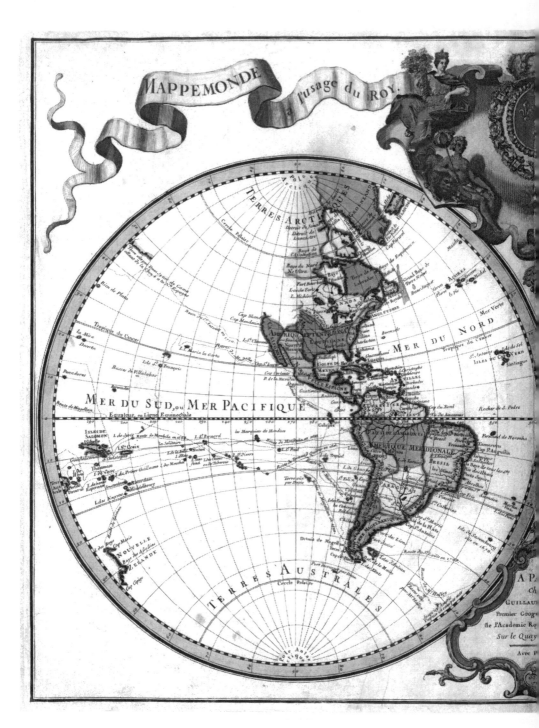

MAPPEMONDE a l'usage du ROY.

TERRES ARCTIQUES

Cercle Polaire

MER DU NORD

Tropique du Cancer

AMERIQUE SEPTENTRIONALE

GOLFE DE MEXIQUE

MER DU SUD, ou MER PACIFIQUE

Equateur, ou Ligne Equinoctiale

ISLES DE SALOMON

AMERIQUE MERIDIONALE

BRESIL

Detroit de Magellan

NOUVELLE ZELANDE

TERRES AUSTRALES

Cercle Polaire

APA
ch
GUILLAU
Premier Geog
fe l'Academie Ro
Sur le Quay

Avec P

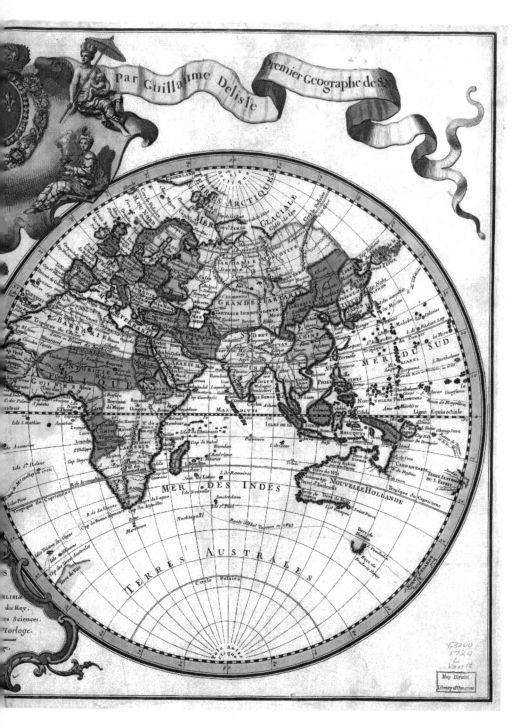

MAP 7. Guillaume Delisle, *Mappemonde a l'usage du roy.* 1720.
Geography and Map Division, Library of Congress, g3200 ct001352

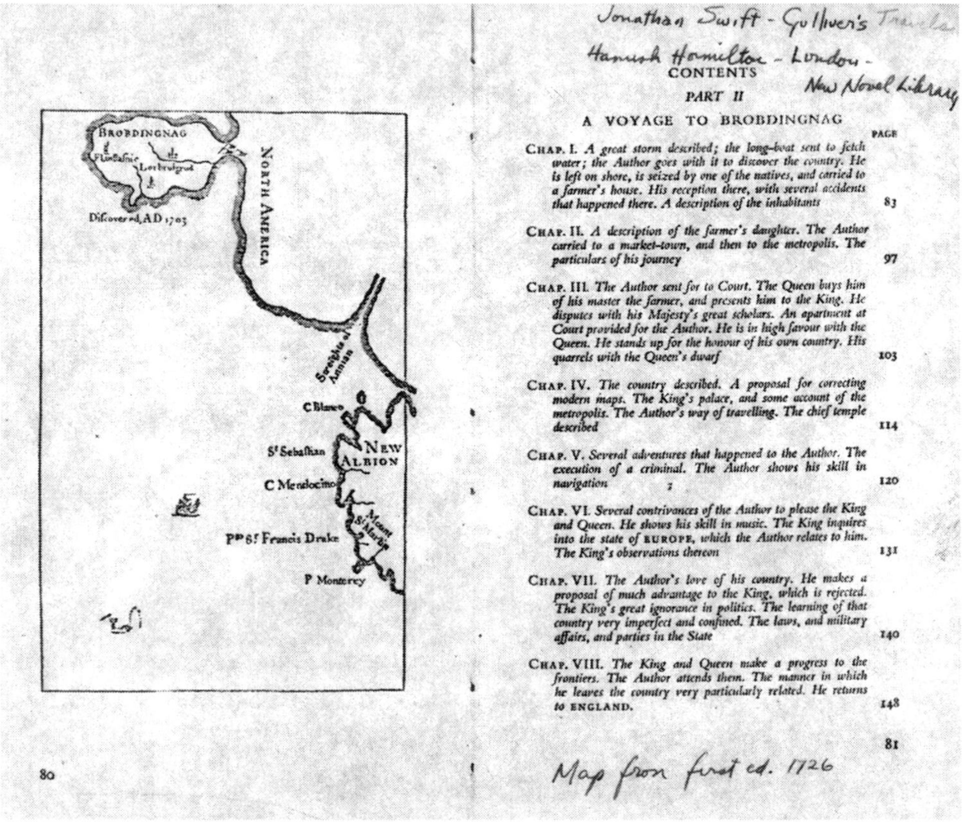

MAP 8. *Brobdingnag.* Jonathan Swift, *Gulliver's Travels* (1726; rpt. London, 1947). Library and Archives Canada, e010771322

phers like Delisle and authors like Swift were generating their western images, the British, French, and Spanish empires were engaged in a series of increasingly violent and decisive struggles for North American dominion. Imperial officials lacked an understanding of western American space sufficient to predict the results of their own and their rivals' decisions concerning the region, but the demands of imperial competition denied European statesmen the luxury of waiting for maps as complete and correct as North American policy was urgent and consequential. Such conditions prompt a question. In a period of increasingly intense imperial rivalry, how did European statesmen make choices involving vast North American territories about which they knew very little?[6]

6. This paragraph and the one preceding it signal an approach placing this study in a different category than numerous early-twenty-first-century treatments of cartographic history and theory.

This question brings to mind and bears on a second one. Surveying 1713 North America, one finds the French, Spanish, and British empires contending for advantage. In 1763, only two of those empires retained continental North American territories. France had surrendered everything, Britain had taken New France and eastern Louisiana, and Spain had extended its recognized American territorial claims north and east to the Mississippi River. Observing these imperial vicissitudes, American historians since Parkman have sought the underlying reasons for them.

Addressing each issue advances understanding of the other. The Seven Years' War furnishes illustrative instances of European officials' making decisions involving the mysterious Far West, and taking that western obscurity into account enables explanation of otherwise puzzling wartime choices.

Historiographical Issues

International rivalry was quite as much a feature of western as of eastern America, even in colonial days, and its story cannot properly be separated from the other. The stage for the contest for the continent was as wide as the hemisphere and its adjacent seas. It was international rivalry that brought into existence as organized communities nearly all the Spanish borderland areas of the Southwest and Pacific coast. These stirring episodes, if treated at all, have been considered only as local history, but they are a part of the general theme. They are no more local history than is the struggle for the St. Lawrence or the Mississippi Valley.

—HERBERT E. BOLTON, 1932

Many of these works follow the late John Brian Harley in criticizing the traditional narrative of cartographic progress and emphasizing the constructed and politically and culturally determined character of early modern maps. See J. B. Harley, *The New Nature of Maps: Essays in the History of Cartography* (Baltimore, 2001). This line of thinking has produced a host of strong books and articles, a number of which will appear in subsequent footnotes. In particular, works inspired by Harley have raised important questions about how cartographers composed maps and how scholars think about cartographic history. They have emphasized the slipperiness of words like "information," "reliable," and "knowledge," noting that different individuals drawing or examining a map can mean equally valid and entirely different things when saying "accurate" or "complete." Scholars have devoted smart volumes to the elusiveness, constructedness, and cultural determinedness of such terms.

This book follows a different tack, starting with distinct questions and being drawn toward dissimilar methods in the attempt to answer them. I concentrate here less on the ideologies and assumptions underlying map construction and more on the way imperial decisions were made with the geographical images and ideas at hand. In this context, the underlying issue for terms like "information," "accuracy," "completeness," and "knowledge" generally involves utility for imperial purposes. More abstractly, a reference to officials' geographic "ignorance" really boils down to their lacking a sense of space and place sufficient for foresight. They knew too little about the West to establish a predictable relationship between actions involving the region and their effects, between the current physical world and future developments.

Where are the manifestos pointing the way to a common colonial history of the continent?

—JOHN MACK FARAGHER, 1994

In combining treatments of western geographic ideas and European imperial rivalry, this book tries to counteract the lingering effects of two tendencies in early American historical writing: Atlantic orientation and topical specialization.[7]

Because of a traditional focus on the Atlantic coast and Basin, the history of early America has often amounted to a history of eastern America. This reflects, in part, the influence of a tenacious argument concerning historical importance. For much of the twentieth century, early American historians wondered whether those seeking to explain the great changes in North America north of Mexico needed to devote sustained attention to the North American West's French, Spanish, and Indian communities. In the United States, at least, the predominant scholarly answer to these questions was no. In the eyes of many, the British Empire and United States' economic, demographic, technological, and political vitality gave them an unrivaled power to transform the North American continent and establish its prevalent cultural characteristics and political institutions. The Spanish borderlands in North America, the French Empire in the Mississippi Valley and western Great Lakes region, and the Indian peoples of the pre-nineteenth-century West seemed colorful but inconsequential historical byways rather than crucial contributors to North American historical evolution. Study of Anglo-America's westward expansion rather than of overwhelmed Indian, French, and Spanish communities was the logical starting point and often the ending point for scholars seeking understanding of major American historical developments.[8]

7. See Bolton's presidential address to the American Historical Association, rpt. in Herbert E. Bolton, "The Epic of Greater America," in Bolton, ed., *Wider Horizons of American History* (New York, 1939), 1–54, esp. 17; John Mack Faragher, comment on James A. Hijiya, "Why the West Is Lost," *WMQ*, 3d Ser., LI (1994), 727–728, esp. 728. An equally prescient and more fully developed call for a more continental approach to early American history can be found in James Axtell, "A North American Perspective for Colonial History," *History Teacher*, XII (1979), 549–562.

8. Canadian historians, acutely aware of the importance of the fur trade, for example, have often been more attentive to far western areas. For an introduction to the issue of western dismissal, see David J. Weber, *The Spanish Frontier in North America* (New Haven, Conn., 1992), 7–8, 355, 359; and see Hijiya's "Why the West Is Lost" and the comments following it, especially the one by Gordon Wood, *WMQ*, 3d Ser., LI (1994), 275–292, 717–754. For examples of scholarly disparagement of the Spanish borderlands' importance, see Earl Pomeroy, "Toward a Reorientation of Western History: Continuity and Environment," *MVHR*, XLI (1955), 579–600, esp. 590; Moses Rischin, "Beyond the Great Divide: Immigration and the Last Frontier," *JAH*, LV (1968), 42–53, esp. 50–51; J. A.

Nonetheless, there have always been scholars wishing to investigate disregarded early North American lands and peoples. Personal interest, the search for understudied topics, and, more recently, cultural and demographic developments in the United States and Canada as well as the inclination to include traditionally marginalized groups in North American historical narratives have led to academic reexamination of historiographically peripheral North American regions. New books are proving successful in their efforts to offer new insights into the West's early history and in their attempts to capture the attention of scholars specializing in other parts of North America. Earlier neglect and the limitations of older approaches have left many gaps in the story of western historical development, and these works are starting to fill them. This book will fill in a few more. The obscurity to which earlier volumes of western history were consigned remains a cautionary tale, however. The ongoing challenge for the authors of early western studies will be sustaining the case for the region's larger significance. They'll need to keep giving those without a personal interest in the area reason to care about what went on there.[9]

Related issues pertain to the place of the Pacific Ocean in North American and Atlantic history. Because much of the eighteenth-century European interest in the Far West revolved around the possible existence of a North American water route to the Pacific, an investigation of European ideas about America's uncharted territories necessarily involves consideration of European inter-

Hawgood, *California as a Factor in World History during the Last Hundred Years* (Nottingham, U.K., 1948), 1.

9. Weber's *Spanish Frontier* and Ramón A. Gutiérrez's *When Jesus Came, the Corn Mothers Went Away: Marriage, Sexuality, and Power in New Mexico, 1500–1846* (Stanford, Calif., 1991) opened the way for a range of excellent studies of the Hispano-Indian marchlands of North America. Richard White's *Middle Ground: Indians, Empires, and Republics in the Great Lakes Region, 1650–1815* (Cambridge, 1991) and Daniel H. Usner, Jr.'s *Indians, Settlers, and Slaves in a Frontier Exchange Economy: The Lower Mississippi Valley before 1783* (Chapel Hill, N.C., 1992) did the same for the Franco-Indian borderlands, regions familiar to many historians from Canada and Louisiana but often obscure for those from other areas. More recent works illuminating the eighteenth-century West include Alan Taylor, *American Colonies* (New York, 2001); Elizabeth A. Fenn, *Pox Americana: The Great Smallpox Epidemic of 1775–82* (New York, 2001); James F. Brooks, *Captives and Cousins: Slavery, Kinship, and Community in the Southwest Borderlands* (Chapel Hill, N.C., 2002); Brett Rushforth, "Savage Bonds: Indian Slavery and Alliance in New France" (Ph.D. diss., University of California, Davis, 2003); Colin G. Calloway, *One Vast Winter Count: The Native American West before Lewis and Clark* (Lincoln, Neb., 2003); Steven W. Hackel, *Children of Coyote, Missionaries of Saint Francis: Indian-Spanish Relations in Colonial California, 1769–1850* (Chapel Hill, N.C., 2005); Kathleen DuVal, *The Native Ground: Indians and Colonists in the Heart of the Continent* (Philadelphia, 2006); Ned Blackhawk, *Violence over the Land: Indians and Empires in the Early American West* (Cambridge, Mass., 2006); Juliana Barr, *Peace Came in the Form of a Woman: Indians and Spaniards in the Texas Borderlands* (Chapel Hill, N.C., 2007); and Pekka Hämäläinen, *The Comanche Empire* (New Haven, Conn., 2008).

est in the South Sea. As with the West, the merits of including discussion of the Pacific Basin in the history of the Atlantic world have, for most scholars, seemed less than self-evident. The most powerful forces shaping the history of the eighteenth-century Atlantic empires were centered on the Atlantic itself. The vast majority of these empires' populations resided on the shores of that ocean; the bulk of their trade and migrations occurred there; the better part of their attention was focused there. Indeed, because Britain and France did not begin systematically exploring the South Sea before James Cook and Louis-Antoine de Bougainville's voyages in the 1760s and 1770s, and because neither nation possessed territory there before the British Empire established a Nootka Sound trading entrepôt and Botany Bay settlement in 1788, many scholars of British and French imperial history have been tempted to dismiss the Pacific's importance for pre-1763 Atlantic world history.[10]

Historians of the Spanish Empire have generally been less prone to disregard the South Sea. The relative paucity of British and French Pacific activities reflected the intensity with which the Spanish Empire tried to reserve the ocean for its own uses. Pacific waters lapped the shores of Peru, Mexico, and the Philippines. Spanish explorers investigated the South Sea long before Cook, and early modern Spanish merchants conducted an annual trans-Pacific Acapulco-Manila trade involving large quantities of silver and silk. Nor were the Spanish always the only Europeans in the ocean. Despite Spain's best efforts, in the years before Cook and Bougainville, French and British pirate, merchant, and naval vessels navigated parts of the South Sea, and the French and British empires sought opportunities to trade with Spain's Pacific colonies or capture the silver-laden Manila galleon. Like early America's western borderlands, however, these European Pacific outposts and forays have often seemed more curious than consequential, and skeptics will need to see more than a few statements about isolated incidents or occasional activities before avowing the historical significance of pre-1763 European engagement with the South Sea.[11]

10. For a recent example of this kind of dismissal, see Nicholas Canny, "Writing Atlantic History; or, Reconfiguring the History of Colonial British America," *JAH*, LXXXVI (1999), 1093–1114, esp. 1111–1113. The works of Glyndwr Williams and John Dunmore offer some of the strongest recent examples of challenges to the traditional neglect of the Pacific in British and French (respectively) imperial historiography.

11. Frequent references to Manila, Chile, Peru, the Pacific, and the Philippines appear in books like Stanley J. Stein and Barbara H. Stein, *Silver, Trade, and War: Spain and America in the Making of Early Modern Europe* (Baltimore, 2000); Guillermo Céspedes del Castillo, *América Hispánica (1492–1898)* (Barcelona, 1994); John Lynch, *Bourbon Spain, 1700–1808* (Oxford, 1989); Geoffrey J. Walker, *Spanish Politics and Imperial Trade, 1700–1789* (Bloomington, Ind., 1979). Relative to the Atlantic, however, the Pacific remains neglected even in Spanish-language historiography. See

The effects of a long-standing Atlantic and east coast orientation are visible in scholarly understanding of the relation between British Pacific encroachments and the American origins of the Seven Years' War. One reason French officials reacted so bellicosely to British Ohio Valley incursions like Washington's was their tendency to view these inroads within a larger context of apparent British designs on Spain's American empire. Pacific projects, like Britain's 1740s expeditions in search of a Northwest Passage from Hudson Bay, helped convince French officials of the reality and immediacy of these grandiose British ambitions. This has been an easy story to miss. Accounts of events leading to the outbreak of Anglo-French hostilities in North America have long focused on the Ohio Valley, Acadia, and the Caribbean. Hundreds of pages of diplomatic documents discuss these regions, and thousands of pages of historical literature have discussed these discussions. Once it was generally understood that historians of the French and British empires in North America considered these the important areas for inquiry, evidence suggesting and even works detailing the importance of European interest in access to the Pacific Ocean could easily be set aside. Scholarship on eastern areas like the Ohio Valley has been so formidable that it has sometimes overshadowed other parts of the continent.[12]

As early American history has often concentrated geographically on the shores of the Atlantic, it has also often tended toward specialization with regard to language, empire, nation, and approach. In the early eighteenth century, four empires and innumerable native communities were active in and around the many territories comprising the Far West. The region, like the continent of which it was a part, was fundamentally multicultural, multinational, and multi-imperial. Such essential diversity notwithstanding, historians — in accordance with the belief that factors internal to individual political, cultural, and local units are sufficient to explain historical change and because of the power of national traditions and the difficulty of mastering multiple languages, source bases, and historiographical traditions — have often chosen to focus on a single political, cultural, or territorial topic. Such concentration can easily create the impression that early American history was the sum of many separate histories taking place on the same continent. In contrast, it was often a snarled struggle for the continent.

Sometimes, a seemingly inconsequential episode — the 1752 arrival of two

---

Carlos Daniel Malamud Rikles, *Cádiz y Saint Malo en el comercio colonial peruano (1698–1725)* (Cadiz, Spain, 1986), 21.

12. This paragraph's treatment of Seven Years' War historiography was inspired by Herbert Butterfield's discussion of Frederick the Great's invasion of Saxony in Butterfield, *The Reconstruction of an Historical Episode: The History of the Enquiry into the Origins of the Seven Years' War* (Glasgow, 1951).

French traders in New Mexico, for instance—can illuminate these struggles. In this case, Spanish reactions to the wayfarers and the complicated diplomatic exchanges that followed can reveal the geopolitical calculations contributing to Spain's bewildering policy of neutrality during the early Seven Years' War. But reaching this conclusion requires attention to three empires' sources detailing Spanish, French, British, and Comanche actions involving localities from Western Europe to Cape Horn to the uncharted upper Colorado River. Much of early American history consisted of "entangled histories" such as this one. Interpreting them entails a wide-ranging approach to early American history.[13]

Such an approach, moreover, should include not just different nations but also different facets of their history. It should relate intellectual developments, particularly in cartography and geography, to imperial policies. It should ask how, or if, the evolving images on maps affected particular decisions in ministries. A good example of why can be seen in accounts of the 1762 French cession of trans-Mississippi Louisiana to Spain. Diplomatic historians treating the cession have often discussed western Louisiana primarily in terms of the Mississippi River areas most densely settled by Europeans and most familiar to French geographers and officials. They have shown less interest in the development and ramifications of French ideas about the unexplored western regions lying within or beyond Louisiana's ill-defined boundaries. Scholars of French exploration and geographic thought, on the other hand, have documented French geographers' frustration with western America's persistent refusal to yield an easy water route to the Pacific, for example, but not the impact of this chagrin on French diplomats wrestling with their British and Spanish counterparts for control of the most valuable North American territories. Diplomatic historians have often overlooked the role of changing European ideas about North American geography, whereas scholars studying these notions have often eschewed detailed consideration of their effects on the geopolitical conceptions of European statesmen. A complete explanation of the 1762 cession requires understanding of both changes in French western ideas over the course of the eighteenth century and the diplomatic maneuvers leading to France's surrender of its colony.

Sources, Topical Boundaries, and Method

Read with questions of geographic uncertainty in mind, the records of Europe's foreign offices enable engagement with the above historiographical issues and

13. See Eliga H. Gould, "Entangled Histories, Entangled Worlds: The English-Speaking Atlantic as a Spanish Periphery," and Jorge Cañizares-Esguerra, "Entangled Histories: Borderland Historiographies in New Clothes," *AHR*, CXII (2007), 764–786, 787–799.

elucidation of the role of the mysterious North American West in international affairs. Europe's foreign ministries have left abundant documentation in the form of correspondence between ministers and ambassadors and reports designed to influence government policy. The functions of foreign ministries included assessing the strategic significance of different areas, analyzing the nature and importance of connections among different geographic regions and political entities, and managing the combustible relations of Europe's competing empires. Explicitly concerned with geopolitics, foreign offices were consequently interested in the implications of changing geographic ideas, and it is possible to reconstruct from the writings of diplomats what the curious images on old maps and the wild notions of expedition promoters meant for the conduct of state.

Documents from three European empires are particularly important. The French, Spanish, and British empires were the great European rivals for North American dominion between 1713 and 1763. Each possessed a European core, established American colonies, and Pacific territories or concerns. Each had to weigh the relative value of its globally scattered interests and consider the connections among its disparate and far-flung imperial objectives. These empires were thinking and acting on the grandest scale, and examination of their perspectives expands our own historical vision. Foreign office records of imperial rivalry make it possible to situate the Far West, the Pacific, and the Seven Years' War on a broad canvas and to establish the historical significance of two neglected regions by relating them to an imperial contest of acknowledged importance.[14]

Of the three empires, that of France merits primary consideration in an investigation of the influence of European ideas about western geography. During the first six decades of the eighteenth century, French geographers and cartographers were generally considered the most advanced in Europe. Canada and Louisiana afforded access to mysterious western lands, and French explorers were trying to extend a European presence into them. Additionally, because the North American territories claimed by France lay between those of Spain and Britain, French officials were thoroughly caught up in the innumerable disputes springing from long and poorly defined imperial boundaries, and they were usefully loquacious in their descriptions of these quarrels.

Less expansionist but more extensive than France's North American territories were Spain's North and South American dominions, extending in the north

14. Lacking American colonies, approaching North America from a different direction, and still having to make the case for its European character, Russia represents a topically related but analytically distinct case, best handled as an influential but peripheral actor in chapters centering on France and Spain.

into Baja California, New Mexico, and Texas. Spanish explorers had reconnoitered parts of the Pacific coast and much of the Southwest in the sixteenth and seventeenth centuries, but the empire conducted little North American exploration between 1713 and 1763. In some cases, records of Spanish discoveries lay unpublicized, concealed, or forgotten in Spain's cavernous imperial archives, and Spanish expeditions had to reinvestigate in the 1770s and 1780s territories sighted or traversed by Spanish explorers in earlier centuries. For the most part, during the decades before 1763, a vulnerable Spanish Empire reacted to the initiatives of its French and British rivals. This only partially diminishes Spain's importance. Its imperial assets, especially the peerless silver mines of Mexico and Peru, constituted one of the great prizes for which Britain and France were competing in their eighteenth-century struggle for colonial and maritime dominance. It was often thought that the addition of Spanish financial resources and military might to the assets of either Britain or France would tip the balance of forces in the fortunate recipient's favor. This made Spain a coveted ally. Furthermore, it was Spanish territory in New Mexico, Mexico, and Peru that French and British explorers were hoping to reach in their futile searches for a Northwest Passage and their more successful journeys westward into North America.[15]

Confined largely to the Atlantic seaboard and a few Hudson Bay outposts, Britain exhibited less interest in the North American Far West than did France and Spain; less, but not none. In the 1740s, Britain launched a series of expeditions in search of a Northwest Passage from Hudson Bay. These efforts elicited an apprehensive French reaction. More generally, the burgeoning population and expansive inclinations of Britain's North American colonies and the Spanish American designs of British merchants and imperialists alarmed French and Spanish officials with the prospect of aggression only the most determined resistance could check. From 1713 to 1763, the British Empire's commercial projects, maritime ventures, demographic vitality, and military victories instigated French and Spanish western policies from a distance, like a looming shadow stirring the suddenly obscured into frightened motion.

These three empires came together as rivals, and they hold together as a subject. They occupy comfortably the central positions in a study of the relation between European geographic ignorance and North American imperial rivalry. Observing the way metropolitan statesmen and diplomats looked out upon a

15. On lost records of Spanish exploration, see Warren L. Cook, *Flood Tide of Empire: Spain and the Pacific Northwest, 1543–1819* (New Haven, Conn., 1973), 3–5; Oakah L. Jones, Jr., "Spanish Penetrations to the North of New Spain," in Allen, ed., *North American Exploration*, II, 7–64, esp. 9, 40–41, 49; Weber, *Spanish Frontier*, 55–56.

distant and obscure North American West enhances understanding of the Seven Years' War and the role of the Far West in it.

It must be admitted, however, that employing this particular approach runs the danger of falling into a classic problem, one perhaps best introduced in a famous passage from Fernand Braudel's preface to his *Mediterranean and the Mediterranean World in the Age of Philip II*:

> When I began it in 1923, it was in the classic and certainly more prudent form of a study of Philip II's Mediterranean policy. My teachers of those days strongly approved of it. For them it fitted into the pattern of that diplomatic history which was indifferent to the discoveries of geography, little concerned . . . with economic and social problems; slightly disdainful towards the achievements of civilization, religion, and also of literature and the arts; . . . shuttered up in its chosen area, this school regarded it beneath a historian's dignity to look beyond the diplomatic files, to real life, fertile and promising. . . .
>
> . . . The historian who takes a seat in Philip II's chair and reads his papers finds himself transported into a strange one-dimensional world, a world of strong passions certainly, blind like any other living world, our own included, and unconscious of the deeper realities of history, of the running waters on which our frail barks are tossed like cockleshells.[16]

Study of pre-twentieth-century diplomacy has never really recovered from Braudel's larger critique, and an investigation of eighteenth-century European statesmen's ideas about western American geography would seem a natural target of it. Perceptual limitations not so different from the blindness Braudel remarked in Philip II constitute the study's main subject. The diplomatic files he found so subject to sterility form its core sources. These papers' authors were more cognizant of larger historical forces' tossing them like cockleshells than Braudel might acknowledge, but diplomats' awareness of these currents made them even more frustrated by their inability to control events. And though diplomatic records partook of a life no less real than any other, Europe's statesmen resided far from the geographic focus of this book and were in many respects more socially and culturally remote from the daily existence of most North American people than they were physically distant from the continent

16. Fernand Braudel, *The Mediterranean and the Mediterranean World in the Age of Philip II*, trans. Siân Reynolds, 2 vols. (New York, 1972), I, 19–21. See also Nathan J. Citino, "The Global Frontier: Comparative History and the Frontier-Borderlands Approach in American Foreign Relations," *Diplomatic History*, XXV (2001), 677–693, esp. 688.

itself. Many of their ideas about the American West and its inhabitants were quite literally outlandish. Examining such antiquated fancies is useful, because the foolish notions of powerful men often weigh on the shoulders of ordinary people, but seeing the Far West through the weak and distorted lens of European diplomatic documents can frustrate readers understandably eager for more immediate apprehension of the region and its inhabitants.

One way to mitigate these problems is to be concerned with matters Braudel saw his masters disdain, to follow lines of inquiry leading away from diplomatic files and Philip's chair — to get out of the office. Another grand old man of Mediterranean historiography, Herodotus, suggests one way of doing this: preceding a focused examination of facets of a great conflict with a broader treatment of pertinent geohistoric, imperial, and cultural issues. In this case, an appropriate and instructive way to begin is by connecting metropolitan geographic ignorance to western circumstances; by responding, that is, to a third question: What was it about the North American Far West and its peoples that made it so hard for Europeans to comprehend the region? Treating this question responds to readers' — and authors' — natural curiosity about the origins of those strange and blinkered notions animating European statecraft, and it also creates an opportunity to introduce material bearing not simply on how the West appeared from Europe but on how it looked to at least some of the Euro-American and Amerindian people in it.

A significant number of documents from the French, Spanish, and British explorers, traders, and missionaries who trod western lands, conferred with their inhabitants, and reported their impressions remain extant. These sources can form a kind of bridge between European perceptions from afar and western conditions underfoot. They include both accounts of what European scouts saw themselves and, critically, some of our only samples of the oral reports and cartographic sketches produced by western Indians. Interpreted and misinterpreted by Europeans, these testimonials and drawings were often the only information about distant lands available to them. Although the documents available are too limited in quantity, perspective, and coverage to satisfy the western enthusiast, they suffice, at least, to point to western probabilities and limit the range of regional possibilities. Moreover, although it is hard to peer beyond what western American sources could see, comparison of documents and events pertaining to an unforthcoming region like the early West with those concerning more yielding areas can illuminate western circumstances. Setting early modern European exploration of the Far West alongside Spanish reconnaissance of sixteenth-century Peru and Mexico and French surveying of eighteenth-century

France, Russia, and China can reveal a good deal about western North America, its peoples, and European difficulties in grasping the region.

Themes and Argument

This book argues that the ambiguity of western Indian geographic information explains European geographic uncertainty regarding much of the continent, and this uncertainty shaped the disposition of North American territory. The mechanism through which the influence of uncertainty operated was the entanglement of doubt about distant lands' contents with worries about rival polities' intentions. The indeterminacy of each magnified unease about the other, complicating statesmen's deliberations. The geographical matter connecting the study's issues and actors was European interest in the potential value of the North American Far West, especially the possibility that some kind of practicable water route extended through the West to the Pacific. Shifting combinations of western geographic apprehension and anticipation contributed to the causes, course, and consequences of the Seven Years' War and thus to the great reconfiguration of North American empire during the second half of the eighteenth century.

The argument develops in five parts, with the parts arranged in loose chronological order and the chapters using an analytical approach to issues of empire and geographic understanding.

The first three parts treat the question of why Europeans cared about, but found it so hard to comprehend, the North American West. Part 1 contrasts *Part 1.* *Spanish* Spanish use of indigenous imperial structures to apprehend large parts of South and Central America with the difficulties the want of such political cohesion created for Spanish reconnaissance in western North America. Part 2 establishes *Part 2* *Silver* the interest in Spanish American silver that made the French and British empires eager to obtain access to Spain's Pacific possessions. It examines Franco-Peruvian commerce during the War of the Spanish Succession and shows how diplomats' efforts to forestall conflicts over this trade led to renewal, in the 1713 Treaty of Utrecht, of the traditional exclusion of non-Spanish shipping from the *Part 3* South Sea. Part 3 introduces optimistic French visions of the West in the first half of the eighteenth century. It contrasts the communicative difficulties and western Indian disunity hindering French exploration during this period with the imperial projects forwarding French geographic undertakings in the North Pacific and East Asia. It finds, in the Far West, Amerindian geographic ambiguity explaining European geographic confusion.

The last two parts concentrate on the implications of European uncertainty about western American geography for the Seven Years' War. They identify unresolved historical problems pertaining to the conflict and use consideration of European notions about the Far West to address them. Part 4 treats the origins and early years of the war. It argues that 1740s and 1750s British Pacific designs in Hudson Bay and elsewhere helped bring France into war with Britain. It contends, further, that simultaneous Spanish concerns about both British Pacific encroachments by way of southern South America and a possible French descent on the Pacific by way of southwestern rivers reinforced the Spanish inclination to remain neutral during the early years of the Seven Years' War, depriving France of crucial assistance in its struggle with Britain. Part 5 considers the vast North American territorial transfers at the end of the Seven Years' War. It argues that, in the years preceding the 1762 French cession of western Louisiana, French officials grew increasingly dubious about the potential value of unexplored North American territories, and this skepticism formed a precondition for the sacrifice of the French colony. Spanish officials accepted the colony because of fear that French diplomats would cede it, and access to the southwestern river route French explorers had been seeking, to Britain. Part 5 closes with the observation that the French and British empires, despite their divergent fortunes at the end of the Seven Years' War, were engaged in a comparable reevaluation of the value of North American territories.

Despite their best efforts to learn from the kind of experiences discussed in this book, Spanish, French, and British officials would all find in the decades following 1763 that no amount of rethinking would generate an easy solution to the dilemmas of eighteenth-century empire.

PART I.

THE SPANISH EMPIRE
AND THE ELUSIVE WEST

1

PEOPLES AND TERRAIN,
DIFFICULTIES AND DISAPPOINTMENTS

Who has yet pretended to define how much of America is included in Brazil, Mexico, or
Peru? It is almost as easy to divide the Atlantic Ocean by a line, as clearly to ascertain the
limits of those uncultivated, uninhabitable, unmeasured regions.
  —SAMUEL JOHNSON, "Observations on the Present State of Affairs," 1756

Spain initiated early modern Europe's engagement with western North American geography, and so it is with the Spanish Empire that a study of the influence
of western geographic ideas properly begins. From the vantage point of the early
twenty-first century, with the results of the last two and a half centuries of geographic investigation close at hand, it is remarkable how little early- and mid-
eighteenth-century Spanish officials knew about the North American continent
their empire had been colonizing since 1519.We can more completely and easily
understand the effects of their uncertainty if we first pay some attention to the
extent of and reasons for this Spanish geographic ignorance.[1]

The two French traders briefly alluded to in the Introduction furnish a good
starting point. On August 6, 1752, Jean Chapuis and Louis Feuilli, two intrepid
but unfortunate travelers from the Illinois country, arrived at the Pecos Mission
on the outskirts of New Mexico. Like the scattering of French scouts who had
preceded them, Chapuis and Feuilli represented an immediate opportunity for
and a potential danger to New Mexico. For the inhabitants of a colony at the
northern extremity of Spain's American empire, French traders offered the possibility of a new source of European goods, perhaps of better quality and lower
price than those lugged by mules up the eighteen-hundred-mile dirt track from
Mexico City. On the other hand, for Spanish officials in a cautious or conscien-
tious frame of mind, French travelers challenged the mercantilist goals of reserv-

1. See Samuel Johnson, "Observations on the Present State of Affairs, 1756," in *The Yale Edition of
the Works of Samuel Johnson, X, Political Writings*, ed. Donald J. Greene (New Haven, Conn., 1977),
184–196, esp. 189.

ing imperial markets for Spanish goods and merchants. More seriously, routes reconnoitered by French explorers one year and employed by French traders the next might carry French soldiers in the future. On this occasion, caution prevailed over opportunism, and New Mexico governor Tomás Vélez Cachupín gave the French visitors an unpleasant welcome. He had Chapuis and Feuilli arrested, relieved of their goods, and escorted to Santa Fe. He questioned the two merchants and then dispatched them to Mexico City. The viceroy of New Spain shipped them to a Cadiz cell in 1754.

Questions Spanish officials put to Chapuis and Feuilli suggest that their arrival was so alarming in part because their successful journey to New Mexico demonstrated a familiarity with North American regions about which the Spanish government knew frustratingly little. When interrogators asked Chapuis and Feuilli the distance "from the presidio of Illinois to this capital of Santa Fe," "the conditions of the journey," and "whether the trade of" New Mexico "with Canada . . . could or could not be made with ease," they were not simply assessing the travelers' intentions and capacities. They were trying also to obtain information about the plains east of New Mexico and north of the Texas missions from Frenchmen who had recently been there. For the Spanish, the region remained, as the 1755 instructions to the new viceroy of New Spain put it, "the unknown country lying between our populated provinces and the western extremity of Louisiana."[2]

2. Questions to Chapuis and Feuilli from "Council of the Indies to His Majesty, Madrid, Nov. 27, 1754," and "Declaration of Juan [Jean] Chapuis, Frenchman," in Alfred Barnaby Thomas, [ed. and trans.], *The Plains Indians and New Mexico, 1751–1778: A Collection of Documents Illustrative of the History of the Eastern Frontier of New Mexico* (Glendale, Calif., 1940), 82–89, 103–106, esp. 83–84, 105. For the viceroy's instructions, see Julian de Arriaga, "Instruccion reservada que trajo el marqués de las Amarillas, recibida del Exmo. Sr. D. Julian de Arriaga, ministro de Indias," June 30, 1755, in *Instrucciones que los vireyes de Nueva España dejaron a sus sucesores. . . .* (Mexico City, 1867), 94–103, esp. 97: "El país incógnito que média entre las provincias que tenemos pobladas y la extremidad occidental de la Luisiana." These instructions are discussed in Council of the Indies, "Expediente sobre la aprehension que Dn Jacinto de Barrios . . . ," Oct. 22, 1756, AGI, Guadalajara 329, 345–366. For other references to the mysterious plains, see "The Diary of Juan de Ulibarri to El Cuartelejo, 1706," "Governor Cuerbó Reports the Return of the Picuríes, 1706," and "Diary of the Campaign of Governor Antonio de Valverde against the Ute and Comanche Indians, 1719," in Alfred Barnaby Thomas, ed. and trans., *After Coronado: Spanish Exploration Northeast of New Mexico, 1696–1727; Documents from the Archives of Spain, Mexico, and New Mexico* (Norman, Okla., 1935), 59, 62, 77, 79, 128; "Don Francisco Cuervo y Valdez to His Majesty, Santa Fé, October 18, 1706," in Charles Wilson Hackett, ed., *Historical Documents Relating to New Mexico, Nueva Vizcaya, and Approaches Thereto, to 1773*, comp. Adolph F. Bandelier and Fanny R. Bandelier, 3 vols. (Washington, D.C., 1923–1937), III, 383–384, esp. 383; and marqués de Altamira, Mexico, Jan. 9, 1751, in José Antonio Pichardo, *Pichardo's Treatise on the Limits of Louisiana and Texas: An Argumentative Historical Treatise with Reference to the Verification of the True Limits of the Provinces of Louisiana and Texas*, ed. and trans. Charles Wilson Hackett, trans. Charmion Clair Shelby, 4 vols. (Austin, Tex., 1931), III, 329–333, esp. 330.

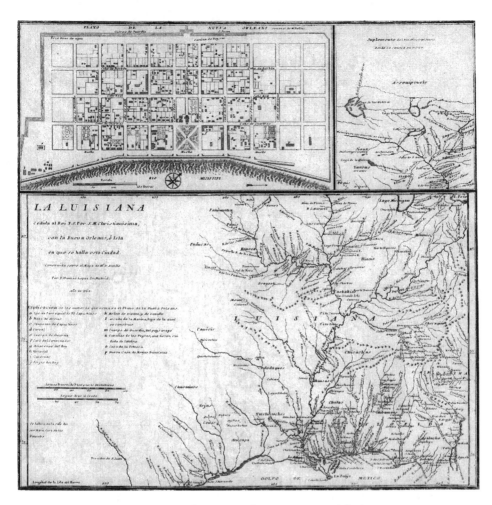

MAP 9. Tomás López de Vargas Machuca, *La Luisiana cedida al rei*. 1762.
Geography and Map Division, Library of Congress, g4010 ar167400

Spanish geographic doubt was not confined to the lands descending from the Rockies to the Mississippi. Uncertainty and error also characterized Spanish ideas about regions north and west of New Mexico. One misconception was that the Pacific or some inland arm of it lay close to the upper Rio Grande Valley or the eastern slopes of the Rockies. In 1750, Vélez Cachupín reported, "From the accounts which the Moaches [Utes] give it is believed that the sea is not very far away" from New Mexico. In 1716, Texas Franciscan missionary Fray Francisco Hidalgo could credibly tell the viceroy of New Spain of indications that "the coast of the South Sea and many vessels" could be seen "from the summit of the range" from which the streams comprising the Missouri flowed. As late as 1775,

Spanish naval explorer Bruno de Hezeta could think little enough of the distance and terrain involved to suggest that a colony at northern California's Trinidad Harbor could "easily deposit . . . trade goods in the interior of New Mexico."[3]

Part of what made the idea of a proximate Pacific credible was the lingering notion that some kind of Northwest Passage — or Strait of Anian, as Spanish geographers often called the elusive water route between the oceans — existed in western North America. New Mexico missionary Fray Alonso de Posada wrote, in a 1686–1687 report on New Spain's northern provinces, of the "gulf" and "Strait of Anian" lying in the region beyond the mountains "east and northeast" of Santa Fe and extending from the North Pacific to Labrador. In 1792, more than a century after Posada's reference to the Strait of Anian and more than a decade after Captain James Cook's exploration had rendered the existence of such a strait unlikely, Spanish captains Dionisio Alcalá Galiano and Cayetano Valdés set out to search inside the Strait of Juan de Fuca for "any communication with the Atlantic by way of the Bays of Hudson, Baffin, et cetera."[4]

More generally, in the mid-eighteenth century, the entire coastline north of modern Oregon's Cape Blanco remained a mystery for Spain. Spanish ships regularly sailed within view of upper and Baja California. Winds occasionally drove them close enough to the lands north of California to see or, in the most dire circumstances, to touch them. Nonetheless, Spanish geographers received too little reliable information from these voyages to form confident or accurate views of the northwest coast. They were uncertain, for instance, how far north familiar landmasses reached. As Jesuit procurator general Gaspar Rodero wrote from Madrid in 1737, "According to modern geographers, the full extent of the Californias is not known." Where charts failed, speculation arose. The great southwestern explorer Father Francisco Eusebio Kino opined in 1710 that northwestern North America would provide a "convenient land route to Asia," sug-

3. Tomás Vélez Cachupín to Juan Francisco de Güemes y Horcasitas [conde de Revilla Gigedo], Mar. 8, 1750, in Pichardo, *Pichardo's Treatise*, ed. and trans. Hackett, trans. Shelby, III, 327; "Fray Francisco Hidalgo to the Viceroy, November 4, 1716," in Mattie Austin Hatcher, trans., "Descriptions of the Tejas or Asinai Indians, 1691–1722, III," *SWHQ*, XXXI (1927), 60; Herbert K. Beals, ed. and trans., *For Honor and Country: The Diary of Bruno de Hezeta* (Portland, Ore., 1985), 72–73.

4. S. Lyman Tyler and H. Darrel Taylor, eds. and trans., "The Report of Fray Alonso de Posada in Relation to Quivira and Teguayo," *NMHR*, XXXIII (1958), 284–314, esp. 305–306; Fray Alonso de Posadas, "Informe á S. M. sobre las tierras de Nuevo Méjico, Quivira y Teguayo," in Cesáreo Fernández Duro, *Don Diego de Peñalosa y su descubrimiento del reino de Quivira* (Madrid, 1882), 53–67, esp. 66; "Instructions from Alejandro Malaspina," and "Instructions from Viceroy Revillagigedo," in John Kendrick, trans., *The Voyage of Sutil and Mexicana, 1792: The Last Spanish Exploration of the Northwest Coast of America* (Spokane, Wash., 1991), 39–48, 49–54, esp. 41, 50. On the origins of the phrase "Strait of Anian," see W. Michael Mathes, "The Early Exploration of the Pacific Coast," in John Logan Allen, ed., *North American Exploration, I, A New World Disclosed* (Lincoln, Neb., 1997), 400–451, esp. 412.

gesting that hazy North Pacific territories referred to as "Jesso" and "Company Land" were sufficiently close or connected to each other and mainland North America to furnish mostly solid footing to "Great Tartary" and "Great China."[5]

More basic than the question of how far "the Californias" extended was that of what the vaguely defined entity designated "California" was. Here, too, misconceptions lingered in the mid-eighteenth century. Attached to the pleasing idea of a fabulous island off North America's west coast, geographers and explorers kept finding ways to separate California from the continent. Though Kino's turn-of-the-eighteenth-century lower Colorado River reconnaissance had confirmed Baja California's peninsularity, the informed and accomplished Spanish explorer and Jesuit missionary Jacobo Sedelmayr was still asserting in the mid-1740s that "our knowledge is not certain" with regard to the question of "whether California be island or continent."[6]

While exploration along the lower Colorado was slowly illuminating the character of Baja California, the upper course of the Southwest's great river remained obscure. In the late 1690s and early 1700s, Kino journeyed along the Gila River upstream from its junction with the Colorado and along the Colorado downstream from this confluence to the sea. He left the upper waters of the Colorado untouched. By 1744, Sedelmayr had ascended the Colorado to its junction with the Williams River and could write with confidence about the entire course of the Gila. For the upper Colorado, he could only repeat and reflect

5. "Gaspar Rodero's Report to Philip V on California, 1532–1736: Description and Exploration of the Country, Account of the Progress of the Enterprise," in Ernest J. Burrus, ed. and trans., *Jesuit Relations: Baja California, 1716–1762* (Los Angeles, Calif., 1984), 184–202, esp. 184–185. For the original Spanish, see "Informe del P. Rodero sobre California (1737)," in Francisco María Picolo, *Informe del estado de la nueva Cristiandad de California, 1702, y otros documentos,* ed. Ernest J. Burrus (Madrid, 1962), 279–300, esp. 280. Eusebio Francisco Kino, *Kino's Historical Memoir of Pimería Alta: A Contemporary Account of the Beginnings of California, Sonora, and Arizona . . . ,* ed. and trans. Herbert Bolton, 2 vols. (Cleveland, Ohio, 1919), II, 259–260. For a comparable French view, see Jean-Bernard Bossu, *Travels in the Interior of North America, 1751–1762,* ed. and trans. Seymour Feiler (Norman, Okla., 1962), 210–213.

6. Kino, *Kino's Historical Memoir of Pimería Alta,* ed. and trans. Bolton, I, 329–354, II, 87–88; Jacobo Sedelmayr, "Relación, 1746," in Peter Masten Dunne, trans., *Jacobo Sedelmayr: Missionary, Frontiersman, Explorer in Arizona and Sonora: Four Original Manuscript Narratives, 1744–1751* ([Tucson], Ariz., 1955), 15–42, esp. 36; Dora Beale Polk, *The Island of California: A History of the Myth* (Spokane, Wash., 1991), 301; "Royal Cédula of King Philip V on the California Missions, 1744," in Charles W. Polzer and Thomas E. Sheridan, eds. and comps., *The Presidio and Militia on the Northern Frontier of New Spain: A Documentary History,* 2 vols. (Tucson, Ariz., 1986–1997), II, part 1, *The Californias and Sinaloa-Sonora, 1700–1765,* 185–193, esp. 186, 190. Ultimately, Jesuit Fernando Consag's 1746 voyage around the Gulf of California persuaded holdouts like Sedelmayr. See Miguèl Venegas and [Andrés Marcos Burriel], *Noticia de la California, y desu conquista temporal, y espiritual hasta el tiempo presente sacada de la historia manuscrita, formada en Mexico año de 1739 por el Padre Miguèl Venegas . . . ,* 3 vols. (Madrid, 1757), I, 3, II, 552, III, 140–195; Polk, *Island of California,* 324–326.

upon Indian reports "that it issues from a great hollow of the earth and carries down with it corn and corncobs." As late as 1773, Spanish explorer Juan Bautista de Anza was writing of the Colorado, "It has been explored only in the neighborhood of its mouth, and that badly."[7]

In the mid-eighteenth century, a sizeable chunk of the New World remained as enigmatic for frustrated Spanish officials as it was for Spain's envious rivals. In the 1755 instructions to New Spain's incoming viceroy, the marqués de Las Amarillas, minister of the Indies Julián de Arriaga requested composition of a map covering the territories from Louisiana to the Pacific. Arriaga observed that maps previously sent to Spain did not allow a full understanding of the terrain and distances involved "because of the small amount of land they comprehend" ["por el corto terreno que comprenden"]; a replacement, he allowed, would be the "work of many years" ["será obra de muchos años"]. Spanish officials trying to protect their empire from British, French, and Indian challenges found their work made much more difficult by confusion about the geography of what New Spain's earlier viceroy the conde de Revilla Gigedo had referred to as "the vast unknown continent of this northern America."[8]

Spanish geographic nescience of such extent at such a late date is remarkable, as the Spanish Empire would seem to have enjoyed the situation, motivation, and duration requisite for filling in the western voids on its North American maps. The same advanced positions making Spanish missions, settlements, and presidios in provinces like New Mexico, Baja California, Sonora, and Texas vulnerable to attack and difficult to supply also made them potential springboards for western exploration, a role they would fulfill after 1763.

Spain would also seem to have had sufficient reason for wanting to investi-

7. Sedelmayr, "Relación, 1746," in Dunne, trans., *Jacobo Sedelmayr*, 20–21, 28; Juan Bautista de Anza to Antonio María de Bucareli y Ursúa, Mar. 7, 1773, in Herbert Eugene Bolton, ed. and trans., *Anza's California Expeditions*, V, *Correspondence* (Berkeley, Calif., 1930), 57–67, esp. 62. Regarding the upper course of the Colorado, Kino could only opine, "Since this Colorado River . . . carries so much water, it must be that it comes from a high and remote land, as is the case with the other large volumed rivers of all the world and terraqueous globe" (Kino, *Kino's Historical Memoir of Pimería Alta,* ed. and trans. Bolton, II, 253).

8. Arriaga, "Instruccion reservada que trajo el marqués de las Amarillas," in *Instrucciones que los vireyes de Nueva España dejaron a sus sucesores,* 97; translations in "The King to the Viceroy, the Marquis of Las Amarillas, Aranjuez, June 30, 1755," in Charles Wilson Hackett, "Policy of the Spanish Crown regarding French Encroachments from Louisiana, 1721–1762," in George P. Hammond, [ed.], *New Spain and the Anglo-American West: Historical Contributions Presented to Herbert Eugene Bolton,* I, *New Spain* (Lancaster, Pa., 1932), 107–145, esp. 138–139. Revilla Gigedo quotation in Aug. 6, 1751, AGI, Guadalajara 137: "El basto incognito Contiente de esta Septentrional America." See also Venegas and [Burriel], *Noticia de la California,* III, 297 ("del vasto Continente desconocido de la America desde la California, hasta el Norte"); Kino, *Kino's Historical Memoir of Pimería Alta,* ed. and trans. Bolton, I, 212, 358, II, 74, 144, 224, 232, 238, 258, 264.

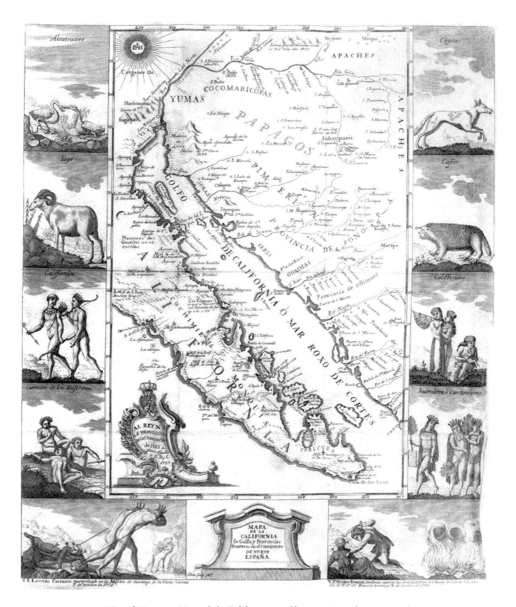

MAP 10. Miguèl Venegas, *Mapa de la California su golfo, y provincias fronteras en el continente de Nueva España*. 1757. Courtesy of the John Carter Brown Library at Brown University

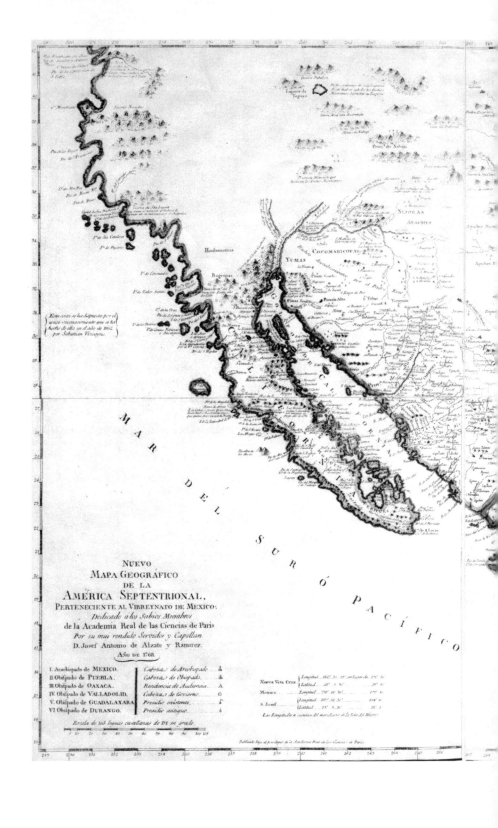

NUEVO
MAPA GEOGRÁFICO
DE LA
AMÉRICA SEPTENTRIONAL,
PERTENECIENTE AL VIRREYNATO DE MEXICO;
Dedicado á los Sabios Miembros
de la Academia Real de las Ciencias de Paris
Por su muy rendido Servidor y Capellan
D. Josef Antonio de Alzate y Ramirez.
AÑO DE 1768.

I. Arzobispado de MEXICO.
II. Obispado de PUEBLA.
III. Obispado de OAXACA.
IV. Obispado de VALLADOLID.
V. Obispado de GUADALAXARA.
VI. Obispado de DURANGO.

MAP 11. José Antonio de Alzate y Ramírez, *Nuevo mapa geográfico de la América septentrional.* 1768. Courtesy of The Bancroft Library, University of California, Berkeley

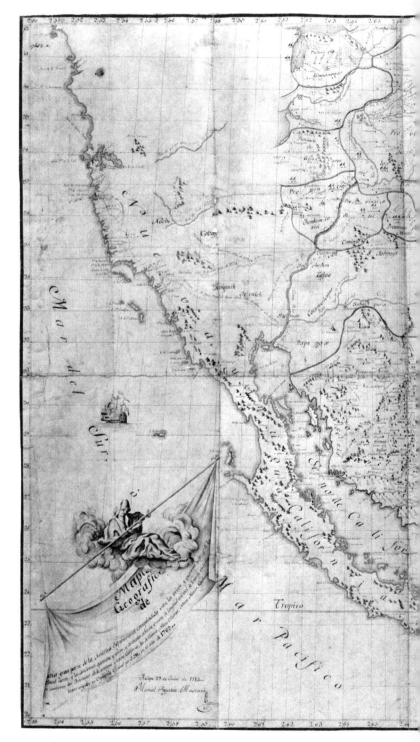

MAP 12. Manuel Agustín
Mascarò, *Mapa geografico
de una gran parte de la
America septentrional*. 1782.
Courtesy of The Bancroft
Library, University of
California, Berkeley

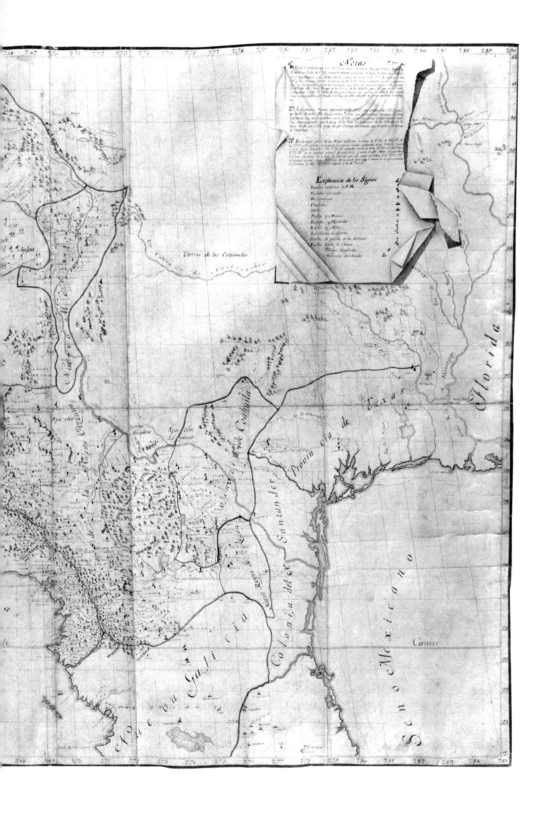

MAP 13. Tomás López de Vargas Machuca, *Mapa de América, sujeto à las observaciónes astronomicas.* c. 1783. Courtesy of the John Carter Brown Library at Brown University

gate the still-uncharted regions of the North American West before 1763. For nervous officials, the possible existence of a Strait of Anian might mean a potential avenue of advance toward the Pacific, the empire's jealously guarded "Spanish Lake." For aspiring conquistadors, the West might offer new realms to conquer. A new Tenochtitlán or an American Tokyo might lie a few hundred leagues beyond Santa Fe or Sonora. Missionaries might find new souls to save and new sites for essaying isolated utopian communities. Slavers might find new bodies to seize, governors new enemies to forestall, traders lucrative sources of or markets for goods.

In addition to apparent motive and suitable position, Spain had also had, in contrast with relative newcomers to the West like France and Britain, time. By the 1750s, Spain had been active in the New World for more than 250 years. The results of its western exploration might have been limited, but the efforts and accomplishments of its explorers had been considerable. Francisco Vásquez de Coronado's (1540–1542) and Juan de Oñate's (1598–1605) *entradas* had investigated the upper Rio Grande Valley, the Grand Canyon, the lower Colorado River, and the trans-Pecos plains. Ships from Juan Rodríguez Cabrillo's and Sebastián Vizcaíno's expeditions might have sailed two thousand miles up the coast from Acapulco in 1543 and 1603. In the seventeenth and early eighteenth centuries, Spanish missionaries and scouts reconnoitered southern Arizona and Texas, and Spanish parties rode into areas comprising the modern states of Colorado, Kansas, and Nebraska. Nonetheless, for want of means or resolve, Spanish explorers before 1763 failed to anticipate the achievements of their Enlightenment successors by following the Pacific coast to Alaska, pushing well beyond Baja California, lower Arizona, New Mexico, and the North Platte and returning with records clear and copious enough to establish and explicate their accomplishments.

This bounded reach of pre-1763 Spanish western reconnaissance contrasts not only with the actual achievements of Spain's later Enlightenment expeditions but also with the exploits of early modern Spanish scouts in other parts of the Americas. In the sixteenth century, Spanish explorers and conquerors threaded the Straits of Magellan, circumnavigated the globe, traversed the Isthmus of Panama, crisscrossed the Andes, strode the causeway to Tenochtitlán, descended the Amazon, and ascended the Paraná. Yet Idaho remained elusive. Consequently, in the centuries after 1492, a Spanish cartographer in Seville or Cadiz or a Spanish king in Madrid, Aranjuez, or the Escorial could sit in a sunny room and direct his gaze from Tierra del Fuego to Cape Blanco and the Rio Grande. He could see, with varying degrees of precision, Mexican and Peruvian territories unimaginable to his medieval predecessors. But his vision failed north of New Mexico's Santa Fe and modern Oregon's Rogue River. What was it about the American West that frustrated the kinds of imperial reconnaissance efforts that succeeded so brilliantly elsewhere and forced Spanish officials to rest weighty decisions on a foundation of dubious information?

The sources most usefully connecting early modern western conditions to Spanish officials' perceptions of the region are the records of Spanish western explorers, missionaries, and officers. These documents display the kinds of information available to early modern Spanish statesmen, and they also disclose a great deal regarding not just how the West looked from Madrid but also how it looked to people who had actually been there. From Cabeza de Vaca's roamings onward, castaways and conquistadors, proselytizers and profiteers, explorers and officials had a great deal to say about western peoples and landscapes.

Such records suffer from grave shortcomings. Their authors had blind spots, biases, and, perhaps worst of all, literary aspirations. Missionaries seeking governmental support for a move north needed to demonstrate natives' susceptibility to conversion. Governors reporting to their superiors wanted to excuse — or obscure — their conduct. Disappointed explorers might have deemed it expedient to tell acquaintances and authorities back home that unendurable privations, insurmountable obstacles, and innumerable Indians had stopped progress rather than admit that surfeit of blisters, longing for loved ones, or lack of resolve had led to ignominious reverse. Spanish scouts could misunderstand what indigenous informants told them, could hear what hope encouraged them to hear, and could see and report what preconceptions and aspirations led them to expect: rumors of villages became reports of cities; a coastline's turn to the northeast became the entrance of a Northwest Passage.

Fortunately, these Spanish western sources do not exist in isolation. Individual Spanish western accounts can be placed alongside one another, as indeed they were by early modern Spaniards, who were no less concerned by evidentiary unreliability than are modern historians. Official investigations often gathered evidence from multiple witnesses. Later explorers and authors criticized their predecessors' wilder claims and unfavorably compared venerable assertions to recent findings. In addition, we can take advantage of the geographic range of early modern Spanish reconnaissance and compare Spanish western writings to those treating other parts of the New World. Though Spanish authors and witnesses from across the Americas often began with similar preconceptions and predispositions, they nevertheless spoke differently about diverse American locales and Amerindian peoples. These variations within the corpus of Spanish sources provide a more reliable indicator of diverse American circumstances than do the words of individual Spanish reports viewed by themselves. In addition to these documentary comparisons, the contentions of Spanish written sources can be checked against the arguments of more than a century of Southwestern and plains archaeological, ethnographic, literary, and historical investigations. Indeed, because existing accounts of Spanish western exploration have done such a nice job of establishing and presenting the course of Spanish western expeditions, this inquiry can pursue an analytical approach, using examples from two and a half centuries of Spanish reconnaissance to illustrate the conditions limiting Spanish western geographic comprehension. The focus will be on pre-1763 exploration, with material from after 1763 adduced when it illuminates issues relevant to earlier decades. With sources in hand, and caveats in mind, it is possible to address the matter of lingering Spanish geographic uncertainty.[9]

9. Excellent overviews of Spanish western exploration include Dennis Reinhartz and Oakah L. Jones, "*Hacia el Norte!* The Spanish *Entrada* into North America, 1513–1549," and Mathes, "Early Exploration of the Pacific Coast," both in Allen, ed., *North American Exploration,* I, 241–291, 400–

Investigation of the limits of Spanish western exploration and geographic comprehension begins most straightforwardly with assessment of the difficulties the West's physical features and indigenous inhabitants presented to Spanish explorers. These obstacles can then be weighed against the inducements attracting Spanish attention and explorers to the region. Throughout, these western challenges, incentives, and opportunities can be considered alongside pertinent aspects of Spanish South and Central American reconnaissance.

Such an inquiry yields a two-layered explanation, the first level of which will appear in this chapter. Most basically, Spanish exploration of western North America lagged because of an unfavorable ratio of possible rewards to manifest difficulties. Early Spanish western explorers found a forbidding coast, perilous plains, and harsh deserts and mountains. They encountered indigenous peoples who were sometimes hostile, occasionally deceptive, and often perplexing. At the same time, Spanish scouts happened upon less in the way of removable wealth, tractable populations, and imposing civilizations than experiences among the Incas, Mayas, and Aztecs had led them to hope for.

Terrain and Peoples

The easiest explanation for this abatement of Spanish western American reconnaissance is that the West's physical characteristics made investigating the region so difficult for explorers. This hypothesis is worth looking into, not least because the relatively comfortable conditions of modern scholarship and travel render the raw physical challenge of the early modern Pacific coast, Mountain West, and Great Plains easily forgotten. European explorers bewailed the land and seascapes so bewitching to modern tourists.

Even today, those parts of the Pacific coast away from the great port cities

---

451; Jones, "Spanish Penetrations to the North of New Spain," and James R. Gibson, "The Exploration of the Pacific Coast," both in Allen, ed., *North American Exploration*, II (Lincoln, Neb., 1997), 7–64, 328–396; Warren L. Cook, *Flood Tide of Empire: Spain and the Pacific Northwest, 1543–1819* (New Haven, Conn., 1973); Bernard DeVoto, *The Course of Empire* (Boston, 1952); and William H. Goetzmann and Glyndwr Williams, *The Atlas of North American Exploration: From the Norse Voyages to the Race to the Pole* (New York, 1992). In addition, David J. Weber, *The Spanish Frontier in North America* (New Haven, Conn., 1992); Colin G. Calloway, *One Vast Winter Count: The Native American West before Lewis and Clark* (Lincoln, Neb., 2003); and Donald Cutter and Iris Engstrand, *Quest for Empire: Spanish Settlement in the Southwest* (Golden, Colo., 1996) are always useful. Such works provide much of the foundation necessary for understanding the role of the mysterious West in eighteenth-century international affairs. In the chapters in this book about Spanish and later French exploration, I have synthesized pertinent insights from existing secondary literature and used comparison between western exploration and the reconnaissance of other parts of the world to yield conclusions going beyond those of earlier works.

remain forbidding as well as beautiful, or perhaps beautiful because forbidding. Long stretches lack safe harbors. Waves are high. Water gives way to rocks, fog conceals them, picturesque capes become lee shores, sudden and strong winds drive ships to destruction. In the early modern period, the currents and winds carrying returning Manila galleons south along the California coast hindered the northward movement of vessels for much of the year. In winter, storms and a countervailing current could carry ships north into unknown waters and east against unforgiving shores. Vessels unable to acquire fresh provisions from an unfamiliar and perilous coast were subject to scurvy.

In such conditions, an exploratory effort like Vizcaíno's 1602–1603 expedition up the California coast could quickly become a catalog of horrors. According to Vizcaíno's diary, by the time the three Spanish ships comprising his flotilla reached Monterey Bay, their "supplies were becoming exhausted because of the length of time" they "had spent in coming." As provisions ran low, men grew ill, making it difficult to work the ships. The officers deemed it necessary to send one vessel back to Mexico with the records of the voyage and the bodies of the infirm. The two remaining ships continued north. As they did so, Spanish sailors found the weather intemperate. The cosmographer attached to Vizcaíno's expedition, Fray Antonio de La Ascensión, wrote of "the severity of winter in this climate, and of the cold" off Cape Mendocino. Worse, a storm struck the ships there and drove them farther north. When healthy sailors were most needed, battering seas and bad food incapacitated them. One ship, the *San Diego*, was reportedly blown to 42° north before a northwest wind made it possible to return. Vizcaíno claimed that, by the time the vessel and its rotting provisions made it back to Monterey, "the mouths of all were sore, and their gums were swollen larger than their teeth, so that they could hardly drink water, and the ship seemed more like a hospital than a ship of an armada." The other ship, the *Tres Reyes*, was impelled into higher latitudes than the *San Diego*, allegedly to the vicinity of Cape Blanco, where, according to the recorded testimony of the boatswain, "the cold was so great that" the Spanish mariners "thought they should be frozen." More significant, at the moment when the coast was trending northeast and the ship had therefore apparently reached the entrance of the Northwest Passage, the crew was unable to proceed. Ascensión averred, "If on this occasion there had been on the captain's ship even fourteen sound men, without any doubt we should have ventured to explore and pass through this Strait of Anian. . . . But the general lack of health and of men who could manage the sails and steer the ship obliged us to turn about toward New Spain, to report what had been discovered and seen, and lest the whole crew should die if we remained longer in that latitude." All told, even though they turned back short of the mythical strait, more than forty men on Vizcaíno's three ships died. Modern historians cannot dismiss the possibility that the mariners involved exaggerated the hardships of the voyage, but early modern Spaniards could not reject the possibility that they had not. Little wonder that more than a century and a half would pass before Spanish ships voluntarily ventured north of Cape Blanco.[10]

Like the Pacific coast, the spectacular terrain of the dry and mountainous Southwest that today invites visitors then repelled explorers. Lands along southwestern rivers like the Gila and Colorado were often lush, cultivated, and populous, and forests often graced higher elevations. But when reconnaissance took Spanish parties away from the Southwest's greener zones, they frequently encountered lands "where the soil is dry and sterile" or the terrain "extremely rough." To journey overland from Tubac (in modern Sonora) to the San Gabriel Mission (the site of modern Los Angeles) in 1774, Juan Bautista de Anza and Father Francisco Garcés required two tries to cross the Colorado Desert and its "horrible sand dunes." Even the famously "phlegmatic" Garcés avowed his "extreme repugnance" at the prospect of another such journey. More spectacularly, in 1540, a party from the Coronado expedition set out to reconnoiter lands northwest of New Mexico. They were blocked by a Grand Canyon so vast they could barely comprehend its dimensions. Scouts failed to reach the chasm's bottom "because of great obstacles they found." Lack of water precluded remaining

10. "Diary of Sebastian Vizcaíno, 1602–1603," in Herbert Bolton, ed., *Spanish Exploration in the Southwest, 1542–1706* (New York, 1916), 52–103, esp. 86, 92–97, 101; Vizcaíno's Diary, in Francisco Carrasco y Guisasola, ed., *Documentos referentes al reconocimiento de las costas de las Californias desde el cabo de San Lucas al de Mendocino* (Madrid, 1882), 95, 99–103, 106; Antonio de La Ascensión, "Brief Report of the Discovery in the South Sea," in Bolton, ed., *Spanish Exploration in the Southwest*, 104–134, esp. 108, 120–121; [Antonio de La Ascensión], "Relacion breve en que se dá noticia del descubrimiento que se hizo en la Nueva-España, en la mar del Sur, desde el puerto de Acapulco hasta más adelante del cabo Mendocino . . . ," in Luis Torres de Mendoza, ed., *Coleccion de documentos inéditos relativos al descubrimiento conquista y organizacion de las antiguas posesiones españolas de América y Oceanía*, VIII (Madrid, 1867), 539–547, esp. 542, 557–558; "The Voyage of Juan Rodriguez Cabrillo," "Father Antonio de La Ascensión's Account of the Voyage of Sebastian Vizcaíno," and "Facsimile of the Account of the Cabrillo Expedition," all in Henry R. Wagner, ed., *Spanish Voyages to the Northwest Coast of America in the Sixteenth Century* (San Francisco, 1929), 72–93, 180–272, 450–463, esp. 92, 244–246, 252–258, 264–265, 461–462; "Relacion, ó diario, de la navegacion que hizo Juan Rodriguez Cabrillo . . . ," in [Buckingham Smith], ed., *Colección de varios documentos para la historia de la Florida y tierras adyacentes* (London, 1857), 173–189, esp. 187–188; "Letter of Sebastián Vizcaíno, Dated at Monterey Bay, 28th December, 1602, Giving Some Account of What He Has Seen and Done during His Exploration of the Coast of the Californias," and "Letters of Sebastián Vizcaíno to the King of Spain, Announcing His Return from the Exploration of the Coast of the Californias, as Far as the Forty-Second Degree of the North Latitude — Dated 23rd May, 1603," both in Donald C. Cutter, ed. and trans., and George Butler Griffin, trans., *The California Coast: A Bilingual Edition of Documents from the Sutro Collection* (Norman, Okla., 1969), 105–109, 111–117, esp. 106, 116.

in the area to investigate further. The Southwest seemed constructed to discourage intruders.[11]

Though less obviously lethal, more generally covered with vegetation, and in many respects more inviting for eventual settlement, the plains proved no less difficult for Spanish explorers. Spanish adventurers in the Southwest marveled at the strikingly varied and precipitous terrain. Spanish scouts east of the Rockies lost themselves on what Coronado referred to as "plains so without landmarks that it was as if we were in the middle of the sea." In this monotony, errant Spanish explorers could easily suffer the same thirst and exhaustion described by figures like Anza and Garcés. In "extreme need," Coronado drank a substance "so bad it contained more mud than water." Navigation on the plains was so challenging that it was not just the alien Spanish who missed their way. Even the Indian inhabitants of the region could go astray. In his diary of a 1706 expedition onto the plains north and east of Taos, Juan de Ulibarri claimed that, although his Indian guide "took especial care," finding "his direction from hummocks of grass placed a short distance apart on the trail by the Apaches, who lose even themselves there," the expedition "became lost entirely." Venturing onto the plains was easy; getting around and back could prove more difficult.[12]

Closely connected to the challenges posed by topography were those presented by human geography. When Spanish expeditions set out into the lands of the American West, the indigenous peoples they encountered could, and frequently did, stop them. The mere threat of attacks discouraged travel in many regions. On the plains, the reception of Spanish parties varied. Some ventured

11. First quotations from Sedelmayr, "Relación, 1746," in Dunne, trans., *Jacobo Sedelmayr*, 18; and "Instruction of Don Thomas Vélez Cachupín, 1754," in Thomas, [ed. and trans.], *Plains Indians and New Mexico*, 129–145, esp. 130. On Anza and Garcés, see Pedro Font, in Herbert Bolton, ed. and trans., *Anza's California Expeditions*, 5 vols. (Berkeley, Calif., 1930), IV, *Font's Complete Diary of the Second Anza Expedition*, 121; Francisco Garcés, "Garcés's Diary of His Detour to the Jalchedunes, 1774," ibid., II, *Opening a Land Route to California: Diaries of Anza, Díaz, Garcés, and Palóu*, 373–392, esp. 391; see also "Anza's Complete Diary, 1774," ibid., 1–130, esp. 57–67, 72–82; and Pedro Font, "Diario Extendido de Font de 1776," or "Expanded Diary of Pedro Font," Dec. 8, 1775, at Web de Anza, http://anza.uoregon.edu/archives.html (accessed June 17, 2009). On the Coronado expedition, see "The Relación de la Jornada de Cíbola, Pedro de Castañeda de Nájera's Narrative, 1560s," in Richard Flint and Shirley Cushing Flint, eds. and trans., *Documents of the Coronado Expedition, 1539–1542* (Dallas, Tex., 2005), 397–398, 451. See also "The Relación del Suceso (Anonymous Narrative)," ibid., 494–507, esp. 499, 504–505.

12. "Coronado's Letter to the King," in Flint and Flint, eds. and trans., *Documents of the Coronado Expedition*, 319–320, 323–324; "Diary of Juan de Ulibarri," in Thomas, [ed. and trans.], *After Coronado*, 66. See also "Diary of the Campaign of Governor Antonio de Valverde," ibid., 110–133, esp. 111; "Gallegos' Relation of the Chamuscado-Rodríguez Expedition," in George P. Hammond and Agapito Rey, eds. and trans., *The Rediscovery of New Mexico, 1580–1594: The Explorations of Chamuscado, Espejo, Castaño de Sosa, Morlete, and Leyva de Bonilla and Humaña* (Albuquerque, N.M., 1966), 67–114, esp. 90–92.

east and returned without incident or after having been the beneficiaries of a friendly reception. On other occasions, even large and well-armed Spanish *entradas* might encounter compelling reasons to beat a hasty retreat. The 1601 Oñate expedition reportedly, and perhaps exaggeratedly, fought a three-to-five-hour "skirmish" with 2,500–5,000 "Escanxaque" Indians, near the Arkansas River in south-central Kansas or north-central Oklahoma. The armored Spanish soldiers apparently suffered no deaths from Indian arrows, but thirty "received slight wounds." Even before the clash, reports of hostile Indians ahead had led Oñate's followers to request that the expedition turn back, and Oñate had reluctantly agreed. Though less lethal than it might have been, the actual combat confirmed the good sense of the fainthearted. Those contemplating future expeditions would have to ruminate on the possibility that the next engagement might turn out worse. A warning by friendly Indians about the hostile Quivira Indians to come sufficed to persuade Captain Alonso Vaca to turn his 1634 expedition around.[13]

In 1720, Pedro de Villasur and his roughly 105-man party turned back too late, and the leader and some 43 of his men never returned to Santa Fe. Determined to ascertain the veracity of reports of Frenchmen moving west toward New Mexico, Villasur appears to have made it to the junction of the Platte and Loup rivers in modern eastern Nebraska, where what was probably a Pawnee-Oto force attacked, inflicting what former New Mexico governor Antonio Valverde called "the most outstanding misfortune that has come to pass in this country." The loss demonstrated most clearly the damage Plains Indians could inflict on even a sizeable and martial Spanish-Indian expedition, particularly when the passage of a century since Oñate's skirmish and the arrival of French traders in the region had dispersed at least some firearms to plains peoples such as those attacking Villasur.[14]

13. For friendly receptions, see Tyler and Taylor, eds. and trans., "Report of Fray Alonso de Posada," *NMHR*, XXXIII (1958), 293–298; Posadas, "Informe á S. M. sobre las tierras de Nuevo Méjico, Quivira y Teguayo," in Duro, *Don Diego de Peñalosa*, 57–61. On Oñate, see "Official Inquiry Made by the Factor, Don Francisco de Valverde, by Order of the Count of Monterrey, concerning the New Discovery Undertaken by Governor Don Juan de Oñate toward the North beyond the Provinces of New Mexico," in George P. Hammond and Agapito Rey, eds. and trans., *Don Juan de Oñate: Colonizer of New Mexico, 1595–1628* (Albuquerque, N.M., 1953), II, 836–871, esp. 848, 858–859, 868; Marc Simmons, *The Last Conquistador: Juan de Oñate and the Settling of the Far Southwest* (Norman, Okla., 1991), 162–164. On the warning to Cabeza de Vaca, see Tyler and Taylor, eds. and trans., "Report of Fray Alonso de Posada," *NMHR*, XXXIII (1958), 298; Posadas, "Informe á S. M. sobre las tierras de Nuevo Méjico, Quivira y Teguayo," in Duro, *Don Diego de Peñalosa*, 60–61.

14. Antonio Valverde to marqués de Valero, Oct. 8, 1720, 162–167 (quotation on 164), Revolledo to Valero, Dec. 9, 1720, 175–177, "Testimony of Aguilar, Santa Fé, July 1, 1726," 226–228, "Testimony of Tamariz," July 2, 1726, 228–230, all in in Thomas, ed. and trans., *After Coronado*; "Charlevoix Visits Wisconsin: His Description of the Tribes," in Reuben Gold Thwaites, ed., State Historical Society

The Villasur disaster points to a more general difficulty. Frontier provinces formed shaky foundations for exploration because soldiers who might have been detached for reconnaissance were instead often engaged in defending existing settlements against outside raiders or in preventing or suppressing revolts by nominally subject populations. In 1735, fifty soldiers at the Corodeguachi Presidio (about forty miles south of modern Douglas, Arizona), the kind of force that might have been used for exploration, were "occupied now in watching over the disturbances among the Pimas Altos, now over those among the Seris, and lastly over anything which might happen in the province which might need the attention of the forces of his Majesty." In the case of New Mexico, the loss of Villasur and so many of his men inhibited Spanish exploration in the following decades. The small colony, nearly destroyed by the Pueblo Revolt forty years before and increasingly beset by raids from its neighbors as the eighteenth century unfolded — "enclosed on all sides by innumerable and warlike nations of heathen enemies," as the Spanish auditor the marqués de Altamira put it in 1753 — could ill afford the loss of forty soldiers in 1720 or the risking of more in the decades that followed.[15]

West and southwest of New Mexico, western Apaches loomed especially large in the consciousness and sometimes the lives of eighteenth-century Spanish officials, soldiers, and missionaries. As Kino put it, aspiring explorers had to cope with "the obstacle of the very difficult passage through the Apaches." Juan Bautista de Anza, the father of the famous California explorer of the same name, noted in a 1735 report that Apaches ambushed journeyers in narrow passes and would "sometimes . . . gather in large numbers to mount a major attack against

of Wisconsin, *Collections*, XVI, *The French Regime in Wisconsin*, I, *1634–1727* (Madison, Wis., 1902), 408–418, esp. 413–414; Weber, *Spanish Frontier*, 169–171; Gottfried Hotz, *Indian Skin Paintings from the American Southwest: Two Representations of Border Conflicts between Mexico and the Missouri in the Early Eighteenth Century*, trans. Johannes Malthaner (Norman, Okla., [1970]). Identification of site and Indian nations is from Weber.

15. "Statement of Don Agustín de Vildosola," 1735, in Donald Rowland, "The Sonora Frontier of New Spain, 1735–1745," in Hammond, [ed.], *New Spain and the Anglo-American West*, I, 147–164, esp. 155; marqués de Altamira, "Opinion," Jan. 14, 1753, in Thomas, [ed. and trans.], *Plains Indians and New Mexico*, 126; see also "Instruction of Don Thomas Vélez Cachupín, 1754," ibid., 129–145, esp. 135; and Robert Ryal Miller, ed. and trans., "New Mexico in Mid-Eighteenth Century: A Report Based on Governor Vélez Capuchin's Inspection," *SWHQ*, LXXIX (1975), 166–181, esp. 169–181. Calloway provides a nice discussion of the general issue of Spanish troops' being tied down by defense burdens in *One Vast Winter Count*, 177–185, 330–331.

The dangers of the plains for New Mexicans might have been long-standing: Gallegos understood Pueblo Indians in 1581 to have said that "they did not want to go" into buffalo country "because the Indians who followed the buffalo were enemies and very cunning; and that the two peoples would kill each other and start trouble." See "Gallegos' Relation," in Hammond and Rey, eds. and trans., *Rediscovery of New Mexico*, 87–88.

travelers, soldiers, or settlements." In such conditions, even a military escort might fail to get a traveler to his destination. Jesuit missionary Ignatius Keller and nine soldiers tried in 1743 to journey north from the Gila River (near modern Coolidge, Arizona) to the Little Colorado Moqui pueblos — settlements formerly visited by Franciscans but lost to the church during the Pueblo Revolt. En route, "Apaches attacked his party," killing one of the soldiers and stealing "most of the horses so that the father was forced to return."[16]

Means other than violence were also available to Indians wishing to check Spanish progress. Locals holding back information about water sources, food supplies, and practicable routes; informants overstating the human and physical dangers to come; and guides leading parties along arduous and indirect paths or simply disappearing when most needed — all these could make the lives of Spanish explorers difficult in the best of circumstances and short in the worst. Franciscan fathers Silvestre de Vélez de Escalante and Francisco Atanasio Domínguez's party had to overcome these kinds of challenges during a 1776 journey through lands comprising the modern states of Colorado, Utah, Arizona, and New Mexico. Ute communities in southwestern Colorado claimed to be unfamiliar with the country through which the Spanish party wished to proceed, warned of hostile Comanches who "would impede" the Spaniards' "passage and even deprive" them of their "lives," and hesitated to "make exchanges for the hoofsore horses" the Spanish were using. Guides withheld route information, "vanished" at inopportune moments, and were rumored to be deliberately choosing the most difficult paths so as to keep the Spanish "needlessly winding around for eight or ten days, to make" them "turn back."[17]

16. Kino, *Kino's Historical Memoir of Pimería Alta*, ed. and trans. Bolton, I, 198; in the same work, see also I, 237, II, 25–32, 171, 254–255; "Juan Bautista de Anza Discusses Apache and Seri Depredations," Aug. 13, 1735, in Polzer and Sheridan, eds. and comps., *Presidio and Militia on the Northern Frontier of New Spain*, II, part 1, 307, 311. Quotations regarding Keller from Sedelmayr, "Relación, 1746," in Dunne, trans., *Jacobo Sedelmayr*, 23, 33. See also the somewhat different account, which doesn't identify the attackers as Apaches, in "Informe anónimo dirigido al provincial (hacia 1753) sobre los acontecimientos bélicos en la región más norteña de las misiones de la Compañía," in Ernest J. Burrus and Félix Zubillaga, eds., *El Noroeste de México: Documentos sobre las misiones jesuíticas, 1600–1769* (Mexico City, 1986), 307–348, esp. 314.

17. Ted J. Warner, ed., *The Domínguez-Escalante Journal: Their Expedition through Colorado, Utah, Arizona, and New Mexico in 1776*, trans. Angelico Chavez (Provo, Utah, 1976), 28, 31–32, 34, 38, 69, 76–77, 100, 147, 149–151, 153, 170, 174–175, 188. See also Francisco Garcés, *A Record of Travels in Arizona and California, 1775–1776*, ed. and trans. John Galvin (San Francisco, 1965), 18–19, 47, 53, 61; Francisco Garcés, *Diario de exploraciones en Arizona y California en los años de 1775 y 1776* (Mexico City, 1968), 28, 53, 56, 64; Joseph P. Sánchez, trans., "Translation of Incomplete and Untitled Copy of Juan María Antonio Rivera's Original Diary of the First Expedition, 23 July 1765," and "Translation of Juan María Antonio Rivera's Second Diary, Oct. 1765," in Sánchez, *Explorers, Traders, and Slavers: Forging the Old Spanish Trail, 1678–1850* (Salt Lake City, Utah, 1997), 137–147, 149–157, esp. 145, 149–150, 153–155.

Indeed, perhaps disappointed by the results or unhappy about the difficulties of exploration, Spanish scouts often suspected that Indian pathfinders actively misled or misinformed them. During his 1684 expedition into Texas, Captain Juan Domínguez de Mendoza claimed that Jumano Indian guide (and expedition instigator) Juan de Sabeata had been generally mendacious and, in particular, had spread false reports of nearby hostile Apaches to halt the Spanish party. More famously, Castañeda's classic account of the Coronado expedition claimed that an Indian captive, "the Turk," used tales of the fabulous city of Quivira to draw the conquistadors away from the pueblos of New Mexico and onto the plains in hopes that Spanish horses would die and Spanish arms weaken there. In Castañeda's tale, when the disappointed Spanish found thatched huts in place of the rich city Indian stories had led them to expect, Coronado "asked El Turco why he had lied and had guided the [people of the expedition] so tortuously. He replied that his land was toward that area, and besides, the [people] of Cicuyc [Pecos] had begged him to get the [Spaniards] lost on the plains. [That was] so that, lacking food supplies, the horses would die. And [the people of Cicuyc] could kill [the Spaniards] without difficulty when they returned, [because they would be] weak. And [they would be able] to avenge what [the Spaniards] had done." Later critics have questioned the reliability of the story, and it is fair to say that its literary qualities suggest that Castañeda might have imposed his own sense of drama on events. At the same time, there is nothing incredible in the account, and Coronado's brutal entrada had given native peoples ample reason to decoy and destroy it.[18]

Less elaborate techniques of obstruction could also prove effective. Indians could merely refuse passage through their territories. The travels of Father Garcés offer one example. He succeeded in 1776 in ascending the Colorado and Little Colorado rivers to the Moqui pueblos. Garcés had passed through the Pima villages downstream on his journey, however, and the Moquis were ap-

18. "Itinerary of Juan Domínguez de Mendoza," in Bolton, ed., *Spanish Exploration in the Southwest*, 320–344, esp. 336. See also Garcés, *Record of Travels in Arizona and California*, ed. and trans. Galvin, 84–85, 89; Garcés, *Diario de exploraciones en Arizona y California*, 83, 85, 89; and "Rivera's Second Diary," in Sánchez, *Explorers, Traders, and Slavers*, 153. Castañeda quotations from "Castañeda de Nájera's Narrative," in Flint and Flint, eds. and trans., *Documents of the Coronado Expedition*, 411, 467. The Flints question the reliability of this account, contending that the Turk was tortured into the admission, that he and his captors probably lacked much in the way of shared vocabulary, and that such tales of betrayal figured frequently in romances of the period. These are doubts worth raising. I find the underlying events of the account credible nonetheless. The Turk had been with the expedition a year, and this seems to me time enough to pick up a few useful phrases. Information extracted under duress need not be false, and I see nothing improbable about a captive trying to lead Coronado and his bloodstained men into trouble. Historical accounts may echo literary texts, but literary texts may also recall actual events.

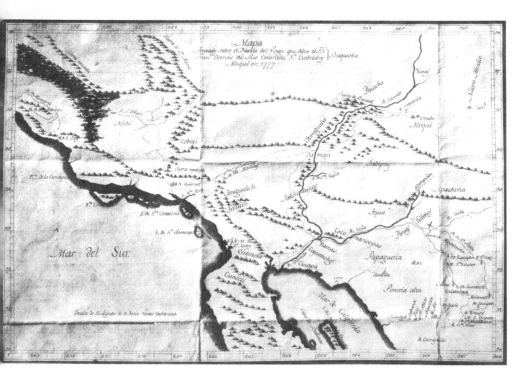

MAP 14. Pedro Font, *Mapa formado sobre el diario del viage que hizo el P. F. Franco Garcès*. 1777. Courtesy of The Bancroft Library, University of California, Berkeley

parently suspicious of a Spanish explorer on good terms with a people they frequently fought. They might also have been wary of any representative of the Spanish imperial authority they had successfully escaped. The Moquis enjoined Garcés to "go back to" his "own land." He did as he was told. The Moqui case is especially vivid, but a variety of Indian groups had ample reasons to hinder Spanish exploration, whether to exclude outsiders from key positions in exchange networks, to keep disruptive and ultimately deadly Spaniards at a distance, or to prevent Spaniards from trading and perhaps allying with enemy nations beyond.[19]

Aspects of the Spanish imperial experience farther south would seem, at first glance, to support an emphasis on the role of harsh terrain or unfriendly Indians in inhibiting Spanish exploration. Remote regions of southern South America — precipitous in the Andes; frigid, desolate, and allegedly giant-haunted east of

19. Garcés, *Record of Travels in Arizona and California*, ed. and trans. Galvin, 75, 77; Garcés, *Diario de exploraciones en Arizona y California*, 76.

them; broken, soggy, and well defended by very real and lethal Araucanians west of them—retained their mysteries into the second half of the eighteenth century. When French mariner Louis-Antoine de Bougainville visited Patagonia in 1767, possible "existence of a race of giants" there remained "a contentious subject," and he felt obligated to mention in his journal that the tallest native he "met attained scarcely a height of 5 ft 9 in." As late as 1790, a Spanish expedition was seeking the "Enchanted City" of the Caesars near Lake Nahuel Huapí (in modern Argentina). Within living memory, the tropical climate, fearsome wildlife, and unwelcoming inhabitants of parts of the Amazon Basin have deterred or frustrated Euro-American interlopers. Parts and peoples of South and Central America long resisted Spanish conquest or cognizance.[20]

Although often making Spanish exploration arduous, however, South American and western North American physical and human obstacles rarely made it impossible. In western North America and the North Pacific, Spanish scouts and sailors would demonstrate this after the Seven Years' War, when foreign encroachments rendered more extensive western American reconnaissance a pressing matter. Spanish ships reached the northwest coast before Cook, and Spanish soldiers and missionaries preceded the United States in California and Utah. More strikingly, in the sixteenth and seventeenth centuries, Spanish explorers' achievements in large parts of South and Central America suggest that virtually no physical or human American obstacle could hold out against determined conquistadors or *marineros*. The North Pacific coast, the Southwest, and the plains might have been forbidding, but no more so than the Straits of Magellan, the breadth of the South Sea, the central and northern Andes, the northern Mexican deserts, the Chaco Boreal, and large parts of the South and Central American rain forests, all of which Spaniards explored, or at least traversed, in the sixteenth and seventeenth centuries. In many cases, these spectacular journeys occurred despite indigenous efforts to mislead and obstruct Spanish explorers and indeed notwithstanding running battles with serried Aztec and Inca armies or dispersed forest hunters and desert raiders. Because Spanish ships coasted the entire perimeter of South and Central America, and Spanish military and missionary expeditions crisscrossed their interiors, Spain was left with pockets of uncertainty interspersed among ribbons of geographic familiarity rather than with an enormous, uninterrupted zone of ignorance comprising roughly two-thirds of North America north of Mexico. Physical geog-

20. Louis-Antoine de Bougainville, *The Pacific Journal of Louis-Antoine de Bougainville, 1767–1768*, ed. and trans. John Dunmore (London, 2002), 11–12. On the Enchanted City of the Caesars, see Edward J. Goodman, *The Explorers of South America* (New York, [1972]), 164–178.

raphy and indigenous resistance contribute to an explanation of the limits of Spanish western exploration but do not complete one.

Disappointments

Ulloa reached California, rounded the cape [San Lucas], and sailed north to the parallel of San Andrés, at a point he named Cape Disappointment [Punta Baja], where he turned back to New Spain because of contrary winds and shortage of provisions. This voyage took him a whole year, and he brought back no word of a good new land. *The game was not worth the candle.*

In the belief that there were some very large and rich islands between New Spain and the Spice Islands, Cortés had thought he might discover a second New Spain on that coast and sea, but in spite of all the ships he had fitted out, and even commanded in person, he accomplished nothing but what I have said.
—FRANCISCO LÓPEZ DE GÓMARA, 1552

All or most of those who went on the expedition that Francisco Vázquez Coronado made in search of the seven *ciudades* . . .

Although they did not find the wealth of which they had been told, they found the beginning of a good land to settle and the wherewithal to search for [wealth] and to go onward. . . . Their hearts weep because they have lost such an opportunity of a lifetime. . . .

. . . I have understood [that] some of those who came back from there would be happy to go back, in order to continue farther so as to recover what was lost.
—PEDRO DE CASTAÑEDA DE NÁJERA, 1560s

To reach a deeper understanding of these limits of Spanish western exploration, it is necessary to consider the sources of the seemingly demonic determination of those demon-fearing sixteenth-century Spanish conquistadors and the consequences when their fonts of motivation ran dry. The discoveries of Spain's first American generations departed so far from expectation as to suggest for a time the need to extend the horizons of the possible. Scripture and romance—as in Columbus's paradisiacal speculations off the coast of Venezuela or Bernal Díaz's epic awe before Tenochtitlán—seemed at times a better guide to New World conditions than medieval Iberian experience. Spain's early conquistadors were willing to endure or inflict almost unimaginable hardship in part because their reward might be quantities of wealth and degrees of status unattainable and almost inconceivable in Europe. As Spanish explorers moved into what is now northern Mexico and the United States, the weight of reality began to drag Spanish fancies down to earth. The early American West disappointed hopes animated by the Central American Isthmus, the Straits of Magellan, Peru, and Mexico. Spanish explorers failed to find comparable riches and dominions north of the Aztec Empire or practicable water passages through North America, and this reduced the incentive for future explorers to push into remaining unknown areas.[21]

A Strait of Anian's existence in the American West, for instance, remained possible but unrealized. La Ascensión had thought that around latitude 43° north, where "the coast and land turns to the northeast," he had seen "the head and end of the realm and mainland of California and the entrance to the Strait of Anian." Neither La Ascensión nor any other Spanish mariner, however, moved far enough into an apparent passage to prove its practicability and value.[22]

This failure to investigate a possible passage's depths was not universally regretted by Spanish officials. For shrewd Spanish statesmen, a Northwest Passage's potential utility might furnish good reason to avoid confirming its existence. For if such a passage was more than chimerical, it might provide a route more useful to British, Dutch, and French interlopers than to Spanish mariners. Francis Drake, who made it at least to and perhaps beyond the northern California coast in 1579, and Thomas Cavendish, who seized a Spanish galleon off Cape San Lucas (in Baja California) in 1587, had alarmed the defenders of Spain's empire by showing how far English raiders might reach even without a passage. Revealing to them an easier northern route to the Pacific might be less than prudent. Ascensión, alarmed by San Diego Bay Indian tales of "a people living inland, of form and figure like our Spaniards, bearded, and wearing collars and breeches" (perhaps members of the Oñate entrada), worried that Dutch or English ships might already have come through the Strait of Anian and established a settlement in western America. His suspicion proved unwarranted, but the dangers a Northwest Passage posed to the Spanish Empire remained.[23]

21. Francisco López de Gómara, *Cortés: The Life of the Conqueror by His Secretary,* ed. and trans. Lesley Byrd Simpson (Berkeley, Calif., 1964), 403–404, brackets and italics in original (for the original Spanish, see Gómara, *Historia de la conquista de México,* ed. Jorge Gurria Lacroix [Caracas, 1979], 312); "Castañeda de Nájera's Narrative," in Flint and Flint, eds. and trans., *Documents of the Coronado Expedition,* 385–386, 436–437, brackets in original.

22. Ascensión, "Brief Report of the Discovery in the South Sea," in Bolton, ed., *Spanish Exploration in the Southwest,* 121; [Ascensión], "Relacion breve en que se dá noticia del descubrimiento que se hizo en la Nueva-España, en la mar del Sur," in Torres de Mendoza, ed., *Coleccion de documentos inéditos relativos al descubrimiento, conquista y organizacion de las antiguas posesiones españolas de América y Oceanía,* VIII, 558: "La costa y tierra da la vuelta al Nordeste, y aquí es la cabeza y fin del reino y Tierra Firme de la California, y el principio y entrada para el estrecho de Anian"; "Father Antonio de La Ascension's Account of the Voyage," in Wagner, ed., *Spanish Voyages to the Northwest Coast of America,* 264–265, 267–268.

23. Ascensión, "Brief Report of the Discovery in the South Sea," in Bolton, ed., *Spanish Exploration in the Southwest,* 117–118; [Ascensión], "Relacion breve en que se dá noticia del descubrimiento que se hizo en la Nueva-España, en la mar del Sur," in Torres de Mendoza, ed., *Colección de documentos inéditos relativos al descubrimiento, conquista y organizacion de las antiguas posesiones españolas de América y Oceanía,* VIII, 554 ("una gente que estaba la tierra dentro, del talle y modo que nuestros españoles, barbados, y con cuellos y valonas"); "Father Antonio de La Ascension's

A less spectacular but more probable western possibility was that of great navigable river systems affording access to the interior or a route to the ocean, as the Paraná and Amazon systems did in South America. River systems comparable to these did, of course, exist in the North American West, but the Spanish Empire failed to exploit them fully. In some cases, along the Pacific coast beyond the Gulf of California, for example, Spanish sailors failed to locate the West's great rivers. The Columbia's mouth lay beyond the known limits of Spanish maritime exploration and the usual route of galleons returning from the Philippines; in any case, it was sufficiently inconspicuous to conceal its significance from Hezeta in 1775 and its existence from Cook in 1778. The Golden Gate into San Francisco Bay and the rivers of the California interior, though along routes sailed by both galleons and explorers, was indistinct enough against the backdrop of the coastal hills to elude verifiable Spanish detection until 1769.[24]

Members of the 1542–1543 Cabrillo-Ferrelo and 1602–1603 Vizcaíno expeditions did see, or at least suspected they saw, other "great," "large," and "copious" rivers between San Diego Bay and Cape Blanco. But they identified no commodity as attractive to Spanish merchants as northwestern sea otter furs would later be to traders from Old and New England. The "fish, game, hazel nuts, chestnuts," and "acorns" with which coastal Indians tried to entice them failed to inspire the avarice of Hernán Cortés and Francisco Pizarro's compatriots. Spanish trading houses remained absent from the mouths of these rivers, and Spanish boats left their upper courses uncharted and their economic or strategic utility undemonstrated.[25]

In addition to practicable passages and navigable rivers, there was also the possibility of wealthy civilizations on the Pacific coast, perhaps comparable to

Account of the Voyage," in Wagner, ed., *Spanish Voyages to the Northwest Coast of America,* 226–227, 268.

24. Samuel Eliot Morison, *The European Discovery of America,* II, *The Southern Voyages, A.D. 1492–1616* (New York, 1974), 675–676.

25. "Voyage of Juan Rodriguez Cabrillo," "Father Antonio de La Ascension's Account of the Voyage," and "Facsimile of the Account of the Cabrillo Expedition," in Wagner, ed., *Spanish Voyages to the Northwest Coast of America,* 72–93, 180–272, esp. 86, 88, 92, 255, 265, 455, 457, 462; "Relacion, ó diario, de la navegacion que hizo Juan Rodriguez Cabrillo," in [Smith], ed., *Colección de varios documentos para la historia de la Florida y tierras adyacentes,* 181, 183, 188; "Diary of Sebastian Vizcaíno," in Bolton, ed., *Spanish Exploration in the Southwest,* 94, 102; Vizcaíno's Diary, in Carrasco y Guisasola, ed., *Documentos referentes al reconocimiento de las costas de las Californias,* 100, 106; Ascensión, "Brief Report of the Discovery in the South Sea," in Bolton, ed., *Spanish Exploration in the Southwest,* 119–121; [Ascensión,] "Relacion breve en que se dá noticia del descubrimiento que se hizo en la Nueva-España, en la mar del Sur," in Torres de Mendoza, ed., *Coleccion de documentos inéditos relativos al descubrimiento, conquista y organizacion de las antiguas posesiones españolas de América y Oceanía,* VIII, 556–558; Mathes, "Early Exploration of the Pacific Coast," in Allen, ed., *North American Exploration,* I, 423.

the Incas or connected to Japan. But Spanish explorers found the California coast unsatisfying, comparisons of it to lands farther south unfavorable. Consider the contrast between the initial indications offered by Peruvian and Californian shores. In 1527, Spanish pilot Bartolomé Ruiz captured a trading raft off the coast of modern Peru or Ecuador containing "blankets of wool and cotton, shirts, . . . and other articles of clothing, all beautifully worked in scarlet, crimson, blue, yellow, and all the other colors with various kinds of designs and figures of birds, animals, fish, and trees." In the same year, Pizarro and his men found on the Gulf of Guayaquil's Santa Clara Island "a great sample of the wealth of the land ahead of them because they found many small gold and silver pieces shaped as hands or a woman's breasts or heads, and a large silver vessel." At Tombes, Pizarro encountered an official ["oréjon"] reporting to Inca emperor Huayna Capac in Quito. The signs pointed to material wealth and imperial grandeur.[26]

The California coast promised less. Cabrillo and Vizcaíno met village rulers rather than imperial officials.[27] Instead of being shown wrought gold and silver by the Indians they came into contact with, Spanish expeditions up the California coast presented European metalwork to its inhabitants. Cabrillo and Vizcaíno's men encountered Indians wearing, not cloth adorned by representations of animals, but instead the skins of the animals themselves. When, on one occasion, coastal California Indians possessed silk, it came from a Spanish "ship . . . driven by a strong wind to the coast and wrecked." This did not encourage Spanish navigation in the area. Spanish vessels moving up North America's Pacific coast observed populous and comfortable towns and wrote enough about the

26. On Ruíz, see "Francisco de Jérez's Account of the Early Expeditions," in John H. Parry and Robert G. Keith, eds., New Iberian World: A Documentary History of the Discovery and Settlement of Latin America to the Early 17th Century, IV, The Andes (New York, 1984), 14–17, esp. 16; "Relación de los primeros descubrimientos de Francisco Pizarro y Diego de Almagro, sacada del códice número CXX de la Biblioteca Imperial de Viena," in Martin Fernandez Navarrete, Miguel Salva, and Pedro Sainz de Baranda, eds., Colección de documentos inéditos para la historia de España, V (Madrid, 1844), 193–201, esp. 197. On Pizarro, see Pedro de Cieza de León, The Discovery and Conquest of Peru: Chronicles of the New World Encounter, ed. and trans. Alexandra Parma Cook and Noble David Cook (Durham, N.C., 1998), 103, 108; Cieza de León, La crónica del Perú, in Carmelo Sáenz de Santa María, ed., Obras completas (1553; rpt. Madrid, 1984), I, 246–247.

27. On village rulers, see "Voyage of Juan Rodriguez Cabrillo," "Father Antonio de La Ascension's Account of the Voyage," and "Facsimile of the Account of the Cabrillo Expedition," in Wagner, ed., Spanish Voyages to the Northwest Coast of America, 72–93, 450–463, esp. 88, 239, 457–458; "Relacion, ó diario, de la navegacion que hizo Juan Rodriguez Cabrillo," in [Smith], ed., Colección de varios documentos para la historia de la Florida y tierras adyacentes, 183; Ascensión, "Brief Report of the Discovery in the South Sea," in Bolton, ed., Spanish Exploration in the Southwest, 118; [Ascensión], "Relacion breve en que se dá noticia del descubrimiento que se hizo en la Nueva-España, en la mar del Sur," in Torres de Mendoza, ed., Coleccion de documentos inéditos relativos al descubrimiento, conquista y organizacion de las antiguas posesiones españolas de América y Oceanía, VIII, 555. Ascensión speaks of a "petty king" ["reyezuelo"].

area's human richness to tantalize later scholars, but they saw little to indicate the material wealth or political structures so alluring to merchants or conquistadors, "no trace whatever of the City of Quivira, notwithstanding every effort."[28]

Inland, the 1540–1542 Coronado entradas into New Mexico, onto the plains, and through the Southwest, like the subsequent 1598–1605 Oñate ventures in the same area, proved less rewarding than anticipated. Coronado and his men found populated areas in both the Southwest and the plains and lands appealing enough to inspire nostalgia in aging conquistadors, but this was less than earlier reports had led them to hope for. When Cabeza de Vaca, the black slave Esteban, and their two companions finally made their way to Spanish settlements in Mexico in 1536, they brought not only tales of suffering but also rumors of

28. On the presentation of metalwork, see Ascensión, "Brief Report of the Discovery in the South Sea," in Bolton, ed., *Spanish Exploration in the Southwest*, 117; [Ascensión], "Relacion breve en que se dá noticia del descubrimiento que se hizo en la Nueva-España," in Torres de Mendoza, ed., *Coleccion de documentos inéditos relativos al descubrimiento, conquista y organizacion de las antiguas posesiones españolas de América y Oceanía*, VIII, 554. On skin clothing, see "Diary of Sebastian Vizcaíno," in Bolton, ed., *Spanish Exploration in the Southwest*, 82, 84; Vizcaíno's Diary, in Carrasco y Guisasola, ed., *Documentos referentes al reconocimiento de las costas de las Californias*, 91, 93, 95; Ascensión, "Brief Report of the Discovery in the South Sea," 120; [Ascensión], "Relacion breve en que se dá noticia del descubrimiento que se hizo en la Nueva-España, en la mar del Sur," in Torres de Mendoza, ed., *Coleccion de documentos inéditos relativos al descubrimiento, conquista y organizacion de las antiguas posesiones españolas de América y Oceanía*, VIII, 557; "Voyage of Juan Rodriguez Cabrillo," "Father Antonio de La Ascension's Account of the Voyage," and "Facsimile of the Account of the Cabrillo Expedition," in Wagner, ed., *Spanish Voyages to the Northwest Coast of America*, 85–86, 88, 236, 256, 455–456, 458; "Relacion, ó diario, de la navegacion que hizo Juan Rodriguez Cabrillo," in [Smith], ed., *Colección de varios documentos para la historia de la Florida y tierras adyacentes*, 179–181, 183.

The wrecked ship was likely the Acapulco-bound galleon *San Agustín*, lost trying to reconnoiter the California coast in 1595. After the ship's demise, a royal decree forbade risking precious and unwieldy galleons on hazardous coastal exploration. See "Diary of Sebastian Vizcaíno," and "Brief Report of the Discovery in the South Sea," both in Bolton, ed., *Spanish Exploration in the Southwest*, 85, 94, 120; Vizcaíno's Diary, in Carrasco y Guisasola, ed., *Documentos referentes al reconocimiento de las costas de las Californias*, 94, 101; [Ascensión], "Relacion breve en que se dá noticia del descubrimiento que se hizo en la Nueva-España, en la mar del Sur," in Torres de Mendoza, ed., *Coleccion de documentos inéditos relativos al descubrimiento, conquista y organizacion de las antiguas posesiones españolas de América y Oceanía*, VIII, 558; "Father Antonio de La Ascension's Account of the Voyage," in Wagner, ed., *Spanish Voyages to the Northwest Coast of America*, 249; "Paragraph of a Letter from the Royal Officials of Acapulco to the Conde de Monterrey, Viceroy of New Spain, Giving Tidings of the Loss of the Ship *San Agustín* — Dated 1st February, 1596," and "Paragraph of a Letter from the Conde de Monterrey, Viceroy of New Spain, to the King of Spain, Giving Notice of the Loss of the Ship *San Agustín* and of Discoveries Made in Her — Dated 19th April, 1596," in Cutter, ed. and trans., and Griffin, trans., *California Coast*, 32–39; William Lytle Schurz, *The Manila Galleon* (New York, 1959), 240; Iris H. W. Engstrand, "Seekers of the 'Northern Mystery': European Exploration of California and the Pacific," in Ramón A. Gutiérrez and Richard J. Orsi, eds., *Contested Eden: California before the Gold Rush* (Berkeley, Calif., 1998), 78–110, esp. 90. Quivira material from "Father Antonio de La Ascension's Account of the Voyage," in Wagner, ed., *Spanish Voyages to the Northwest Coast of America*, 252, 255, 264.

wealthy and populous cities to the north. These rumors reinforced Indian merchant reports of gold, silver, and big pueblos. Esteban failed to survive a 1539 effort to investigate these rumors, but Fray Marcos de Niza, the leader of the Spanish reconnaissance effort, did. He wrote of "seven *ciudades*" and claimed to have seen the first of them, Cíbola, which was "grander than the *Ciudad de México.*" When Coronado and his men ventured north looking for these cities, they found instead "seven small towns," the first of which was "a small *pueblo* crowded together and spilling down a cliff." In place of the "wealth of gold or rich jewels" they had imagined, they found "two bits of emerald . . . in some paper, also some very worthless, small, red stones." New Mexico fell short of the old.[29]

Coronado retained hopes for better prospects in the lands beyond the Rio Grande Valley. He claimed in a letter to Charles V that Indian informants had led him to believe that, on the plains, the entrada would encounter "much grander towns and buildings" than those of New Mexico, with "lords who ruled them" and "ate out of golden dishes." Instead, the Spanish expedition trudged to a Quivira with houses made "of thatch," where the conquistadors found "neither gold nor silver" nor "news of it." The closest they came was a "lord" wearing "a copper medallion [suspended] from his neck." Oñate's later expeditions led to a Spanish colony's establishment in New Mexico but did little to dispel the inauspicious impressions Coronado left of the regions to the east and west of it or to inspire governmental support for "new discovery."[30]

As Coronado and Oñate's entradas failed to find a Cibola or Quivira com-

29. For Marcos de Niza, see "The Viceroy's Instructions to Fray Marcos de Niza, November 1538, and Narrative Account by Fray Marcos de Niza, August 26, 1539," in Flint and Flint, eds. and trans., *Documents of the Coronado Expedition*, 75, 87. For the reactions of Coronado and his men, see "Coronado's Letter to the Viceroy, August 3, 1540," and "Castañeda de Nájera's Narrative," ibid., 258–259, 267, 393–394, 446–447.

30. On Coronado's hopes and disappointments, see "Coronado's Letter to the King, October 20, 1541," in Flint and Flint, eds. and trans., *Documents of the Coronado Expedition*, 319–320, 323–324: "muy mayores pueblos y casas meJores . . . señores que los mandavan y que se sirVian en VasiJas de oro"; "hay . . . de paJa." "No se Vio entre aquella gente oro ni plata ni notiçia de ello" ("Castañeda de Nájera's Narrative," 411, 467). For other references to thatched roofs in the same volume, see "Relación de Suceso" and "Juan Jaramillo's Narrative," 501, 506, 517, 523. On the copper medallion, see "Castañeda de Nájera's Narrative," 411, 467 (brackets in original): "El señor traya a el cuello Una patena de cobre." On Oñate's expeditions, see "Opinion of the Audiencia of Mexico as to the Continuation of the Conquest and Discovery of New Mexico, May 14, 1602," "Letter from the Fiscal of the Audiencia of Mexico, May 14, 1602, Regarding the Discovery of New Mexico," "Summary of the Five Discourses Presented by the Viceroy concerning the Situation in the Territory That Has Been Pacified and Settled by Adelantado Don Juan de Oñate in the Provinces of New Mexico . . . ," and "Discussion and Proposal of the Points Referred to His Majesty concerning the Various Discoveries of New Mexico," 1602, in Hammond and Rey, eds. and trans., *Don Juan de Oñate*, II, 895–924, esp. 895.

mensurate with their desires, Oñate and others of Coronado's successors were also frustrated in their quest for another celebrated will-o'-the-wisp in the lands north of the Valley of Mexico. In the sixteenth century, Spanish missionaries, scholars, and explorers heard and saw expressions of a vague and shifting tradition holding that major peoples of Mexico such as the Mexicas, Toltecas, and Tlaxcoltecas had migrated from someplace in the north denominated variously by names such as Aztlan, Whiteness, the Seven Caves, Chicomoztoc, Colhuacan, Teguayo, the Land of Herons, or the Lake of Copala.[31]

In at least some versions of the Mexican origin story, the physical and perhaps magical distance of Aztlan from the Valley of Mexico, the austerity of Aztlan's inhabitants, and the hardships of its emigrants made it seem an unlikely objective for Spanish exploration. It was not clear from the legends that Edenic Aztlan had been especially rich in the worldly treasures Spaniards sought nor that it could be reached by earthly means. In the most well-known version of the Mexican origin story—that from missionary-ethnographer Fray Diego Durán's 1581 *History of the Indies of New Spain*—the inhabitants of Colhuacan, living as they did "poorly and simply," knew "nothing about" "the wealth" possessed by the later Aztecs. Rich or poor, moreover, Aztlan sounded hard to get to. At one point, Durán suggests that the distance to Aztlan was "so short that it" could "be covered in a month." In a subsequent, Arabian Nights–like account of Aztec emperor Montezuma I's (reigned 1440–1469) attempt "to seek out the place where his ancestors had dwelt," however, Durán has the monarch's advisor Tlacaelel warn that after the Mexicas' ancestors had "departed from their home everything turned into thorns and thistles. The stones became sharp, . . . the bushes . . . prickly and the trees . . . thorny. . . . Everything there turned against them, . . . so they could not return there." In such circumstances, it was best to send "wizards, sorcerers, magicians, who with their enchantments and spells" could "discover that place" rather than a conventional expedition of conquest or reconnaissance. Perhaps fittingly, centuries later, what or where Aztlan was remains uncertain.[32]

31. Diego Durán, *The History of the Indies of New Spain*, trans. Doris Heyden (Norman, Okla., 1994), 3–4, 9–21, 212–222; Durán, *Historia de las Indias de Nueva España e islas de la Tierra Firme* (Mexico City, 1967), II, 13, 18–29, 215–224. For a secondary introduction to this subject, see S. Lyman Tyler, "The Myth of the Lake of Copala and Land of Teguayo," *Utah Historical Quarterly*, XX (1952), 313–329; Tyler and Taylor, eds. and trans., "Report of Fray Alonso de Posada," *NMHR*, XXXIII (1958), 285–314.

32. Durán, *History of the Indies*, trans. Heyden, 12, 212–222; Durán, *Historia de las Indias*, II, 21, 215–222: "esas riquezas que traéis, no usamos acá de ellas, sino de pobreza y llaneza"; "siendo tan poco el camino que en un mes se anda"; "enviar a saber en qué lugares habían habitado sus antepasados"; "después que de allí salieron, todo se volvió espinas y abrojos; las piedras se volvieron puntiagudas para lastimarlos y las yerbas picaban, los árboles, espinosos: todo se volvió contra ellos,

This was not for lack of searching. In the late sixteenth and early seventeenth centuries, Spanish expeditions proceeded north, sometimes hearing what sounded like more encouraging reports en route. In one appealing variation of this tradition, they heard that somewhere to the northwest of Mexico and New Mexico lay a *"Laguna de Oro"* surrounded by rich and populous settlements. Spanish expeditions led by Francisco de Ibarra in the mid-1560s and Antonio de Espejo in 1582–1583 sought this Lake of Copala or Lake of Gold, and might have made it as far as the Gila Valley or the Casas Grandes region of Chihuahua (Ibarra) and Zuñi pueblo and the Colorado River (Espejo). One member of the Espejo expedition, Fray Bernaldino Beltran, reportedly heard from New Mexican Indians "of a large lake, with many towns and inhabitants, where the people rode in canoes bearing large, bronze-colored balls in the prows." Espejo wrote of hearing at Zuñi pueblo in 1583 about "a large lake where the natives claimed there were many towns. These people told us that there was gold in the lake region, that the inhabitants wore clothes, with gold bracelets and earrings, that they dwelt at a distance of a sixty days' journey from the place where we were." The 1604–1605 Oñate land expedition to the lower Colorado also picked up reports of "a lake on whose shores lived people who wore on their wrists bands or bracelets of a yellow metal" at a distance of "nine or ten days' travel." (They also heard fabulous tales of peoples sporting monstrously long ears or "virile members," sleeping in trees or underwater, and subsisting solely on the smell of food.)[33]

---

para que no supiesen ni pudiesen volver allá"; "brujos o encantadores y hechiceros, que, con sus encantamientos y hechicerías, descubriesen estos lugares."

33. On the *Laguna de Oro,* see Tyler, "Myth of the Lake of Copala," *Utah Historical Quarterly,* XX (1952), 313–329; Tyler and Taylor, eds. and trans., "Report of Fray Alonso de Posada," *NMHR,* XXXIII (1958), 304–305; Posadas, "Informe á S. M. sobre las tierras de Nuevo Méjico, Quivira y Teguayo," in Duro, *Don Diego de Peñalosa,* 65; Alonso de Benavides, *Fray Alonso de Benavides' Revised Memorial of 1634: With Numerous Supplementary Documents Elaborately Annotated,* ed. and trans. Frederick Webb Hodge, George P. Hammond, and Agapito Rey (Albuquerque, N.M., 1945), 40–41. On the Ibarra and Espejo expeditions, see Tyler, "Myth of the Lake of Copala," *Utah Historical Quarterly,* XX (1952), 316–317; George P. Hammond and Agapito Rey, trans., *Obregón's History of 16th Century Explorations in Western America, Entitled, Chronicle, Commentary, or Relation of the Ancient and Modern Discoveries in New Spain and New Mexico* (Los Angeles, Calif., 1928), 43, 50, 152–216; Hammond and Rey, eds. and trans., *Rediscovery of New Mexico,* part 2, 153–242. For Beltran, see "Brief and True Account of the Discovery of New Mexico by Nine Men Who Set out from Santa Bárbara in the Company of Three Franciscan Friars," ibid., 142–143; "Relacion breve y verdadera del descubrimiento del Nuevo Mexico," in Torres de Mendoza, ed., *Coleccion de documentos inéditos relativos al descubrimiento, conquista y organizacion de las antiguas posesiones españolas de América y Oceanía,* XV (Madrid, 1871), 146–150, esp. 149. For the information from Zuñi pueblo, see "Report of Antonio de Espejo," in Hammond and Rey, eds. and trans., *Rediscovery of New Mexico,* 213–232, esp. 225, 227; "Relacion del viage, que yo, Antonio Espejo, . . . hize con catorce soldados y un relijioso de la orden de San Francisco, á las provincias y poblaciones de la Nueva México," in *Coleccion*

What did these northern tales of a fabulous lake signify? Polite or mischievous Indian informants might have been indulging the wishes or testing the gullibility of their Spanish audience. Or Spaniards might have been hearing what they wanted to hear. Desperately hoping for an El Dorado in North America surpassing that of the continent to the south, and in keeping with a tendency of many North American explorers, they might have translated their Indian interlocutors' words into a description compatible with Spanish greed or fancy. They might have taken Indian references to nonprecious metals as evidence of gold or silver. Copper bells formed an important article of southwestern trade, and lacustrine or riparian peoples possessing bells or other ornaments of this metal could easily have generated the kinds of reports Europeans were hearing. Father Francisco de Escobar, who kept a diary on Oñate's journey to the Gulf of California, alluded to the danger of such confusion when, after much describing, gesturing, pointing, and showing on the part of Colorado River Indians, he confessed, "They almost convinced me beyond all doubt that there were both yellow and white metals in the land, though there is no proof that the yellow metal is gold or that the white is silver, for of this my doubts are still very great."[34]

One other possibility is that eager Spaniards like Beltran and Espejo were misunderstanding imperfect rumors of some actual area or combination of areas approximating the characteristics appearing in tales. The Puget Sound–Inside Passage region of the Pacific Northwest fits parts of the descriptions. The inland Pacific extensions in this area resemble lakes in many respects and have long been dotted with populous settlements. The inhabitants Europeans observed there in the late eighteenth century exhibited a remarkable facility with canoes as well as striking cloaks, hats, and copper ornaments. A long chain of trading relations connecting the pueblos of New Mexico and Arizona to the distant

---

de documentos inéditos relativos al descubrimiento, conquista y organizacion de las antiguas posesiones españolas de América y Oceanía, XV, 118, 180–181 (volume includes two slightly different versions). On the Oñate expedition, see "Fray Francisco de Escobar's Diary of the Oñate Expedition to California, 1605," in Hammond and Rey, Don Juan de Oñate, II, 1012–1031, esp. 1019, 1025–1026. Zárate Salmerón's 1626 secondhand account of the Oñate entrada placed the alluring lake and its gold-wearing inhabitants 400 hundred leagues from Mexico City, "14 days' journey" beyond the Colorado River, northwest of New Mexico ([Gerónimo de] Zárate Salmerón, Relaciónes: An Account of Things Seen and Learned by Father Jerónimo de Zárate Salmerón from the Year 1538 to Year 1626, trans. Alicia Ronstadt Milich [New York, 1966], 67, 69, 75–76, 90–94). For the original Spanish, see Zárate Salmerón, Relaciónes de todas las cosas que en el Nuevo México se han visto y sabido . . . , in Documentos para servir a la historia del Nuevo México, 1538–1778 (Madrid, 1962), 113–204, esp. 165, 167, 174–175, 190–195.

34. On copper bells, see Brian M. Fagan, Ancient North America: The Archaeology of a Continent, 3d ed. (New York, 2000), 314, 319, 323, 332, 348. For Escobar, see "Fray Francisco de Escobar's Diary of the Oñate Expedition to California, 1605," in Hammond and Rey, eds. and trans., Don Juan de Oñate, II, 1018–1020.

Pacific Northwest might have produced the kinds of stories Spanish explorers were hearing, especially if southwestern Indians were describing peoples they had heard about from others but had not seen themselves. Puget Sound and the Inside Passage might conceivably lie two hard months from Zuñi pueblo. A less distant explanation is that Spaniards were hearing and perhaps conflating accounts of other western bodies of water such as the Colorado Delta, intermittent Lake Cahuilla (in the Salton Trough in today's southern California), the Santa Barbara Channel, or the larger lakes in what is now Utah.[35]

In any case, though Spanish explorers heard tales of Aztec migration and observed monuments purportedly left by Aztec ancestors (such as the Casa Grande ruins in south-central Arizona), they failed to find a terrestrial realization of the desired Lake of Gold. This weakened or even discredited one impulse for exploration. By the late seventeenth and early eighteenth century, though still mentioning without elaboration "Gran Teguayo" as an object of exploration, an explorer like Kino could number "a lake . . . of gold" among the "great errors and falsehoods imposed upon us by those who have delineated this North America with feigned things which do not exist."[36]

These early Spanish failures to find promised riches amid the challenging terrain and peoples of the West provide an almost satisfactory explanation for the slowing of Spanish exploration after the first decade of the seventeenth century. Only almost, because the mineral, agricultural, scenic, and human bounties found by later explorers make it difficult to accept even the most reasonable explanation for Spanish inertia. Almost, also, because seventeenth- and eighteenth-century French explorers found reasons and means to move toward western North American regions repelling Spanish advances, thereby throwing Spanish inertia into high relief.

35. Conversations with historian Peter Wood made me much more attentive to the possibility of connections between the northwest coast and other parts of the Far West. Good introductions to Northwest Coast Indians include Philip Drucker, *Cultures of the North Pacific Coast* (San Francisco, 1965); Erna Gunther, *Indian Life on the Northwest Coast of North America: As Seen by the Early Explorers and Fur Traders during the Last Decades of the Eighteenth Century* (Chicago, 1972); and Kenneth M. Ames and Herbert D. G. Maschner, *Peoples of the Northwest Coast: Their Archaeology and Prehistory* (London, 1999). On connections between Zuñi pueblo and other parts of the Southwest, see T. J. Ferguson and E. Richard Hart, *A Zuni Atlas* (Norman, Okla., 1985), 52–55.

36. Kino, *Kino's Historical Memoir of Pimería Alta*, ed. and trans. Bolton, I, 91, 127–129, 213, 359 (quotation), II, 243, 258, 264. For other eighteenth-century references to Teguayo, see "Information Which I, Fray Carlos Delgado, Give Your Reverence of El Gran Teguayo, Which Is between West and North: It Is Distant Two Hundred Leagues, More or Less, from This Custodia [1745?]," in Hackett, ed., *Historical Documents Relating to New Mexico*, comp. Bandelier and Bandelier, III, 415–416; "Rivera to Casa Fuerte, Presidio del Paso del Río del Norte, September 26, 1727," in Thomas, ed. and trans., *After Coronado*, 211.

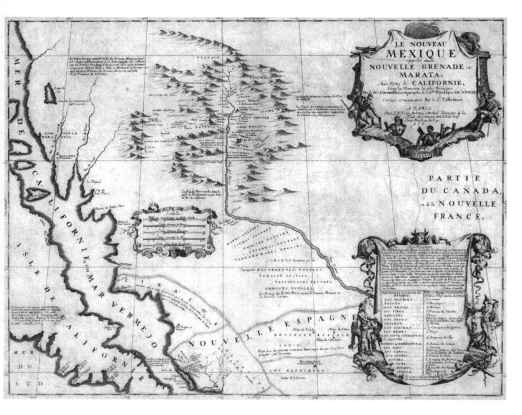

MAP 15. Vincenzo Coronelli, *Le nouveau Mexique.* c. 1685.
Geography and Map Division, Library of Congress, g4300 ct001821

A basic Spanish difficulty arose from the North American West's geographic relation to the rest of the Spanish Empire. When seventeenth- and eighteenth-century French explorers moved west and south toward the Pacific, Mexico, and New Mexico, they were moving toward known concentrations of wealth in Peru, New Spain, and, ultimately, the East Indies. The intervening lands and waters were uncertain, but the galleons returning from Spanish America and the East India ships coming back from Asia repeatedly verified the value of the final objective. In contrast, when Spanish explorers moved north from Mexico up the Pacific coast, into the mountainous Southwest, or onto the western and southern plains, they were, in one sense, going the wrong way. They were moving away from the New World's established Peruvian and Mexican centers of gravity, away from what some historians have called "Nuclear America," and into areas of probable risk and uncertain reward or toward French and British colonies whose economic value lagged for many years behind those of Spain.

From an imperial perspective, this was a questionable diversion of men and money away from the protection of Spain's core assets; from the entrepreneurial point of view of a private expedition financier, a dubious proposition.

This did not stop entirely the extension of Spanish colonization. From the 1547 discovery of silver deposits at Zacatecas, silver strikes pulled New Spain's mining operations and support activities northward. Beyond this mining frontier, regions such as New Mexico, Baja California, Texas, and Florida became — like Chile, Paraguay, and northern Argentina — the sites of Spanish missions and settlements. Their utility as venues for Christian conversion efforts, for the supply or security of more vital mining regions and bullion routes, or for the generation of a modest livelihood for a small number of colonists sufficed to bring missions and settlements into existence and sometimes to sustain them. But this somewhat meager importance failed to generate rapid, thorough, and effective efforts to explore the regions beyond.

French western exploration possessed the additional advantage of a well-developed fur trade. Because this commerce formed a central part of the economy of New France, and because the American West furnished pelts in large quantities, French explorers had both an obvious motive to move into western lands and a possible means of financing their operations. Moreover, because Indian cooperation was critical for the fur trade and because the commerce in pelts was crucial for cementing the Indian alliances necessary for the survival of the underpopulated French colonies, French explorers and officials were more likely than their Spanish counterparts to think in terms of collaborating with select native peoples rather than dominating them. French explorers could move a well-developed and essential economic activity into profuse new terrain, and they benefited from a justification for their roamings less dependent on the location of mineral wealth or the subjugation of settled native peoples.

The Spanish Empire, in contrast, having acquired the precious metals and native labor its rivals coveted, had less need to develop the kind of alternative enterprises sustaining the French, Dutch, and British North American colonies. New Mexico provides an exception proving the rule. As David J. Weber has noted, lacking the economic opportunities blessing Peru and Mexico, needy New Mexican settlers sent slaves, hides, and coarse furs from their Indian neighbors to markets farther south. A combination of factors, however, kept these activities small in scale and limited in appeal. These included the high cost of transport, limited demand for furs in Mexico, measures prohibiting — with varying degrees of success — Indian enslavement, meager interest among southwestern Indians in trapping and trading in small mammals, the labor demands of New Mexican agriculture, and the dangers of Indian attacks. Consequently, settlers already in New Mexico trafficked in furs and slaves, but such commerce was insufficiently lucrative to draw many opportunity-seeking Spaniards from Mexico City to Santa Fe and beyond. New Mexico and its trades in animals and people remained marginal affairs. Before 1763, the colony did not serve, as Canada and the Illinois country did for France, as a jumping-off point for extensive and more-or-less sanctioned western exploration.[37]

When, moreover, Spanish officials were looking for new American revenue sources, other regions often appeared more promising than the North American West. For the French American Empire, the region west of the Great Lakes and Mississippi constituted one of the only areas relatively open for expansion: south from the Saint Lawrence and east from the Mississippi led to British entanglements, the islands of the Caribbean by their very nature left little room for growth, and France failed before the end of the Seven Years' War to establish a presence in South America substantial and lasting enough to open hinterlands for exploration. For the Spanish Empire, in contrast, many areas presented themselves to curious geographers and potential explorers. Why set out from Mexico into North American deserts or Pacific coastal rocks when the Amazon and Orinoco basins stood within reach of Peru and the Spanish Main? This question was particularly pointed when the tropics seemed more likely to offer plants and other resources of value than did the dishearteningly barren landscapes covering much of the plains and Southwest. By the mid-eighteenth century, reform-minded Spanish ministers were already contemplating the possibility of using the natural riches of South America's tropical forests to revive the Spanish imperial economy. It was not yet clear how the arid North American Southwest could contribute immediately to the same end.[38]

37. David J. Weber, *The Taos Trappers: The Fur Trade in the Far Southwest, 1540–1846* (Norman, Okla., 1971), 12–21. Unsanctioned exploration from New Mexico is a possibility worth considering. To give one example, the Spanish Empire had its illicit and therefore usually anonymous western traders. As Weber has noted in his study of New Mexico's fur and hide trade, Spanish traders hoping to acquire slaves and pelts from Ute Indians were certainly moving beyond the colony's frontiers into the Great Basin and the southern Rockies after 1765 and might very well have been doing so decades earlier. Such activities would have familiarized at least some Spaniards with lands and peoples of great interest to imperial officials. Because "a royal order prohibited trade in Indian lands," however, these wide-ranging New Mexicans and their collaborators had every reason to keep such activities secret. We are left to speculate about what unauthorized and therefore undocumented wanderers might have been saying among themselves but keeping from metropolitan authorities (Weber, *Taos Trappers*, 23–26). Along these lines, Kino mentioned having evidence that, before the Pueblo Revolt, New Mexican Spaniards were journeying as far as the junction of the Gila and San Pedro rivers (in southeastern Arizona) to trade with the Pimas Sobaiporis (Kino, *Kino's Historical Memoir of Pimería Alta*, ed. and trans. Bolton, II, 257). See also Warner, ed., *Domínguez-Escalante Journal*, trans. Chavez, 90; Sánchez, *Explorers, Traders, and Slavers*, 17–39.

38. On tropical forests and the Spanish economy, see Demetrio Ramos Pérez, "La época de la

Indeed, in the early modern period, it was in many respects South America and its Incas, rather than North America and its fallen Aztecs, that presented the most plausible and enticing mirages to aspiring discoverers. This was especially true of the comparatively well-grounded reasons for searching around the lengthy perimeter of the Inca Empire. The Inca royal family furnished the most evident of these. Between 1537 and 1572, a fleeing Manco Inca and his successors had managed to construct and maintain a capital and state in exile northwest of Cuzco in the Vilcabamba region, where varied and difficult terrain and climate offered the Incas some protection against Spanish invaders. Manco and his followers carried considerable wealth with them, and when a Spanish expedition finally destroyed Vilcabamba, it gained a new supply of Inca treasure. These were verifiable events yielding tangible rewards, and, as J. H. Parry has remarked, actual Inca flight and the muddled Spanish notions to which it gave rise made more plausible the idea that other escapees from the Inca Empire might have found shelter in impenetrable thickets of the Amazon jungle or inaccessible corners of the Andes. These dreams lingered long after the Inca Empire had fallen. Not only did an Indian (at least in some versions of the tale) metropolis like the Enchanted City of the Caesars remain a plausible fancy into the late eighteenth century, but lost cities of the Incas and the Amazon Basin have been sought and found in the twentieth century—and the twenty-first is still young.[39]

Moreover, the many manifest antagonisms created by the bruising and wide-ranging character of Inca expansion suggested to Spaniards that, in addition to wealthy Inca escapees, it might also be reasonable to seek rich Inca enemies. These likely foes could be connected to other tales of South American treasure, such as the persistent legend of El Dorado. A good example is the story of the celebrated Ancoalli, a leader of the Chancas, an Andean people who fought, served, and fled the Incas in the mid-fifteenth century. Peruvian cosmographer and historian Pedro Sarmiento de Gamboa wrote in his 1572 *History of the Incas* of Ancoalli and the Chancas' decision to flee the treacherous designs of an Inca Empire jealous of Chanca martial prowess. The Chancas agreed "to seek a rugged and mountainous land where the Incas, even if they sought them, would not be able to find them." Other sixteenth-century Spanish authors, such as the

nueva Monarquía," in Luis Navarro García, ed., *Historia general de España y América*, XI, *América en el siglo XVIII: Los primeros Borbones*, 2d ed., part 1 (Madrid, 1989), xi–xli, esp. xxxvii–xxxix.

39. On Vilcabamba, see John Hemming, *The Conquest of the Incas* (New York, 1970), 230–231, 250–261, 425–440. On the implications of Inca flight, see J. H. Parry, *The Discovery of South America* (New York, 1979), 261; Henry J. Bruman, "Sovereign California: The State's Most Plausible Alternative Scenario," in Bruman and Clement W. Meighan, *Early California: Perception and Reality* (Los Angeles, Calif., 1981), 1–41, esp. 6–8. On later South American exploration, see Goodman, *Explorers of South America*, 57–64, 170–178; Hemming, *Conquest of the Incas*, 474–500.

celebrated Pedro Cieza de León, linked this fifteenth-century Chanca exodus to another vigorous South American legend, El Dorado. Cieza de León placed the "descendants of the famous chieftain Ancoallo" in the general area of Moyobamba (in what is now north-central, trans-Andean Peru) and indicated on the basis of Indian tales that, "having crossed the Andes Mountains," the Chancas "came to a great lake, which I believe must be the site of the tale they tell of El Dorado, where they built their settlements and have multiplied greatly."[40]

Together and separately, the story of fleeing Ancoalli and the legend of splendid El Dorado inspired Spanish exploration. A relative of Francisco Pizarro, Diego Pizarro de Carvajal, reportedly asked permission to lead an expedition to seek the errant Chanca chief. Francisco de Orellana's fantastic 1541–1542 descent of the Amazon occurred as a by-product of Gonzalo Pizarro's search not only for cinnamon trees in "the province of La Canela" but also of his quest for "[the region around] Lake El Dorado . . . a very populous and very rich land" east of Quito.[41]

Like the lost cities of the Incas sought and found by sixteenth-century Spaniards and twentieth-century Yale men, these stories of fleeing peoples, local spices, and golden lakes rested on a factual foundation. Though Spaniards never met him, Ancoalli existed, and sixteenth-century Spaniards observed the preserved skins of members of the Chanca nation he led. A kind of cinnamon tree grew east of the Andes, and, though reality fell short of legend, scholars have noted that there does seem to have been in the fifteenth century an Indian ceremony on Lake Guatavita (in modern Colombia) involving a chief's being coated in gold dust. Nearby Lake Siecha has yielded a golden statuette of what appears

40. Pedro Sarmiento de Gamboa, *History of the Incas*, trans. Clements Markham (1907; rpt. Mineola, N.Y., 1999), 115–117; Sarmiento de Gamboa, *Historia Indica por Pedro Sarmiento de Gamboa*, in P. Carmelo Sáenz de Santa Maria, ed., *Obras completas del Inca Garcilaso de la Vega*, 4 vols. (Madrid, 1960–1965), IV, 189–279, esp. 243. See also Garcilaso de la Vega, *Royal Commentaries of the Incas and General History of Peru*, trans. Harold V. Livermore, 2 vols. (Austin, Tex., 1966), I, 300–302; Pedro de Cieza de León, *The Incas of Pedro de Cieza de León*, ed. Victor Wolfgang von Hagen, trans. Harriet de Onis (Norman, Okla., 1959), 100, 130–131; Cieza de León, *Crónica del Perú*, in Sáenz de Santa María, ed., *Obras completas*, I, 104, 199: "los descendientes del famoso capitán Ancoallo"; "pasando por la montaña de los Andes, caminó por aquellas sierras hasta que llegaron, según también dicen, a una laguna muy grande, que yo creo debe ser lo que cuentan del Dorado, adonde hicieron sus pueblos y se ha multiplicado mucha gente."

41. Pizarro de Carvajal's expedition was "abandoned" around 1536 "for lack of supplies" (Cieza de León, *Discovery and Conquest of Peru*, ed. and trans. Cook and Cook, 421; Cieza de León, *Crónica del Perú*, in Sáenz de Santa María, ed., *Obras completas*, I, 349). On the Orellana expedition, see "Letter of Gonzalo Pizarro to the King" (brackets in original), in José Toribio Medina, ed., *The Discovery of the Amazon*, trans. Bertram T. Lee and H. C. Heaton (1934; rpt. New York, 1988), 245–251, esp. 245; see also "Oviedo's Description of Gonzalo Pizarro's Expedition to the Land of Cinnamon . . . ," ibid., 390–404, esp. 391–392, 402; Parry, *Discovery of South America*, 261–262; Goodman, *Explorers of South America*, 66–68.

to be a raft-borne royal ceremony. South America gave and continues to give good reasons to explore and reexplore its vast and difficult landscapes.[42]

Conditions north of the Valley of Mexico offered less sustenance for conquistador dreams. One reason was the absence of the notion that rich royals from or wealthy enemies of the Aztec Empire had sought refuge by fleeing north. No equivalent of Vilcabamba took shape in Arizona, and Spanish explorers lacked therefore one type of concrete example on which they could rest their conjectures. The passage of time also militated against North American entradas. The South American stories motivating exploration possessed a basis not only in reality but in recent reality. Vilcabamba was a city of the sixteenth century; Chanca flight and golden lacustrine ceremonies, happenings of the fifteenth. Aztlan, in contrast, was distant history, perhaps timeless mythology. To the extent that Durán's account can be rendered consistent with itself and conventional chronology, the ancestors of the Indian nations of Mexico seem to have left Aztlan in the ninth century. The intervening years made the reality of Mexican Indian origins that much more difficult to discern or believe. Finally, though Rocky Mountains promised silver veins, western shores sustained prosperous communities, and sierra streams would reveal golden nuggets, scholarly investigations have not yet uncovered an original of the Lake of Gold as closely connectable to its derivative legends as lakes Guatavita and Siecha are to El Dorado.[43]

Between the first decade of the seventeenth century and the years after the Seven Years' War, western North America gave prospective Spanish explorers good reason to stay home—and insufficient incentive to push Spanish reconnaissance more than incrementally or occasionally forward.

42. For the flayed Chancas, see Cieza de León, *Discovery and Conquest of Peru*, ed. and trans. Cook and Cook, 317; Cieza de León, *Crónica del Perú*, in Sáenz de Santa María, ed., *Obras completas*, I, 314. On the El Dorado legend, see John Hemming, *The Search for El Dorado* (New York, 1978), 104–105, 195–198; Goodman, *Explorers of South America*, 66–68, 77, 380; Parry, *Discovery of South America*, 260–261.

43. Durán, *History of the Indies*, trans. Heyden, 12; Durán, *Historia de las Indias*, II, 21.

2

EXPLOITING INDIGENOUS
GEOGRAPHIC UNDERSTANDING

The Indians [from the California coast to New Mexico] may very well have with each other some sort of communication and commerce, extending to considerable distances, although they do not maintain it very directly because they do not like to go very far from their native lands. It is true, however, that from hand to hand and from neighbor to neighbor they exchange the things which some have in abundance and others lack, and in these interchanges give reciprocal notices of their lands.

—MIGUEL COSTANSÓ, 1772

From this square [in Cuzco] four highways emerge; the one called Chinchay-suyu leads to the plains and the highlands as far as the provinces of Quito and Pasto; the second, known as Cunti-suyu, is the highway to the provinces under the jurisdiction of this city and Arequipa. The third, by name Anti-suyu, leads to the provinces on the slopes of the Andes and various settlements beyond the mountains. The last of these highways, called Colla-suyu, is the route to Chile. Thus, just as in Spain the early inhabitants divided it all into provinces, so these Indians, to keep track of their wide-flung possessions, used the method of highways.

—PEDRO DE CIEZA DE LEÓN, 1553

Consideration of the inhibiting effects of western exploratory difficulties and disappointments, of the Far West's geographic position, and of the competitive allure of other potential targets of investigation yields a fair explanation for the pre-1763 Spanish failure to explore the better part of the North American West. Such consideration leaves open, however, the question of why, if the physical and human geography of the North American West rendered its thorough exploration by Spaniards themselves arduous and unappealing, they could not avail themselves of Indian information, utilizing native geographic knowledge as a substitute for Spanish sails and feet. When Vizcaíno sailed the coast, when Kino trudged along the Colorado, when Vélez Cachupín ransomed "Indian children" taken from and by "all nations," why not inquire about what lay beyond the frontiers of Spanish rule and reconnaissance? To give a more pointed example, Villasur's death in 1720 demonstrated the dangers inherent in pushing Spanish expeditions onto the plains. In the same year, New Mexico captain Don Felix Martínez reported speaking to "captives in New Mexico" from the same Pawnee nation that had attacked Villasur and his men. Talking to Pawnees in New Mexico might have furnished an easier means of acquiring information than exchanging gunshots in Nebraska.[1]

Spanish investigators did, in fact, try to take advantage of Indian geographic awareness. Fray Posada, while serving as a missionary "on the more remote frontiers" of New Mexico in the 1650s, "acquired information about the lands from the infidel Indians." He talked at Pecos with "Apacha Indians" holding "some captive Indian children from Quivira" and with an Indian from the Jemez pueblo "who told him . . . of having been captive in Teguayan provinces for a period of two years." Kino, in 1701, at a Colorado River crossing "a day's journey" from the river's mouth, asked "Quiquima," "Cutgana," and "Hogiopa" Indians "about everything farther on, particularly toward the west and south." Members

1. Miguel Costansó to Antonio María de Bucareli y Ursúa, Sept. 5, 1772, in Herbert Eugene Bolton, ed. and trans., *Anza's California Expeditions*, V, *Correspondence* (Berkeley, Calif., 1930), 8–11, esp. 9; Pedro de Cieza de León, *The Incas of Pedro de Cieza de León*, ed. Victor Wolfgang von Hagen, trans. Harriet de Onis (Norman, Okla., 1959), 144; Cieza de León, *La crónica del Perú*, in Carmelo Sáenz de Santa María, ed., *Obras completas* (1553; rpt. Madrid, 1984), I, 117: "Desta plaza salían cuatro caminos reales; en el que llamaban Chinchasuyo se camina a las tierras de los llanos con toda la serranía, hasta las provincias de Quito y Pasto; por el segundo camino, que nombran Condesuyo, entran las provincias que son sujetas a esta ciudad y a la de Arequipa. Por el tercero camino real, que tiene por nombre Andesuyo, se va a las provincias que caen en las faldas de los Andes, y a algunos pueblos que están pasade la cordillera. En el último camino destos, que dicen Collasuyo, entran las provincias que llegan hasta Chile. De manera que, como en España los antiguos hacían división de toda ella por las provincias, así estos indios, para contar las que había en tierra tan grande, lo entendían por sus caminos."

"Indian children" quotation from "Revilla Gigedo to Marqués de Ensenada, Mexico, June 28, 1753," in Alfred Barnaby Thomas, [ed. and trans.], *The Plains Indians and New Mexico, 1751–1778: A Collection of Documents Illustrative of the History of the Eastern Frontier of New Mexico* (Glendale, Calif., 1940), 111–114, esp. 111. On New Mexico's acquisition of Indian captives, see also Juan Joseph Lobato to Tomás Vélez Cachupín, Aug. 28, 1752, ibid., 114–117; and "Declaration of Fray Miguel de Menchero, Santa Bárbara, May 10, 1744," in Charles Wilson Hackett, ed., *Historical Documents Relating to New Mexico, Nueva Vizcaya, and Approaches Thereto, to 1773*, comp. Adolph F. Bandelier and Fanny R. Bandelier, 3 vols. (Washington, D.C., 1923–1937), III, 395–413, esp. 401. Martínez quotation from "Declaration of Martínez, México, November 13, 1720," in Alfred Barnaby Thomas, ed. and trans., *After Coronado: Spanish Exploration Northeast of New Mexico, 1696–1727; Documents from the Archives of Spain, Mexico, and New Mexico* (Norman, Okla., 1935), 170–172, esp. 171. Fray Miguel de Menchero wrote of speaking at a 1731 trade fair with a "Ponnas" Indian who averred that "his country" was 110 "days, or suns" away ("Declaration of Fray Miguel de Menchero, Santa Bárbara, May 10, 1744," in Hackett, ed., *Historical Documents Relating to New Mexico*, comp. Bandelier and Bandelier, III, 401).

of the Alarcón, Espejo, Oñate, Rivera, and Portolá expeditions mentioned looking at maps drawn by the Indians they met.[2]

The results of such interrogations proved as disappointing in many respects as those from other forms of Spanish information-gathering, as we have seen. What made it so difficult for Spanish western explorers and officials to use indigenous informants to help them peer into territories beyond the Spanish frontier?

Comparison offers one path to an answer. Spanish difficulties in making use of native geographic reports in the North American West can be set alongside Spanish acquisition and employment of indigenous geographic knowledge within the Aztec and Inca empires. Similarities and contrasts among these different geographic endeavors can bring into sharper relief the factors most significant in limiting and facilitating Spanish apprehension of western topography. They can do more still. We can use the relative ease or inability of Spanish information-gathering in different areas not only to learn something about Spanish reconnaissance but also to suggest something about the circumstances and geographic horizons of western Indian nations.

When Spanish western investigations came up short, and quizzical Spaniards turned to native sources of geographic intelligence, they missed in the lands north of Mexico the indigenous imperial expertise and infrastructures fostering

2. For the Posada quotation, see S. Lyman Tyler and H. Darrel Taylor, eds. and trans., "The Report of Fray Alonso de Posada in Relation to Quivira and Teguayo," *NMHR*, XXXIII (1958), 285–314, esp. 288, 301, 305; for the original Spanish, see Fray Alonso de Posadas, "Informe á S. M. sobre las tierras de Nuevo Méjico, Quivira y Teguayo," in Cesáreo Fernández Duro, *Don Diego de Peñalosa y su descubrimiento del reino de Quivira* (Madrid, 1882), 53–67, esp. 53, 62, 65. For Kino, see Eusebio Francisco Kino, *Kino's Historical Memoir of Pimería Alta: A Contemporary Account of the Beginnings of California, Sonora, and Arizona . . .* , ed. and trans. Herbert Bolton, 2 vols. (Cleveland, Ohio, 1919), I, 317–318. For the other expeditions, see "Narrative of Alarcón's Voyage," in Richard Flint and Shirley Cushing Flint, eds. and trans., *Documents of the Coronado Expedition, 1539–1542* (Dallas, Tex., 2005), 185–222, esp. 204; "Report of Antonio de Espejo," in George P. Hammond and Agapito Rey, eds. and trans., *The Rediscovery of New Mexico, 1580–1594: The Explorations of Chamuscado, Espejo, Castaño de Sosa, Morlete, and Levyva de Bonilla and Humaña* (Albuquerque, N.M., 1966), 213–231, esp. 230; Testimony of Juan Rodríguez, in "Investigation Made by Order of His Lordship concerning the Discovery of New Mexico," "Questioning of the Indian Miguel Brought from New Mexico by the Maese de Campo, Vicente de Zaldívar," and "Fray Francisco Escobar's Diary of the Oñate Expedition to California, 1605," in George P. Hammond and Agapito Rey, eds. and trans., *Don Juan de Oñate: Colonizer of New Mexico, 1595–1628* (Albuquerque, N.M., 1953), II, 837–871, 871–877, 1012–1031, esp. 867, 872–875, 1024; Joseph P. Sánchez, trans., "Translation of Juan María Antonio Rivera's Second Diary, October 1765," in Sánchez, *Explorers, Traders, and Slavers: Forging the Old Spanish Trail, 1678–1850* (Salt Lake City, Utah, 1997), 149–157, esp. 153; "The Portolá Expedition of 1769–1770: Diary of Miguel Costansó," Academy of Pacific Coast History, *Publications*, ed. Frederick J. Teggart, 4 vols. (Berkeley, Calif., 1911), II, 161–327, esp. 184–185.

h geographic comprehension of the great polities of Meso- and Andean
ca. In the North American West, precise and far-reaching indigenous geo-
c understanding proved less accessible and understandable — and seem-
ugly less extant. As a result, unable or unwilling to reconnoiter large parts of
the American West themselves, and unsuccessful in their efforts to acquire clear
geographic information regarding distant lands from native peoples, Spanish investigators were left to speculate about the lands and waters to the north. Spanish officials struggling to assess the implications of American geography for imperial policy would have to make do.

Communicating Geographic Information

Indigenous linguistic diversity formed the most basic problem. It was often difficult for Spanish scouts to understand what they were hearing. An early modern Spaniard walking from the southern plains to the Pacific could very well
come across languages from the Caddoan, Kiowa-Tanoan, Uto-Aztecan, Eyak-
Athabaskan, and Cochimí-Yuman language families. If he lingered in California,
he would have to request additional paper and ink to record the bewildering
variety of tongues. Spanish explorers commented often on the remarkable array
of languages they encountered. In 1540, Hernando de Alarcón heard from the
"principal" of one of the villages he passed that the Colorado's banks were "inhabited by [speakers of] twenty-three languages. . . . And [he said] that besides
these twenty-three languages, there were more, also on the river, that he did not
know." Coronado reported in a 1541 letter to Charles V that "in each town" on
the plains "they speak their own" language and that he "suffered" from "having
lacked someone who understands them."[3]

One native response to this linguistic variety was sign language. Spaniards
noted its extensive use and had uneven results in their attempts to understand
it. Castañeda described the Coronado expedition's meeting a plains "people very
skillful with [hand] signs, such that it seemed as if they were talking. And they
made a thing understood so that there was no longer need for an interpreter."
Vizcaíno's expedition, in contrast, picked up from the Indians of Santa Catalina Island (off the coast of southern California between Los Angeles and San
Diego) what they thought were reports "by signs" of shipwrecked Spaniards but
gave up efforts to verify the testimony, "as we could not understand their language, all was confusion and there was little certainty as to what they said." Sign

3. For Alarcón, see "Narrative of Alarcón's Voyage," in Flint and Flint, eds. and trans., *Documents
of the Coronado Expedition,* 195 (brackets in original). For Coronado, see "Vásquez de Coronado's
Letter to the King," ibid., 317–325, esp. 321, 324.

language was probably better than nothing for most Spaniards. At the same time, its widespread use might have borne mute testimony to the difficulty even for native peoples of understanding one another's tongues.[4]

Trying to understand signs was one way Spanish investigators could attempt to communicate with native peoples. Another was to find a spoken language familiar, perhaps as a second or third rather than mother tongue, to Indians over a large area. Spaniards might learn such a language, or native interpreters knowing both it and Spanish could assist communication with many peoples. Kino hoped Pima would serve this purpose and was sanguine about the extent to which both local Indians and Jesuit missionaries could manage the challenges of the Colorado and Gila's plethora of idioms. He asserted that the "Pima language" extended "more than two hundred leagues into the interior, even among the other and distinct nations of the Cocomaricopas, Yumas, and Quiquimas." In his view, it was neither the sole tongue of the region nor the natal speech of all its peoples, but "in all places" were "found intermingled some natives who speak both languages, that of the nation where they are and our Pima tongue, and therefore everywhere we have plenty of good interpreters." Such wide dispersal of Pima could aid the efforts of Kino and others in the Colorado–Gila River region, but it is worth noting that, even allowing for the inevitable vagueness of Kino's optimistic estimate of distance, his description of Pima suggests its value for about five to six hundred miles. This is a substantial distance, but beyond it—and much of the West lay more than six hundred miles from the lower Colorado—the Spanish would have to find new interpreters of new tongues. Pima would not carry them to Puget Sound.[5]

4. "The Relación de la Jornada de Cíbola, Pedro de Castañeda de Nájera's Narrative, 1560s," in Flint and Flint, eds. and trans., Documents of the Coronado Expedition, 408, 463 (brackets in original). See also "Inquiry concerning the New Provinces in the North, 1602," in Hammond and Rey, eds. and trans., Don Juan de Oñate, II, 836–877, esp. 836, 854–855; Francisco Garcés, A Record of Travels in Arizona and California, 1775–1776, ed. and trans. John Galvin (San Francisco, 1965), 89; Garcés, Diario de exploraciones en Arizona y California en los años de 1775 y 1776 (Mexico City, 1968), 89. For Vizcaíno, see "Diary of Sebastian Vizcaíno," in Herbert Bolton, ed., Spanish Exploration in the Southwest, 1542–1706 (New York, 1916), 86; Vizcaíno's Diary, in Francisco Carrasco y Guisasola, ed., Documentos referentes al reconocimiento de las costas de las Californias desde el cabo de San Lucas al de Mendocino (Madrid, 1882), 95. Costansó had much the same to say about communication with bands of Indians around San Francisco Bay in November 1769: "Some of the men asked them various questions by means of signs, in order to obtain from them information they desired, and they were very well satisfied with the grimaces and the ridiculous and vague gestures with which the natives responded— a pantomime from which, truly, one could understand very little, and the greater part of the men understood nothing"; "The scouts arrived at night, . . . undeceived in regard to the information of the natives, and their signs which at last they confessed were quite unintelligible" ("Diary of Miguel Costansó," in Teggart, ed., Academy of Pacific Coast History, Publications, II, 270-271).

5. Kino, Kino's Historical Memoir of Pimería Alta, ed. and trans. Bolton, II, 269-270. Soldier Hernán Gallegos, who accompanied and chronicled the 1581-1582 Chamuscado-Rodríguez expedi-

Kino's optimism notwithstanding, although Spanish missionaries and traders learned Indian tongues, Spanish explorers often found that, even when knowing the same general language as the people they met or being accompanied by translators of it, comprehension proved elusive. Father Escalante records a 1776 encounter with a group who spoke "Yuta," but "so differently from all the rest" that neither Escalante and Father Domínguez's Ute-speaking Spanish interpreter nor their "Laguna" (Uintah) Ute companion succeeded in communicating clearly with them. In a similar case, even Father Francisco Garcés, despite his considerable linguistic acumen, found in 1775 that he was "not able to explain" himself "so well to" one group, the "Jalliquamais," "because although their language seems to be the same as that of the Cajuenches the difference is marked." With so many languages and dialects dispersed throughout the Southwest, even the most linguistically gifted Iberians found it difficult to learn enough of them to communicate with the peoples of a really large area.[6]

In addition to difficulties arising from the diversity of languages, Spanish scouts also coped with the even more basic problem of the ambiguity of words. This can be seen in misunderstandings regarding the relation between North American lands and North Pacific waters beyond Cape Blanco. One reason figures like Hidalgo and Vélez Cachupín thought some part of the Pacific lay close to New Mexico or the Front Range of the Rockies was because they thought Indians were telling them this. Part of the confusion might have come from not knowing exactly what was meant by various words understood to suggest "near" or "far." In the diary of his 1706 journey to Apache *rancherías* at El Cuartelejo (probably north of the Arkansas River in western Kansas), Sergeant Major Juan

tion to New Mexico, noted Pueblo Indian trade with Plains buffalo hunters and recorded Pueblos saying that, through this trade, "by communicating with one another, each nation had come to understand the other's language." This suggests both initial Pueblo-Plains communicative difficulties and the eventual capacity to overcome them. See "Gallegos' Relation of the Chamuscado-Rodríguez Expedition," in Hammond and Rey, eds. and trans., *Rediscovery of New Mexico*, 67–114, esp. 87. The same pattern is visible in Garcés's description of the Yumas and "Jamajabs": "su lengua es distinta, pero con la comunicación se entienden bastante" (Garcés, *Record of Travels in Arizona and California*, ed. and trans. Galvin, 34; Garcés, *Diario de exploraciones en Arizona y California*, 42).

6. Ted J. Warner, ed., *The Domínguez-Escalante Journal: Their Expedition through Colorado, Utah, Arizona, and New Mexico in 1776*, trans. Angelico Chavez (Provo, Utah, 1976), 81, 177. See also Sánchez, trans., "Translation of Incomplete and Untitled Copy of Juan María Antonio Rivera's Original Diary of the First Expedition, 23 July 1765," in Sánchez, *Explorers, Traders, and Slavers*, 137–147, esp. 145: "The variety of languages is so great that there are some tribes that do not understand each other." For Garcés, see his *Record of Travels in Arizona and California*, ed. and trans. Galvin, 19; and Garcés, *Diario de exploraciones en Arizona y California*, 29. For other references to the difficulties multiple languages posed for Garcés and the Indians he met and to the conditions and techniques enabling them to overcome these communicative obstacles, see pages 9, 17, 21, 24, 28, and 34 in the translated volume, and 20, 27, 31, 33, 36, and 42 in the Spanish original.

de Ulibarri reported a conversation with Apaches "about the seas to the north and east, they said that they know about them only from other tribes that the seas were not very far, and before coming to that of the north, there are three days of long journeys beyond a tribe whom they call the Pelones." If these northeastern seas were the Great Lakes, as seems possible, then this gives some indication of what "far" could mean. In fact, if the Great Lakes were not considered far from western Kansas, then something like the Pacific, the distance to which was not immeasurably greater, might not seem especially remote either. On the other hand, if Ulibarri's Apaches were speaking of northern and eastern bodies of water closer to western Kansas and smaller than the Great Lakes (a stretch of the Missouri or the Platte, for example) then some large lake or river to the northwest (Salt Lake, for instance) might also be going under a designation Spanish listeners were interpreting as "sea." In either case, Spaniards could easily get a vague and perhaps inaccurate understanding of key features of western geography. The bodies of water in question and the distance to them was unclear.[7]

Spanish investigators might hope that descriptions of the people living on these seas, rivers, or lakes would help clarify matters, but questions of oceanic identification became tangled instead with tricky questions of ethnic classification. What was understood to be the sea was often reported to have white or Spanish inhabitants, but what "white" or "Spanish" meant was less than evident. The 1720 testimony of Santa Fe soldier Bartolomé Garduño furnishes a good example. Garduño indicated that Indians at El Cuartelejo had informed him that the Pawnees living "on the other side of the river which they call Jesús María [probably the South Platte]" were "living with Spaniards who are well clothed, and with white women, and go about on the sea in their houses of wood, giving him to understand they are small craft, and that in that region the sea is close." It is highly unlikely that Spaniards resided beyond the South Platte, but French traders whom some Indian groups might have denominated "Spaniards" were moving west from the Great Lakes and Mississippi in the late seventeenth and early eighteenth centuries. Spanish testimony lends some credence to this possibility. In his 1706 diary, Juan de Ulibarri noted that the Apaches around El Cuartelejo spoke of "the other Spaniards who live farther toward the east." In 1726, Yldefonso Rael de Aguilar, a Santa Fe resident and Villasur expedition survivor, testified that "Spaniards" was the "name which they [the residents of the village that attacked Villasur] give to all those who are white." Just

7. "The Diary of Juan de Ulibarri to El Cuartelejo, 1706," in Thomas, ed. and trans., *After Coronado*, 59–77, esp. 74.

as Euro-Americans used general terms like "Ute," "Apache," and "Pawnee" to cover loosely a complicated and diverse human reality, Plains Indians appear from Spanish sources to have been using terms like "French" or "Spanish" to denominate what eighteenth-century Europeans would have considered separate peoples.[8]

Another possibility is that Garduño could have been hearing of a nation like the Hidatsas, Mandans, or Arikaras boating on the Missouri or one of its tributaries. Other scraps of evidence point in this direction. In 1716, Fray Francisco Hidalgo wrote that the "Caynigua and Panni Indians" were "white Indians" who traded "in clothing, French guns, trinkets, and many other things." He indicated further that on one of the rivers flowing into the Missouri "there is a large city. . . . Indians do not live in this large city, but it is inhabited by white people. They must be either Tartars or Japanese from beyond the watershed." A large Japanese or Tartar city there was not, but, as will be discussed at greater length in subsequent chapters, large towns of trading people often described as "light-skinned" and often confused with Europeans dotted the upper Missouri and its tributaries. From a distance, through the medium of Indian perceptions and languages, it was difficult for Spaniards to determine who was being talked about.[9]

The coming together on the plains of a range of interpretive problems can be seen in the case of an Indian prisoner known by the Spanish as Miguel. A cap-

8. "Diary of Juan de Ulibarri," "Declaration of Garduño, México, November 15, 1720," and "Testimony of Aguilar, Santa Fé, July 1, 1726," in Thomas, ed. and trans., *After Coronado*, 59–77, 172–174, 226–228, esp. 73, 227. Identification of the South Platte River is from Thomas, 37, map (260–261).

9. On the "bull boats" employed by Missouri Valley groups, see Donald J. Lehmer, "Plains Village Tradition: Postcontract," in William C. Sturtevant, ed., *Handbook of North American Indians*, XIII, *Plains*, part 1, ed. Raymond J. DeMallie (Washington, D.C, 2001), 245–255, esp. 253. For Hidalgo, see Mattie Austin Hatcher, trans., "Descriptions of the Tejas or Asinai Indians, 1691–1722, III," *SWHQ*, XXXI (1927), 50–62, esp. 60. Interrogated by New Mexican authorities in 1750, Frenchman Pedro Latren (or Satren) claimed that the "Canes (or Kances) nation" were "white Indians, very proficient in the use of firearms" and that they were the ones who had defeated the 1720 Villasur expedition. See Tomás Vélez Cachupín, Santa Fé, Mar. 5, 1750, in José Antonio Pichardo, *Pichardo's Treatise on the Limits of Louisiana and Texas: An Argumentative Historical Treatise with Reference to the Verification of the True Limits of the Provinces of Louisiana and Texas . . .*, ed. and trans. Charles Wilson Hackett, trans. Charmion Clair Shelby, 4 vols. (Austin, Tex., 1941), III, 312–320, esp. 315. Word of light-skinned trading peoples was making its way to Spanish missionaries in the East Texas region where Hidalgo was picking up his information. In 1722, Fray Isidro Félix de Espinosa reported receiving information from a Missouri River Indian nation about the forty-eight pueblos of the Arikaras. See Espinosa, *Crónica de los Colegios de Propaganda Fide de la Nueva España*, ed. Lino G. Canedo, 2d ed. (Washington, D.C., 1964), 689: "Y de una [nación] que está poblada por el río Missuri corriente arriba, hay noticia de la nación arricará, que son cuarenta y ocho pueblos en término de diez leguas." Peter Wood has suggested in a different context the possibility that Northwest Coast Indians, rather than Japanese or Tartar merchants, might have been crossing the Continental Divide to trade with upper Missouri peoples.

tive of the Escanxaques, Miguel was taken by the Oñate expedition during the 1601 battle mentioned in the previous chapter. He accompanied Oñate's men on their return to New Mexico and then journeyed to Mexico City, where he offered tantalizing information during official questioning in 1602. Miguel "explained by signs the place where he was born" and "how he was taken prisoner and carried away by the [Escanxaque] enemy to other lands where he grew up." These signs were made more intelligible by his making a "picture map" for his interrogators, marking on paper "some circles resembling the letter 'O,'" and explaining "in a way easily understood . . . what each circle represented"; "he drew lines, some snakelike and others straight, and indicated by signs that they were rivers and roads." Giving further details, "he said that from the pueblo of Tancoa, where he was born, to the place where he was taken captive and grew up, it was twenty-two days' travel. . . . From the Great Settlement where the Spaniards took him prisoner to the place where he was held captive by the Indians, it was fifteen days' travel. From his pueblo of Tancoa to Encuche, where gold is found, it was forty-four days." If these figures for trip length are accurate, and if an average day's travel by foot consisted of somewhere between, say, five and twenty-five miles, Miguel might very well have been personally familiar with or well informed about thousands of square miles of plains terrain. This would fit well with scholarly arguments from archaeological, folkloric, and European documentary evidence that not only trade goods but also at least some people were covering remarkable distances across the plains and prairies, perhaps especially before the disruptive effects of European diseases, Spanish interlopers, and Spanish slave traders were fully felt.[10]

10. On the taking of Miguel, see testimony of Baltasar Martínez, Juan de León, and Juan Rodríguez, in "Investigation Made by Order of His Lordship concerning the Discovery of New Mexico," and "Questioning of the Indian Miguel," in Hammond and Rey, eds. and trans., *Don Juan de Oñate*, II, 847–849, 860–861, 868–877. For discussions and photographic reproductions of the Miguel map, see Mark Warhus, *Another America: Native American Maps and the History of Our Land* (New York, 1997), 9, 24–32; G. Malcolm Lewis, "Maps, Mapmaking, and Map Use by Native North Americans," in David Woodward and G. Malcolm Lewis, eds., *Cartography in the Traditional African, American, Arctic, Australian, and Pacific Societies* (Chicago, 1998), 51–182, esp. 125–127. For the quotations, see "Questioning of the Indian Miguel," in Hammond and Rey, eds. and trans., *Don Juan de Oñate*, II, 873–874. In 1694, "the captain of the Llanos Apaches" reported "twenty-five to thirty days" journeys from his ranchería to "the first settlements" of "the kingdom of Quivira." See "Diego de Vargas, Campaign Journal, 30 April–7 May 1694, DS," in John L. Kessell, Rick Hendricks, and Meredith D. Dodge, eds., *Blood on the Boulders: The Journals of Don Diego de Vargas, New Mexico, 1694–97*, 2 vols. (Albuquerque, N.M., 1998), I, 217–222, esp. 220. For discussion of plains and prairie movement, see Colin G. Calloway, *One Vast Winter Count: The Native American West before Lewis and Clark* (Lincoln, Neb., 2003), 53–65, 277; Brian M. Fagan, *Ancient North America: The Archaeology of a Continent*, 3d ed. (New York, 2000), 130–134, 146–156; Gary Clayton Anderson, *The Indian Southwest, 1580–1830: Ethnogenesis and Reinvention* (Norman, Okla., 1999), 13–29.

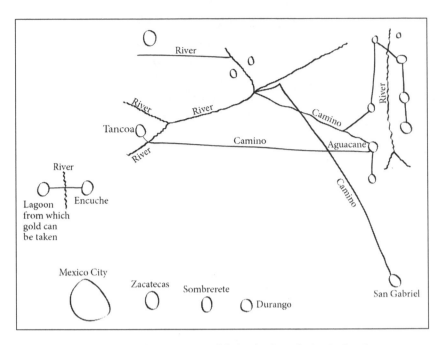

MAP 16. Miguel map. 1602. Simplified and redrawn by Jim DeGrand

Interpreting Miguel's testimony was tricky for early-seventeenth-century Spaniards, however. Part of the problem was linguistic. Miguel and his Spanish interlocutors were relying on signs rather than a common verbal language. The map he drew was highly suggestive, but Spanish examiners had great difficulty matching the symbols he drew on paper to the terrain and settlements of early North America. There was also the danger that the experiences bringing Miguel before Spanish officials might be shaping the information he was giving them. Some of what appeared to be indigenous geographic information communicated by him to his questioners might be instead improvisations calculated to please them or ideas gleaned from Spaniards on the long journey from the plains to Mexico City. One member of Oñate's expedition, Baltasar Martínez, suspected as much, testifying that Miguel told "what the soldiers and those who questioned him wanted to hear." There would be nothing unreasonable about Miguel's prevaricating to mollify his violence-prone and gold-loving captors. Miguel is a fascinating case for modern historians, because he hints at a pre-European plains world of vast horizons, but he was a frustrating case for seventeenth-century Spaniards because they could make neither sense nor certainty of what he was telling them. This was the larger problem confronting

Spanish investigators trying to comprehend the lands east of New Mexico and north of coastal Texas. Information seemed to be coming from great distances, but it was difficult to determine what reports meant.[11]

Spaniards also encountered difficulties acquiring and interpreting intelligence in the physically and culturally very different areas north and west of New Mexico. Despite tough terrain, the dry and mountainous Southwest was certainly a region where information moved. Spanish observers noted Indian reports of distant Spanish activities. As Ascensión seems to have heard rumors of the Oñate expedition from San Diego Indians in 1602, Cabrillo's 1542 expedition up the California coast and Alarcón's 1540 probe up the Colorado River picked up repeated reports of "men like us" tromping around the interior, perhaps parties from the Coronado entrada or groups of Spaniards active in northern Mexico. Similarly, in the late 1760s and early 1770s, Spanish explorers like Anza, Garcés, and Miguel Costansó (engineer on the 1769 Portolá expedition) surmised that Indians south of the Colorado and at the junction of the Gila and Colorado rivers were hearing about Spanish expeditions on the California coast and Spanish activities in New Mexico and that Indians on the Santa

11. For Martínez's suspicions, see "Inquiry concerning the New Provinces in the North, 1602," in Hammond and Rey, eds. and trans., *Don Juan de Oñate*, II, 849. Miguel's testimony was, moreover, often contradicted by itself or by the testimony of others, and this made him less credible as an informant. He seems to have indicated, for example, both that he had and had never returned to his natal village; and the testimony of different Plains Indians challenged his statements. One instance concerns clothing. When "shown a Mexican Indian wearing an ordinary white blanket," Miguel was understood by his Spanish questioners to have indicated "that he had never seen or heard of people anywhere who dressed like that, but that the Indians of his land, and of other places he knew about, all dressed in skins of deer or buffalo." Other Indians from the community from which Miguel was taken, however, though possessing neither blankets nor cloth of cotton or a similar fabric, seem at least to have heard of such things. Baltasar Martínez remembered seeing "an Escanxaque Indian with a cotton thread tied to the end of his bow as a great decoration. When the Spaniards asked him by signs where he had obtained it, he replied, also by signs, that he got it at some pueblos in the direction of San Gabriel and the province where the Spaniards had been" (845, 876).

To keep the difficulties of interpreting Miguel's testimony in perspective, it is worth noting that Spanish officials also found Oñate's reports less than clear. New Spain's Viceroy Monterrey observed in 1602 that it was difficult to determine where precisely Oñate had gone during his 1601 expedition onto the plains because his "papers" were "so confused." In particular, the viceroy noted, "One does not find in them any record of latitude," nor did Oñate "have anyone along who knew how to take it." See "Viceroy Monterrey to the King," in "Discussion and Proposal of the Points Referred to His Majesty concerning the Various Discoveries of New Mexico," 1602, in Hammond and Rey, eds. and trans., *Don Juan de Oñate*, II, 906–924, esp. 917; "Inquiry concerning the New Provinces in the North, 1602," ibid., 870–871; Viceroy Monterrey, "Discurso y proposicion que se hace á vuestra magestad de lo tocante á los descubrimientos del Nuevo México por sus capítulos de puntos diferentes," in *Coleccion de documentos inéditos relativos al descubrimiento, conquista y organizacion de las antiguas posesiones españolas de América y Oceanía* (Madrid, 1871), XVI, 56.

Barbara Channel were receiving goods and information originating in the Rio Grande Valley. Such reports indicate rapid movement, across hundreds of miles of difficult terrain, of what must have seemed to Indian communities like highly interesting information. They also raise two questions: What was the relation between the circulation of reports and the movement of people? And what was the outer range of these southwestern nations' cognizance and mobility?[12]

The earliest Spanish reports record lengthy Indian journeys. Alarcón met one man who claimed he had made a thirty-day journey to "Cíbola" and that he traveled "customarily . . . along a trail that followed the river, [and] it consumed forty days." Another informant mentioned journeys to "Cíbola" normally lasting "two complete moon cycles." Personal familiarity with peoples and things far from the lower Colorado River seems to have gone along with these journeys. Alarcón met individuals familiar with buffalos and possessing not only precise but also seemingly eyewitness information about inland Spanish expeditions such as those of Esteban and Coronado.[13]

Eighteenth-century evidence generally sustains Alarcón's 1540 impressions of regional horizons. Contemplating connections among the lower Colorado and Gila people he met and their neighbors beyond, Kino estimated in the first decade of the eighteenth century that peoples and messengers in the area were making roughly 500- to 600-mile journeys, whereas goods were moving perhaps 750 to 900 miles. The "messages from the Pimas were from one hundred and seventy leagues' journey, and those from the Quiquimas were from two hundred leagues; and the blue shells, which occur only on the opposite coast of Cali-

12. "Narrative of Alarcón's Voyage," in Flint and Flint, eds. and trans., *Documents of the Coronado Expedition*, 193, 199–201; "Voyage of Juan Rodriguez Cabrillo," "Facsimile of the Account of the Cabrillo Expedition," in Henry R. Wagner, ed., *Spanish Voyages to the Northwest Coast of America in the Sixteenth Century* (San Francisco, 1929), 72–93, 450–463, esp. 85–86, 88, 454–455, 458; "Relacion, ó diario, de la navegacion que hizo Juan Rodriguez Cabrillo . . .," in [Buckingham Smith], ed., *Colección de varios documentos para la historia de la Florida y tierras adyacentes* (London, 1857), 173–189, esp. 179–181, 183; Juan Bautista de Anza to Bucareli y Ursúa, May 2, 1772, Costansó to Bucareli y Ursúa, Sept. 5, 1772, José Antonio de Areche to Bucareli y Ursúa, Oct. 12, 1772, and Matheo Sastre to Bucareli y Ursúa, Oct. 19, 1772, all in Bolton, ed. and trans., *Anza's California Expeditions*, V, 3–7, 8–11, 12–23, 33–40, esp. 4–6, 9–10, 14, 38; Garcés, *Record of Travels in Arizona and California*, ed. and trans. Galvin, 24, 27, 34, 81, 87; Garcés, *Diario de exploraciones en Arizona y California*, 33, 36, 42, 80, 87.

13. "Narrative of Alarcón's Voyage," in Flint and Flint, eds. and trans., *Documents of the Coronado Expedition*, 197–204 (brackets in original), esp. 197, 203. Sedelmayr would later echo Alarcón's remark about lower Colorado Indian familiarity with animals more commonly associated with the plains. The Yumas and Cocomaricopas had "knowledge of the buffalo, though the animals do not roam their country" (Jacobo Sedelmayr, "Relación, 1746," in Peter Masten Dunne, trans., *Jacobo Sedelmayr: Missionary, Frontiersman, Explorer in Arizona and Sonora: Four Original Manuscript Narratives, 1744–1751* [(Tucson), Ariz., 1955], 15–42, esp. 26).

fornia and the South Sea, came almost three hundred leagues." In 1776, Garcés found the same general pattern Alarcón and Kino had observed on the Gila and lower Colorado holding true farther up the latter river. The Indian nations he encountered there evinced familiarity with a range of peoples and landscapes to the east and west. At Santo Angel, northwest of the Colorado's intersection with the Bill Williams Fork and south of modern Needles, Garcés met members of the "Chemevet nation" who "gave much information" about "the Yuta nation, . . . as also of the Comanches" to the east and northeast. The "Jamajabs" along the Colorado north and east of the Chemevets "had heard of" the Spanish "Fathers living near the sea" and "offered to accompany" Garcés on his journey to the San Gabriel Mission, to which they "knew the way." The Jaguallapais on the eastern side of the Colorado, with whom Garcés later "spoke at length . . . about the distance to Moqui and New Mexico," gave him "information about the whole area." Such distant horizons are noteworthy, especially in light of the difficulty of the southwestern climate and terrain. Spanish explorers seem to have been picking up indications of a regional network of trade and information exchange.[14]

What Spaniards did not discern among the southwestern Indians they met before 1763 was information bearing clearly on areas much beyond southern California and the Colorado, Rio Grande, and Great basins. Spanish observers were encountering the extent and limits of individual Indian nations' geographic familiarity. In 1540, the "Cumana" Indians could give Alarcón "information about many peoples" and knew that the Colorado rose well beyond what Alarcón had seen, but he understood that they "were not familiar with where its origin was because it came from very far away." Similarly, though many southwestern peoples had heard reports of many Spanish activities, not everyone had heard of all of them. Garcés found Colorado River Indians familiar with the San Gabriel Mission in what is now Los Angeles. Escalante and Domínguez, in contrast, failed to find among the peoples of what is now western Utah and northwestern Arizona "any reports about the Spaniards and padres of the said Monterey" in northern California. Although pre-1763 Spaniards were able to pick up southwestern reports concerning areas now included in the states of California, Nevada, Arizona, Utah, Colorado, and New Mexico, it is not clear that they were hearing firsthand accounts of lands and peoples much beyond them.[15]

14. Kino, *Kino's Historical Memoir of Pimería Alta,* ed. and trans. Bolton, I, 59, 317–318, II, 186 (quotation), 246–247, 268–269; Garcés, *Record of Travels in Arizona and California,* ed. and trans. Galvin, 32, 34, 60; Garcés, *Diario de exploraciones en Arizona y California,* 40, 42, 63.

15. "Narrative of Alarcón's Voyage," in Flint and Flint, eds. and trans., *Documents of the Coronado Expedition,* 197–204 (brackets in original); Warner, ed., *Domínguez-Escalante Journal,* trans. Chavez, 70, 87, 171, 180.

On occasion, even nearer southwestern terrain features might have eluded the familiarity of southwestern peoples. A possible example concerns Mohave Indians and Lake Cahuilla. One puzzling Spanish notion about western geography derived from what Spaniards thought were Indian references to a branch of the Colorado River flowing to the central or northern California coast. Anza claimed that, in 1774, "an Indian of the Soiopa [Mohave] tribe" told him "that three days' travel up the Colorado . . . it divides" and "that the smaller branch turned to the north to join another river larger than this Colorado." Anza indicated that the Soiopas were "ignorant" of this Colorado offshoot's "disembogue-ment into the sea." Mexican viceroy Antonio María de Bucareli y Ursúa specu-lated that the Colorado branch in question might be the river "which runs to the port of San Francisco." This is a peculiar idea, because large rivers normally gain tributaries rather than losing offshoots, and, to the extent that California state civil engineers allow it to do so, the modern Colorado flows into the Gulf of California without natural branches diverting themselves to the Pacific. One plausible explanation for the old Spanish notion is that the Mohaves were refer-ring to the old Colorado River's periodic tendency to overflow its usual banks and flow into the Salton Sink (or Trough). When this occurred, a lake of vary-ing dimensions, known as Lake Cahuilla, formed in the basin now occupied by the Salton Sea. The 1774 diary of Fray Juan Díaz furnishes details making this interpretation a reasonable possibility. Díaz noted that the "Soyopa Indians" living "about forty leagues above" the Gila-Colorado junction indicated that it was "when the [Colorado] river" was "in flood" that "a very large branch" sepa-rated and ran "west." If this interpretation is correct, and Indians were reporting this intermittent offshoot of the Colorado, what is surprising is not that Spanish geographers muddled southwestern geography but that Mohave Indians were unaware of this occasional river's ultimate destination in the relatively nearby Salton Trough. Perhaps Anza misunderstood, and the Mohaves recognized that a branch of the Colorado sometimes flowed into Lake Cahuilla but were unclear on the relation of Lake Cahuilla to the great ocean to the west. Perhaps frequent warfare made lands west of the Colorado too dangerous for at least some of the peoples living along the river. Jacobo Sedelmayr offered an observation support-ing this possibility, claiming that Yumas he spoke with in 1749 near the conflu-ence of the Gila and Colorado rivers had "blue shells from the seacoast which is beyond the mouth of the" Colorado. This sea was "to the north." Or so they had "heard it reported. They said they do not go there for fear of their enemies." More specifically, Garcés reported that the Mohaves had been on unfriendly terms with the "Jalcheduns" to the south of them, and it is possible, if these hostilities were long-standing, that they blocked Mohave cognizance of the area including the Salton Trough.[16]

If peoples like the Mohaves were unfamiliar with the termination of a river in Lake Cahuilla, Indian tales of monstrous beings to the northwest of the lower Colorado, such as those recorded in the Escobar account of Oñate's California expedition, appear in a different light. Most likely, Oñate's men were the butt of an Indian joke, or they misunderstood metaphorical language. But it is also possible that they were hearing about the Colorado Indian equivalent of the sorts of oddities enriching works like the *Odyssey* and the *Argonautica,* with memories of the way lands and seas not yet rendered familiar by regular travel and contact were represented as sites of strange and dangerous beings.[17]

That southwestern Indian geographic horizons would be limited, even if such limits were hundreds of miles away from a particular village, is not surprising. It would not be unusual, either. One could easily fill a volume with examples of remarkably circumscribed geographic vision. It is still arresting to observe how

16. "Anza's Complete Diary, 1774," in Bolton, ed. and trans., *Anza's California Expeditions,* II, *Opening a Land Route to California: Diaries of Anza, Díaz, Garcés, and Palóu* (Berkeley, Calif., 1930), 1–130, esp. 46–47; "Diario de Juan Bautista de Anza," Feb. 9, 1774, Anza diary, at Web de Anza, http://anza.uoregon.edu/archives.html (accessed June 17, 2009): "hablé con un Indio de Nacion Soyopa, y me dio la noticia, de que tres dias de camino Rio Colorado arriba donde el viaje, se partia; que lo mas caudaloso era este, y el menor brazo tiraba para el Norte á juntarse con otro rio maior, que este Colorado, y que aquel en realidad es su agua mas colorada que la de este, cuio desague al mar dice lo ignoran." For Bucareli y Ursúa's view, see Anza to Bucareli y Ursúa, Feb. 9, 1774, and Bucareli y Ursúa to Julián de Arriaga, May 27, 1774, in Bolton, ed. and trans., *Anza's California Expeditions,* V, 113–116, 169–172, esp. 114, 170. See also Garcés, "Garcés's Diary of His Detour to the Jalchedunes, 1774," in Bolton, ed. and trans., *Anza's California Expeditions,* II, 373–392, esp. 380; and Garcés, "Diario de Garcés de 1774," at Web de Anza, http://anza.uoregon.edu/archives.html. See Salton Sea Authority, http://www.saltonsea.ca.gov/about/history.htm; and Jefferson Reid and Stephanie Whittlesey, *The Archaeology of Ancient Arizona* (Tucson, Ariz., 1997), 122–123, for discussions of historic and prehistoric Lake Cahuilla. For Díaz, Sedelmayr, and Garcés, see "Díaz's Diary from Tubac to San Gabriel, 1774," in Bolton, ed. and trans., *Anza's California Expeditions,* II, 245–290, esp. 263–264; Jacobo Sedelmayr, "Trek to the Yumas," in Dunne, trans., *Jacobo Sedelmayr,* 55–66, esp. 60; Garcés, *Record of Travels in Arizona and California,* ed. and trans. Galvin, 29–34, 92; Garcés, *Diario de exploraciones en Arizona y California,* 37–42, 92.

17. One might make similar comments about the tales Rivera picked up in his 1765 journeys into Ute and Paiute country. He heard of one tribe "that kills people solely with the smoke that they make"; of a "large trench which is so broad that trade is made without people crossing it. The people throw what they want to trade"; of people who "ate their children" because of "poor hunting"; and of a "people who are like rocks." Rivera's incomprehension probably accounts for the fanciful quality of these reports — "people, very white with hair the color of straw," for instance, might simply have been people wearing fibrous hats — but it is also possible that Rivera was passing on the kind of distortions occurring when informants were discussing things they had heard about but not seen. See Sánchez, trans., "Rivera's Original Diary of the First Expedition" and "Rivera's Second Diary," in Sánchez, *Explorers, Traders, and Slavers,* 137–147, 149–157, esp. 145, 155.

little Herodotus could say about northern Europe. Southwestern Indians were likely more adapted to landscape and climate than incoming Spaniards. They were better versed in local languages, customs, and traditions and more likely to benefit from well-established alliances based on regular and long-standing, face-to-face contacts. But these were relative rather than absolute advantages of indigenous peoples over newcomers, and even these advantages likely diminished with distance. After hundreds of miles, southwestern Indians, too, would likely have been following unfamiliar trails, encountering outlandish peoples, and hearing strange tongues.

They might also have been running into hostile peoples. Spanish sources mention repeatedly both the role of the inevitable difficulties of strangeness and distance in constraining movement as well as that of the immediate hazards of conflict. Spanish writers kept talking not only about their own fears of and fights with hostile Indian peoples but also about the fears and fights between Indians. Before taking these Spanish sources at face value, it is always worth recalling Spanish writers' reasons for emphasizing Indian violence. Internecine Indian hostilities provided a serviceable justification for the establishment of missions or imposition of ostensibly benevolent rule. Spaniards whose brutality quickly became the subject of a Black Legend might have found it soothing or expedient to portray others as more ferocious than they. More subtly, the kinship ties, rituals, and alliances fostering peaceful relationships might have been harder to discern and less interesting to relate than obvious and arresting instances of hostile villages clubbing each other.

It might also have been true that some of the violence Spaniards were observing was taking place precisely because Spaniards—accompanied by their viruses, horses, and guns—had moved into the area. Though eighteenth-century Spaniards in particular often thought of themselves as peacemakers rather than conquistadors, their presence might have increased rather than diminished hostilities among Indian nations. Spanish slave-raiding and captive-purchasing furnishes one reason why this might have been so. Impelled by motives religious, humanitarian, and selfish, Spaniards in the Colorado River region, like Oñate's men and New Mexico's governors, often bought captives taken by the Indian groups with whom they were trading. Depending on the character of the purchaser and perspective of the observer, souls were saved, civilization was spread, cheap labor or enforced companionship was acquired. As in the case of Miguel, a potential source of geographic information was also obtained. Sedelmayr believed mercury deposits existed "along the upper Colorado" because, upon seeing mercury, a "Nijora Indian girl . . . sold by the Pimas" indicated to her Spanish master by "pointing to her country" and "by her gestures and whole attitude that there was much of the same substance in her own land." At the same time, Spanish slave-raiding and the captive trade might very well have worsened the Indian conflicts that often checked Spanish reconnaissance. Fray Pedro Font, in his diary of the 1775–1776 Spanish expedition to Monterey, claimed that warfare among the Yumas and other peoples of the lower Colorado often involved capturing "a few children in order to take them out to sell in the lands of the Spaniards." This "commerce" in captives, he contended, was "the reason why they have been so bloody in their wars." If Font was correct, the Spanish arrival might have intensified conflict in the Southwest and on the plains, and one possible hypothesis is that the process of Spanish expansion created conditions impeding Spanish exploration. On the other hand, Font was not around to observe the Colorado River valley in 1450, and it may be that the conduct he witnessed partook more of continuity than novelty.[18]

Questions of authorial bias and of change, transformation, or introduction notwithstanding, what Spaniards wrote about violence among southwestern Indian peoples has to be considered. The evidence is too potentially valuable, too abundant, and too consistent to ignore, and what multiple authors wrote in many different decades was that the movements and often lives of many southwestern Indians were frequently constrained and channeled by hostile neighbors.

Observers like Alarcón, Kino, Sedelmayr, and the various participants in Anza's California expeditions furnish numerous indications of people's or reports' traveling hundreds of miles. They also give many instances of journeys foregone or curtailed because of fear of enemies. Alarcón mentioned multiple indications of long-standing warfare among the Colorado peoples he encountered and of the ways in which moving upriver could be hazardous. Leaving the land "of the lord of Quicoma," he was warned that he was "nearing those who were their enemies," and he "found lookouts were posted on guard on their borders." Kino, Sedelmayr, and the members of Anza's expeditions found not only evidence of hostilities but examples of those hostilities' impeding free move-

18. Sedelmayr, "Relación, 1746," in Dunne, trans., *Jacobo Sedelmayr*, 40. Font quotation in Bolton, ed. and trans., *Anza's California Expeditions*, IV, *Font's Complete Diary of the Second Anza Expedition* (Berkeley, Calif., 1930), 102; Pedro Font, "Diario extendido de Font de 1776," Dec. 7, 1775, at Web de Anza, http://anza.uoregon.edu/archives.html. For a recent account asserting the novelty of European disruptive influences in the Southwest, see Ned Blackhawk, *Violence over the Land: Indians and Empires in the Early American West* (Cambridge, Mass., 2006). Books on the highly controversial topic of pre-European southwestern violence include Steven LeBlanc, *Prehistoric Warfare in the American Southwest* (Salt Lake City, Utah, 1999); and Christy G. Turner II and Jacqueline Turner, *Man Corn: Cannibalism and Violence in the Prehistoric American Southwest* (Salt Lake City, Utah, 1998).

ment. Kino, for instance, when discussing at Mission San Xavier del Bac (the site of modern Tucson) "what means there might be whereby to penetrate to the Moquis of New Mexico, . . . found that by going straight north the entry would be very difficult, since these Pimas were on very unfriendly terms with the Apaches who live between." Sedelmayr reported that along the Gila River east of the Yuma villages "a broad belt of intermediate territory" separated the warring Yumas and Cocomaricopas. Fray Juan Díaz wrote that, in 1774, five days after crossing the Colorado at the Colorado-Gila Junction, the "natives who were with" the Spanish party "said they could not go with us any farther, because the natives who lived in that country were their enemies, and their minds were so possessed with fear that we were not able even by means of urging, presents, or promises to induce them to go any further."[19]

It was not that everybody was at war with everybody all the time, but everybody the Spanish met seemed to be at war with somebody some of the time. Garcés furnished a kind of summary view of this situation for the mid-1770s. He noted that the Gila and lower Colorado peoples were presently at peace, but he was not sure such tranquillity would last. He compiled a list of ten area Indian groups, the "Cucapás," "Jalliquamais," "Yumas," "Jalcheduns," "Jamajabs," "Hopis," "Yavipais," "Tejua Yavipais," "Chemeguavas," and "Indians of the Gila River." According to his observations, each of these was friendly with anywhere from one to nine or more Indian peoples and had been hostile toward two to six others. This combination of enmities and amities may help to explain observed instances both of multi-day journeys and fearful immobility. Logically, travel would be easier in lands dominated by traditional friends or enjoying periods of peace, more hazardous in territories rife with longtime enemies or suffering intervals of bloodshed. Such a pattern would also allow information and goods to circulate. Although it is likely that reluctance to endure hardship and risk violence inhibited many southwestern Indian journeys, goods and news could still often have passed from village to village and nation to nation, as Costansó surmised in the 1772 quotation from the beginning of this chapter.[20]

19. "Narrative of Alarcón's Voyage," in Flint and Flint, eds. and trans., *Documents of the Coronado Expedition*, 198; Kino, *Kino's Historical Memoir of Pimería Alta*, ed. and trans. Bolton, I, 237; Sedelmayr, "Trek to the Yumas," in Dunne, trans., *Jacobo Sedelmayr*, 60; "Díaz's Diary from Tubac to San Gabriel, 1774," in Bolton, ed. and trans., *Anza's California Expeditions*, II, 245–290, esp. 270–271. For comparable examples from the same set of Spanish expeditions, see Sastre to Bucareli y Ursúa, Oct. 19, 1772, in Bolton, ed. and trans., *Anza's California Expeditions*, V, 32–40, esp. 35, 37; "Anza's Complete Diary, 1774," and "Díaz's Return Diary, 1774," ibid., II, 1–130, 291–306, esp. 7–9, 126–127.

20. Garcés, *Record of Travels in Arizona and California*, ed. and trans. Galvin, 91–92; Garcés, *Diario de exploraciones en Arizona y California*, 91–93. See also "Rivera's Original Diary of the First Expedition," in Sánchez, *Explorers, Traders, and Slavers*, 137–147, esp. 145.

Imperial Comparisons

Lingering Spanish geographic frustrations in the North American West contrast with salient aspects of Spain's spatial and human conquests farther south. These Peruvian and Mexican triumphs went hand in hand with, and indeed depended on the rapid acquisition of, native geographic information and the effective use of indigenous transportation infrastructure. Seeing what made this possible in parts of South and Central America helps to explain why it was so difficult in the North American West.

If Spanish investigators were to exploit indigenous geographic materials, native communities needed to possess some sense of distant terrain and peoples, and those ideas needed to exist in a form cultural outsiders could comprehend. Significant differences between the way geographic information circulated, collected, and was processed within the imperial polities of South and Central America and the non-state communities of the North American West made this much easier for Spaniards in greater Peru and Mexico than in the lands later comprising California, Arizona, New Mexico, and Texas.

A basic advantage for Spaniards in the Aztec and Inca empires involved language. Indeed, perhaps the most straightforward way to measure the linguistic predicament confronting Spaniards on the southern plains and in the Southwest is through comparison with the communicative circumstances they encountered elsewhere. It was not just that Spanish explorers and missionaries north of Mexico complained about the challenges of understanding their Indian interlocutors but that they often found the difficulties of linguistic diversity more intractable there than farther south. Kino's Piman hopes notwithstanding, Spanish scouts failed to find a verbal language in the Colorado Basin or southern plains playing the role of lingua franca over a vast area. They found no single translator as significant as the famous La Malinche / Doña Maria had been for Cortés because they encountered no language as widely and strikingly useful as Nahuatl had been within the Aztec Empire.

The role of Quechua, or, more precisely, "Southern Peruvian Quechua," within the Inca Empire allows for a still more arresting, if perhaps less familiar, comparison. Nahuatl had been widely spoken in central Mexico before the arrival of the Mexicas and the establishment of their dominions. It had been the language of the rulers of the Toltec Empire, and had Europeans arrived in Mexico centuries before 1519, they would already have found knowledge of Nahuatl very useful. In contrast, whereas various Quechuan languages seem to have dispersed "across the central Andes" before Inca expansion, Southern Peruvian Quechua, the official tongue of the Inca Empire, appears to have spread with the empire as a matter of policy. Spanish observers suggested that Inca rulers believed a successful empire needed an imperial language to bind it together, that travel and administration were too difficult without a widely understood tongue. Cieza de León remarked that Inca empire builders realized "how difficult it would be to travel the great distances of their land where every league and at every turn a different language was spoken, and how bothersome it would be to have to employ interpreters to understand them." If the usually astute Cieza de León was correct, the Spanish were not the only American imperialists frustrated by linguistic diversity, and Inca experience may serve as an additional indicator of the kind of communicative problems Spaniards were grappling with north of Mexico. The Incas reportedly responded to the communications challenge by, in the words of the translator of Inca traditions Juan de Betanzos, ordering "the lords and natives of the provinces to join in learning the general language of Cuzco so that it would be possible to understand them." Despite the difficulties of enforcing such a mandate, it appears to have been reasonably effective. Native tongues remained vigorous and varied, and later scholars indicate that, somewhat contrary to the impression conveyed by Betanzos, the Incas do not seem to have hoped for southern Peruvian Quechua's mass adoption by subject peoples. Instead, knowledge of it was a requirement for local officials, and southern Peruvian Quechua came to form what linguist Bruce Mannheim has called a "thin overlay" above other tongues throughout the Inca polity. One result was the utility of a single language on a scale well beyond anything the Spanish found north of Mexico. Kino lauded the value of the Pima language for a distance of 200 leagues. Cieza de León observed that Quechua "was known and used in an extension of more than 1,200 leagues." Consequently, Spanish conquerors accompanied by one translator, or missionaries who had learned one language, could communicate with many peoples over a vast area. With Quechua, the Spanish "could go everywhere."[21]

21. Bruce Mannheim, *The Language of the Inka since the European Invasion* (Austin, Tex., 1991), 6, 9–10, 16–18, 33–34; Cieza de León, *Incas of Pedro Cieza de León*, ed. von Hagen, trans. Onis, 169–171; Cieza de León, *Crónica del Perú*, in Sáenz de Santa María, ed., *Obras completas*, I, 174–175 ("Y entendido por ellos cuán gran trabajo sería caminar por tierra tan larga y adonde a cada legua ya cada paso había nueva lengua, y que sería gran dificultad el entender a todos por intérpretes"; "se sabía y usaba una lengua en más de mil y doscientas leguas"; "pues podían con ella andar por todas partes"); Juan de Betanzos, *Narrative of the Incas*, ed. and trans. Roland Hamilton and Dana Buchanan (Austin, Tex., 1996), 107; Betanzos, *Suma y narración de los Incas*, ed. María del Carmen Martín Rubio (Madrid, 2004), 153; Pedro Sarmiento de Gamboa, *History of the Incas*, trans. Clements Markham (1907; rpt. Mineola, N.Y., 1999), 121; Sarmiento de Gamboa, *Historia Indica por Pedro Sarmiento de Gamboa*, in P. Carmelo Sáenz de Santa Maria, ed., *Obras completas del Inca Garcilaso de la Vega*, 4 vols. (Madrid, 1960–1965), IV, 189–279, esp. 245. See also Kendall A. King and

Beyond the issue of the tongue in which information was expressed, there was also the matter of the speed and distance information moved. In this case, it is not immediately obvious that conditions favored Spaniards in Peru and Mexico over those to the north. News seems to have traveled far and fast among both the imperial and nonimperial peoples the Spanish encountered. When expeditions under Juan de Grijalva and Cortés landed on the Mexican coast in 1518 and 1519, reports of their arrival quickly reached Montezuma in Tenochtitlán. When Pizarro marched and sailed along the Pacific coast of modern Ecuador and Peru in the late 1520s and early 1530s, descriptions passed rapidly to the Inca courts in Quito and Cuzco. Such a rapid flow of reports within these empires was striking, but not clearly unique. Figures like Alarcón and Ascensión seem to have been witnessing comparably speedy and lengthy information circulation farther north.[22]

Despite this apparent similarity, several features of imperial information flow seem to have distinguished it from that among the decentralized communities of the Southwest. One was the pattern of reports. Following sources like Costansó and other Spanish observers, information seems to have proceeded in the Southwest as though along many threads of many sizes tied together at multiple points, with each knot representing an exchange of news among traders or warriors, or between friends and neighbors or captors and captives. In such cases, information appears to have moved as the result of a series of autonomous individual or small-group decisions.

Nancy H. Hornberger, "Quechua as a Lingua Franca," *Annual Review of Applied Linguistics*, XXVI (2006), 177–194, esp. 180–182.

22. On Cortés, see Hernan Cortés, *Letters from Mexico*, ed. and trans. Anthony Pagden (New Haven, Conn., 1986), 55, 86; Cortés, *Cartas de relación*, ed. Manuel Alcalá (Mexico City, 1988), 34–35, 52; Bernal Díaz [del Castillo], *The Conquest of New Spain*, trans. J. M. Cohen ([New York, 1963]), 35, 91; Díaz, *Historia verdadera de la conquista de la Nueva España*, ed. Miguel León-Portilla, 2 vols. (Madrid, 1984), I, 97, 162–163; Francisco López de Gómara, *Cortés: The Life of the Conqueror by His Secretary*, ed. and trans. Lesley Byrd Simpson (Berkeley, Calif., 1964), 58; López de Gómara, *Historia de la conquista de México*, ed. Jorge Gurria Lacroix (Caracas, Venezuela, 1979), 47; "Book Twelve of the Florentine Codex," in James Lockhart, ed. and trans., *We People Here: Nahuatl Accounts of the Conquest of Mexico* (Berkeley, Calif., 1993), 56–65. On Pizarro, see Betanzos, *Narrative of the Incas*, ed. and trans. Hamilton and Buchanan, 184; Betanzos, *Suma y narración de los Incas*, ed. Carmen Martín Rubio, 235; Pedro de Cieza de León, *The Discovery and Conquest of Peru: Chronicles of the New World Encounter*, ed. and trans. Alexandra Parma Cook and Noble David Cook (Durham, N.C., 1998), 113, 126, 159, 197; Cieza de León, *Crónica del Perú*, in Sáenz de Santa María, ed., *Obras completas*, I, 248, 252, 262, 273; Sarmiento de Gamboa, *History of the Incas*, trans. Markham, 186–187; Sarmiento de Gamboa, *Historia Indica*, in Sáenz de Santa Maria, ed., *Obras completas del Inca Garcilaso de la Vega*, IV, 273; Titu Cusi Yupanqui, *An Inca Account of the Conquest of Peru*, trans. Ralph Bauer (Boulder, Colo., 2005), 59–65; Alessandra Luiselli, ed., *Instrucción del Inca don Diego de Castro Titu Cusi Yupanqui* (Mexico City, 2001), 31–38.

In the Inca and Aztec empires, likely in addition to the many-strand information transfer evident in the Southwest, reports of imperial interest also moved from points throughout the empire to its court and capital, as if following the spokes of a wheel to its hub. A sense of obligation often seems to have impelled this sending of reports, as when local officials or communities sensed the court would wish to know of the arrival of strange outsiders. On other occasions, an imperial directive was involved, as when Montezuma or Atahualpa sent representatives to observe the Spaniards and report back. In either case, and in both empires, the flow of certain kinds of information was organized and directed from a central point. Communities might pass information to one another as they saw fit, but there was also a sense among at least some communities, and certainly on the part of imperial envoys, that news of imperial interest must move quickly to their rulers. Although information likely dispersed, it also collected at particular points and in the hands of particular individuals. Inca emperors in Cuzco and Aztec rulers in Tenochtitlán could catalog the resources of their dominions from their palaces.[23]

Their detailed knowledge of the tribute capacity of their possessions provided the most striking illustration of the results of such information flow. Cortés marveled at the Aztec tribute lists detailing for Montezuma the products of every province and attesting to imperial power capable of drawing items hundreds of leagues to the capital. Farther south, Cieza de León remarked upon the intelligence-gathering capacities of the Inca state: "When the Lord-Inca wished to learn what all the provinces between Cuzco and Chile . . . were to contribute, he sent out . . . persons who enjoyed his confidence, who went from village to village observing the attire of the natives and their state of prosperity, and the fertility of the land, and whether they had flocks, or metals, or stores of food, or the other things which they valued and prized. After they had made a careful survey, they returned to report to the Inca about all this." In short, the kind of detailed, precise, and comprehensive information Spaniards would have loved to possess for the North American West had, for the territories dominated by the Inca and Aztec empires, already been marshaled by Spain's conquering predecessors. Spanish western North American explorers found no political center

23. For a sense of obligation, see Cieza de León, *Discovery and Conquest of Peru*, ed. and trans. Cook and Cook, 108; Cieza de León, *Crónica del Perú*, in Sáenz de Santa María, ed., *Obras completas*, I, 247; Yupanqui, *Inca Account of the Conquest of Peru*, trans. Bauer, 63; Luiselli, ed., *Instrucción del Inca don Diego de Castro Titu Cusi Yupanqui*, 35. For imperial directives, see *Discovery and Conquest of Peru*, 159, 192–195, and *Crónica del Perú*, I, 262, 272–273; John Hemming, *The Conquest of the Incas* (New York, 1970), 30–31; Miguel Leon-Portilla, ed., *The Broken Spears: The Aztec Account of the Conquest of Mexico*, trans. Angel Maria Garibay K. and Lysander Kemp (Boston, 1962), 16–17; "Book Twelve of the Florentine Codex," in Lockhart, ed. and trans., *We People Here*, 86–87.

like Tenochtitlán or Cuzco where multiple instances of local expertise had been gathered and organized to form a far-reaching whole.[24]

This helps to explain the difficulty Spanish scouts ran up against when trying to identify distant peoples mentioned by their western Indian interlocutors. It is part of what made the problem of ethnic classification especially intractable on the plains and in the regions visited by Spaniards beyond the Rockies. Western trading centers like Pecos, where plains hunters and pueblo farmers came together, were impressive and important in their own right, but they were a much smaller affair than the great imperial capitals farther south. When Spaniards in the Colorado and Rio Grande valleys and on the southern plains heard reports of distant and strange peoples, they could often do little more than listen. Unable to see many of the peoples in question, Spaniards had difficulty determining the exact referents of these often vague and fanciful rumors. In Tenochtitlán and Cuzco, in contrast, Spaniards could enjoy a fair tour of the human geography of the Aztec and Inca empires by simply walking the streets of and countryside surrounding the capital city. This was not only because merchants and tribute-bearers from so many lands brought their goods to the imperial centers but also because imperial decrees required many provincial nobles to reside there at least part of the time and, in the case of the Inca Empire, parts of subject populations to relocate there. A sort of human inventorying had already taken place, and Spaniards could learn from it without wearing out their shoes.[25]

24. Micheal D. Coe, *Mexico: From the Olmecs to the Aztecs,* 4th ed. (London, 1994), 171–172; Cortés, *Letters from Mexico,* ed. and trans. Pagden, 109; Cortés, *Cartas de relación,* 66–67; Cieza de León, *Incas of Pedro de Cieza de León,* ed. von Hagen, trans. Onis, 162; Cieza de León, *Crónica del Perú,* in Sáenz de Santa María, ed., *Obras completas,* I, 166: "Pues como el señor quisiese saber lo que habían de tributar todas las provincias que había del Cuzco hasta Chile, . . . mandaba salir . . . personas fieles, y de confianza, las cuales iban de pueblo en pueblo mirando el traje de los naturales y posibilidad que tenían, y la grosedad de la tierra, o si en ellas había ganados, o metales, o mantenimientos, o de las demás cosas que ellos querían y estimaban; lo cual mirado con mucha diligencia, volvían a dar cuenta al señor de todo ello." See also Betanzos, *Narrative of the Incas,* ed. and trans. Hamilton and Buchanan, 90–91, 100, 109, 163, 165, 180; Betanzos, *Suma y narración de los Incas,* ed. Carmen Martín Rubio, 136–137, 146, 155–156, 213, 215, 231; Sarmiento de Gamboa, *History of the Incas,* trans. Markham, 132, 146, 149–150; Sarmiento de Gamboa, *Historia Indica,* in Sáenz de Santa Maria, ed., *Obras completas del Inca Garcilaso de la Vega,* IV, 249–250, 255, 256–257.

25. Betanzos, *Narrative of the Incas,* ed. and trans. Hamilton and Buchanan, 120, 170; Betanzos, *Suma y narración de los Incas,* ed. Carmen Martín Rubio, 166–167, 220; Cieza de León, *Incas of Pedro de Cieza de León,* ed. von Hagen, trans. Onis, 46, 62–63, 109, 148 157–158; Cieza de León, *Second Part of the Chronicle of Peru,* ed. and trans. Clements R. Markham (New York, [1964]), 180; Cieza de León, *Crónica del Perú,* in Sáenz de Santa María, ed., *Obras completas,* I, 106, 117–118, 162, 172–173, 206–207; Cortés, *Letters from Mexico,* ed. and trans. Pagden, 107, 472; Cortés, *Cartas de relación,* 65; Sarmiento de Gamboa, *History of the Incas,* trans. Markham, 120; Sarmiento de Gamboa, *Historia Indica,* in Sáenz de Santa Maria, ed., *Obras completas del Inca Garcilaso de la Vega,* IV, 244–245.

One consequence of these means of Aztec and Inca imperial information collection was that, if Spaniards could secure the willing or coerced cooperation of indigenous imperial rulers or officials, they could learn a great deal about American geography. Indigenous monarchs and functionaries could advise Spaniards of distant lands and conduct parties to them. When Diego de Almagro wanted to investigate Chile, the Inca royal brother Paullu and those under his authority could show the way. Cortés wrote that, while in Tenochtitlán, he was "finding out many of the secrets of Mutezuma's lands and of those which bordered on them and those of which he had knowledge." When Cortés inquired about a Gulf port for his ships, Montezuma's servants promptly brought Cortés "a cloth with all the coast painted on it." Such maps often proved opaque to Europeans, especially before later native cartographers modified them to suit Spanish eyes, but Montezuma furnished more than a picture. He told Cortés "to decide whom to send, and he would provide the means and the guide to show" the coastal regions in question to them. Cortés's men immediately received a guided tour. Like Cortés, Spanish explorers north of Mexico often heard reports of what sounded like gold or silver; they, however, had great difficulty confirming the value or ascertaining the location of such metals. Within the Aztec Empire, when Cortés wanted more information about reported gold mines, Montezuma sent his own servants with Spaniards selected by Cortés to point out the ore-bearing rivers and bring back samples.[26]

This Inca and Aztec guidance, along with the indigenous hub-and-spoke pattern of information circulation, hint at another distinction between the imperial and non-state polities the Spaniards encountered, one concerning the directness, verification, and clarification of information. Whereas Indian communities in the Southwest and on the plains appear frequently to have been interpreting and transmitting secondhand news, imperial courts in Quito, Cuzco, or Tenochtitlán seem to have received and demanded less mediated reports. Montezuma reportedly heard of the arrival of the Spanish from a common man who had seen them. Manco Inca similarly was reported to have heard of the Spaniards

26. On Almagro, see Hemming, *Conquest of the Incas*, 228, 244. For the Cortés quotations, see Cortés, *Letters from Mexico*, ed. and trans. Pagden, 94, 112–113; Cortés, *Cartas de relación*, 57, 69; López de Gómara, *Cortés*, ed. and trans. Simpson, 181–182; López de Gómara, *Historia de la conquista de México*, 143. For the sending out of sample and information gatherers, see Cortés, *Letters from Mexico*, ed. and trans. Pagden, 92–93; Cortés, *Cartas de relación*, 56–57; López de Gómara, *Cortés*, ed. and trans. Simpson, 179–180; López de Gómara, *Historia de la conquista de México*, 141–142. For a similar example from the Spanish reconnaissance of the Inca Empire, see "Cieza de León's Account of the Early Voyages," in John H. Parry and Robert G. Keith, eds., *New Iberian World: A Documentary History of the Discovery and Settlement of Latin America to the Early 17th Century*, 5 vols. (New York, 1984), IV, *The Andes*, 18–39, esp. 38–39.

from coastal eyewitnesses. Such firsthand reports were probably less prone to distortion.[27]

Even under the best of circumstances, however, as generations of lawyers, police officers, and defendants can attest, verbal reports and eyewitness testimony can be imprecise, inaccurate, or incomprehensible. To counter these inevitable imperfections of information transmission, the Amerindian empires employed various techniques to confirm, refine, and translate the material flowing through imperial channels. Most basically, they dispatched officials to check reports. When a coastal commoner recounted to Montezuma a tale of strange men, or of a "round hill" in the water, off went an official to interrogate other locals and to insist upon seeing for himself. When ominous warnings about the Spaniards tromping around Peru reached Atahualpa, envoys set out to "learn their intentions and their ways." Nor were even these additional verbal reports from handpicked agents deemed sufficient. In Bernal Díaz's account, Montezuma's representatives instructed "skilled painters . . . to make realistic full-length portraits of Cortes and all his captains and soldiers, also to draw the ships, sails, and horses, Doña Marina and Aguilar, and even the two greyhounds. The cannon and cannon-balls, and indeed the whole of our army, were faithfully portrayed, and the drawings were taken to Montezuma." Inca leaders went beyond pictorial representations of the newcomers, seeking to acquire flesh-and-blood Spaniards. Varying accounts report Huayna Capac's requesting Spaniards be brought to him for examination and two Spaniards' being taken to Atahualpa for the same purpose. In the North American West, Indian peoples heard and were curious about Spaniards but did not evince the same capacity to command and enhance information about them. Within the Aztec and Inca empires, when Spaniards were taken to see places, peoples, or things, they were replicating processes of information-gathering and checking already in use.[28]

27. On Montezuma, see Diego Durán, *The History of the Indies of New Spain,* trans. Doris Heyden (Norman, Okla., 1994), 495–496; Durán, *Historia de las Indias de Nueva España e islas de la Tierra Firme* (Mexico City, 1967), II, 505–506; Leon-Portilla, ed., *Broken Spears,* trans. Garibay and Kemp, 16–17. On Manco Inca, see Yupanqui, *Inca Account of the Conquest of Peru,* trans. Bauer, 60, 63–64; Luiselli, ed., *Instrucción del Inca don Diego de Castro Titu Cusi Yupanqui,* 31–32, 35–37.

28. On Montezuma's dispatch of an official, see Durán, *History of the Indies of New Spain,* trans. Heyden, 495–496; Durán, *Historia de las Indias,* II, 505–506 ("cerro redondo"); Leon-Portilla, ed., *Broken Spears,* trans. Garibay and Kemp, 16–17. On Atahualpa's envoys, see Cieza de León, *Discovery and Conquest of Peru,* ed. and trans. Cook and Cook, 159, 191–195; Cieza de León, *Crónica del Perú,* in Sáenz de Santa María, ed., *Obras completas,* I, 262, 272–273. On Montezuma's painters, see Díaz, *Conquest of New Spain,* trans. Cohen, 91; Díaz, *Historia verdadera,* 162–163; López de Gómara, *Cortés,* ed. and trans. Simpson, 58; López de Gómara, *Historia de la conquista de México,* 47. On Inca leaders seeking Spaniards, see Cieza de León, *Discovery and Conquest of Peru,* ed. and trans. Cook and Cook, 113, 126; Cieza de León, *Crónica del Perú,* in Sáenz de Santa María, ed., *Obras completas,*

When Spanish explorers and their Aztec and Inca guides set out to see and report, they benefited also from developed administrative and infrastructural systems. Beyond New Mexico, Spanish explorers like Kino, Garcés, Escalante, and Domínguez had to coax horses, guides, and meager rations from a succession of communities. In Mexico, local grandees often furnished Cortés not only with provisions for hundreds but also with hundreds of bearers to carry them. In western Utah in 1776, Escalante and Domínguez found themselves "in great distress, without firewood and extremely cold." Seeing the surrounding mountains "covered with snow," they feared that, before reaching them, "the passes would be closed to us," and the party might be trapped "two or three months in some sierra where there might not be any people or the wherewithal for our necessary sustenance." In contrast, when crossing one mountain pass on their journey to Tenochtitlán, Cortés and his men found at the summit "more than a thousand cartloads of firewood, all very well stacked." Spanish travelers in the Andes marveled at well-constructed, closely managed, marvelous, and nerve-wracking bridges. Those in the Southwest spoke of fords, rafts, and swimming. The issue in the Andes was fear of heights; in the Southwest, of drowning. The terrain was no more difficult in the American West than it was in parts of South and Mesoamerica, but the great polities of those more southerly regions had built and organized on a large scale the kinds of physical and human infrastructure needed to make movement easier.[29]

One of these, the Inca road system, merits particular and additional attention because of its value as an index of Inca geographic conceptions. The geographical knowledge of the peoples in and around the Aztec Empire was impressive in its detail, precision, and beauty of expression, but it is not clear how far be-

I, 248, 252; Yupanqui, *Inca Account of the Conquest of Peru*, trans. Bauer, 60; Luiselli, ed., *Instrucción del Inca don Diego de Castro Titu Cusi Yupanqui*, 32.

29. On provisions and bearers in Mexico, see Díaz, *Conquest of New Spain*, trans. Cohen, 109, 134; Díaz, *Historia verdadera*, 182–183, 221; López de Gómara, *Cortés*, ed. and trans. Simpson, 94, 135, 346, 355; López de Gómara, *Historia de la conquista de México*, 75, 106, 269, 276. On Escalante and Domínguez, see Warner, ed., *Domínguez-Escalante Journal*, trans. Chavez, 70–71, 170–171. On prepared firewood in Mexico, see Cortés, *Letters from Mexico*, ed. and trans. Pagden, 55 (quotation); Cortés, *Cartas de relación*, 35; López de Gómara, *Cortés*, ed. and trans. Simpson, 95; López de Gómara, *Historia de la conquista de México*, 75. On Andean bridges, see, for instance, Miguel de Estete's account of Hernando Pizarro's 1533 journey, reproduced in J. H. Parry, *The Discovery of South America* (New York, 1979), 185–191. On the absence of bridges in the Southwest, note Kino's description of crossing the lower Colorado in 1701 in a "large basket" atop a "good raft" (*Kino's Historical Memoir of Pimería Alta*, ed. and trans. Bolton, I, 316–317), and Garcés and Escalante's frequent references to difficult, bridgeless, and wet river crossings. See, in particular, Garcés, *Record of Travels in Arizona and California*, ed. and trans. Galvin, 48, 86; Garcés, *Diario de exploraciones en Arizona y California*, 54, 85; Warner, ed., *Domínguez-Escalante Journal*, trans. Chavez, 94–101, 185–188; "Rivera's Original Diary of the First Expedition," in Sánchez, *Explorers, Traders, and Slavers*, 145.

yond the range of that empire and the Mesoamerican culture area it extended. Despite probable trade connections between Mesoamerica and the Southwest, despite the spread of crops and customs north from Mexico, and despite the reconnaissance activities of the famous Pochteca merchant-spies, information picked up in the Valley of Mexico and the participation of Mexican Indians in expeditions like that of Coronado seems to have done little to disabuse Spanish explorers of their wilder notions about regions to the north. The sections of the Florentine Codex concerned with peoples and animals do not present anything clearly recognizable as a buffalo or an Indian from the territories currently comprising the United States. The geographic horizons of Miguel or of Alarcón's informants might have extended as many miles as those of Montezuma.[30]

The Incas were another matter, and their roads are the most eloquent expression of the difference. When Spanish scouts began probing the Inca Empire in the 1520s, the main Inca highland road running through Quito and Cuzco to modern Argentina's Mendoza River, and thus across some of the most precipitous terrain on earth, was roughly 3,513 miles long—some 700 miles more than the modern driving distance from Los Angeles to New York. Another parallel road ran along the coast, and lateral routes connected the primary roads to each other and to individual towns, increasing the total mileage to "at least . . . 14,000 miles." Imperial mandates and local compliance ensured that distance markers, copious storehouses, and comfortable shelters graced the roads at regular intervals. For purposes of this discussion of geographic thought, however, what was most interesting about the road system was not the technical skill required to construct it, nor even the state authority and labor mobilization needed to build and maintain it. Most striking, instead, was that the existence and operation of the road system constituted a physical manifestation of the scale and precision of Inca geographic conceptions. Officials could think of individual locations in terms of their relation to the road system, and the sum of these localities and relations could be imagined as a single entity. The road system even made it possible to physically experience them as such. Before his death in the mid-1520s, the twelfth Inca Huayna Capac had made use of the road system to set imperial foot in lands stretching from central Chile to southern Colombia.[31]

30. Bernardino de Sahagún, *Florentine Codex: General History of the Things of New Spain*, trans. Charles E. Dibble and Arthur J. O. Anderson, 13 vols. ([Salt Lake City], Utah, 1950–1982), IX, *The Merchants*, X, *The People*, and XI, *Earthly Things*.

31. Quotation and mileage figures from John Hyslop, *The Inka Road System* (New York, 1984), xiii, 223. See also Betanzos, *Narrative of the Incas*, ed. and trans. Hamilton and Buchanan, 174–177, 181; Betanzos, *Suma y narración de los Incas*, ed. Carmen Martín Rubio, 225–228, 232; Cieza de León, *Second Part of the Chronicle of Peru*, ed. and trans. Markham, 199–219; Cieza de León, *Crónica del Perú*, in Sáenz de Santa María, ed., *Obras completas*, I, 212–219; Sarmiento de Gamboa, *History of the*

Spanish and later explorers found what they often called roads in the American West. Domínguez and Escalante, for example, reported traveling on innumerable trails, paths, and roads ["veredas," "sendas," "caminos"] of varying size, quality, and degree of usage, and their reliance on established routes was so habitual that the inability to find or follow one, or the necessity of traveling without one, merited mention. Twenty-first-century scholars and enthusiasts are still marveling at the network of roads around Chaco Canyon. Indeed, from the accounts of a variety of explorers, not to mention the work of generations of archaeologists, anthropologists, and historians, it appears that a lattice of trading and raiding routes, many probably of immemorial antiquity, covered the length and breadth of North America north of Mexico. As noteworthy as these routes and roads were, European explorers encountered nothing in the North American West combining the vastness of scale, degree of development, and centralization of control evident in the Inca road network. More significant, though, than the absence of the kind of physical and administrative infrastructure present in the Inca Empire was Spanish explorers' failure to locate in the American West the kind of conceptual superstructure accompanying Inca roads and rule. The Inca road system was valuable not simply as the fruit of human design but because the humans designing it were available for questioning. Spanish explorers met many western North American individuals who could tell them much about local or regional geography, but not rulers and officials who could reliably and comprehensibly describe a good part of the continent.[32]

Indeed, one advantage for Spanish conquistadors in the Inca, and to a lesser extent Aztec, empires was that their Indian collaborators and captives often proved far more useful than those available to Spanish explorers farther north. Spanish *adelantados* in the North American West obtained information from members, captives, and observers of communities, and even from leaders of nations, but not from rulers of an array of peoples spread over a vast area. From emperors and their officials, Spaniards could obtain rooms filled with precious metal as well as information metaphorically placed on a silver platter.

Comparisons among Spanish experiences in different parts of the Americas — between state and non-state and imperial and decentralized communities as well as between the South and North American mountain wests — return us, in fact, to the meaning of a phrase like "undiscovered West." "Undiscovered" obviously cannot mean "uninhabited," for many thousands of people clearly

Incas, trans. Markham, 158–166; Sarmiento de Gamboa, *Historia Indica*, in Sáenz de Santa Maria, ed., *Obras completas del Inca Garcilaso de la Vega*, IV, 260–264.

32. Warner, ed., *Domínguez-Escalante Journal*, trans. Chavez, 5, 8–9, 135, 137.

lived in the western North American regions unseen by pre-1763 Europeans. It cannot mean "unfrequented," because many western Indians were making long journeys through its territories. It cannot mean "never discovered," because recent migrants or the ancestors of entrenched peoples had, at some point, to have seen landscapes for the first time.

It could mean that a process of reconnaissance, familiarization, and organization comparable to what had happened in South America's Andean West was not visible to Spanish and other European investigators of early modern western North America. If the impression conveyed by sixteenth-century interpreters and expounders of Inca traditions is correct, the expansion of the Inca Empire and the extension of Inca geographic horizons went hand in hand. Initially conversant with a circumscribed area around Cuzco, the Incas would receive reports of more distant zones suitable for and vulnerable to conquest; would scout, invade, and subjugate them; and would then consolidate their rule over and refine their understanding of them. When Spanish scouts began probing the Pacific coast, parts of South America extending from Ecuador to Chile had already been discovered by an Inca Empire formerly unfamiliar with them and could be, to use an antiquated but useful meaning of the word, "discovered" again, in the sense of being revealed by Inca nobles and functionaries to conquistadors.[33]

It is possible that western Indians at someplace like the Dalles, a Red River Comanche camp, or that continental spire where the Green, Snake, and Missouri rivers originate, possessed a conception of western North America as vast as the Inca view of the Andes; but pre-1763 Spaniards failed to find, understand, and exploit such geographic mastery.

THIS LEFT THE HARRIED mid-eighteenth-century Spanish diplomat in a difficult position as he struggled to assess the degree of danger posed by French and British interlopers. The foundation for decisions informed by understanding of western American geography had not been laid. The difficulties of exploring the West and the disappointments of the expeditions that tried had prevented

33. On Inca conquests, see Betanzos, *Narrative of the Incas*, ed. and trans. Hamilton and Buchanan, 81, 93, 113, 119–120, 124–125, 148–152; Betanzos, *Suma y narración de los Incas*, ed. Carmen Martín Rubio, 126, 139, 159, 165–166, 171–172, 197–201; Cieza de León, *Second Part of the Chronicle of Peru*, ed. and trans. Markham, 147, 168–169, 183, 195, 208–209, 213; Cieza de León, *Crónica del Perú*, in Sáenz de Santa María, ed., *Obras completas*, I, 197, 203, 207, 211, 215–217; Sarmiento de Gamboa, *History of the Incas*, trans. Markham, 129, 133–134, 145, 157, 166; Sarmiento de Gamboa, *Historia Indica*, in Sáenz de Santa Maria, ed., *Obras completas del Inca Garcilaso de la Vega*, IV, 248, 250–251, 255, 260, 264.

sustained and successful Spanish reconnaissance of the region. The challenges of acquiring comprehensible geographic information from western Indians had vitiated one means of compensating for the shortcomings of Spanish exploration.

The question that remains is whether this was about degree of comprehension or about transmission of understanding. Europeans might have found it easier to acquire geographic information from indigenous empires than from non-state communities, but that does not prove that imperial eyes saw farther than those of decentralized villages and bands. Inca roads and Mexican cartographers might simply have made geographic conceptions more evident than did the ephemeral sketches and elusive traditions of the peoples stretching between the western Spanish-Indian borderlands and the Arctic Ocean. Still, the question arises: how far could geographic horizons extend in the absence of a durable, large-scale, centralized government capable of suppressing local conflicts and channeling violent impulses; of constructing, maintaining, and policing transportation routes; and of gathering and organizing geographic information? To what degree could geographic conceptions be refined? To what extent, that is, did intellectual mastery of geography depend on political control of territory?

Although most of this book is concerned with the way empires dealt with complications arising from the continuation of geographic uncertainty and the presence of imperial rivals, this discussion of the Spanish Empire suggests that lingering ignorance of western geography can be ascribed in part to an earlier absence of empire. This proposition will be explored in greater detail in subsequent consideration of the limits of eighteenth-century French geographic understanding.

Before turning to French reconnaissance of the North American West, however, it is necessary to consider the goal toward which much French and British eighteenth-century exploration was directed: the Pacific Ocean.

PART II.

SOUTH SEA INTERLUDE

3

THE ALLURING PACIFIC OCEAN

The most celebrated goal of early modern French, British, and Anglo-American western exploration was to find some kind of Northwest Passage to the Pacific. French scouts looking for a river route to the South Sea pushed west of lakes Superior and Winnipeg in the 1730s, 1740s, and 1750s. British ships sought a Northwest Passage from Hudson Bay in the 1740s. Thomas Jefferson, famously, hoped Lewis and Clark would find a passage to India, not the Bitterroot Mountains they actually encountered. The persistence of this quest, long after the hard experience of European explorers had dispelled Columbus's vision of a short westward voyage to the Indies, has raised many questions. One of these is why enthusiasts kept finding reasons to hope for the existence of a transcontinental water route to the Pacific. More basic still is the question of why they cared about finding one. To understand the durability, difficulty, and implications of the Anglo-French search for a Northwest Passage, it is necessary first to consider where a navigable water route through North America would have led.

During the half century from 1713 to 1763, and indeed in the decades before and after, the still largely uncharted South Sea offered much to draw the attention, inspire the imagination, and direct the actions of European officials. They could be almost certain the Pacific contained undiscovered islands and, the dimensions of the ocean being ample, it might hold unknown continents. Should a water passage through North America exist, the Pacific might still provide a way to East Asian commerce less difficult than that around the Cape of Good Hope. The South Sea promised access, moreover, to Spanish silver needed to finance trade with China or wars in Europe. Whether this silver remained in the hands of Spanish Chilean and Peruvian subjects, ascended South America's Pacific coast to Panama, or crossed from Acapulco to Manila to support the Spanish colony in the Philippines, it enticed eighteenth-century Europeans who often retained the avarice, but rarely enjoyed the opportunities, of Pizarro and Cortés. Together, geographic mysteries, commercial opportunities, and financial importunities rendered the South Sea and the possibility of novel and superior routes to and

across the Pacific subjects of great interest to European officials. Such considerations would draw pirates into the South Sea in the late seventeenth century and
merchants and privateers in the early eighteenth. Paradoxically, they would contribute also to the official exclusion of most non-Spanish ships from the Pacific
after the War of the Spanish Succession. This restoration of Spanish claims to a
closed Pacific would limit possible French avenues of approach to the American
West and sow the seeds of future conflicts among western Europe's maritime
powers.

Uncharted Waters, Imagined Lands

The lingering obscurity of the eighteenth-century Pacific arose in part from geographic fundamentals. Reaching the South Sea required Europeans to sail thousands of miles and to round one of two proverbially difficult capes or to find their
way across the unforgiving land barriers of Eurasia or the Americas. Once Europeans entered the Pacific, its immeasurable magnitude portended scurvy, ship-
wreck, and desolation to mariners unfamiliar with its waters and unacquainted
with the navigational techniques developed over the centuries by South Sea
islanders. In 1492, fewer than five weeks out from the Canaries, Columbus's men
"could . . . bear no more." Five weeks was a third of what Ferdinand Magellan
and his crew endured on their 1520–1521 Pacific crossing. In Antonio Pigafetta's
account,

> They were three months and twenty days without eating anything (*i.e.*,
> fresh food), and they ate biscuit, and when there was no more of that
> they ate the crumbs which were full of maggots and smelled strongly of
> mouse urine. They drank yellow water, already several days putrid. . . .
> The gums of some of the men swelled over their upper and lower teeth,
> so that they could not eat and so died. . . . In these three months and
> twenty days they went four thousand leagues in an open stretch in the
> Pacific Sea. . . . And they saw but two uninhabited islands, where they
> saw no other things than birds, and trees. . . . If God had not given them
> fine weather, they would all have perished of hunger in this very vast sea.
> And they were certain that such a voyage would never again be made.[1]

1. The account of the Pacific that follows rests on a sturdy foundation of scholarly literature.
J. H. Parry provides a short introduction to the Pacific and the eighteenth-century European empires in *Trade and Dominion: The European Overseas Empires in the Eighteenth Century* (New York,
1971), 235–256. O. H. K. Spate gives a more comprehensive overview of European exploration of
the Pacific in the series The Pacific since Magellan, I, *The Spanish Lake* (Minneapolis, 1979), and II,

Such voyages were made again, and many sailors survived them, in part because of the "fine weather" gracing Magellan and the Pacific, more fundamentally because of the favorable winds blowing west and east across the South Sea. But the winds that made crossing the Pacific possible made surveying it difficult. Those carrying ships across the Pacific to Asia blew vessels north of the majority of the Pacific's islands, and the westerlies propelling Manila galleons back to the Americas drove them closer to the Aleutians than to Tahiti. Such was the tyranny of winds and distance that the Hawaiian Islands, which lay between the northern and southern routes of the Spanish galleons, were apparently never seen, or at least never reported by them. Even when Europeans descried land amid the South Sea's watery solitudes, their inability to accurately determine longitude at sea often left them unable to fix the discovery on their charts. The Solomon Islands afford one example. Alvaro de Mendaña sighted them in 1568. Thereafter, these islands were sometimes seen but rarely recognized by European mariners, and they wandered about European maps more like Sinbad's whale than bodies of land. Only in the late eighteenth century did they achieve the stationary position on paper merited by their immobility in the Pacific.[2]

Beyond these matters of geography and technology, Europeans from outside Spain had to reckon with politics. In the centuries between Vasco Núñez de Balboa's 1513 Pacific sighting and the 1763 Treaty of Paris, the South Sea was, for Europeans, a "Spanish Lake," a mare clausum. The reasons for this dated back to the early decades of the sixteenth century. As part of the ongoing attempt to establish boundaries between the territories claimed by Spain and Portugal, the 1529 Treaty of Saragossa established, at least in the minds of the Iberian empires, that the Moluccas (or "Spice Isles") would be exclusively Portuguese

Monopolists and Freebooters (Minneapolis, 1983). Glyndwr Williams offers an excellent treatment of English interest in the ocean in The Great South Sea: English Voyages and Encounters, 1570–1750 (New Haven, Conn., 1997), 1–12, 48–75. William Lytle Schurz treats the "Spanish Lake" in the context of his account of The Manila Galleon (New York, 1959). And John Dunmore introduces French activities in the South Sea in French Explorers in the Pacific, 2 vols. (Oxford, 1965–1969), I, The Eighteenth Century, 1–53. For the quotations, see "The Atlantic Crossing and the Discovery of the West Indies, as Related by Columbus," in John H. Parry and Robert G. Keith, eds., New Iberian World: A Documentary History of the Discovery and Settlement of Latin America to the Early 17th Century, 5 vols. (New York, 1984), II, The Caribbean, 22–58, esp. 29; Antonio Pigafetta, The Voyage of Magellan: The Journal of Antonio Pigafetta, trans. Paula Spurlin Paige (Englewood Cliffs, N.J., 1969), 24–25.

2. Parry, Trade and Dominion, 236–238, 245, 247; Schurz, Manila Galleon, 228–229; Spate, Spanish Lake, 108–109, 119–127; Derek Hayes, Historical Atlas of the North Pacific Ocean: Maps of Discovery and Scientific Exploration, 1500–2000 (Seattle, 2001), 17–19, 35; Thomas Suárez, Early Mapping of the Pacific: The Epic Story of Seafarers, Adventurers, and Cartographers Who Mapped the Earth's Greatest Ocean (Singapore, 2004), 22, 56, 60–67, 70–72, 77, 82, 84, 88, 96, 100, 104, 112, 152, 154–155, 177, 180; Williams, Great South Sea, 9–11, 55.

and the Pacific solely Spanish. Spanish laws of 1540, 1558, 1560, 1563, and 1692 gave renewed expression to Spain's claims to an exclusive right to sail Pacific waters. Although other European nations never formally accepted this Iberian treaty, and although British, Dutch, and French ships appeared occasionally in South Sea waters—Francis Drake, for example, sailed up the Pacific coast to California and then west across the ocean and back to England on his famous 1577–1580 voyage—Spain's rivals established neither regular Pacific routes nor lasting South Sea settlements. In contrast, the Spanish Empire moved quickly from Pacific exploration to American and even Asian colonization. Spain began establishing cities in western Central America (Panama) by 1519, Peru (Lima) by 1535, Chile (Santiago) by 1541, western Mexico (Acapulco) by 1550, and the Philippines (Manila) by 1571.[3]

Though Spanish cities arose on the Americas' Pacific rim, and though Spanish ships commenced an annual Mexico-Philippines commerce, Spanish explorers exhausted their efforts before they had completed investigation of the Pacific's seventy million square miles and innumerable islands. This is not to say that the achievements of Spanish navigators were inconsiderable, nor always as evanescent as Mendaña's experience with the Solomons. Spanish mariners established a serviceable roundtrip route from Acapulco to Manila between 1527 and 1571. An expedition led by Pedro de Quiros and Luis Vaez de Torres located Vanuatu in the New Hebrides and sailed through what came to be called the Torres Strait between New Guinea and Australia in 1606. But Spanish voyages failed to turn up the Old Testament's golden Ophir and the wealthy islands visited by Inca legend. They did not locate Marco Polo's "7459 Islands" in the "Eastern Sea of Chin," "12,700 Islands" in the "Sea of India," and Locac (Lucach, or "Beach") with its "gold in incredible quantity." When Spanish officials found that maintenance of existing Indies possessions and prosecution of inescapable European enterprises already expended more resources than the Spanish Empire could generate, they ceased risking valuable ships on always dangerous and usually fruitless exploratory ventures. In the eighteenth century, when Spanish imperial power was declining in many respects relative to that of rivals France and Britain, foreign observers like John Campbell (in his revised 1744–1748 edition of John Harris's *Complete Collection of Voyages and Travels*) suggested that Spain shied away from support for Pacific reconnaissance also because of a fear that Spanish discoveries might be seized and exploited by others. Whatever the

3. Max Savelle, *The Origins of American Diplomacy: The International History of Anglo-America, 1492–1763* (New York, 1967), 12–16; Schurz, *Manila Galleon*, 20, 287–288, 297; Spate, *Spanish Lake*, 94–96.

reasons, the incompleteness of Spanish Pacific cognizance left open intriguing possibilities for speculation and, it later turned out, islands and continents for investigation.[4]

One looming South Pacific possibility, nicely expounded by historians of Pacific exploration, was some kind of southern continent. The Greco-Egyptian geographer Ptolemy (A.D. c. 100–170) had posited the existence of such a land-mass, imagining that it connected southern Africa and southeastern Asia and enclosed thereby the southern oceans. One sign of the Renaissance rediscovery of Ptolemy's work was the circulation from the 1470s of European maps displaying this vision. Much of Ptolemy's geographic thought would not, however, survive the voyages of exploration it helped to inspire. The journeys of Bartolomeu Dias and Magellan demolished parts of the Ptolemaic conception of the antipodes by demonstrating that one could sail into the Indian Ocean by rounding the Cape of Good Hope or by traversing the Pacific and the Malay Archipelago. The idea of some form of southern continent survived nonetheless, and sixteenth-century maps from leading cartographers such as Abraham Ortelius (1527–1598) included a kind of inflated Antarctica pushing up from the South Pole.[5]

One reason Europeans continued to entertain notions of a southern continent was because they could rest their surmises not only on Ptolemy's ancient authority but also on more contemporary reason and faith. Gerardus Mercator (1512–1594) and many who followed him suggested that planetary stability necessitated a southern landmass to act as equipoise to the familiar northern continents. The theologically minded opined that the symmetrical imperatives of a divine creation required the presence of lands in the Southern Hemisphere comparable to those in the north.[6]

A still more fundamental lifeline for belief in the existence of a southern continent was provided by Europeans' failure to sail far enough south in the Indian

4. On the Acapulco-Manila route, see Schurz, *Manila Galleon*, 216–283; Spate, *Spanish Lake*, 91–106. On the Quiros-Torres expedition, see Spate, *Spanish Lake*, 132–141; Williams, *Great South Sea*, 56–58. The Polo quotations are in Marco Polo, *The Book of Ser Marco Polo, the Venetian: Concerning the Kingdoms and Marvels of the East*, ed. and trans. Henry Yule, 2 vols. (1903; rpt. London, 1921), II, 264, 276, 424. On Spanish disappointments and exploratory caution, see Pedro Sarmiento de Gamboa, *History of the Incas*, trans. Clements R. Markham (1907; rpt. Mineola, N.Y., 1999), 135; Sarmiento de Gamboa, *Historia Indica*, in P. Carmelo Sáenz de Santa Maria, ed., *Obras completas del Inca Garcilaso de la Vega*, 4 vols. (Madrid, 1960–1965), 4 vols., 189–279, esp. 251; Suárez, *Early Mapping of the Pacific*, 61, 77–79, 84–85, 103; Parry, *Trade and Dominion*, 235; Schurz, *Manila Galleon*, 228–240; Williams, *Great South Sea*, 8–9; John Harris, ed., *Navigantium atque Itinerantium Bibliotheca; or, A Complete Collection of Voyages and Travels . . .*, 2 vols. (London, 1744), I, 65.

5. Parry, *Trade and Dominion*, 238–239; Williams, *Great South Sea*, 1–12; Suárez, *Early Mapping of the Pacific*, 35–38, 64–65, 69.

6. Williams, *Great South Sea*, 1–12; Suárez, *Early Mapping of the Pacific*, 35–38, 64–65.

and Pacific oceans to disprove the existence of landmasses there. Dutch navigator Abel Tasman sailed along Australia's northern and northwestern coast, Tasmania's southern shore, and New Zealand's northwest littoral in 1642 and 1644. Piratical English naturalist William Dampier observed northern and western Australia, northern New Guinea, and western New Britain between 1688 and 1700. Retired Batavian lawyer Jacob Roggeveen happened upon Easter Island and the Samoan group during his 1721–1722 circumnavigation. However impressive they might have been in their own right, these early modern European voyages left a vast area of obscurity between Chile and New Zealand where a symmetrically disposed divinity might have deposited one more continental creation. The musings of British privateer and circumnavigator Woodes Rogers in his 1712 *Cruising Voyage round the World* offer a nice illustration of the kinds of reflection stimulated—or necessitated—by the limited reach of European Pacific exploration:

> I have often admir'd that no considerable Discoveries have yet been made in South Latitude from America to the East Indies: I never heard the South Ocean has been run over by above three or four Navigators, who varied little in their Runs from their Course, and by consequence could not discover much. I give this Hint to encourage our South Sea Company, or others, to go upon some Discovery that way, where for ought we know they may find a better Country than any yet discover'd, there being a vast Surface of the Sea from the Equinox to the South Pole, of at least 2000 Leagues in Longitude that has hitherto been little regarded, tho' it be agreeable to Reason, that there must be a Body of Land about the South Pole, to counterpoise those vast Countries about the North Pole. This I suppose to be the Reason why our antient Geographers mention'd a Terra Australis Incognita.[7]

In the middle decades of the eighteenth century, interest in a southern continent and southern islands began to increase, and so did recommendations that they be sought. A 1731 memoir by Arthur Dobbs, the persistent and influential promoter of British searches for a Northwest Passage from Hudson Bay, spoke of a "New and beneficial Commerce . . . among many Islands and Countrys in the Great Southern Ocean not yet fully discover'd Such as the Islands of Solomon etc." John Campbell encouraged the exploration of New Guinea, advocated a British settlement on the island of Juan Fernández (off the coast of

7. Parry, *Trade and Dominion*, 239–241; Williams, *Great South Sea*, 57–75, 106–132, 206; Suárez, *Early Mapping of the Pacific*, 113–115. Quotation from Woodes Rogers, *A Cruising Voyage round the World . . .* (London, 1712), 324–325.

Chile) to facilitate trade and navigation, and averred, "There is . . . in all Probability, another Southern Continent." In 1756, Charles de Brosses, already a classicist of note and soon to be an important linguist and historian of religion, produced a two-volume *Histoire des navigations aux terres australes*. Brosses accepted the need for large bodies of land in the Southern Hemisphere to balance those in the north and, like Campbell, recommended establishments in New Britain and Juan Fernández to facilitate Pacific commerce.[8]

The existence of a continent or islands in the southern Pacific constituted one possibility; the existence of as-yet undiscovered islands or peninsulas in the North Pacific, another. Well into the eighteenth century, Europeans knew too little about the South Sea's northern waters to be certain what they contained. Campbell could still declare in 1744, "We are, in a manner, ignorant of what lies between America and Japan, and all beyond that Country lies buried in Obscurity." Brosses could still aver that it was simply not plausible that the waters between Japan and North America could fail to yield islands "rich in spices." Furthermore, just as some saw the need for a southern Pacific continent to balance known northern landmasses, others saw the need for lands farther north in the Pacific to balance the Eurasian and American continents on either side of the South Sea. Henry Ellis, who sailed with the 1746–1747 British expedition to Hudson Bay, argued in 1750, "We have a kind of moral Certainty, that there must be either a great Continent, or considerable Islands to the Westward of America, in order to constitute the Equipose, and they too must lie to the North, that is under those Parallels of Latitude, between California and Japan."[9]

One often-sought body of land in this North Pacific murk was "Gama Land."

8. William Barr and Glyndwr Williams, eds., *Voyages to Hudson Bay in Search of a Northwest Passage, 1741–1747*, 2 vols. (London, 1994–1995), I, 34; Harris, ed., *Navigantium atque Itinerantium*, I, 264, 331–335, 341; Charles de Brosses, *Histoire des navigations aux terres australes: Contenant ce que l'on sçait des moeurs et des productions des contrées découvertes jusqu'à ce jour . . .*, 2 vols. (Paris, 1756), I, 2, 13–18, II, 362–369. See also [Pierre Louis Moreau] de Maupertuis, *Lettre sur le progrès des sciences* (n.p., 1752), 7–26; Dunmore, *French Explorers in the Pacific*, I, 45–49; Spate, *Monopolists and Freebooters*, 267.

9. Harris, ed., *Navigantium atque Itinerantium*, I, 336; Brosses, *Histoire des navigations aux terres australes*, I, 4: "Il n'est pas possible qu'entre le Japon et l'Amérique il n'y ait, dans le vaste ocean pacifique, un grand nombre d'isles riches en épiceries." See also Rogers, *Cruising Voyage*, 325; J[oseph]-N[icholas] Delisle to comte de Maurepas, May 25, 1729, in H. Omont, "Lettres de J.-N. Delisle au comte de Maurepas et à l'abbé Bignon sur ses travaux géographiques en Russie (1726–1730)," in *Bulletin de la section de géographie*, XXXII (1915), 134–148, esp. 144: "Cette mer Orientale, par où l'on peut parvenir des terres de la Russie au Japon et sur laquelle se trouve la terre d'Yeço, la terre de la Compagnie, la Corée, etc., cette mer, dis-je, est la plus inconnue de toute la terre." Ellis quotation from Henry Ellis, *Considerations on the Great Advantages Which Would Arise from the Discovery of the North West Passage* (1750; rpt. San Francisco, 1959), 4. Ellis was elected a Fellow of the Royal Society in 1748 and would go on to serve as governor of the colonies of Georgia and Nova Scotia.

The idea owed its origin to a 1589 or 1590 voyage from Macao to Acapulco, during which Joao de Gama believed he had spotted land in the northwestern Pacific and after which Gama Land appeared frequently on North Pacific charts. Gama Land became the target of eighteenth-century exploration. German scholar Gerhard Friedrich Müller, a participant in Vitus Jonassen Bering's 1741 North Pacific voyage, spoke of going in search of "land seen by Don Jean Gama" because it appeared on a 1731 North Pacific chart, and Bering and his officers believed that "the author of the map would not have represented anything on uncertain ground." After subsequent investigation convinced the party "of the nonexistence of the Land of Gama," and after the separation, shipwrecks, scurvy, and death consequent from vainly looking for mythical lands in stormy and foggy North Pacific waters, Bering's men came to "believe that they were misled to a useless navigation."[10]

The "Company Land" and Yedso ("Jesso") mentioned by Kino in Chapter 1 provided two other instances of geographic confusion. Company Land dated to a 1643 Dutch expedition led by Maerten Gerritsz Vries (or Fries). Vries observed parts of the Kuril Islands and mistakenly believed one of them, which he named after the Dutch East India Company, was connected to the continent. Company Land appeared in various sizes and North Pacific locations thereafter, often as an Asian or North American peninsula. Yedso, whose variant spellings might have outnumbered its population, derived from imperfect ideas of Hokkaido. It appeared as an island or as an American or Asian continental extension and, because of its northern location, seemed to Arthur Dobbs in 1731 a favorable site for a "New Trade for our [Britain's] Woollen Manufactures."[11]

10. On Gama Land, see [Joseph-Nicholas Delisle], "Memoir Presented to the Senate with Map Which Bering Used in Going to America," in F. A. Golder, *Russian Expansion on the Pacific, 1641–1850: An Account of the Earliest and Later Expeditions Made by the Russians along the Pacific Coast of Asia and North America* . . . (Cleveland, Ohio, 1914), 301–313, esp. 302–309; Hayes, *Historical Atlas of the North Pacific*, 58–62, 67, 75, 85; Suárez, *Early Mapping of the Pacific*, 27, 46–47; Spate, *Spanish Lake*, 108; Glyn[dwr] Williams, *Voyages of Delusion: The Northwest Passage in the Age of Reason* (London, 2002), 204–205, 240, 244. On Müller and Bering's 1741 expedition, see Gerhard Friedrich Müller, *Bering's Voyages: The Reports from Russia*, ed. and trans. Carol Urness (Fairbanks, Alaska, 1986), 35, 38, 79, 99–101, 176. See also the Dec. 7, 1741, "Report from Captain Aleksei I. Chirikov to the Admiralty College concerning His Observations and Explorations along the Coast of America," in Basil Dmytryshyn, E. A. P. Crownhart-Vaughan, and Thomas Vaughan, eds. and trans., *To Siberia and Russian America: Three Centuries of Russian Eastward Expansion; A Documentary Record*, 3 vols. ([Portland, Ore.,] 1988), II, *Russian Penetration of the North Pacific Ocean: 1700–1797*, 139–158, esp. 139; Georg Wilhelm Steller, *Journal of a Voyage with Bering, 1741–1742*, ed. and trans. O. W. Frost, trans. Margritt A. Engel (Stanford, Calif., 1988), 51, 85–86; Spate, *Monopolists and Freebooters*, 228–251.

11. [Delisle], "Memoir Presented to the Senate," in Golder, *Russian Expansion*, 308–313; Hayes, *Historical Atlas of the North Pacific*, 6, 8, 34, 36–38, 48–49, 54–55, 58–62, 67, 75, 111; Barr and Williams,

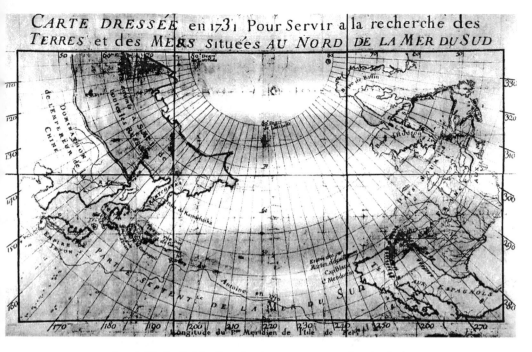

CARTE DRESSÉE en 1731 Pour Servir a la recherche des
TERRES et des MERS Situées AU NORD DE LA MER DU SUD

MAP 17. Joseph-Nicolas Delisle, *Carte dressée en 1731 pour servir a la recherche des terres et des mers situées au nord de la mer du Sud*. 1731. Library and Archives Canada, NMC 6909

What made these reported territories especially interesting were rumors that islands north or east of Japan were rich in gold and silver. This notion was rendered somewhat plausible by the claim of Marco Polo that in "Chipangu" "gold is abundant beyond all measure," by reports of a sixteenth-century Portuguese ship's having found islands east of Japan *"rricas de plata y muy pobladas,"* and by the fact that Japan was, after Spanish America, the main early modern silver producer. In the first decade of the seventeenth century, interest in the islands "Rica de Oro" and "Rica de Plata" distracted Spanish attention from projects to establish a California naval station. The 1643 Dutch expedition mentioned above, as well as its 1639 predecessor, were searching for these islands. The possibility of North Pacific mineral wealth continued to pique European interest in the eighteenth century. In 1730, a Spanish royal directive instructed the governor of the Philippines to find and occupy Rica de Oro and Rica de Plata. After Martin Spanberg, as part of Bering's Second Kamchatka expedition, conducted three ships to the Kuril Islands and Japan in 1739, the February 27, 1740, *Gazette de France* com-

---

eds., *Voyages to Hudson Bay*, I, 34; Spate, *Monopolists and Freebooters*, 39–41, 236; Williams, *Voyages of Delusion*, 49, 204–205, 240, 244.

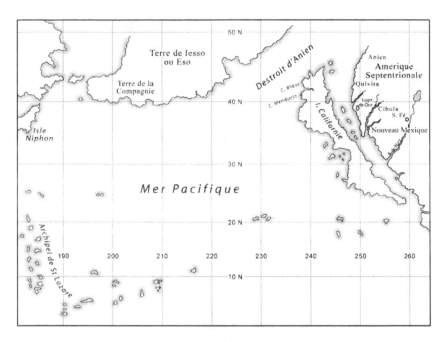

MAP 18. The North Pacific. Drawn by Jim DeGrand, after Pierre Du Val, *Carte universelle du monde.* 1684. For a photographic reproduction of a detail of the original map at the Library of Congress, see Derek Hayes, *Historical Atlas of the North Pacific Ocean: Maps of Discovery and Scientific Exploration, 1500–2000* (Seattle, 2001), 6

mented on the "large quantities of gold and copper money" possessed by the inhabitants of the "thirty-four islands" Spanberg was reported to have discovered.[12]

Just as unknown or unexplored lands might ring or dot the North Pacific, advanced and as-yet uncontacted cultures might grace the North American Pacific coast, making a Northwest Passage a route to lucrative commerce or conquest.

12. First quotations are from Polo, *Book of Ser Marco Polo,* ed. and trans. Yule, II, 253; and "Letter of Fray Andrés de Aguirre to the Archbishop of Mexico Giving an Account of Some Rich Islands Inhabited by Civilized People, Discovered by a Portuguese Trader, and Situated in Latitude 35° to 40° North—Written in 1584–85," in Donald C. Cutter, ed., *The California Coast: A Bilingual Edition of Documents from the Sutro Collection,* trans. Cutter and George Butler Griffin (Norman, Okla., 1969), 9–17, esp. 14–15. On Japanese silver production, see Ward Barrett, "World Bullion Flows, 1450–1800," in James D. Tracy, ed., *The Rise of Merchant Empires: Long-Distance Trade in the Early Modern World, 1350–1750* (Cambridge, 1990), 224–254, esp. 245–247. On Spanish island interest, see Schurz, *Manila Galleon,* 231–238; Hayes, *Historical Atlas of the North Pacific,* 31, 34, 111; and W. Michael Mathes, ed., *Californiana,* I, *Documentos para la historia de la demarcación comercial de California, 1583–1632* (Madrid, 1965), docs. 2, 71, 84, 88–90, 92–93, 99, 103, 105, 107, 109–110, 113, 121. On reports after Spanberg's expedition, see "Documents Bearing on the Voyage of Captain Spanburg from Kamchatka to Japan in June, July, and August, 1739: Discovery of Thirty-Four Islands in the North Pacific Ocean," in Golder, *Russian Expansion,* 330–333; Müller, *Bering's Voyages,* ed. and trans. Urness, 38, 90.

Some authors emphasized the Asiatic origin of or influence on these peoples. Arthur Dobbs spoke of "a Strong presumption that there are Many civiliz'd Nations" in northwestern North America and asserted that, since it was "highly probable that North America at least was peopled from the Eastern Regions near Japan, there is more Reason to believe they are better civiliz'd in those Countrys, the nearer We approach to the Asiatick Coast." Others authors pointed to especially accomplished Amerindian civilizations. Daniel Coxe, in the 1727 edition of his *Description of the English Province of Carolana,* wrote of the lands beyond a range of hills north of New Mexico. A western river descended from these hills to a large lake and from that lake to the Pacific. On this lake lived "Two or Three Mighty Nations, under Potent Kings, abundantly more civiliz'd, numerous, and warlike, than their Neighbours, differing greatly in Customs, Buildings, and Government, from all the other Natives of this Northern Continent: ... they are cloathed, and build Houses, and Ships, like Europeans, having many of great Bigness, in length 120 or 130 Foot, and carry from 2, to 300 Men, which navigate the great Lake, and it is thought the adjacent Parts of the Ocean." These coastal nations might, he suggested, trade with ships from China or Japan.[13]

Europeans did not find a new Canton, Nagasaki, Tenochtitlán, or Cuzco in the Pacific Northwest. The possibility of wealthy civilizations there, like that of Terra Australis Incognita, Ophir, Locac, Gama Land, Company Land, and Yedso in the Pacific, enticed the credulous, but because their existence was uncertain, their power to shape the conduct of European officials was inconsistent. The value of other Pacific lands was more sure, their significance more stable.

China

Ainsi à parler en général, ce n'est plus qu'avec de l'argent qu'on peut trafiquer utilement à la Chine.

— JEAN-BAPTISTE DU HALDE, *Description géographique, historique, chronologique, politique, et physique de l'empire de la Chine et de la Tartarie chinoise,* 1735

One of these Pacific lands, as would be expected from the discussion above, was Japan, and Japan did attract European attention. Dobbs spoke of a Northwest

---

13. Dobbs quoted in Barr and Williams, eds., *Voyages to Hudson Bay,* I, 34–35. By his own account, Coxe's idea of an easy passage across western mountains and down to the Pacific came from the French author Lahontan; that of Indian civilizations trading with Asia, from a variety of Spanish authors. For Coxe, see his *Description of the English Province of Carolana, by the Spaniards Call'd Florida, and by the French La Louisiane* (1727; rpt. San Francisco, 1940), microfilm, 28–29, 49–51. See also [Louis Armand de Lom d'Arce, baron de] Lahontan, *Oeuvres complètes,* ed. Réal Ouellet and Alain Beaulieu, 2 vols. (Montreal, 1990), I, 392–423.

Passage offering an "easy and short way to Japan," and Spanberg succeeded in making a difficult and short journey there from Kamchatka. Despite the promise of the "four-cornered, gold coins" Spanberg's men received there, eighteenth-century Japan would generally provide little to Europeans. As had been the case since the 1630s, Japan remained officially closed to outsiders and the Japanese themselves confined to their home islands. A limited Dutch trade at Nagasaki constituted the only European access. Even the obstacle-ignoring Dobbs allowed that, to initiate the trade, he envisioned it would be necessary to "force Japan into a Beneficial Treaty of Commerce." This would have to wait for the nineteenth century.[14]

Instead of insular Japan, imperial China—object of comparison, admiration, and disdain for Enlightenment philosophers, source of refined goods for European consumers and stylistic models for European craftsmen, and cause of increasing disquiet for some European manufacturers and statesmen—was the primary Asian-Pacific trading partner for eighteenth-century Europeans. As Edward Gibbon observed in his eighteenth-century history of Rome, the western European purchase of Chinese luxury goods dated back to the century of Trajan. What set Gibbon's century apart from those of his predecessors was that the European demand for Chinese goods such as silk, porcelain, and tea was increasing, the volume of Sino-European trade expanding, and the number of participants in this once limited exchange multiplying. Jean-Baptiste Du Halde, author of the monumental 1735 *Description géographique, historique, chronologique, politique, et physique de l'empire de la Chine* declared that the river at Canton, the main site of Sino-European commerce, looked like a forest because of the quantity of ships there. The value of tea imported into the British Empire on such ships grew from roughly £8,000 in 1699–1701 to £848,000 in 1772–1774. This growth in trade reflected not merely the upper-class craze for chinoiserie that left western Europe littered with incongruous pagodas and Chinese-themed wallpaper but also the fact that Chinese products that had been traditionally, or initially, luxuries to be enjoyed by the wealthy were becoming items of mass consumption. Merchant and philanthropist Jonas Hanway claimed in his 1757 *Essay on Tea*, for example, that the English consumption of tea began with its introduction by two aristocrats in 1666 and then, because of the tendency of the lower classes to imitate the privileged, "descended to the *Plebæan* order" around 1700. A similar trend occurred on the other side of the Atlantic. On his extensive

14. J[ean]-B[aptiste] Du Halde, *Description géographique, historique, chronologique, politique, et physique de l'empire de la Chine et de la Tartarie chinoise,* 4 vols. (Paris, 1735), II, 173; Barr and Williams, eds., *Voyages to Hudson Bay,* I, 34; Müller, *Bering's Voyages,* ed. and trans. Urness, 90; Schurz, *Manila Galleon,* 99–128.

and often disapproving 1744 journey among the British American colonies from Maryland to Massachusetts, Scottish doctor Alexander Hamilton was shocked both by the absence of cultivation among the lower classes he disdained and by the presence of refinement among those seemingly incapable of affording it. Amid the wilds of New York, he met a family that, though mired in poverty, hoped to be steeped in tea: it possessed "a set of stone tea dishes, and a tea pot," a "tea equipage" Hamilton considered "quite unnecessary" for a family with so little else.[15]

The implications of the taste for Chinese goods were economic as well as social and cultural. The tea to be drunk, the Chinese porcelain on which to serve it, and the silk that adorned its imbibers had to be paid for. The problem for traders responding to European consumer demand for Chinese goods was that, before Captain Cook's crew discovered the value of Pacific Northwest sea otter pelts and before the British Empire began exploiting the commercial possibilities of opium, Chinese consumers desired European products less than Europeans wanted Chinese ones. Europeans had little to offer that Chinese merchants wanted to buy.[16]

The exception was silver. It was China's official means of exchange and the currency in which the Chinese government insisted taxes be paid. Moreover, stimulated in part by the introduction of crops such as potatoes, sweet potatoes, and maize, China's already huge population grew larger still over the course of the eighteenth century, going from very roughly 100 million in 1685 to about 210 million in 1767. This burgeoning population needed increasing amounts of silver to finance its private and public transactions.[17]

The vast majority of the silver needed to satisfy the material and monetary

15. On Canton, see Du Halde, *Description géographique*, II, 173: "La riviere paroît comme une grande forêt, par la multitude des Vaisseaux qui s'y trouvent." On tea importation, see Jacob M. Price, "The Imperial Economy, 1700–1776," in W[illiam] Roger Louis, ed., *The Oxford History of the British Empire*, 5 vols. (Oxford, 1998–1999), II, *The Eighteenth Century*, ed. P. J. Marshall and Alaine Low, 78–104, esp. 83. On the growing consumption of Chinese goods, see David L. Porter, "Monstrous Beauty: Eighteenth-Century Fashion and the Aesthetics of the Chinese Taste," *Eighteenth-Century Studies*, XXXV (2002), 395–411, esp. 396. On tea drinking, see [Jonas Hanway], *An Essay on Tea . . .*, in Hanway, *A Journal of Eight Days Journey from Portsmouth to Kingston upon Thames: Through Southampton, Wiltshire, etc. with Miscellaneous Thoughts . . .*, 2 vols. (London, 1757), II, 21–22; Alexander Hamilton, *The Itinerarium of Dr. Alexander Hamilton*, in Wendy Martin, ed., *Colonial American Travel Narratives* (New York, 1994), 173–328, esp. 216.

16. Price, "Imperial Economy," in Louis, ed., *Oxford History of the British Empire*, II, 80; Porter, "Monstrous Beauty," *Eighteenth-Century Studies*, XXXV (2002), 401.

17. Jonathan D. Spence, *The Search for Modern China* (New York, 1990), 20, 46, 93–95; K. N. Chaudhuri, *The Trading World of Asia and the English East India Company, 1660–1760* (Cambridge, 1978), 176, 200; Dennis O. Flynn and Arturo Giráldez, "Cycles of Silver: Global Economic Unity through the Mid-eighteenth Century," *JWH*, XIII (2002), 391–427.

wants of the populations of the growing Chinese Empire and the greater European world originated in the Americas. Indeed, as contemporary observers were well aware, the enormous increase in European trade with Asia and in the European consumption of Asian goods in the seventeenth and first half of the eighteenth centuries was possible only because New World silver acquired legally or illegally from the Spanish Empire gave Europeans the means to pay for their purchases. When an East India Company ship left the British Isles for Asia, bullion typically constituted 80 percent or more of its cargo. Montesquieu noted, "America furnished Europe with the material for its commerce in the vast part of Asia called the East Indies." Adam Smith concurred: "The silver of the new continent seems in this manner to be one of the principal commodities by which the commerce between the two extremities of the old one is carried on, and it is by means of it, in a great measure, that those distant parts of the world are connected with one another."[18]

Spanish American Silver

It was this Spanish silver—drawn from Peruvian and Mexican mines, necessary for trade with China, useful as a means of exchange in Europe, and serviceable in diplomacy and war—that helps to explain the striking extent of European interest in Spanish Pacific settlements. In the sixteenth, seventeenth, and eighteenth centuries, estimates of Spanish American silver production range from roughly 100,000 to 150,000 tons. This constituted about 85 percent of world silver production for the period. For the eighteenth century alone, one estimate

18. On the role of Spanish silver, see Artur Attman, *American Bullion in the European World Trade, 1600–1800,* trans. Eva Green and Allan Green (Göteborg, Sweden, 1986), 4–7, 31, 33, 56, 74; Fernand Braudel, *Civilization and Capitalism: 15th-18th Century,* trans. Siân Reynolds, 3 vols. (New York, 1982–1984), II, *The Wheels of Commerce,* 198–199; Chaudhuri, *Trading World of Asia,* 2–3, 6–10, 200, 456; Allan Christelow, "Contraband Trade between Jamaica and the Spanish Main, and the Free Port Act of 1766," *HAHR,* XXII (1942), 309–343, esp. 311; Christelow, "Great Britain and the Trades from Cadiz and Lisbon to Spanish America and Brazil, 1759–1783," *HAHR,* XXVII (1947), 2–29, esp. 22–23; Catherine Manning, *Fortunes à Faire: The French in Asian Trade, 1719–48* (Hampshire, U.K., 1996), 33; Henri Sée and Léon Vignols, "L'envers de la diplomatie officielle de 1715 à 1730: La rivalité commerciale des puissances maritimes et les doléances des négociants français," *Revue belge de philologie et d'histoire,* V (1926), 471–491, esp. 478; Geoffrey J. Walker, *Spanish Politics and Imperial Trade, 1700–1789* (Bloomington, Ind., 1979), 7. On bullion in East India ships, see P. J. Marshall, "The British in Asia: Trade to Dominion, 1700-1765," in Louis, ed., *Oxford History of the British Empire,* II, 488. For the quotations, see [Charles de Secondat, baron de] Montesquieu, *The Spirit of the Laws,* ed. and trans. Anne M. Cohler, Basia C. Miller, and Harold S. Stone (Cambridge, 1989), 354–355, 392; Adam Smith, *An Inquiry into the Nature and Causes of the Wealth of Nations,* ed. R. H. Campbell, A. S. Skinner, and W. B. Todd, 2 vols. (Indianapolis, 1981), I, 225.

has Spanish silver accounting for roughly 90 percent of global output, or 51,000 out of 57,000 tons.[19]

The most celebrated early modern silver source was the ore-filled mountain of Potosí (in modern Bolivia), the riches of which the Spanish first began to discover in 1545. For centuries thereafter, mules carried the better part of Potosí's production to the Pacific coast at Arica, and ships then carried it to Callao and Panama City. In Panama, the silver was disembarked and sent overland to the Atlantic city of Portobello, where it was then reloaded on ships for transport to Spain. After a period of decline in the seventeenth century, silver production at Potosí appears to have tripled between the 1720s and 1780s, with the annual value of output in the century as a whole ranging between three and ten million pesos.[20]

Potosí formed one of the two great Spanish American silver sources, northern Mexico's mines the other. Though less renowned than Potosí, these Mexican mines yielded more silver in the 1700s than those of Peru, ultimately accounting for 67 percent of New World production. Moreover, their rate of output increased more quickly, quadrupling over the course of the eighteenth century. Combining the figures for Peruvian and Mexican production reveals that Spain's mines poured forth twice as much silver between 1686 and 1810 as they did between 1560 and 1685.[21]

The better part of the silver of Peru and Mexico moved eastward across the Atlantic to Spain and the rest of Europe. A significant portion, however, moved in the opposite direction, from Acapulco to Manila and Manila to China. From

19. Barrett, "World Bullion Flows," in Tracy, ed., *Rise of Merchant Empires*, 224–256, esp. 224–226, 237; Richard L. Garner, "Long-Term Silver Mining Trends in Spanish America: A Comparative Analysis of Peru and Mexico," *AHR*, XCIII (1988), 898–935, esp. 898.

20. On silver's itinerary, see Walker, *Spanish Politics and Imperial Trade*, 7–8; Peter T. Bradley, *The Lure of Peru: Maritime Intrusion into the South Sea, 1598–1701* (Hampshire, U.K., 1989), 2; E. W. Dahlgren, *Les relations commerciales et maritimes entre la France et les côtes de l'Océan pacifique (commencement du XVIIIe siècle)*, I, *Le commerce de la mer du Sud jusqu'à la paix d'Utrecht* (Paris, 1909), 61; Carlos Daniel Malamud Rikles, *Cádiz y Saint Malo en el comercio colonial peruano (1698–1725)* (Cadiz, 1986), 132. On silver production, see Peter Bakewell, "Mining in Colonial Spanish America," in Leslie Bethell, ed., *The Cambridge History of Latin America*, 11 vols. (Cambridge, 1984–2008), II, *Colonial Latin America*, 105–151, esp. 148; D. A. Brading, "Bourbon Spain and Its American Empire," ibid., I, *Colonial Latin America*, 389–439, esp. 422; Garner, "Long-Term Silver Mining Trends," *AHR*, XCIII (1988), 899, 903, 908, 910.

21. Brading, "Bourbon Spain and Its American Empire," in Bethell, ed., *Cambridge History of Latin America*, I, 420–421; Garner, "Long-Term Silver Mining Trends," *AHR*, XCIII (1988), 899–901, 934–935; John Lynch, *Bourbon Spain, 1700–1808* (Oxford, 1989), 154–156; Alexander von Humboldt, *Political Essay on the Kingdom of New Spain*, ed. Mary Maples Dunn, trans. John Black (Norman, Okla., 1972), 145–183, 239.

1565 to 1815, Spain conducted an annual Mexico-Philippines galleon trade involving the exchange of vast quantities of Spanish silver for Asian goods. Generally, one or two ships, ranging from a few hundred to a few thousand tons in size, made, "in thirsty scorbutic squalor," the three-month journey from Acapulco to Manila and the six-month return voyage to Mexico. Chinese silks were the most important commodity shipped from Manila, but porcelain, Indian cottons, Persian and Chinese rugs, Indonesian spices, and a host of other items also made the annual eastward journey in the "*nao de China*." Because merchants violating Spanish trade regulations often tried to keep the amount of silver they sent east on the galleon secret, it is difficult to state precisely the quantities of New World silver passing from Acapulco to Manila. Estimates range from an average of 2–3 million pesos a year to a low of about 600,000 pesos. Many signs in Mexico attested to the prominence of this trans-Pacific commerce. An ocean away from Asia, New Spain's inhabitants called the path from Acapulco to Mexico City the "China Road." Alexander von Humboldt, who had originally planned to take the galleon from Mexico to the Philippines after completing his early-nineteenth-century researches in Spanish America, reported that because of "the frequent communication between Acapulco and the Philippine Islands many individuals of Asiatic origin, both Chinese and Malays, have settled in New Spain."[22]

Man and nature took their toll on the ponderous galleons trying to cross the world's largest ocean. A ship laden with American or Asian riches tempted raiders. In 1587, the English freebooter Thomas Cavendish captured the Manila galleon on its return voyage to Acapulco. In 1709, Woodes Rogers took the smaller of two ships on their way back to Mexico. Even in the absence of these plunderers, the Pacific passage was difficult, and the ocean and its shores claimed more than thirty ships between 1565 and 1815. For years, beachcombers collected beeswax from the cargo of one that crashed near present-day Tillamook, Ore-

22. The scurvy quotation is from Parry, *Trade and Dominion*, 237. The high bullion estimate comes from Schurz, *Manila Galleon*, 189–190; the low is from Humboldt, *Political Essay on the Kingdom of New Spain*, ed. Dunn, trans. Black, 182. But note that Humboldt's low estimate was the average since the sixteenth-century commencement of the trade. Later in his book (206–207), Humboldt estimated that, at the time of his own visit to New Spain, approximately 1–1.3 million piasters were passing annually from Acapulco to Manila. Barrett ("World Bullion Flows," in Tracy, ed., *Rise of Merchant Empires*, 248–249), estimates that about fifteen tons of silver went annually to the Philippines in each of the three quarter-century periods between 1700 and 1775. Earlier, on page 237, Barrett suggests that 20,000 of the 130,000–150,000 tons of silver produced in the Americas between 1500 and 1800 went to the Philippine trade and its subsidies. See also Dahlgren, *Les relations commerciales et maritimes*, I, 69; and William Schell, Jr., "Silver Symbiosis: ReOrienting Mexican Economic History," *HAHR*, LXXXI (2001), 89–133, esp. 89, 91. On Asia in America, see Humboldt, *Political Essay on the Kingdom of New Spain*, ed. Dunn, trans. Black, 45; Schurz, *Manila Galleon*, 27–33, 63, 362, 384–386.

gon. In the journals of the Lewis and Clark expedition, Clark reported meeting on the Oregon coast "with the party of *Clât Sops* who visited us last . . . a man of much lighter Coloured than the nativs are generaly, . . . freckled with long duskey red hair, about 25 years of age, . . . Certainly . . . half white at least," who might have descended from survivors of both Spanish and British shipwrecks.[23]

Some of Spanish America's silver crossed the Pacific to the Philippines. Some crossed the Atlantic to Spain. Some found its way to the bottom of both oceans. A significant quantity passed through the hands of Spanish America's inhabitants, and this last fact made eighteenth-century Spanish America an important market for European goods. Silver gave Spanish American consumers the means to pay for such goods, and, in the eighteenth century, a large population meant that the demand for them was substantial. Once the regions comprising Spain's American empire had recovered from the initial demographic collapse triggered by the arrival of pestilential Old Worlders, their inhabitants became numerous. In 1800, a year for which good comparative statistics are available, the United States' population was about 5.3 million; Spanish America's, between 13 and 17 million.[24]

Spanish American demand for European goods derived not only from the size of America's populations and the output of its mines but also from Spanish governmental policies. For much of the sixteenth century, the Spanish government had tried to stimulate American agriculture and manufactures to meet the needs of Spain's American subjects. In the latter part of that century, imperial policy began to move in a different direction. As the colonies came to be seen in Spain primarily as sources of precious metals, the government sought ways to encourage the production of bullion and its transfer to the peninsula. It was hoped that discouraging agriculture and industry in the Spanish colonies would make them dependent on peninsular Spain for consumer goods, cloth, and agricultural produce. The colonies would have to mine more gold and silver to pay for these Spanish imports, and American demand would stimulate peninsular industry and farming. As a result of such policies, millions of Spanish American

23. On captured galleons, see Rogers, *Cruising Voyage*, 293, 300–303; Williams, *Great South Sea*, 41–42, 152–155. On wrecked galleons, see Schurz, *Manila Galleon*, 15, 193–195, 256, 263, 281, 303. On Lewis and Clark on the Pacific, see Gary E. Moulton, ed., *The Journals of the Lewis and Clark Expedition*, VI, *Nov. 2, 1805–Mar. 22, 1806* (Lincoln, Neb., 1990), 147–148; Warren L. Cook, *Flood Tide of Empire: Spain and the Pacific Northwest, 1543–1819* (New Haven, Conn., 1973), 31–40.

24. On Spanish America as a market, see Allan Christelow, "French Interest in the Spanish Empire during the Ministry of the Duc de Choiseul, 1759–1771," *HAHR*, XXI (1941), 515–537, esp. 516–517. On Spanish America's population, see Humboldt, *Political Essay on the Kingdom of New Spain*, ed. Dunn, trans. Black, 31–33, 37, 238; Nicolás Sánchez-Albornoz, "The Population of Colonial Spanish America," in Bethell, ed., *Cambridge History of Latin America*, II, 3–35, esp. 34.

customers became reliant for many of their material wants on producers on the other side of the Atlantic.[25]

In some cases, these wants inclined toward extravagance. European travelers devoted pages to descriptions of Peruvian upper-class luxuries. Antonio de Ulloa wrote at length about the clothing worn in Lima. He commented, in the somewhat free rendition of his eighteenth-century English translator, "It may be said without exaggeration, that the finest stuffs . . . are more generally seen at Lima than in any other place; vanity and ostentation not being restrained by custom or law. Thus the great quantities brought in the galleons and register ships notwithstanding they sell here prodigiously above their prime cost in Europe, the richest of them are used as cloaths, and worn with a carelessness little suitable to their extravagant price." Four pages followed describing the lavish dress and striking appearance of Peruvian women.[26]

Spain's mercantile system failed to function as its architects intended, and many of the goods purchased by Spain's American subjects originated outside of Iberia. Although Spanish officials hoped to reserve Spanish American markets for Spanish producers, they could not succeed in doing so. Spain never produced enough to supply both its own and its American consumers. In addition, economic mismanagement, the high taxes required to finance Spain's innumerable European wars, and inflation encouraged by peninsular importation of gold and silver tended to make Spanish products more expensive than those of other European nations.

Spanish America's fabled riches also inflamed the ambitions of non-Spanish merchants. The actual production of mines such as Potosí was impressive

25. Walker, *Spanish Politics and Imperial Trade,* 1–4; Christelow, "Contraband Trade between Jamaica and the Spanish Main," *HAHR,* XXII (1942), 311.

26. Jorge Juan and Antonio de Ulloa, *A Voyage to South America,* trans. John Adams (New York, 1964), 196–200. The Spanish text from Antonio de Ulloa, *Viaje a la América meridional,* ed. Andrés Saumell, 2 vols. (Madrid, 1990), II, 71, reads, "Todos visten con mucha ostentacion, y puede decirse sin exageracion que las telas que se fabrican en los paises donde la industria trabaja para conseguir sus invenciones se lucen en *Lima* mas que en ninguna otra parte por la mucha generalidad con que se gastan, siendo esto causa de que tengan consumo las muchas que llevan las armadas de *galeones* y *registros.* Y aunque su costo es allí tan subido que no se puede comparar con el que tienen en *Europa* los mismos generos, no embaraza este ni para que dexen de vestirse de las mejores ni para no usarlas con desenfado y generosidad, sin poner aquel cuidado en su conservacion que parece corresponiente á su mucho costo." Ulloa became well known in Europe because of his participation in a Franco-Spanish expedition (1735–1744) seeking to measure an arc of the meridian near Quito, part of a larger effort to determine the earth's exact shape. Other authors seconded these kinds of observations. See William Betagh, *A Voyage round the World: Being an Account of a Remarkable Enterprize, Begun in the Year 1719, Chiefly to Cruise on the Spaniards in the Great South Ocean . . .* (London, 1728), 262–263; Amédée-François Frézier, *A Voyage to the South-Sea and along the Coasts of Chili and Peru, in the Years 1712, 1713, and 1714* (London, 1717), 218–219.

enough, but distance, wishful thinking, the awestruck reports of travelers such as Ulloa, and the embellishments of storytellers made American mines and Spanish colonies appear even richer to envious Britons and Frenchmen than they actually were. Potosí and Peru became bywords for wealth, and, as scholars have remarked, the name of the British South Sea Company (founded 1711) bore witness to the importance of Spain's Pacific possessions in the European imagination. French, British, and Dutch merchants saw that they could profitably supply Spain's American consumers with better and cheaper goods than their Spanish competitors, either by selling goods in Spain for reshipment to America or by sending goods directly and illegally to Spain's colonies. Ultimately, the better part of the cargo on the annual Spanish fleets to the Indies would come from outside Spain, and smugglers would haunt Spanish America's coasts. Ulloa, in a 1749 *Discourse and Political Reflections on the Kingdoms of Peru* he coauthored with fellow Spanish officer and Peruvian observer Jorge Juan, declared, "In order to deal with the evils of illicit trade in the Indies, one must assume that every port, city and town is infected by the malady. The only difference is in degree: smuggling is more widespread in some areas than in others." Assessing the implications of this smuggling, the seventeenth-century German legal scholar Samuel Pufendorf more colorfully and famously joked, "Spain kept the cow, and the rest of Europe drank the milk."[27]

Mere cupidity would explain the desire of merchants to ply their wares in these Spanish markets, but more than individual avarice was involved. Spanish silver possessed a global economic significance of which commercial-minded European governments were acutely conscious. By the eighteenth century, Europe's economies and societies had developed such that they needed American silver as much as they wanted it. In this period, western Europe suffered from a recurring trade imbalance not only with China but also with the Baltic

27. On the enticing Pacific, see Bradley, *Lure of Peru*, 2–4; Christelow, "Contraband Trade between Jamaica and the Spanish Main," *HAHR*, XXII (1942), 314; Dahlgren, *Les relations commerciales et maritimes*, I, 62–66; Charles Frostin, "Les Pontchartrain et la pénétration commerciale française en Amérique espagnole (1690–1715)," *Revue historique*, CCXLV (1971), 307–336, esp. 319; Humboldt, *Political Essay on the Kingdom of New Spain*, ed. Dunn, trans. Black, 83; Richard Pares, *War and Trade in the West Indies, 1739–1763* (Oxford, 1936), 5; Schurz, *Manila Galleon*, 362–366; Stanley J. Stein and Barbara H. Stein, *Silver, Trade, and War: Spain and America in the Making of Early Modern Europe* (Baltimore, 2000), 136. On smuggling, see Dahlgren, *Les relations commerciales et maritimes*, I, 63–66; Walker, *Spanish Politics and Imperial Trade*, 11–14. Ulloa quotation from Jorge Juan and Antonio de Ulloa, *Discourse and Political Reflections on the Kingdoms of Peru: Their Government, Special Regimen of Their Inhabitants, and Abuses Which Have Been Introduced into One and Another, with Special Information on Why They Grew up and Some Means to Avoid Them*, ed. and trans. John J. TePaske, trans. Besse A. Clement (Norman, Okla., 1978), 42. Pufendorf quoted in Christelow, "Great Britain and the Trades from Cadiz and Lisbon," *HAHR*, XXVII (1947), 3.

(East Prussia, Poland, and Russia) and the Levant. The Baltic region supplied commodities such as grain, lumber, hemp, flax, tallow, wax, bar iron, leather, skins, furs, and potash. For a naval power such as Britain, Norwegian timber, Swedish tar, and Russian hemp proved indispensable for the masts and ropes needed by the fleet, efforts to obtain these stores from the British colonies in North America having proved unsatisfactory. Britain could not sell enough of its own products to cover its purchases of Baltic commodities, and, consequently, large quantities of bullion flowed out of Britain to pay for them.[28]

Silver

Silver had, moreover, other uses besides financing branches of international commerce. It served as a useful means of exchange in Europe. Its importance in this regard is visible in phenomena such as the apparent synchronization of the velocity of exchange in France with the arrival of the silver-laden galleons in Cadiz and dispersal of part of their bullion in France. Precious metals such as silver also backed European paper currency. This was especially important during wars, when the unpredictability of events contributed to the uncertainty of paper money's value.[29]

More generally, the period's reigning mercantilist ideas emphasized bullion's importance for national power. Although Gibbon noted in 1776, "It has been observed, with ingenuity, and not without truth, that the command of iron soon gives a nation the command of gold," it was also observable that the command of silver could give nations the command of iron. Silver paid soldiers, built ships, and bought cannons. It also purchased allies. Fernand Braudel noted in *The Wheels of Commerce* that, in 1756, in exchange for Frederick the Great's allegiance in the Seven Years' War, Britain sent thirty-four wagonloads of silver coins bouncing along the roads to Berlin. Jonas Hanway explained the larger point:

> With gold and silver we can engage armies, and maintain fleets to fight
> our battles, and save our country: but without them we cannot even
> carry on a defensive war in our own country. It would be a difficult task
> to persuade a soldier, native or foreigner, to accept a bit of tin or lead in

28. Attman, *American Bullion*, 4–7, 74; Barrett, "World Bullion Flows," in Tracy, ed., *Rise of Merchant Empires*, 250; Daniel A. Baugh, "Maritime Strength and Atlantic Commerce: The Uses of 'a Grand Marine Empire,'" in Lawrence Stone, ed., *An Imperial State at War: Britain from 1689 to 1815* (London, 1994), 185–223, esp. 198; Patrick K. O'Brien, "Inseparable Connections: Trade, Economy, Fiscal State, and the Expansion of Empire, 1688–1815," in Louis, ed., *Oxford History of the British Empire*, II, 53–77, esp. 74.

29. On synchronization, see Louis Dermigny, "Circuits de l'argent et milieux d'affaires au XVIIIe siècle," *Revue historique*, CCXII (1954), 239–278, esp. 239–240. D. W. Jones presents a thorough and perspicacious discussion of the difficulties of wartime finance in *War and Economy in the Age of William III and Marlborough* (Oxford, 1988), 14–20, 95–97.

the place of gold or silver. . . . We have been hitherto enabled to support great fleets, and upon emergencies great armies also: we have checked the encroachments of France whose extent of dominion, and number of inhabitants, are so much greater than ours. But how have we been able to do this? Not by the force of valor only, but of money.[30]

Because American silver formed a crucial component of state power, domination of the sources of silver by a single vigorous nation could threaten the independence of other European states. Such had been the case with Habsburg Spain in the sixteenth and early seventeenth centuries. In the eighteenth century, a situation in which trade gave the different European powers roughly equal access to Spanish silver was a necessary condition for a balance of power among them. The attempt by one of their number to dominate this Spanish trade must elicit a vigorous response. The War of the Spanish Succession provided the first illustration of this.[31]

30. Edward Gibbon, *The History of the Decline and Fall of the Roman Empire*, ed. David Womersley, 3 vols. (London, 1994), I, 247; O'Brien, "Inseparable Connections," in Louis, ed., *Oxford History of the British Empire*, II, 55; Baugh, "Maritime Strength and Atlantic Commerce," in Stone, ed., *Imperial State at War*, 186; Braudel, *Civilization and Capitalism*, trans. Reynolds, II, 196; [Hanway], *Essay on Tea*, 172–173.

31. Chaudhuri, *Trading World of Asia*, 153; Stein and Stein, *Silver, Trade, and War*, 41, 119. For a general and insightful discussion of the relation between state finances and power in the early modern period, see chaps. 2 and 3 of Paul Kennedy, *The Rise and Fall of the Great Powers: Economic Change and Military Conflict from 1500 to 2000* (New York, 1987).

4

THE PACIFIC OCEAN AND THE
WAR OF THE SPANISH SUCCESSION

Le principal objet de la guerre présente est celui du commerce
des Indes et des richesses qu'elles produisent.
—LOUIS XIV, February 18, 1709

Tho' some religious-headed people fancy that money got by privatiering won't prosper, yet
I may venture to say the St. Malo men are as rich and florishing as any people in France.
—WILLIAM BETAGH, *A Voyage round the World*, 1728

On November 16, 1700, Louis XIV proclaimed, in accordance with the will of
Charles II of Spain, that Louis's grandson Philip, duc d'Anjou, would inherit
the Spanish throne. The ensuing War of the Spanish Succession (1702–1714)
would determine which European powers would profit from the riches of Spanish America. Louis and his ministers hoped that French commercial penetration of the Spanish Empire would make Spanish resources available to French
merchants and those merchants' income available to royal tax collectors. In the
years after 1700, the French government directed its diplomatic, military, and
commercial policies to this end. William III of England and the Dutch Republic
saw that control of Spanish American resources might give Louis the capacity
to achieve the European dominance he had long sought and others had long
feared. William and other European leaders felt they had to resist Louis's Spanish design.[1]

1. Louis XIV to Michel-Jean Amelot, Feb. 18, 1709, in Baron de Girardot, ed., *Correspondance de
Louis XIV avec M. Amelot, son ambassadeur en Espagne, 1705–1709*, 2 vols. (Paris, 1864), II, 121 ("The
principal object of the present war is that of the commerce of the Indies and the riches they produce"); William Betagh, *A Voyage round the World: Being an Account of a Remarkable Enterprize,
Begun in the Year 1719, Chiefly to Cruise on the Spaniards in the Great South Ocean* ... (London, 1728),
308; E. W. Dahlgren, *Les relations commerciales et maritimes entre la France et les côtes de l'Océan pacifique (commencement du XVIIIe siècle)*, I, *Le commerce de la mer du Sud (jusqu'à la paix d'Utrecht)*
(Paris, 1909), I, 561–562, 580–581, 633, 636, 659; Charles Frostin, "Les Pontchartrain et la pénétration commerciale française en Amérique espagnole (1690–1715)," *Revue historique*, CCXLV (1971),

During the War of the Spanish Succession, there transpired a comm
innovation of great significance for the study of eighteenth-century Eur
interest in the undiscovered North American West: direct commerce be
French ports and the Spanish settlements in Peru and Chile. By giving F
merchants and statesmen extensive experience with the profits available from
Spanish Pacific trade—and painful familiarity with the difficulty of reaching
Chile and Peru by way of Cape Horn—this commerce demonstrated the potential value of an easier western North American route to the South Sea.[2]

The antecedents to direct French trade with Peru and Chile in the first decades of the eighteenth century lay in the last three decades of the seventeenth.
From the 1670s to the early 1690s, simultaneously pushed by imperial efforts to
curtail Caribbean piracy and pulled by the lure of lightly guarded Spanish cities
and treasure ships, buccaneers had journeyed into the Pacific by crossing the
Isthmus of Panama or sailing around South America. They plagued the Americas' western shores, burning towns, taking ships, disrupting trade, and forcing
the Spanish Empire to divert scarce pesos to defend a region usually protected
by isolation. Ulloa remarked in his 1748 *Voyage to South America* that the city
of Guayaquil (in modern Ecuador) still suffered from being pillaged by pirates
in 1686. After sustaining numerous such costly attacks, Spain's Pacific colonies
finally succeeded in the 1690s in driving the pirates away.

Some of the buccaneers made it back to Europe, bearing tales of Spanish
wealth that excited the avarice of their countrymen in French port towns like
Saint Malo, in much the same way that Spanish conquistadors some two centuries before had inspired their Extremaduran brethren with tales of Aztec and

---

307–336, esp. 310, 319–320; D. W. Jones, *War and Economy in the Age of William III and Marlborough*
(Oxford, 1988), 2, 4; Henry Kamen, *The War of Succession in Spain, 1700–15* (London, 1969), 125–127,
135, 143; Luis Navarro García, "La política indiana," and Lucio Mijares Pérez, "Política exterior:
La diplomacia," both in Navarro García, ed., *Historia general de España y América*, XI, *América en
el siglo XVIII: Los primeros Borbones*, part 1, 2d ed. (Madrid, 1989), 3–64, 65–100, esp. 3–5, 65–66;
Curtis Nettels, "England and the Spanish-American Trade, 1680–1715," *JMH*, III (1931), 1–32, esp.
19; James Pritchard, *In Search of Empire: The French in the Americas, 1670–1730* (Cambridge, 2004),
360; Stanley J. Stein and Barbara H. Stein, *Silver, Trade, and War: Spain and America in the Making
of Early Modern Europe* (Baltimore, 2000), 6, 121; Geoffrey J. Walker, *Spanish Politics and Imperial
Trade, 1700–1789* (Bloomington, Ind., 1979), 19.

2. Scholarly knowledge of this French Pacific commerce can be traced back to the work of the
Swedish historian Erik W. Dahlgren, director of the Royal Library in Stockholm. Working in
the first decades of the last century, Dahlgren produced one book and three articles that remain
the necessary basis for an understanding of early-eighteenth-century French Pacific activity. In
addition to the already-cited *Les relations commerciales et maritimes*, see "Le comte Jérôme de Pontchartrain et les armateurs de Saint-Malo (1712–1715)," *Revue historique*, LXXXVIII (1905), 225–263;
"Voyages français à destination de la mer du Sud avant Bougainville (1695–1749)," *Nouvelles archives
des missions scientifiques et littéraires*, XIV (1907), 423–554; "L'expédition de Martinet et la fin du
commerce français dans la mer du Sud," *Revue de l'histoire des colonies françaises*, I (1913), 257–332.

Inca treasure. We can, perhaps, get the flavor of these boozy yarns by examining published versions of pirate exploits. One French buccaneer, Raveneau de Lussan, published an account of his South Sea adventures in 1689. According to his narrative, Lussan returned with pearls, gold, and jewels worth thirty thousand pieces of eight (gained, admittedly, more from gambling with other pirates than from looting the Spanish himself). One reason Lussan carried his wealth in precious stones and gold was that he considered silver too bulky for transport over the Isthmus of Panama. Whatever his preferences in booty, he affirmed the quantities of silver available in Peru: "This rich mineral is so abundant in this country that most of the things made in France of steel, copper, and iron are made out here of silver." So ubiquitous (and heavy) was silver that Lussan indicated that he and his companions would sometimes not even bother to chase down collections of thousands of pieces of eight. At least on the pages of filibuster memoirs, the Spanish Pacific offered easy money. In Lussan's optimistic view, "To succeed out here and amass a considerable fortune without taking extreme risks or suffering, all that is required is a good boat supplied with enough provisions to last for a long period."[3]

Such tales of wealth, whether recounted in Saint Malo dives or set to paper by Paris publishers, could not fail to attract the attention of French government officials. Louis XIV, for reasons a few moments' reflection on the cost of building Versailles and fighting Europe make apparent, had long been interested in gaining access to Spanish imperial wealth. During the 1690s, when France and Spain were fighting on opposite sides in the War of the League of Augsburg, Louis and French ministers such as the head of the Department of the Navy, Jérôme Phélypeaux de Maurepas, comte de Pontchartrain (1699–1715), were eager to advance on Spain's Pacific possessions and markets.[4]

3. Peter T. Bradley, *The Lure of Peru: Maritime Intrusion into the South Sea, 1598–1701* (Hampshire, U.K., 1989), 103–195; Guillermo Céspedes del Castillo, "La defensa militar del istmo de Panama a fines del siglo XVII y comienzos del XVIII," *Anuario de estudios americanos*, IX (1952), 235–275; Dahlgren, *Les relations commerciales et maritimes*, I, 73–74, 89–98; John Dunmore, *French Explorers in the Pacific*, 2 vols. (Oxford, 1965–1969), I, *The Eighteenth Century*, 10; Peter Gerhard, *Pirates of the Pacific, 1575–1742* (Lincoln, Neb., 1990), 135–194, 240–242; Jorge Juan and Antonio de Ulloa, *A Voyage to South America*, trans. John Adams (New York, 1964), 81, 84; Ulloa, *Viaje a la América meridional*, ed. Andrés Saumell, 2 vols. (Madrid, 1990), I, 232, 236. For a discussion of the way in which a handful of pirate attacks on one region could produce large and lasting effects, see Kris E. Lane, "Buccaneers and Coastal Defense in Late-Seventeenth-Century Quito: The Case of Barbacoas," *CLAHR*, VI (1997), 143–173. The translated quotations are from Marguerite Eyer Wilbur, ed. and trans., *Raveneau de Lussan: Buccaneer of the Spanish Main and Early French Filibuster of the Pacific* (Cleveland, Ohio, 1930), 227–228, 252–253, 256. The original French can be found in Raveneau de Lussan, *Journal du voyage fait à la mer du Sud avec les Flibustiers de l'Amérique, en 1684 et années suivantes* (Paris, 1690), 202, 229.

4. His father held the title from 1690 to 1699, but Jérôme was effectively acting as secretary from

French officials contemplated a variety of routes to the South Sea. As the buccaneers had recently shown, the Isthmus of Panama provided one possible road to the Pacific, and, moving in that direction, French forces succeeded in taking Cartagena in 1697. More innovatively, Louis and Pontchartrain discussed the possibility of reaching the Pacific by means of a Northwest Passage from Hudson Bay. In a 1698 memoir concerning the French strategy for negotiating Anglo-French boundaries in the Hudson Bay region, Louis and Pontchartrain spoke of the "importance" of obtaining all of Hudson Bay and of the extent to which its loss would be detrimental to France ["de quelle importance il est d'obtenir l'entiere possession de la Baye du Nord, et combien la perte en seroit prejudiciable a la nation"]. This importance arose not only because of the top-quality furs to be had in the icy lands inland from the bay but also because of the likelihood that ascending the rivers discharging into it would lead to silver and gold deposits and "a passage to the Pacific Ocean which would shorten considerably the route to the East Indies."[5]

On a more immediately practical level, Louis also authorized French attempts to reach the Pacific by sailing around Cape Horn. In 1695, inspired by tales of pirate loot, French naval officer Jean-Baptiste de Gennes left La Rochelle for Peru with six ships. He failed to get farther than the western extremity of the

---

1696. See John C. Rule, "Jérôme Phélypeaux, Comte de Pontchartrain, and the Establishment of Louisiana, 1696–1715," in John Francis McDermott, ed., *Frenchmen and French Ways in the Mississippi Valley* (Urbana, Ill., 1969), 179–197, esp. 179.

5. The original French for the quotation reads, "L'apparence quil y a de trouver en remontant les Rivieres qui se dechargent dans cette baye, des mines d'or, et d'argent, et un passage a la mer pacifique qui abregeroit de beaucoup la route des Indes orientalles." See "Mémoire pour servir d'instruction au Sieur Comte de Tallard . . . son ambassadeur extraordinaire en Angleterre, et au Sieur Phelypeaux d'Herbaut . . . intendant de la marine et des armées navales de France," July 7, 1698, "Signé Louis, et plus bas Phelipeaux," CP, Angleterre, 208, 29r–29v. French and English forces had fought for control of the bay in the 1680s and 1690s, but the 1697 Treaty of Ryswick had left unresolved their disputes about control of the bay and ownership of the forts and territories around it. See E. E. Rich, "The Hudson's Bay Company and the Treaty of Utrecht," *Cambridge Historical Journal*, XI (1954), 182–203, esp. 188; Max Savelle, *The Diplomatic History of the Canadian Boundary, 1749–1763* (New Haven, Conn., 1940), xi; Savelle, *The Origins of American Diplomacy: The International History of Anglo-America, 1492–1763* (New York, 1967), 116–120; Nellis M. Crouse, *Lemoyne d'Iberville: Soldier of New France* (Ithaca, N.Y., [1954]), 14–66, 90–117, 139–154. Pontchartrain took cartography quite seriously. He acquired an extensive private collection of maps and globes and, in 1695–1696, began gathering charts within a ministry of marine bureau. He frequently consulted geographers working there, collaborating closely with members of the Delisle family. See Lucie Lagarde, "Le passage du Nord-Ouest et la mer de l'Ouest dans la cartographie française du 18e siècle, contribution à l'etude de l'oeuvre des Delisle et Buache," *Imago mundi*, XLI (1989), 19–43, esp. 20–23; Rule, "Jérôme Phélypeaux," in McDermott, ed., *Frenchmen and French Ways in the Mississippi Valley*, 179, 182–184, 197; Dale Miquelon, "Les Pontchartrain se penchent sur leurs cartes de l'Amérique: Les cartes et l'impérialisme, 1690–1712," *Revue d'histoire de l'Amérique française*, LIX (2005), 53–71.

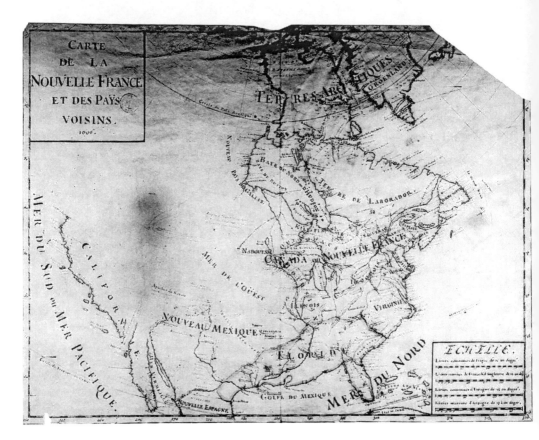

MAP 19. Guillaume Delisle, *Carte de la nouvelle France et des paÿs voisins*. 1696. Library and Archives Canada, NMC 6351

Straits of Magellan, and the expedition returned to France in 1697. The lust for gain proved stronger than the fear of Cape Horn, and French efforts to sail to Chile and Peru continued. In 1698, the French Compagnie royale de la Mer pacifique was founded, with Pontchartrain as its presiding officer and with a thirty-year exclusive right to trade with all the islands and coasts of the Pacific not occupied by a European power as its privilege. The clause limiting trade to areas unoccupied by other European powers enabled French diplomats to claim that France was not officially violating the Spanish Empire's commercial integrity, but French merchants quickly demonstrated they had no intention of allowing the provision to hinder their trade with Spanish possessions. The company's first expedition, consisting of the *Phélipeaux,* the *Comte de Maurepas,* and two smaller ships, left La Rochelle in December 1698. After requiring more than six months to pass the Straits of Magellan, the two larger ships reached the Pacific, sold their goods in Peru and Chile—despite the hostility of the Spanish viceroy—and returned to La Rochelle in August 1701. The expedition failed to turn a profit, but it gave French sailors more experience with the navigational challenges of reaching the Pacific. It also showed that Spanish American consumers would pay high prices in silver for French goods and thus convinced French merchants of the wisdom of sending more ships in the future.[6]

Once the War of the Spanish Succession began in 1702, Spain's naval forces proved insufficient either to protect Spanish commerce from English and Dutch raiders or to supply Spain's distant imperial possessions. Peru remained disconnected from Spain for much of the conflict, and the galleons normally sailing from Spain to South America every year did so only once between 1695 and 1713. One French observer, Amedée-François Frézier, an army engineer dispatched by Louis XIV to survey the coasts and assess the defenses of Chile and Peru, referred to the "Scarcity of Merchandizes there was in the Country, by reason of the Stoppage of the Trade of the Galeons." French merchants rushed to take advantage of these circumstances by sailing directly to Peru and Chile. Between 1702 and 1713, 98 French ships arrived in the Pacific. To give one example of the density of the French presence, Frézier reported, "There assembled at La Conception [in January 1714] 15 Sail of French, great and small, and about 2600 Men." The merchants of Saint Malo took the most active role in this trade, and the sailing of their ships through the South Atlantic gave the Malvinas (Falkland) Islands one of their two names.[7]

These French trading voyages placed the French and Spanish governments in a quandary. French officials knew that Spanish silver was especially important for the French economy and government in time of war. A 1705 instructive memoir for Michel-Jean Amelot, France's ambassador to Spain, provides an example of this French interest in Spanish silver: "The commerce that is conducted in

6. On Gennes's expedition, see Dahlgren, *Les relations commerciales et maritimes*, I, 87–88, 98–102; Carlos Daniel Malamud Rikles, *Cádiz y Saint Malo en el comercio colonial peruano (1698–1725)* (Cadiz, 1986), 48. Louis authorized the Compagnie royale de la Mer pacifique to "faire seule le commerce pendant trente années, à l'exclusion de nos autres sujets, depuis le cap de Saint-Antoine, sur la côte déserte, sur les côtes des détroits de Magellan, Le Maire et Browars, et sur les côtes et dans les îles de la mer du Sud ou Pacifique, non occupies par les puissances de l'Europe" (Dahlgren, *Les relations commerciales et maritimes*, I, 120). See also Dahlgren, "Voyages français à destination de la mer du Sud," *Nouvelles archives des missions scientifiques et littéraires*, XIV (1907), 423–554, esp. 425. On the company's early expeditions, see *Les relations commerciales et maritimes*, I, 120–144.

7. On Peruvian commercial isolation, see Dahlgren, "L'expédition de Martinet," *Revue de l'histoire des colonies françaises*, I (1913), 261; Kamen, *War of Succession*, 140, 177; Walker, *Spanish Politics and Imperial Trade*, 21–23, 69, 137. Quotations from Amédée François Frezier, *A Voyage to the South-Sea and along the Coasts of Chili and Peru, in the Years 1712, 1713, and 1714* (London, 1717), 201, 280. The French edition of Frézier's account appeared in 1716 (Dunmore, *French Explorers in the Pacific*, I, 27). "Ninety-eight ships" estimate from Malamud Rikles, *Cádiz y Saint Malo*, 62.

Spain . . . produces the better part of the silver that goes out into Europe, and it is certain that the more merchandise one carries to the Spanish, the more gold and silver bullion and specie one brings back. This is why Mr. Amelot must take particular care to maintain and augment the commerce of the nation by all the means that he will judge the most suitable and that the merchants will be able to suggest to him." Direct trade with Peru provided an effective means of obtaining this Spanish silver, and it also furnished the colonies of France's Spanish ally with products Spain was temporarily unable to offer. On the other hand, Louis and his officials did not wish to offend their Spanish allies' sensibilities with blatant encroachments on traditionally protected markets.[8]

French officials and French policy vacillated, but under the stress of war, the need for silver usually trumped other considerations. With regard to the Pacific, therefore, French policy, especially in the early phase of the war, generally continued officially forbidding French trade with Spain's Pacific colonies while official neglect and legal loopholes tacitly permitted it. In 1701, for example, the French government forbade the Compagnie royale de la Mer pacifique from trading with Spanish South Sea colonies. In the same year, Pontchartrain granted French ships permission to search "uninhabited lands" ["terres inhabitées"] in the Pacific Basin for mineral deposits, and in 1705, Louis authorized his subjects to make discoveries in the Pacific ["aller aux découvertes"], though not to trade there. Pontchartrain and Louis had not, strictly speaking, flouted Spain's commercial regulations, but one could not doubt violation would follow.[9]

As the war continued, some French officials became increasingly skeptical about the value of direct trade with Spain's colonies. They grew sensitive to the

8. The original French of the quotation reads, "Le commerce qui se fait en Espagne est d'autant plus considérable, qu'il produit la plus grande partie de l'argent que se répand dans tous les autres États de l'Europe, et il est certain que plus on porte de marchandises aux Espagnols, plus on rapporte de matières et d'espèces d'or et d'argent. C'est pourquoi le sieur Amelot doit avoir une attention particulière à maintenir et à augmenter le commerce de la nation par tous les moyens qu'il jugera les plus convenables et que les négociants pourront lui suggérer" (from Dahlgren, *Les relations commerciales et maritimes*, I, 330). On page 423 of the same volume is a reproduction of a 1706 letter from Chamillart, the *contrôleur général des finances*, to Pontchartrain: "The king is so convinced, Monsieur, of how important it is that the kingdom continue to receive new hard-currency [or silver] financial support from the return of vessels which carry merchandise to the Indies and into the places from which one can obtain it, in order to replace a part of that which has been going out for several years for the subsistence of the armies His Majesty has been obliged to maintain outside of his territories" ["Le Roi est tellement convaincu, Monsieur, de l'importance dont il est que le Royaume continue à recevoir de nouveaux secours d'argent par le retour des vaisseaux qui portent des marchandises aux Indes et dans les endroits d'où l'on en peut tirer, pour remplacer une partie de celui qui sort depuis plusieurs années pour la subsistance des armées que Sa Majesté a été obligée d'entretenir hors de ses États"].

9. Dahlgren, *Les relations commerciales et maritimes*, I, 183–185, 197, 337–338, 400, 586; Malamud Rikles, *Cádiz y Saint Malo*, 55, 75.

complaints of the unlucky merchants uninvolved in unmediated Peruvian and Chilean commerce. Many of these querulous French traders had been accustomed before the war to selling their goods in Cadiz for shipment to America on the annual Spanish galleons. They found, during the conflict, that Saint Malo merchants going directly to South America's west coast had preempted their best customers. Pontchartrain, although actively involved in extending French trade into the Pacific before the War of the Spanish Succession, became vigorously opposed to it during the war. He worried not only about the disruption of the traditional French Cadiz trade and about complaints from Spanish officials but also that the lure of profits to be gained from trade with Peru and Chile would cause French sailors and merchants to abandon commerce with France's own colonies. Other French officials worried about increasing Spanish protests. In 1708, Amelot, having served four years as French ambassador to Spain, warned Louis that the flood of French goods was ruining traditional Peruvian and New Spanish commerce, provoking continual Spanish complaints and jeopardizing Franco-Spanish relations. Frézier, having observed firsthand the effects of French trade in Chile and Peru, would echo this sentiment in his 1717 volume: "The French resorting thither without Measure, have carry'd many more Goods than the Country could use; that Plenty has obliged them to sell the said Goods at very low Rates, and has ruin'd the Spanish Merchants, and consequently the French for several Years." Amelot recommended that Louis restrict the activities of French merchants.[10]

As Frézier's report suggests, attempts by officials such as Pontchartrain to curb France's wartime Pacific trade were largely unsuccessful. Pontchartrain's indecisive and sometimes ineffectual personality, his awkward — in some eyes, "loathsome" — appearance, and his related inability to make his inferiors obey his commands formed one reason. Another was connivance in the trade by other French officials: Michel Chamillart, the French secretary of war from 1701 to 1709, for example, felt that France must obtain silver regardless of the conse-

10. A natural conflict existed between French merchants who wished to trade directly with Peru and those who preferred to continue to sell their goods in Spain for shipment to the colonies. The Cadiz merchants complained bitterly to the French government about the disruption of their customary commerce. They had good reason, as the French Pacific traders happily undercut their rivals. In 1705, for example, French merchants active in the contraband trade in the Pacific sought to delay the sailing of the Spanish fleets from Cadiz, because the continued absence of the fleets would increase demand for the goods carried by French interlopers to Peru. See Dahlgren, *Les relations commerciales et maritimes*, I, 273–274; Walker, *Spanish Politics and Imperial Trade*, 28, 60. For French officials' concerns and Frézier's observations, see Pontchartrain to Amelot, July 15, 1705, CP, Espagne, 153, 26r–27r; Feb. 20, 1709, CP, Espagne, 195, 126r–126v; Amelot to Louis XIV, Mar. 31, 1708, CP, Espagne, 179, 188r–189r; Frostin, "Les Pontchartrain et la pénétration commerciale française," *Revue historique*, CCXLV (1971), 321–323; Frezier, *Voyage to the South-Sea*, 201.

quences. Lack of consistent opposition to the trade on Louis's part added a third. Even a resolute French governmental policy banning Pacific trade might have done little. As Swedish historian Erik W. Dahlgren observed in his pioneering account of this early-eighteenth-century Franco-Spanish South Sea trade, although Spanish officials, impressed by the power and efficiency of Louis's absolutist government, doubted that French merchants could successfully disobey its commands without tacit support, Saint Malo merchants proved quite willing to disregard official prohibitions in quest of fantastic profits. Valdivia was a long way from Versailles.[11]

Spanish officials found that French voyages to Peru and Chile also posed a dilemma for the Spanish Empire. The Spanish wanted to maintain the traditional exclusion of foreigners from imperial commerce, but they recognized the need for French wartime assistance in supplying and protecting Spain's overseas possessions. Spain's Philip V, moreover, was bound to Louis not only by family ties but also by the need for French help in his war against the other contender for the Spanish throne, Archduke Charles of Austria. Spanish sentiments remained opposed to French trade with the empire, but Spanish policy wavered. In September 1700, Charles II of Spain had forbidden French commerce with Spanish America. In January 1701, Philip allowed French ships to enter Spanish American ports, though not, in theory, to trade there. French ships ignored the theory and enjoyed a thriving contraband trade. Then, in June 1705, spurred by American reports of this illicit French trade, Philip strengthened regulations enjoining the seizure of merchandise from ships involved in illegal trade. In 1706, however, Philip, concerned about Archduke Charles's intrigues to gain the loyalty of some of Spain's American colonies for himself but lacking an available Spanish ship, asked Louis to send a French ship to the Pacific bearing orders counteracting the archduke's efforts. In 1710, Philip dispatched passports authorizing French ships to journey to the Pacific (it is not clear why, but probably to increase either the revenues of Spain or of the financially straitened Spanish ambassador in France who issued them). Philip's policies regarding French ships in the Pacific followed no single course.

In addition to the challenge of managing his overbearing French ally and

11. Dahlgren, Les relations commerciales et maritimes, I, 272–273, 278–280, 319, 336–339, 358–361, 589, 660–661; Dahlgren, "Voyages français à destination de la mer du Sud," Nouvelles archives des missions scientifiques et littéraires, XIV (1907), 425–426; Dahlgren, "Le comte Jérôme de Pontchartrain," Revue historique, LXXXVIII (1905), 225–263, esp. 226–227, 242, 249; Dahlgren, "L'expédition de Martinet," Revue de l'histoire des colonies françaises, I (1913), 262; Kamen, War of Succession, 149–155; Gaston Rambert, "Marseille et le commerce 'interlope' en mer du Sud (1700–1723)," Provence historique, XVII (1967), 32–60, esp. 35–38; Walker, Spanish Politics and Imperial Trade, 20, 25–26, 50–52. The quotation is from Pritchard, In Search of Empire, 238–239.

grandfather, Philip also faced difficulties in securing the compliance of his colonial officers. Peruvian viceroys, thousands of miles from Madrid and acutely aware both of local shortages of goods and of personal and viceregal want of funds, sometimes granted trading authorizations to French merchants contravening imperial regulations. The viceroy from 1707 to 1710, the marqués de Castelldosríus, provides the most egregious example (one described nicely by the historian Geoffrey Walker). Castelldosríus combined an appreciation for French culture with a tolerance of French smuggling. He had previously served as Spanish ambassador in France, where he had become an intimate of Louis XIV and where, by borrowing 6,300 pistoles from the Compagnie royale de la Mer pacifique, he had become cozy with France's South Sea trading interests. It was Castelldosríus who had declared, *"Il n'existe plus de Pyrénnées"* when informing Louis of the news that Philip of Anjou had inherited the throne of Spain. Upon arriving in Peru, he acted as if he had also said, *"Il n'existe plus de* Spanish Lake." He placed his cronies in posts throughout the viceroyalty, and in exchange for a healthy commission, Castelldosríus and his henchmen cooperated in French merchants' attempts to sell their goods in the viceroyalty. He "openly espouses their Interest, and encourages them," claimed Woodes Rogers. Castelldosríus then condemned the trade publicly while profiting from it privately.[12]

To understand the tenacity with which merchants and officials clung to direct trade with Peru and Chile, and to appreciate the lingering effects of this trade on French interest in the North American West, it is necessary to consider the profits made by French merchants. Though it is impossible to arrive at precise figures, the quantities of bullion returning to France were enormous. Dahlgren estimated conservatively that French ships involved in the South Sea trade in the early decades of the eighteenth century brought at least two hundred million livres of silver back to France. To put this in perspective, estimates put the total value of silver coinage in France around 1700 at no more than five hundred million livres. The treasure returns of the year 1709 alone made possible a re-coinage of the French currency. More recently, Carlos Daniel Malamud Rikles has suggested that France extracted roughly fifty-five million pesos from the Spanish Pacific in the first quarter of the eighteenth century. He avers that these French merchants engrossed at least 68 percent of Peru's foreign trade. Stanley J. Stein

<hr/>

12. Woodes Rogers, *A Cruising Voyage round the World* . . . (London, 1712), 196; "Memoire contenant le commerce de la mer du Sud presenté a mylord compte Doxfort grand tressorier d'Agleterre [sic] par le Captaine Alexandre Lion," Sept. 3, 1712, Add. Mss. 70,163; Dahlgren, *Les relations commerciales et maritimes*, I, 401–402, 614–615; Kamen, *War of Succession*, 143–145; Malamud Rikles, *Cádiz y Saint Malo*, 241–242, 254; Walker, *Spanish Politics and Imperial Trade*, 22, 34–45, 61–62, 137–138.

and Barbara H. Stein have mentioned a peso estimate at the higher end of the spectrum, suggesting that Saint Malo merchants might have returned up to ninety-nine million pesos to France between 1703 and 1715, a quantity sufficient to double the French money supply. To get a sense of the magnitude of these peso figures, consider that Alexander Hamilton estimated the total value of both specie and paper money in the thirteen colonies on the eve of the American Revolution at about thirty million pesos. Direct trade with Peru and Chile, in short, formed one of the most lucrative branches of eighteenth-century European commerce.[13]

The Pacific and the Utrecht Settlement

French trade with Peru and Chile could not go unnoticed across the channel. Just as late-seventeenth-century pirate tales inspired French royal interest in South Sea commerce, early-eighteenth-century French South Sea gains attracted envious English attention. English merchants, statesmen, and privateers wanted the same silver Saint Malo ships were coming home with. Some of the most vigorous expressions of these jealous and covetous sentiments can be seen in documents produced around the 1711 founding of the British South Sea Company, an organization formed explicitly to take advantage of what seemed to be fabulous opportunities for trade with the Spanish Empire.

Prominent in these papers were French profits derived from Pacific trade and the idea that Britain should emulate French methods of obtaining them. An example comes from Robert Allen, a survivor of an ill-starred 1698–1700 effort to found a Scottish colony in Darien (in modern Panama) with subsequent experience working in Panama, Quito, and Guayaquil. In "An Essay on the Nature and Methods of Carrying on a Trade to the South Sea," part of a package of papers solicited by the South Sea Company, Allen spoke of the "immence Treasure" French traders had lifted from Chile and Peru and noted, "This is so well known that in the news Papers we have from time to time had it publish'd how many Millions they have brought and are daily bringing from the South Seas." In Allen's view, British merchants could gain South Sea Spanish American markets for themselves. British goods, he asserted, cost less than those of France, and

13. Dahlgren, "Voyages français à destination de la mer du Sud," *Nouvelles archives des missions scientifiques et littéraires*, XIV (1907), 426, 430–432; Maurice Filion, *Maurepas, ministre de Louis XV (1715–1749)* (Montreal, 1967), 139; Malamud Rikles, *Cádiz y Saint Malo*, 65, 67, 80–81, 90, 280; John J. McCusker, *Money and Exchange in Europe and America, 1600–1775: A Handbook* (Chapel Hill, N.C., 1978), 7; Stein and Stein, *Silver, Trade, and War*, 113. In referring to a doubling of the French "monetary stock," the Steins do not specify whether they mean specie or all currency.

Spanish American consumers lacked any particular affection for French traders. If British ships could sail to the South Sea and trade directly with its Spanish inhabitants rather than rely on indirect trade by way of Spain and the Spanish Main, British traders could compete for bullion-rich Pacific littoral customers.[14]

Should the Spanish Empire prove unwilling to offer commercial access to British merchants, there was always the possibility of using force. Important in this regard was the long-standing idea that southern South American Indians — many of whom had successfully resisted Spanish rule — would happily join British invaders in a fight against Spain. Former East India merchant and energetic South Sea Company advocate Thomas Bowrey wrote in 1711, "The Indians about Baldivia are numerous, Valliant, and utter Enemies to the Spaniards, therefore may probably joyn with us against them." Similarly, barber, surgeon, author, and veteran of 1680s Pacific buccaneer expeditions Lionel Wafer spoke of both the mineral and human prospects Chile offered. The Indians around both Coquimbo and Valdivia were "inveterate Enemies and were att Continual war with the Spaniards when I was there. They are a valiant brave generous and warlike People and very popolous, theire Contry abounds with all sorts of Riches as Gold Silver etc." French trade had shown that Pacific South America had much to offer, and war or trade could help Britons get it.[15]

14. Robert Allen, "An Essay on the Nature and Methods of Carrying on a Trade to the South Sea," Add. Mss. 28,140, 25v–28r. His essay, which appears also among Robert Harley's papers in Add. Mss. 70,163, was written sometime late in the War of the Spanish Succession and submitted to the South Sea Company during winter 1711–1712. On Allen and his writings, see Glyndwr Williams, *The Great South Sea: English Voyages and Encounters, 1570–1750* (New Haven, Conn., 1997), 172. For documents expressing similar ideas, see Thomas Bowrey, Marine Square, "Proposall for Takeing Baldivia in the South Seas by Thomas Bowrey," "Copy Deliverd to the Lord High Treasurer 1711," Sept. 10, 1711, Add. Mss. 28,140, 31r–31v; Bowrey, Marine Square, "Proposal for a Setlement in the Way to the South Seas," "Copy Deliverd to the Lord High Treasurer 1711," Sept. 10 or 11, 1711, Add. Mss. 28,140, 31v–32v (copies of both Bowrey documents appear in Add. Mss. 70,163, "Official Papers of Robert Harley"); "Captain Rogers's Account How a Trade May Be Carried on to the South Sea," Add. Mss. 28,140, 30r–30v.

15. For a concise recent discussion of Araucanian resistance to the Spanish Empire, see David J. Weber, *Bárbaros: Spaniards and Their Savages in the Age of Enlightenment* (New Haven, Conn., 2005), 54–68. Bowrey quotation from Bowrey, "Proposall for Takeing Baldivia," Sept. 10, 1711, Add. Mss. 28,140, 31r–31v (see also "Proposal for a Setlement in the Way to the South Seas," Sept. 10 or 11, 1711, Add. Mss. 28,140, 31v–32v). For Wafer's report, solicited by Tory leader and South Sea Company promoter Robert Harley, see Lionel Wafer, n.d., Add. Mss. 70,163. Such notions retained their hold on the British imagination into the years following the Treaty of Utrecht and the South Sea Bubble. Writing of Chile around 1726, Woodes Rogers stated, "If any armament hence gets possession of any Port in Chili, the Indians woud certainly join our forces and distress all parts of the South Sea, if not reduce the Citty of Lima, and the Sea ports of Peru to submit to whatever terms are demanded within six months after such force is in those parts." See "Papers Relating to Trade and Revenue," "Capt. Rogers . . . of the South Sea," Add. Mss. 19,034, 70r–70v. Date from Williams's discussion in *Great South Sea*, 202.

South Sea enthusiasts were, in fact, casting longing glances not just at Chile and Peru but also at territories on the eastern side of the Andes. Thomas Pindar's "Short Scheme for Improvement of the America Company" argued that the inland territories of what is today Argentina would not only afford access to the "richest mynes" of the Spanish but might also prove to be the sites of new mines in the future. An anonymous memoir from the same collection of documents pertaining to British trade as Pindar's went beyond claims of mineral wealth, speaking rosily of Argentina's "bounty both of the Climate and Soyle" and contending that "without Exageration no place under Heaven can with Justice be prefer'd before it in both these respects." It indicated further that the Potosí mining region was accessible from Buenos Aires and that the city's "Province is the Southernmost of the whole Kingdom of Peru, so Consequently all South America may be Supply'd with Goods [from it]." The implication was that acquisition of a foothold in southeastern South America could open a road to Spanish riches.[16]

Avarice alone would suffice to explain merchant interest in Spanish America, but more than greed was at issue. Discussion of Spanish wealth reflected also an awareness of the fragility of British public finances. England had expended vast sums and incurred huge debts in the Nine Years' War and the War of the Spanish Succession, and many officials and pundits felt the kingdom needed to acquire territories or commercial concessions to help restore its fiscal health. The anonymous 1711 "Observations on the South Sea Trade and Company" noted, "The nation has for near twenty years, sustained the weight of two Burdensom Expensive wars, the second of which is yet unfinished and must, of necessity, be carried on with the utmost vigor." The author warned of "insuperable Difficulties in Satisfying Publick Debts." He contended, moreover, that Britain had so far missed the wartime gains that might compensate for its financial expenditures: "Hither to we have been Fighting, like Knights Errant, for Honours sake, and to redress publick grievances with out any regard to our private Interest. . . . We have even neglected the pursuit of those advantageous Conditions on which we enter'd into the grand Alliance, viz. That we should remain Masters of the Conquests we should make in the West-Indies." The author advanced the South Sea Company as the solution to these difficulties, arguing that trade with the Spanish Pacific would "open such a Vein of Riches, will return such wealth as in a few Years, will make us more than Sufficient Amends for the vast Expences we have

16. Thomas Pindar, "To the Right Honourable the Earle off [sic] Oxford Lord High Treasurer of Great Brittain," "A Short Scheme for Improvement of the America Company," n.d., Add. Mss. 70,164. The anonymous document is also in Add. Mss. 70,164. It lacks title, author, or date. It does have an internal pagination system; I'm referring to pages 2–4.

been at, since the Revolution." Such commerce would, in the author's view, open markets to British woolens, irons, and manufactures, offering employment to poor Britons at home and at sea.[17]

These kinds of ideas shaped the British approach to the negotiations ending the War of the Spanish Succession. Like their French counterparts, British statesmen saw the connection between maritime commerce and state power and sought to promote overseas trade and protect it from rivals. British officials and merchants recognized and envied the extent of France's Pacific commerce and profits. In fact, they saw these French gains as threatening, attributing France's ability to finance the War of the Spanish Succession to acquisition of Peruvian silver and imputing the wartime decline of Jamaica's trade with Spanish colonies to French usurpations of it by way of Spanish Pacific ports. Depriving France of the South Sea trade formed one of Britain's main wartime objectives.[18]

British diplomats went beyond the negative aim of excluding France from the Pacific. Military victories had put British negotiators in a position to ask much from their adversaries, and designs on the Spanish Empire and demands for Spanish commercial concessions appeared not only among the productions of South Sea Company promoters but also in the papers of British diplomats. An example came on July 21, 1711, when Britain's poet-diplomat Matthew Prior arrived at the French palace at Fontainebleau bearing Britain's preliminary peace proposal. He and the marquis de Torcy, the French minister of foreign affairs, met twice, on July 22 and 29. Prior requested four sites for British commercial posts in the Spanish Indies, two on the Atlantic coast and two on the long stretch of Pacific coastline between California and the Straits of Magellan (the "si grande estendüe de terre depuis la Californie jusqu'au detroit de Magellan"). Prior indicated that France and Britain could keep this concession secret from the Dutch, who might otherwise ask for comparable privileges, and that Britain would not oppose a similar grant to Louis from his Spanish royal grandson.[19]

17. "Observations on the South Sea Trade and Company," 1711, Add. Mss. 70,163. In my transcription of the document, I have "Knight Errants."

18. Stein and Stein, Silver, Trade, and War, 154; Dahlgren, Les relations commerciales et maritimes, I, 570, 578, 636, 729; Jones, War and Economy, 40; Nettels, "England and the Spanish-American Trade," JMH, III (1931), 29–30; Rogers, Cruising Voyage, ix; A Letter to a Member of the P——t of G——t-B——n, Occasion'd by the Priviledge Granted by the French King to Mr. Crozat (London, 1713), 20–21.

19. "Mr Secretary St John to the Queen," Sept. 20, 1711, SP 105/266, 21r; marquis de Torcy, July 22, 1711, CP, Angleterre, 233, 52v, 53v, 56v–57r (quotation); Louis XIV and Torcy, "Instruction pour le sieur Mesnager, chevalier de l'Ordre de St. Michel, député au Conseil de Commerce," Aug. 3, 1711, rpt. in Paul Vaucher, ed., Recueil des instructions données aux ambassadeurs et ministres de France depuis les traités de Westphalie jusqu'à la Révolution française, XXV, Angleterre, III, 1698–1791 (Paris, 1965), 76–102, esp. 84–85; L. G. Wickham Legg, ed., "Torcy's Account of Matthew Prior's

Desperate to end the war, French officials considered this idea. Later documents, such as a 1727 instructive memoir for French negotiators in Spain, recalled that, in the deliberations leading to the Peace of Utrecht, French diplomats contemplated having Spain cede the Chilean city of Concepción to the British Crown. It had only been, the memoir noted, because France was in circumstances demanding peace at any price that a concession of such importance had been under discussion ["Il faloit une Extremité Comme Celle ou lon étoit en Ce tems-la, d'obtenir la paix a Tout prix, pour Sacrifier un article de Cette Consequence"]. French officials were open to the necessity of similar concessions elsewhere. In January 1712, Pontchartrain's instructions to the French plenipotentiaries at Utrecht authorized them to cede all of Hudson Bay to Britain if such a transfer constituted the price of a peace settlement. He added that such a sacrifice should include the provision that Britain would grant France free navigation of any Northwest Passage discovered in the bay. French diplomats did later feel compelled to give up Hudson Bay, though, interestingly, they omitted the provision regarding a Northwest Passage.[20]

Negotiations at Fontainebleau in July 1711," *EHR*, XXIX (1914), 525–532; Williams, *Great South Sea*, 170–171. For a detailed account of Prior's role in the negotiations ending the War of the Spanish Succession, see Legg, *Matthew Prior: A Study of His Public Career and Correspondence* (Cambridge, 1921), 144–218.

20. "Memoires utils dans la conjoncture des préliminaires de la paix, d'un futur Congrés, et d'une ambassade en Espagne, par raport au comerce de France en 1727," CP, Espagne, 352, 18r–18v (see also Richard Wall to José de Carvajal y Lancaster, May 5, 1749, Estado 4503, AGS). The French text regarding a Northwest Passage reads, "En reservant cependt aux François la liberté d'y naviguer si par la suite on decouvroit un passage pour faire le tour du monde." See Pontchartrain, "Memoire concernant les colonies, le commerce, et la navigation, pour Messrs. les plenipotentiaires du roy," Jan. 2, 1712, MD, Amérique, 24, 36r–36v (other copies in MD, Amérique, 22, 25r, and MD, France, 1426, 81r–81v). I have not been able to find post–January 1712 references to a Northwest Passage in the correspondence among Louis, Pontchartrain, and the French Utrecht plenipotentiaries nor an explanation for the lack of a provision mentioning it. Perhaps French officials thought better of stimulating British interest in seeking a passage from bay to ocean; perhaps they had begun to doubt its existence or utility. By 1716–1717, memoirs to and from the Conseil de la marine were speaking of a passage too far north to be useful. See Crouse, *Lemoyne d'Iberville*, 102–103; "Mémoire joint à la lettre de MM. de Vaudreuil et Bégon du 12 Novembre 1716, pour être porté à monseigneur le duc d'Orléans, délibéré par le conseil, le 3 février 1717," and "Nécessité d'établir trois postes pour parvenir de la á la découverte de la mer de l'Ouest," in Pierre Margry, ed., *Découvertes et établissements des Français dans l'ouest et dans le sud de l'Amérique septentrionale (1614–1754): Mémoires et documents originaux*, VI, *Exploration des affluents du Mississipi et découverte des montagnes Rocheuses (1679–1754)* (Paris, 1886), 495–498, 498–503. For a thoughtful discussion of the French approach to North America at Utrecht, see Dale Miquelon, "Envisioning the French Empire: Utrecht, 1711–1713," *FHS*, XXIV (2001), 653–677. Miquelon emphasizes the maritime orientation of French policy and its general disdain for the American interior. French interest in a Northwest Passage, either from Hudson Bay or in the form of river networks farther south, is pertinent to this argument because, if such a passage existed, parts of the North American interior would suddenly become maritime territory. If Pontchartrain stopped mentioning a passage because of growing skepticism

They thought better of the idea of having Spain grant Britain a Pacific port. Mindful of the profits French merchants gained from direct trade with Peru and Chile during the war and from indirect trade with the Spanish colonies by way of Cadiz before it, French officials had long feared the consequences of allowing British or Dutch ships in Pacific waters. Even before the War of the Spanish Succession began, Louis had intended to exclude the English and Dutch from trade with Spain's colonies. Later, in July 1704, an instructive memoir for France's ambassador to Spain, the duc de Gramont, warned that, if English and Dutch merchants gained access to the South Sea, they could easily flood the Peruvian market with goods carried across the Pacific from the East Indies. This would ruin the traditional commerce in which European goods from countries such as France and Spain traveled across the Atlantic to Spain's Pacific colonies.[21]

Related concerns remained strong in the war's latter years. In response to Prior's 1711 proposal, Antoine Pecquet (the elder), one of Torcy's functionaries at the foreign office, strenuously opposed allowing Britain to obtain any kind of establishment off the Americas' Pacific coast. In Pecquet's view, the most negligible possession that could be offered to the British was the Isle of Juan Fernández. The island was currently uninhabited, he noted, and produced only trees and goats. In British hands, Pecquet worried that it would become a well-populated and thriving trading entrepôt for Asian and European goods the British would use to supply Peruvian and Mexican markets ["le plus grand entrepot du monde des manufactures d'Europe et d'Asie, dont les Anglois fourniroient les Royaumes du Perou, et du Mexique"]. The wealth of Spain's colonies would make Britain's Pacific outpost rich, whereas France would lose the profits it had obtained by selling products such as cloth to Spanish colonial markets. Pecquet concluded, therefore, that the English must be convinced it was in the general interest to exclude all nations but Spain from the Pacific.[22]

---

about its existence, this would reinforce the argument about French disparagement of the American interior. If, on the other hand, references to a passage stopped because of concerns about drawing British attention to a water route that might exist, this would suggest a more favorable French assessment of American interior possibilities. Miquelon (676) and Mapp would both like to know more about what Pontchartrain was thinking.

21. On Louis XIV's interest in commercial exclusion, see Mark A. Thomson, "Louis XIV and the Origins of the War of the Spanish Succession," in Thomson, *William III and Louis XIV*, ed. Ragnhild Hatton and J. S. Bromley (Liverpool, 1968), 140–161, esp. 148. For Gramont's warning, see "Memoire sur le commerce d'Espagne aux Indes, pour Mr. Le Duc de Gramont ambassadeur du roy en Espagne," July 1704, CP, Espagne, 139, 35r, 40v.

22. See Dahlgren, *Les relations commerciales et maritimes*, I, 632–635. The Juan Fernández cluster would soon contribute to a great work of literature: Alexander Selkirk, the prototype for Robinson Crusoe, was marooned by buccaneers on one of the islands there in 1704. The British privateer Woodes Rogers rescued him in 1709. See Rogers, *Cruising Voyage*, 125–131. Pecquet was a *premier commis*, a sort of foreign office bureau chief, responsible for receiving and redacting documents,

Louis also consulted Nicholas Mesnager, the French representative in London. Mesnager's views were similar to those of Pecquet. He opined in a letter to Torcy that the presence of such British establishments in Peru, "amidst the riches of the world," would cause frequent disputes among rival European nations and had therefore to be avoided. Mesnager suggested that, in the place of such concessions, France should cede to Britain — or, more precisely, should have Spain cede to Britain — a 15 percent decrease in the duties paid on British goods entering Spain and the privilege for thirty consecutive years of supplying slaves to the Spanish Empire. Louis followed the recommendations of his advisors, and the French government rejected Prior's proposal.[23]

Britain's negotiators wanted peace, and they were willing to share Spain's Pacific trade with France, but they preferred to continue hostilities rather than let France have the Pacific trade to itself. French officials feared that any British Pacific presence would ruin French trade and exacerbate Anglo-French commercial rivalry, triggering new wars. French diplomats came to see that they could secure both an end to the war and the exclusion of British merchants from the Pacific only if French ships also stayed out of the South Sea. The only point both empires could ultimately agree upon was that Spain alone should enjoy the right to sail Pacific waters. A prohibition of non-Spanish Pacific shipping formed a part of the 1713 Utrecht settlement. In article 6 of the Anglo-French treaty of April 11, 1713, Louis renounced any French effort to seek alterations in the Spanish trading system prevailing during the reign of Charles II (1665–1700). Since Spain had forbidden French Pacific commerce during Charles II's reign, acceptance of this provision precluded future licit French trade there. In its agreements with Britain and the Dutch Republic, Spain extended the same prohibition to them.[24]

---

and thus a personage of considerable potential influence (Antoine Pecquet, "Memoire pour repondre aux demandes que les Anglois font avant l'ouverture des conferences pour la paix," late summer 1711, CP, Angleterre, 233, 75r–77r). On Pecquet and his office, see Hamish Scott, "Diplomatic Culture in Old Regime Europe," in Hamish Scott and Brendan Simms, eds., *Cultures of Power in Europe during the Long Eighteenth Century* (Cambridge, 2007), 63; Camille Piccioni, *Les premiers commis des affaires étrangères au XVIIe et au XVIIIe siècles* (Paris, 1928), 179–183; Jean-Pierre Samoyault, *Les bureaux du secrétariat d'état des affaires étrangères sous Louis XV* (Paris, 1971), 35–41, 301.

23. Nicholas Mesnager to Torcy, Aug. 25, 1711, CP, Angleterre, 233, 202r–203r; Legg, "Torcy's Account of Matthew Prior's Negotiations," *EHR*, XXIX (1914), 531; Frances Gardiner Davenport, ed., *European Treaties Bearing on the History of the United States and Its Dependencies*, 4 vols. (Gloucester, Mass., 1967), III, *1698–1715*, 148–150; John G. Sperling, *The South Sea Company: An Historical Essay and Bibliographical Finding List* (Boston, 1962), 11–13.

24. Pontchartrain to Torcy, Aug. 23, 1713, CP, Espagne, 223, 29r–31v; Pontchartrain, "Mémoire concernant le commerce maritime, la navigation et les colonies pour servir d'instruction a M. d'Iberville Envoyé extraordre. de Sa Mté en Angre," Nov. 10, 1713, CP, Angleterre, 250, 56r; "Mémoires utils dans la conjoncture des préliminaires de la paix," CP, Espagne, 352, 16v; Dahlgren, "Voy-

Continuing Efforts to Exclude French Shipping from the Pacific after the Treaty of Utrecht

Of course, signing a treaty is one thing; ensuring compliance with it, another. French and international laws might forbid direct commerce between the Spanish Pacific and French ports, but French traders still wanted the profits of this commerce and were indeed eager to get what they could before enforcement of trade restrictions curtailed their activities. When, moreover, hostilities with Britain and the Dutch Republic ended, French traders were emboldened by the absence of Dutch and English privateers. French ships kept sailing. Louis XIV issued the first prohibition of South Sea trade on January 18, 1712. Ten French ships left for the South Sea in the same year. Thirteen to fifteen more left in 1713. The Spanish king complained of French merchants' sailing into the South Sea in violation of Louis's prohibitions, and Louis reissued his strictures on July 31, 1713, on this occasion threatening violators with the galleys. Despite the renewed edicts, sixteen French ships sailed for the Pacific in 1714. In that year, British secretary of state (and future philosopher) Henry Saint John, Viscount Bolingbroke, warned France's ambassador that continued French Pacific trade could give those in Britain who opposed peace with France a pretext for protests. France's regency government declared on January 29, 1716, that French captains flouting the law would receive the death penalty. Spanish complaints continued, and in December 1716, in an attempt to render trade prohibitions more effectual, an official expedition of four French ships and 1,500 French and Spanish sailors under the French officer J. N. Martinet left Cadiz for the Pacific and seized six French ships. Martinet's expedition temporarily halted French South Sea commerce, but a brief resurgence occurred in 1719 and 1720, when French merchants hoped to take advantage of France and Spain's being on opposite sides during the War of the Quadruple Alliance.[25]

---

ages français à destination de la mer du Sud," *Nouvelles archives des missions scientifiques et littéraires,* XIV (1907), 427; Dahlgren, *Les relations commerciales et maritimes,* I, 563, 674–676, 715, 725; Malamud Rikles, *Cádiz y Saint Malo,* 77. Utrecht treaties reprinted in Davenport, ed., *European Treaties Bearing on the History of the United States,* III, 210–211: "Sa Majesté Tres Chrestienne demeure d'accord et s'engage que son intention n'est pas de tacher d'obtenir ny mesme d'accepter a l'avenir que pour l'utilité de ses sujets il soit rien changé, ny innové dans l'Espagne ny dans l'Amerique Espagnole, tant en matiere de commerce, qu'en matiere de navigation, aux usages pratiquez en ces pays sous le regne du feu Roy d'Espagne Charles Second."

25. Pontchartrain to Amelot, June 14, 1714, CP, Espagne, 235, 208r–208v; Pontchartrain to Torcy, Aug. 23, 1713, CP, Espagne, 223, 29r–31v; Pontchartrain, "Mémoire concernant le commerce maritime," Nov. 10, 1713, CP, Angleterre, 250, 56r; Pierre Lemoyne d'Iberville to Pontchartrain, July 2, 1714, CP, Angleterre, 257, 10r–10v; Sept. 4, 1714, CP, Angleterre, 258, 132r–132v; "To the R Honorable James Stanhope Esquire, His Majesty's Principal Secretary of State," Add. Mss. 25,559, 49r–50r;

Underlying French efforts to halt French Pacific trade were concerns about the reactions it inspired. French officials feared that continued French commerce in the South Sea could give British merchants an excuse to initiate their own trade. This worry is evident in a French comment on Britain and Spain's December 14, 1715, signing of a new commercial agreement. The pact confirmed the 1667 Anglo-Spanish Treaty of Madrid as the basic document regulating trade, and both nations agreed to eliminate commercial innovations made since then. In 1716, a French foreign office memoir worried that "innovations" might refer to the French ships continuing to trade in the Pacific and that England would use the treaty as a justification to send warships into the South Sea to remove French interlopers. The memoir also indicated (incorrectly) that the *asiento* would allow England to dispatch ships into the Pacific to supply the Spanish colonies with slaves. The presence of all these ships in the South Sea, in turn, would allow England to dominate the commerce of Peru and Chile.[26]

A weightier danger was that a continuation or resurgence of French South Sea commerce would incite new Anglo-French hostilities. Fears of such conflict were apparent in reactions to a 1721 proposal by, ironically, Martinet, that the French government request permission to send French merchant ships to Peru. French memoirs commenting on this proposal fretted that its acceptance would bring commercial competition, diplomatic complications, and armed conflict.

Matthew Prior to Charles Townshend, Jan. 11, 15, 22, 1715, SP 78/159, 322r, 330r–331r, 347v–348r; James Wishart to Viscount Bolingbroke, "From Her Majesties Ship the Rippon in the Bay of Cadiz," Apr. 27, 1714, esp. no. 4, "French Ships Which Have Gone Lately to the South Sea," and "Minutes Taken by the Command of Their Excys the Lords Justices on the Representation of the Directors of the South Sea Company, concerning the Difficulty's Which Remain upon the Assiento and Assignment Thereof," [August 1714?], both in SP 94/82; Jan. 14, 1715, SP 105/27, 130r–130v; Oct. 1, 1720, MD, Amérique, 6, 287r–288v; Dahlgren, *Les relations commerciales et maritimes*, I, 231, 684; Dahlgren, "L'expédition de Martinet," *Revue de l'histoire des colonies françaises*, I (1913), 264–268, 282–283, 325; Dahlgren, "Voyages français à destination de la mer du Sud," *Nouvelles archives des missions scientifiques et littéraires*, XIV (1907), 423–554, esp. 427–428; Dunmore, *French Explorers in the Pacific*, I, 22–24; Léon Vignols and Henri Sée, "La fin du commerce interlope dans l'Amérique espagnole," *Revue d'histoire économique et sociale*, XIII (1925), 300–313, esp. 300–302; Malamud Rikles, *Cádiz y Saint Malo*, 148–149; Walker, *Spanish Politics and Imperial Trade*, 142.

26. Anglo-Spanish commercial agreement rpt. in Davenport, ed., *European Treaties Bearing on the History of the United States*, III, 253–255: "As there may have been innovations in commerce, his Catholic Majesty promises to use all possible endeavors on his part for abolishing them, and for the causing them to be by all means avoided in the future. In like manner his Britannic Majesty promises to use all possible endeavors for abolishing all innovations on his part, and for avoiding them by all means in the future." For the 1716 French memoir, see "Observations sur le Tté. de Commerce signé à Madrid le 14e. xbre. 1715. entre l'Espe. et l'Angre," Feb. 1, 1716, CP, Espagne, 246, 164r–165r, 168r, 175r–176v. There are actually two memoirs here with the same title and apparently the same author. I am treating them as a single work. See also Dahlgren, "L'expédition de Martinet," *Revue de l'histoire des colonies françaises*, I (1913), 283–284, 308–309.

One warned that establishment of a direct commerce with Peru would cause France's competitors to send their own ships to the South Sea, and this would be a perennial source of discord ["semence intarissable de discorde"]. Another memoir discussing Martinet's proposal recommended that, instead of France's sending its own ships to the Pacific, the nations of Europe should act in concert to prohibit the entry of any non-Spanish ships into that ocean.[27]

French officials worried not just about English but also about Spanish reactions to French ships in the Pacific. Redoubled Spanish complaints accompanied the flurry of French voyages in 1719 and 1720. Early in 1722, Cardinal Guillaume Dubois, the French minister of foreign affairs (1718–1723), received a letter from José de Grimaldo, Spain's leading minister from 1719–1724, decrying the numerous French merchant ships that had docked in Callao with the intention of conducting illicit trade with Peru's inhabitants. Grimaldo demanded that the French government prevent any more ships from leaving for the Pacific, and he warned that failure to do so would damage the union, mutual understanding, and economic well-being of France and Spain.[28]

Dubois took these Spanish grievances seriously. After receiving Grimaldo's protest, he promptly wrote a letter to Amelot asking for advice. In his missive, Dubois spoke of the issue raised by Grimaldo as "important" and indicated that careful consideration of a French response was *"d'une extrême conséquence"* because of the possible repercussions for French commerce and Franco-Spanish relations. In an April response, Amelot recalled the *"trésors immenses"* that France had acquired in its South Sea commerce but noted also that such commerce generated complaints from Spanish citizens and officials. No matter how close the union of France and Spain, he thought, the Spanish would never relax their prohibition of French Pacific trade. If, by some miracle, the Spanish

27. Martinet, July 21, 1721, 67r–95r, "Observations sur le memoire de M. Martinet joint a sa lettre du 21e Juillet 1721," 99r–100r, anonymous comment on Martinet's memoir, July 21, 1721, 101r–103r, all in CP, Espagne, 303.

28. "La continuation du commerce forcé dans la mer du Sud pourra inquiéter et troubler les peoples de ces royaumes et avoir d'autres conséquences peu favorables à l'union, à l'intelligence et à l'opulence des deux nations" ("Extract d'une lettre de M. le marquis de Grimaldo au cardinal Dubois," in Vignols and Sée, "La fin du commerce interlope dans l'Amérique espagnole," *Revue d'histoire économique et sociale*, XIII [1925], 304). For a reiteration of Grimaldo's complaint from the Spanish ambassador in France, see D. Patricio Laules to Dubois, Mar. 3, 1723, CP, Espagne, 327, 353r–354r. For another example of a Spanish complaint of this kind, see "Memoire des vaisseaux qui s'arment dans les ports de la France et destinés pour aller *a la mer du Sud*," 1720, CP, Espagne, 296, 13r–13v. British agents in Paris were also following and reporting news of renewed French voyages to the South Sea. See Daniel Pulteney to James Craggs, Dec. 22, 1719, 92r, Pulteney, Dec. 24, 1719, 94r, "In Mr. Pulteney's of Feb 10, ns 1719/20," SP 78/166, 131r–131v, 136r; Robert Sutton to Craggs, Oct. 30, 1720, SP 78/169, 88v–89v.

government did allow such commerce, Amelot felt that the English and Dutch would oppose it with *"toutes leurs forces,"* throwing the French government into exactly the kind of awkward diplomatic situation it was trying to avoid. Dubois accepted Amelot's reasoning, conceding in a letter that French ships should not be allowed to sail into the Pacific in peacetime and affirming the validity of the agreements forbidding such navigation. In 1724, the French government renewed its 1716 decree forbidding French merchants from carrying on unlawful commerce with the Spanish colonies. Direct French trade with the Pacific was largely over.[29]

These efforts to end French South Sea trade were one manifestation of larger trends in French policy after the Treaty of Utrecht. France had suffered heavily from the human and financial costs of Louis XIV's incessant wars. In the decades between the 1713 close of the War of the Spanish Succession and the 1739 outbreak of the War of Jenkins' Ear, French leaders such as the duc d'Orléans (regent for Louis XV from 1715 to 1723) and Cardinal André-Hercule de Fleury (Louis XV's principal minister from 1726 to 1743) usually sought to avoid jeopardizing peaceful European conditions and French economic recovery by antagonizing Britain or Spain separately or, more critically, simultaneously. Despite continued French interest and merchant efforts, the French government generally saw the diplomatic necessity of getting and keeping French traders out of the South Sea. The movement of French ships into the Pacific constituted a flagrant violation of the terms of the Utrecht settlement and could easily involve France in the kind of diplomatic controversies, and perhaps hostilities, with Britain and Spain that French policy as a whole sought to avoid. Despite recollections of the gains to be made from direct trade in the Spanish Lake, the French government tried, for the most part, to confine French navigation to other oceans.[30]

29. "Lettre du Cardinal Dubois à Amelot," "Mémoire sur l'envoie qui a été fait de quelques vaisseaux français a la mer du Sud," in Vignols and Sée, "La fin du commerce interlope," *Revue d'histoire économique et sociale,* XIII (1925), 304–308. For Dubois's acceptance of Amelot's reasoning, see Dubois, "Memoire servant de réponse a celuy que a esté presenté le 3. mars 1722. par M. Dellaules Ambassadeur d'Espagne au sujet des vaisseaux de la Compagnie des Indes de retour de la mer du Sud," "remis copie de ce memoire à Mr. Laules à l'audience le 9e. Mars *1723,*" CP, Espagne, 327, 399r–401v. The year 1724 also saw a new Peruvian viceroy taking office and launching a vigorous effort to clean up any remaining French Pacific commerce. See Dahlgren, "Voyages français à destination de la mer du Sud," *Nouvelles archives des missions scientifiques et littéraires,* XIV (1907), 429; Dahlgren, "L'expédition de Martinet," *Revue de l'histoire des colonies françaises,* I (1913), 332; Dunmore, *French Explorers in the Pacific,* I, 24; Malamud Rikles, *Cádiz y Saint Malo,* 149–151; Vignols and Sée, "La fin du commerce interlope," *Revue d'histoire économique et sociale,* XIII (1925), 302; Walker, *Spanish Politics and Imperial Trade,* 152–154, 164. There were a few French voyages to Spain's Pacific colonies during the War of the Austrian Succession.

30. For discussions of Anglo-French and Franco-Spanish relations between 1713 and 1739, see

French merchants and government had profited from American resources and Spanish imperial trade during the War of the Spanish Succession. The desire to continue to profit remained after Utrecht. But Utrecht and its aftermath had closed off one site of and route to such gains. French officials and traders might retain designs on Spain and interest in American wealth, but they were going to have to find means other than direct commerce with the Spanish Pacific to realize their ambitions. The North American Far West would offer one alternative.

Arthur McCandless Wilson, *French Foreign Policy during the Administration of Cardinal Fleury, 1726–1743: A Study in Diplomacy and Commercial Development* (Cambridge, Mass., 1936); Derek McKay and H. M. Scott, *The Rise of the Great Powers, 1648–1815* (London, 1983), 94–158; Jeremy Black, *European International Relations, 1648–1815* (Hampshire, U.K., 2002), 138–157.

PART III.

FRANCE AND THE ELUSIVE WEST AFTER THE TREATY OF UTRECHT

5

VISIONS OF WESTERN LOUISIANA

"You know there are mines in these mountains; I have often heard you say that you
believed in their existence."
"Reasoning from analogy, Richard, but not with any certainty of the fact."
"You have heard them mentioned, and have seen specimens of the ore, sir; you
will not deny that! and, reasoning from analogy, as you say, if there be mines in South
America, ought there not to be mines in North America, too?"
— JAMES FENIMORE COOPER, *The Pioneers*

In the history of the United States, and in an account of the agreements ending
the Seven Years' War, Louisiana figures most prominently as the colony France
gave away. But before France yielded Louisiana, it had to acquire claims to the
great Mississippi Valley — twice. To understand the first of France's Louisiana
cessions, it is necessary to examine the reasons why French officials once found
the colony so desirable.

In 1699, one year after founding the French Compagnie royale de la Mer pacifique, Pierre Lemoyne d'Iberville's landing on the Mississippi Delta — followed
by his and his brother Jean-Baptiste de Bienville's erection of forts and nego-
tiation of alliances with local Indian nations — initiated the French colony of
Louisiana. French empire and resources were preoccupied during the War of
the Spanish Succession, and the nascent colony developed little during the first
years of the eighteenth century. The desirability of maintaining claim to it was
open to question. In late 1711 and early 1712, Louis XIV and Pontchartrain discussed ceding the infant Louisiana colony to Spain in exchange for the Spanish portion of the Caribbean island of Saint-Domingue. They ultimately decided against this course, and Louis instead gave a commercial monopoly for

An earlier version of material from this chapter and Chapter 13 appeared in Paul Mapp, "French
Geographic Conceptions of the Unexplored American West and the Louisiana Cession of 1762,"
in Bradley G. Bond, ed., *French Colonial Louisiana and the Atlantic World* (Baton Rouge, La., 2005),
134–174. The author thanks Louisiana State University Press for allowing him to make use of his
previous work.

Louisiana to financier Antoine Crozat in 1712. This grant initiated a five-decade effort — admittedly, a highly uneven effort — to cultivate a French presence in the Mississippi Valley.[1]

If, after the War of the Spanish Succession, the French Empire could not go south by sea to Cape Horn, the Straits of Magellan, Chile, and Peru or to the Pacific through a Northwest Passage from Hudson Bay, it could perhaps go by land and navigable rivers into whatever lay beyond the Mississippi, the Great Lakes, and the great bay to their north. The empire's harbingers could proceed from either the old colony of New France or the new colony of Louisiana. Succeeding chapters will treat the continuation of French westward investigation from Canada; this one will focus on French visions of Louisiana's western reaches.

The territories comprising the Louisiana of French claims and imagination extended far beyond the boundaries of the later American state, dwarfing not only France's insular Caribbean colonies but also France itself. The grandest such claims encompassed the Mississippi basin and took France to the doorstep of the mysterious regions beyond it. Indeed, part of what rendered French Louisiana policy so difficult — and to historians, so interesting — was the connection between the consequences of French decisions regarding the trans-Mississippi West and the character of a region French scouts had failed to explore and French officials were unable to comprehend. Those western reaches of Louisiana might contain nothing but inhospitable deserts, impassable mountains, impenetrable forests, and Indians too impoverished to engage in trade and too distant and intractable to enlist against Britain. Then again, the Far West might hold a Northwest Passage, a new Potosí, fertile soils, and inhabitants more like those of Japan or Peru than the Saint Lawrence and Great Lakes. Future bounties might compensate for Louisiana's poor beginnings. In assessing the significance of their new colony, French statesmen would have somehow to estimate the value of the unknown.

To many of these statesmen, the unknown looked pretty appealing. Such seems to have been the case with many of the personnel working in or writing for the French foreign office. Because of the role they played in negotiating

1. James Fenimore Cooper, *The Pioneers* (New York, 1964), 304. On discussions of ceding Louisiana, see Jérôme Phélypeaux de Maurepas, comte de Pontchartrain, "Memoire concernant les colonies, le commerce, et la navigation, pour Messrs. les plenipotentiares du roy," Jan. 2, 1712, MD, France, 1425, 85r–86r; Pontchartrain to Jean-Baptiste Colbert, marquis de Torcy, Dec. 15, 1711, 236r–236v, Pontchartrain to Nicholas Mesnager, Jan. 2, 1712, 258r–258v, both in CP, Hollande, 231; Marcel Giraud, *Histoire de la Louisiane française*, 2 vols. (Paris, 1953–1958), I, *Le règne de Louis XIV (1698–1715)*, 229; Dale Miquelon, "Envisioning the French Empire: Utrecht, 1711–1713," *FHS*, XXIV (2001), 659.

treaties like those ending the War of the Spanish Succession—in determining, that is, what would be gained and lost at the bargaining table—diplomats' ideas about a region's merits and deficiencies could be especially consequential. One reason to pay particular attention to foreign office papers bearing on western Louisiana from the first half of the eighteenth century is because comparison to documents from the years preceding the 1762 Louisiana cession can elucidate the relation between changing French geographic ideas and the sacrifice of France's last mainland North American colony. Documents collected within the French foreign ministry archives suggest that, from roughly 1712 to 1747, many of the conceptions concerning western Louisiana within the French diplomatic community were quite positive. Indeed, for the French foreign office, the years between the end of the War of the Spanish Succession and the end of the War of the Austrian Succession are best labeled the period of the presumptive and central value of unknown North America. Great significance was attached to the unexplored regions west of the Mississippi. The tone of pertinent foreign office memoirs was optimistic. They spoke of great size as a desirable colonial feature. They predicted that western lands would be fertile and rich and that navigable rivers would facilitate their exploitation. They expected that trade or conquest would deliver Spain's colonial North American wealth to France. Authors filled the unknown with features extrapolated from the familiar. Some posited the existence of a Sea and River of the West linking western Louisiana to the South Sea, and even writers skeptical about particular predicted western features looked forward to finding something of value in mysterious trans-Mississippi territories. The future looked bright for France's new Louisiana colony.[2]

One reason was the colony's size. Authors of foreign ministry memoirs in the first half of the eighteenth century held that Louisiana and the unexplored portions of the continent beyond it were enormous. A 1716 report by Mobile missionary François Le Maire to the regent duc d'Orléans's council called Louisiana "a vast and immense extent of land." An unsigned 1739 memoir on the French and British North American colonies stated, "This continent is of a prodigious

2. By foreign office sources, I mean documents produced or collected by foreign office personnel and retained within the foreign ministry archives. Some of these documents are particular to the foreign ministry, whereas others can be found in multiple official collections. Louis XIV ordered the establishment of the foreign ministry archives in 1709, and archival documents discussing Louisiana become abundant beginning around 1711. Louis XIV's death in 1715 stimulated production of additional foreign ministry papers discussing Louisiana. The regent following the Sun King, Philip, duc d'Orléans, solicited a wide range of opinions as part of a post-Louis reexamination of French government, and Louisiana policy formed one topic of interest. Louisiana's place in imperial negotiations and policy led to the continued production and gathering of documents pertaining to the colony in later decades.

extent. There is a large part of it that one has never penetrated and that is entirely unknown to us. The French possess Canada and Louisiana there" ["Ce continent est d'une prodigieuse etendüe. Il y en a vue grande partie où l'on n'a jamais penetré et qui nous est entierement inconnue; Les Francois y possedent le Canada et la Loüisiane"].[3]

Louisiana's great size derived partly from uncertainty concerning the colony's boundaries. The 1712 grant to Crozat included all the lands watered by rivers flowing into the Mississippi below its junction with the Illinois. A circa 1717 memoir from Louisiana's royal finance officer (ordonnateur), Marc Antoine Hubert (1716–1720), to French Conseil des finances president the duc de Noailles spoke vaguely of Louisiana's western boundaries as the lands of the Sioux and the Missouri River to its source. A 1718 memoir from (three-time) Louisiana governor Jean-Baptiste de Bienville to the Conseil de la marine conceded that Louisiana remained without definite boundaries ["Bornes fixes"] but talked of New Mexico and Sioux territory as approximate western limits. The Abbé Raguet, who was named ecclesiastical director of the French Compagnie des Indes in 1724, referred to the Rio Grande in New Mexico and the mountains to the east of it as a western boundary and Hudson Bay as a northern one. Because the Missouri's sources and New Mexico's northern and Sioux territory's southern extent remained imprecisely determined, and because Hudson Bay lay so far north, these boundary markers left a great deal of room for Louisiana to extend west. In fact, if there were a gap between New Mexico or the Missouri headwaters in the south and Hudson Bay or the Sioux country in the north, these definitions allowed Louisiana to stretch to the Pacific. The space between New Mexico and the "Nation du Serpent" on Jacques-Nicolas Bellin's 1755 map of North America displays a possible route for this westward reach.[4]

3. François Le Maire, "Memoire sur la colonie de la Louisiane," 1716, MD, Amérique, 1, 52r; "Memoire sur les colonies francoises et angloises de l'Amerique septentrionale 1739," MD, France, 1990, 241r. See also Abbé Raguet, writing sometime between 1729 and 1740, in "Du domaine et des limites de la Loüisiane," MD, France, 1991, 43v. In this chapter, translations from French are mine unless otherwise noted.

4. "Seconde lettre d'un ministre espagnol, de la cour de Madrid a un de ses amis aux Pais-Bas," September 1712, CP, Espagne, 218, 487v–488r (document also in "Lettres-patentes du roy, qui permettent au Sieur Crozat secretaire du roi de faire seul le commerce dans toutes les terres possédées par le roy et bornées par le Nouveau Méxique et autres," Sept. 14, 1712, CP, États-Unis, supplément 6, 10r); Marc Antoine Hubert, "Memoire au sujét de l'etablissement de la colonie de la L'ouisianne envoyé par ordre de Monseig'neur le d'uc de Noailles," c. 1717 (date from Waldo G. Leland, John J. Meng, and Abel Doysié, *Guide to Materials for American History in the Libraries and Archives of Paris*, II, *Archives of the Ministry of Foreign Affairs* [Washington, D.C., 1943], 870), 139r, and Jean-Baptiste de Bienville to Conseil de la marine, June 10, 12, 1718, 198v–199r, both in MD, Amérique, 1; Raguet, "Du domaine et des limites de la Loüisiane," MD, France, 1991, 43r, 61v–63r. For descriptions of Louisiana authors like Hubert in action in the colony, see Jean-Baptiste Bénard de La

Authors opined, moreover, that Louisiana's great size made it a better possession. In part, this was because Louisiana's territories were thought to contain navigable rivers enabling settlement and commerce and rich landscapes making cultivation and trade worthwhile. The "vaste et immense étenduë de Terre" Le Maire spoke of in his 1716 memoir to the Regency Council was watered, in his view, by innumerable rivers. It enjoyed fertile soils suitable for the production of all manner of fruits and abounded in a variety of commercially valuable wildlife. An unsigned 1733 memoir on French and Dutch American commerce argued that the vast "extent" of Louisiana would be suitable for all sorts of agricultural production, including, for example, tobacco and silk. It contended further that Louisiana possessed navigable rivers that could aid colonization and that it was likely a river would be found leading southwest to the Pacific. Many French authors foresaw a profitable Franco-Indian commerce in western animal products. Trans-Mississippi furs would substitute for those no longer available in trapped-out territories farther east. Regions too warm to generate top-quality furs could still produce saleable hides. Geographer John Logan Allen has referred to this recurring French emphasis on the richness and utility of western lands as the "motif of the Garden." It was an inviting notion, and it might have sustained early-eighteenth-century French authors' assumption that possessing more North American territory would be advantageous. Because measureless westward extent was the colony's most striking characteristic, assessments of colonial value viewing size as a positive attribute tended to designate Louisiana a prize possession.[5]

More generally, the reasoning techniques French authors employed to esti-

Harpe, *The Historical Journal of the Establishment of the French in Louisiana,* ed. Glenn R. Conrad, trans. Joan Cain and Virginia Koenig (Lafayette, La., 1971); Richebourg Gaillard McWilliams, ed. and trans., *Fleur de Lys and Calumet: Being the Pénicaut Narrative of French Adventure in Louisiana* (Baton Rouge, La., 1953).

5. On Louisiana's magnitude, see Le Maire, "Memoire sur la colonie de la Louisiane," 1716, 52r, 54v, "Troisieme avis a la Compagnie des Indes touchant la decouverte des mines et des nouvelles terres de la Louïsiane," 1719, 460r, and Marc Antoine Hubert, "Memoire sur la riviere, les terres et les sauvages du Missoury," [June 1718?] (date from Leland, *Archives of the Ministry of Foreign Affairs,* 871), 221r–222v, all in MD, Amérique, 1; "Sur le commerce des Hollandois et des François en Amerique . . . ," 1733, MD, France, 1996, 113v–115r; "Mémoire sur l'etat de la colonie de la Louisiane en 1746," MD, Amérique, 2, 204r; French Company of the Indies, "Colonie de la Louisiane," 1730, MD, Amérique, 7, 346r. On the Indian trade, see Bienville, "Memoire sur la Loüisiane," 44r–44v, 46v, 50v, Hubert, "Memoire sur la colonie de la L'ouisianne," 137r, "Memoire pour faire connoitre dequelle importance il est de conserver la colonnie de la Loüisiane," c. 1717 (date from Leland, *Archives of the Ministry of Foreign Affairs,* 870), 188v, and "Memoire sur la Loüisiane," 1717, 154r, all in MD, Amérique, 1. On the "motif of the Garden," see John Logan Allen, *Passage through the Garden: Lewis and Clark and the Image of the American Northwest* (Urbana, Ill., 1975), 2–4. For an especially clear example of this motif in operation, see McWilliams, ed. and trans., *Fleur de Lys and Calumet,* 255.

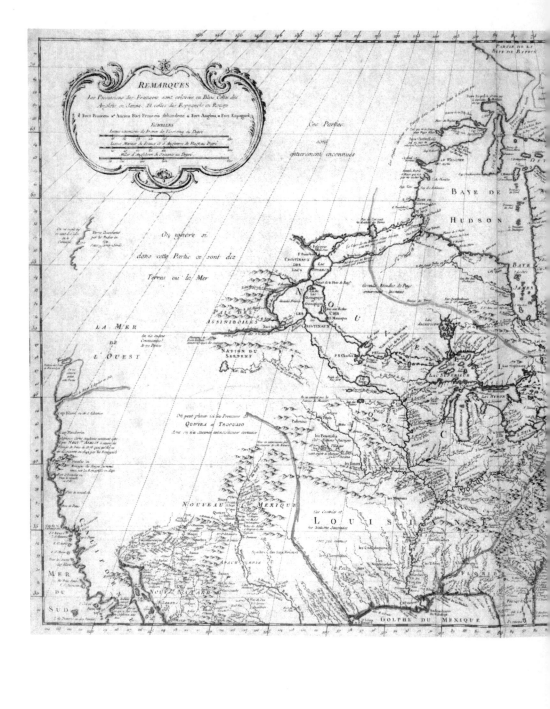

MAP 20. Jacques-Nicolas Bellin, *Carte de l'Amerique septentrionale depuis le 28. degré de latitude jusqu'au 72.* 1755. Special Collections, University of Virginia Library

mate western territorial value heightened unexplored Louisiana's perceived importance. To predict the features of trans-Mississippi regions, these authors extrapolated from what they knew, or thought they knew, about more familiar American areas. And not just any areas: the bases of conjecture were often territories of proved, or at least reputed, worth rather than those of lackluster performance — Mexico rather than Canada. Le Bartz, one of Crozat's Louisiana agents, claimed in a circa 1716 memoir that "one could not doubt" that mines would be found in Louisiana as rich as those the Spanish were thought to possess in New Mexico, because the mountain chain from which the Spanish drew so much gold and silver extended into the French colony. As additional evidence, Le Bartz claimed that Missouri River Indians were supplying gold under the name of copper to the Spanish; he hoped use of Mississippi River transport and appealing French merchandise would allow France to take over this trade. In a 1721 memoir, Charles Legac inferred like Le Bartz, or like a character from a James Fenimore Cooper novel, that the same mountain chain would provide the same ores as in New Mexico, but he added his belief that it was "incontestable" that Louisiana mountains would also contain lead, copper, and iron deposits. A 1727 memoir that appears to have been either written or approved by French minister of marine Jean-Frédéric Phélypeux, comte de Maurepas (1723–1749), agreed that Louisiana possessed gold and silver deposits because the mountains yielding them in Mexico reached into the "immense extent" ["etenduë immense"] of Louisiana.[6]

These authors might have been credulous, but they were not foolish in surmising that unassayed trans-Mississippi territories might hold ore deposits. Not only had Spaniards been scraping fortunes out of Mexican silver mines for centuries, but Brazilians were eking lucre out of massive gold deposits located during the 1680s and 1690s. From the vantage point of the early eighteenth century, a new El Dorado had recently been found, and others might await discovery. Of most immediate significance, the technique of extrapolating from characteristics of familiar and reputedly wealthy lands lent itself, in French foreign office

6. Le Bartz, "Mémoire," 161, and "Memoire de Charles Legac cy devant directeur pour la Compagnie des Indes à la Loüisianne," Aug. 25, 1721, 124r, both in MD, Amérique, 1; "Remis par M. du Maurepas en Novembre 1727," "Memoire pour le partage de l'administration de la Compagnie des Indes entre le controolleur general et le secretaire d'etat ayant le Department de la marine," MD, France, 1991, 211v–212v. See also "Lamothe Cadillac to Pontchartrain," Oct. 26, 1713, in Dunbar Rowland and Albert Godfrey Sanders, eds. and trans., *Mississippi Provincial Archives: French Dominion*, II, *1701–1729* (Jackson, Miss., 1929), 162–204, esp. 175–180. Authors used similar reasoning to argue that Louisiana would be suitable for silk production because it was at the same latitude as China. See "Memoire sur la Loüisiane," 1717 (date from Leland, *Archives of the Ministry of Foreign Affairs*, 870), MD, Amérique, 1, 147v.

memoirs, to a rosy assessment of the potential value of regions whose true character remained unknown. To this way of thinking, western North America was a treasure house. Too late to instruct eighteenth-century diplomats, the California Gold Rush would prove this surmise correct.

Though not every French author shared these optimistic views of trans-Mississippi territory, from 1712 to 1747, even figures questioning particular aspects of the region's supposed bounty found other characteristics substantiating western Louisiana's value. In a 1717 memoir, Crozat expressed fear of over-extending Louisiana west, as he felt the Spanish had done with their Peruvian and Mexican empire. He foresaw little value in a Pacific port. He felt no need to encourage exploration of the unknown regions west of Louisiana, and he advised readers to verify the accounts of "Canadiens" who claimed to have done so. He argued that predictions of easy and lucrative trade with New Mexico lacked a solid foundation, noting that Spanish governors were generally less than welcoming to foreign merchants ["Les Commandans Espagnols sont extrememenr jaloux du commerce de leurs colonies"] and that the rough country and polyglot Indian nations west of Louisiana would challenge any traveler. In the same memoir, however, Crozat argued that French Louisiana played the valuable role of barrier between the British and Spanish colonies. This implied, in contradiction of his earlier statement, that, despite the difficult terrain and hostile Indians allegedly impeding Franco-Spanish commerce, practicable routes to New Spain and New Mexico were available for British discovery and use. Although dubious about routes to the Pacific, Crozat accepted the notion of rich mines in New Mexico's mountains and at Louisiana's latitude. He believed that mineral lodes must exist in the range he imagined extending east into French Mississippi Valley territory, and, though questioning certain explorers' tales, he accepted reports of western gold and copper deposits, of New Mexicans trading Spanish hardware for Indian lucre, and of Louisiana's "metallicité."[7]

Other figures displayed attitudes similar to Crozat's. In a 1725 memoir, Louisiana governor Bienville disparaged other authors' accuracy and honesty. He called "bizarre" the idea of opening commerce with New Mexico by a river, the Missouri, the ascent of which began hundreds of leagues from the Spanish colony and led away from it. He opined, however, that the Missouri flowed through lands rich in mineral deposits, animals, and potential Indian allies or trading partners. Another anonymous circa 1717 memoir discussing the establishment of Louisiana shared Crozat and Bienville's skepticism about French

7. Antoine Crozat, May 14, 1717, MD, Amérique, 1, 227r–227v, 230r, 235r–235v, 238r. A circa 1718 memoir by Joseph Le Gendre (262r–271r) is similar in its reasoning.

fur-traders' yarns but suggested these raconteurs were disclosing less than they had seen rather than inventing more. It further suggested that voyageurs were concealing valuable discoveries for fear that the Compagnie des Indes would use its commercial monopoly to seize the fruits of their wanderings. In this case, skepticism enhanced the perceived value of Louisiana and the unexplored West rather than diminished it.[8]

Though authors recognized Louisiana's many difficulties, they retained an attitude emphasizing the colony's future rather than current value. In 1733 and 1746, writers acknowledging the colony's disappointing economic performance to date envisaged nonetheless its ultimate success. Étienne de Silhouette, whose later career included stints as French comptroller general of finances and as a member of the 1750s Anglo-French boundary commission, argued in a 1747 memoir about English commerce, finances, and navigation that it was "essentiel . . . de conserver et d'augmenter" Louisiana: "One should not by any means evaluate the value of the colonies that France possesses by the profits that it has derived from them up to now. What renders their conservation of extreme interest for the state is . . . [in addition to the danger they would pose in English hands] the benefit that France can eventually derive from them." Skepticism about aspects of Louisiana's value in this period remained compatible with a forward-looking view of a worthwhile colony.[9]

Proximity to New Mexico enhanced Louisiana's apparent value. French geographers had received descriptions of the Spanish colony. Some came, in the 1670s and 1680s, for example, from an exiled New Mexico governor, Don Diego de Peñalosa (1661–1664). New Mexico appeared on French maps. The descriptions French geographers had heard were often inaccurate, however. French maps differed significantly from New Mexican reality (see the upper course of the Rio Grande on Guillaume Delisle's 1718 "Carte de la Louisiane et du cours

8. Bienville, "Memoire sur la Loüisiane," 1725, MD, Amérique, 1, 7r, 27r, 46r–46v, 50v (see also [Bienville], "Memoir on Louisiana," 1725 or 1726, in Dunbar Rowland and Albert Godfrey Sanders, eds., *Mississippi Provincial Archives: French Dominion*, III, *1704–1743* [Jackson, Miss., 1932], 499–539, esp. 518, 534); "Mémoire au sujet de l'etablissement de la nouvelle colonie de la Loüisiane," [1717?] (date from Leland, *Archives of the Ministry of Foreign Affairs*, 870), 193v–194r. For an optimistic view of the Missouri, see "D'Artaguette to Pontchartrain," June 20, 1710, in Rowland and Sanders, eds., *French Dominion*, II, 55–60, esp. 59.

9. "On ne doit point évaluer le prix des Colonies que la France possede dans l'Amerique Septentrionale par le profit qu'elle en a retiré jusqu'icy. Ce qui en rend la conservation extremement interessante pour l'etat, c'est: . . . 2° L'utilité que la France en peut retirer dans la suite pour elle même" (Étienne de Silhouette, "Observations sur les finances la navigation et le commerce d'Angleterre," 1747, MD, Angleterre, 46, 48v, 72v–73r). See also "Sur le commerce des Hollandois et des François en Amerique," MD, France, 1996, 113r; "Mémoire sur l'etat de la colonie de la Louisiane en 1746," MD, Amérique, 2, 220r.

du Mississipi"), and a comparison of French accounts of New Mexico with the actual state of the territory demonstrates that the Spanish colony remained beyond the horizon of early-eighteenth-century French geographic mastery. For Spanish officials and colonists, New Mexico suffered from poverty and isolation, endured Indian revolts and raids, and wanted significant bullion deposits. In French foreign office memoirs, New Mexico signified wealth. Le Maire's 1716 memoir to the regent's council declared that it possessed the "richest mines of the Spanish" ["les plus riches mines des Espagnols"]. An unsigned memoir from sometime between 1717 and 1720 claimed that New Mexico enjoyed more working silver mines than old Mexico. Other memoirs, dating from 1717 and around 1733, expressed similar opinions.[10]

French authors saw New Mexico, moreover, as not just rich but also accessible. Le Maire argued that west of Louisiana lay "a country little known up to now, which leads by land to New Mexico, and to the new kingdom of Leon, and to other provinces of the Spanish" ["un pays peu connu jusqu'à présent et qui conduit par terre au nouveau Mexique, au nouveau Royaume de Leon, et a d'autres provinces des Espagnols"]; he spoke of the "certitude" of an "advantageous commerce" with the mine-owning Spanish ["la certitude d'y faire un commerce avantageux"]. The unsigned 1733 memoir "Sur le commerce des Hollandois et des François en Amerique" averred that all agreed on Louisiana's happy situation, the colony bordering on the New Mexican localities richest in mines, areas with which Louisiana communicated by navigable rivers. The unsigned 1739 memoir on the French and British American colonies spoke of the "ease . . . of establishing commercial relations between Louisiana and New Mexico by land" ["Facilité . . . de Lier un Commerce de La Loüisiane avec Le Nouveau Mexique par Terre"]. Many other memoirs articulated the same confidence in the practicality of profitable Louisiana–New Mexico trade. And should trade prove unsatisfactory, the roads leading to New Mexican commerce could facilitate the colony's conquest. The "Projet abregé d'une entreprise de la Louisiane

10. On the role of Peñalosa, see William Brandon, *Quivira: Europeans in the Region of the Santa Fe Trail, 1540–1820* (Athens, Oh., 1990), 104; David J. Weber, *The Spanish Frontier in North America* (New Haven, Conn., 1992), 131, 148–149; Peter H. Wood, "La Salle: Discovery of a Lost Explorer," *AHR*, LXXXIX (1984), 294–323, esp. 313–317. For early-eighteenth-century French views of New Mexico, see Le Maire, "Memoire sur la colonie de la Louisiane," 1716, MD, Amérique, 1, 53r; "Projet abregé d'une entreprise de la Louisiane sur le nouveau Mexique," MD, France, 1991, 119r (another copy of this memoir in MD, Amérique, 7, c. 1717–1720, 207r–211v) (dating of 1717 to 1720 from document's past-tense references to 1716 and proposal of a 1720 invasion); "Memoire pour faire connoitre dequelle importance il est de conserver la colonnie de la Loüisiane," c. 1717, 188r–188v, and "Sur le commerce des Hollandois et des François en Amerique," 1733, 113v, both in MD, Amérique, 1; see also "Duclos to Pontchartrain," Oct. 9, 1713, in Rowland and Sanders, eds., *Mississippi Provincial Archives*, II, 79–143, esp. 139.

MAP 21. Guillaume Delisle, *Carte de la Louisiane et du cours du Mississipi.* 1718. Special Collections, University of Virginia Library

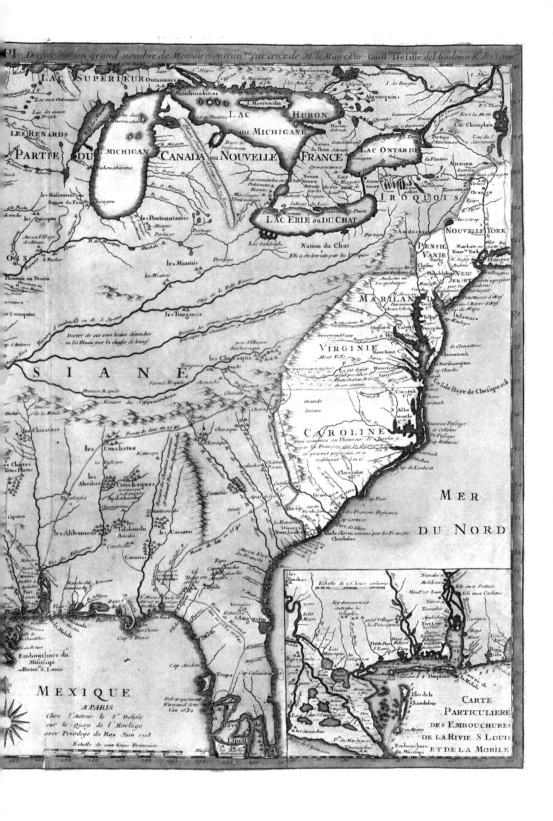

sur le nouveau Mexique," playing on a theme present from the earliest days of Louisiana, argued that the Red, Missouri, Arkansas, and Rio Grande rivers enabled an easy French invasion of New Mexico. By means of trade or conquest, New Mexico would serve as a conduit through which Spanish silver would flow to France.[11]

The West's Indian inhabitants would forward this and other French projects. Indian guides, it was expected, would show French explorers and traders routes to New Mexico and other western destinations. Indian nations might, suggested Le Maire in 1716, serve as middlemen in the Franco-Spanish trade, their presence perhaps alarming the Spanish less than would subjects of one of Spain's imperial rivals. More likely and more often mentioned was the idea that Indian nations could be induced to join French parties in attacks on Spanish possessions. Foreign office memoirs presumed that Spanish cruelty had antagonized the West's native peoples, who would therefore welcome French assistance against a common enemy. French trade goods, guns among them, would sweeten the deal. Upper Missouri River Indians, ordonnateur Hubert suggested, held the Spanish "en horreur," and merchandise and gifts could bring them to act in accordance with French interests. As western Indians could serve as partners in raids on Spanish territory, they could also help defend Louisiana against British aggression, Indian warriors counterbalancing Anglo-American numbers. With Indian alliances secured, Le Maire opined, Louisiana need fear neither Spanish distrust nor English hatred ["Quand nous aurons étendu et assuré nos alliances avec les nations sauvages et que les particuliers y trouveront leur interet, nous n'aurons plus à craindre ni la deffiance des Espagnols ni la haine des Anglois"].[12]

11. Quotations from Le Maire, "Memoire sur la colonie de la Louisiane," 1716, MD, Amérique, 1, 52r, 56v; "Sur le commerce des Hollandois et des François en Amerique," 1733, 113r–114r, "Memoire sur les colonies francoises et angloises de l'Amerique septentrionale," 1739, 307v, both in MD, France, 1996. See also "Memoire sur la Loüisiane," c. 1717, 147r–147v, "Observations particulieres sur l'utilité de la colonie du Mississipi . . . ," c. 1717, 287r, Compagnie d'Occident, ". . . de la Compagnie du Sud d'Angleterre avec la comp. d'occident qu'on propose d'establir en France," c. 1717 (date from Leland, *Archives of the Ministry of Foreign Affairs,* 871), 319v, and "Memoire de Charles Legac cy devant directeur pour la Compagnie des Indes à la Loüisianne," 1721, 125r, all in MD, Amérique, 1; Company of the Indies, "Colonie de la Louisiane," 1730, MD, Amérique, 7, 349; "Projet abregé d'une entreprise de la Louisiane sur le nouveau Mexique," MD, France, 1991, 120r–122r (also in MD, Amérique, 7, 207r–211v); Nellis M. Crouse, *Lemoyne d'Iberville: Soldier of New France* (Ithaca, N.Y., 1954), 194.

12. On Indian guides, see the following from MD, Amérique, 1: Le Maire, "Memoire sur la colonie de la Louisiane," 1716, 56v–57v; Hubert, "Memoire sur la riviere, les terres et les sauvages du Missoury," 222v (and see "Memoir of D'Artaguette to Pontchartrain on Present Condition of Louisiana," May 12, 1712, in Rowland and Sanders, eds., *Mississippi Provincial Archives,* II, 60–67, esp. 65); on Indian middlemen, see Le Maire, "Memoire sur la colonie de la Louisiane," 1716, 66v. On Indians' aiding France against Spain, see Hubert, "Memoire sur la colonie de la L'ouisianne," 136r; Le Maire, "Memoire sur la colonie de la Louisiane," 67r–67v; "Memoire pour faire connoitre

Even more than mineral deposits, fertile lands, Spanish colonies, and Indian allies, the most spectacular western geographic features French authors imagined were a "Sea of the West" extending inland from and offering access to the Pacific, and some kind of "River of the West" flowing into this sea or the ocean itself. The illustrious French cartographer Guillaume Delisle offered a 1717 memoir discussing this possibility. He argued that the Sea of the West communicated with the Pacific through a strait between latitudes 43 and 45°; that this strait would give France a new route to China and Japan; that the wealthy and sophisticated Amerindian city of Quivira lay on the shores of the Sea of the West, about eighty to ninety leagues west of New Mexico; and that an east Asian nation was trading with North America's west coast. He suggested the Missouri would provide a convenient westward water route and that a height of land from which a river drained into the Pacific must exist west of Louisiana.[13]

Delisle based his argument on reports from Indian informants, Spanish geographers, English navigators, and French explorers. He concluded it was unlikely that so many different sources could evince such a high degree of consistency if no Sea of the West existed. He felt also that interested nations such as England had probably tried to conceal their knowledge of the Sea of the West and had thereby contributed to the uncertainty surrounding the matter. Delisle accepted that his work was conjectural but argued that the benefits that would arise from Pacific access justified an aggressive search for routes to the ocean. Though his ideas may seem far-fetched today, Delisle's contemporaries respected his opinions. He served as young Louis XV's geography tutor and received in 1718 the title *premier géographe du roi*.[14]

---

dequelle importance il est de conserver la colonnie de la Loüisiane," c. 1717, 190v; Hubert, "Memoire sur la riviere, les terres et les sauvages du Missoury," 223v; "Observations particulieres sur l'utilité de la colonie du Mississipi," c. 1717, 287v. See also "Duclos to Pontchartrain," Oct. 9, 1713, in Rowland and Sanders, eds., *Mississippi Provincial Archives*, II, 79–143, esp. 96–97. On Indians' aiding France against Britain, see the following from MD, Amérique, 1: Le Maire, "Memoire sur la colonie de la Louisiane," 67r–67v; Bienville, "Memoire sur la Loüisiane," 50v; "Mémoire sur l'etat de la colonie de la Louisiane en 1746," 212v–215r.

13. Delisle, "Conjectures sur l'existence d'une mer dans la partie occidentale du Canada, et du Mississipi," MD, Amérique, 1, 241r–253r. John Logan Allen discusses the distant origins and effects on Anglo-American exploration of this idea in "Geographical Knowledge and American Images of the Louisiana Territory," in James P. Ronda, ed., *Voyages of Discovery: Essays on the Lewis and Clark Expedition* (Helena, Mont., 1998), 39–58, esp. 44–45, 56.

14. Michel Antoine, Louis XV ([Paris], 1989), 73–74; Joseph D. Castle, "The Cartography of Colonial Louisiana," in Light Townsend Cummins and Glen Jeansonne, eds., *A Guide to the History of Louisiana* (Westport, Conn., 1982), 139–148, esp. 141; Marcel Giraud, *Histoire de la Louisiane française*, II, *Années de transition (1715–1717)* (Paris, 1958), 13–24; Lucie Lagarde, "Le passage du Nord-Ouest et la mer de l'Ouest dans la cartographie française du 18e siècle, contribution à l'etude de l'oeuvre des Delisle et Buache," *Imago mundi*, XLI (1989), 19–43, esp. 20–30.

Other writers echoed Delisle. Hubert transmitted to the Conseil de la marine a 1717 memoir speaking of a western sea and a river leading to it, of a single mountain containing the sources of this river and the Missouri, of New Mexico's proximity to the Pacific, and of possible lucrative trade with China and Japan that would justify the initial cost of western exploration. The French Compagnie des Indes expressed a similar position in a 1719 memoir, arguing for systematic exploration of the West because the region could yield, among other benefits, a river route to the Pacific that would reduce the cost of French East Asian trade. Belief in the possible existence of these legendary North American attributes weighed heavily on the question of Louisiana's importance. Recent experience had demonstrated the value of direct trade with Peru, European commerce with China was increasing, and Japan might yet be induced to open its ports. Getting there, however, was most of the difficulty rather than half the fun. Cape Horn and the Straits of Magellan frightened the sensible, the Panama Canal lay in the future, and those taking the route around Africa might grow old before touching Cathay. Should the River and Sea of the West exist, they would give the French Empire a relatively easy means of reaching the South Sea, a substitute for the Hudson Bay passage Louis and Pontchartrain had envisaged in 1698 and French diplomats had given up in 1713. A practicable river route to the Pacific would mean that North America was permeable rather than impenetrable, Louisiana central instead of marginal. In the fullness of time, when French relations with Spain and Britain allowed or French finances required, it might permit a reopening of the trade with Peru and Chile that Utrecht had closed.[15]

Such ideas about western geography were attractive, but were they influential? The presence of various western and Louisiana documents among French foreign office papers suggests their likely interest for ministry figures, and the reports indicate the sorts of ideas about Louisiana available to curious diplomats. One additional technique for assessing the reception and influence of these memoirs is to compare them with ministry correspondence to see whether French diplomacy was conducted in accordance with memoirs' sentiments. It was. The 1720 instructions to the marquis de Maulevrier, negotiator of an alliance with Spain, and a contemporary letter that appears to be from minister of foreign affairs Dubois to the French regent, for example, ruled out any possibility of exchanging Louisiana for the Spanish half of Saint-Domingue. An additional unsigned 1721 paper contended that Louisiana lay close to New Mexico and could easily trade with it by way of the West's navigable rivers. Another

15. MD, Amérique, 1: Hubert, "Memoire sur la colonie de la L'ouisianne," 1717, 134r, 136v–137v; "Premier avis a la Compagnie des Indes . . . ," 1719, 460v–462v.

anonymous set of observations for French negotiators in Spain in 1727 spoke of the great potential value of Louisiana and of the agricultural bounty and New Mexican commerce the British would obtain if they should ever gain possession of the French colony. For these reasons, "It is of great consequence to preserve this country for France" ["il Est de grande Consequence a la France de Conserver Ce Paÿs"]. In 1741, Silhouette, acting as a French agent in London, sent a report to foreign minister Jean-Jacques Amelot de Chaillou warning that British conquest of Veracruz and New Mexico would give France's imperial rival the wealth required to play a more significant role in Europe than ever before. In a 1747 letter, Maurepas expressed similar fears of British aggression toward New Mexico and observed that British possession of Louisiana would advance these hostile ambitions. These letters shared the assumptions of the memoirs: that Louisiana was a colony of vast extent, actual or latent worth, and strategic import because of the access it afforded to a wealthy New Mexico.[16]

For the French foreign office in the first half of the eighteenth century, Louisiana was the stuff of dreams. Memoirs spoke, not of a great American desert, but of rivers running through fertile country to a New Mexican silver bonanza and Asian-Pacific commercial opportunities. Authors speculated and extrapolated freely. Lands west of the Rockies, and much territory east of them, remained unknown to French officials, and one could still speak, as an author did in 1746, of the importance of ascending the Missouri and of "discovering the continent of the West of which we have only a very imperfect idea" ["decouvrir le continent de l'ouest dont nous n'avons qu'une idée très imparfaite"]. The working assumption was that America's boundless and unknown western lands would contain something of great value to France. Louisiana as a colony demanded development, reconnaissance, patience, and continued French dominion.[17]

16. "Mémoire pour servir d'instruction au marquis de Maulevrier, lieutenant général des armées du roi, ... allant a Madrid en qualité d'envoyé extraordinaire de sa majesté auprès du roi d'Espagne," Sept. 9, 1720, in A. Morel-Fatio and H. Léonardon, eds., Recueil des instructions données aux ambassadeurs et ministres de France depuis les traités de Westphalie jusqu'à la Révolution française, XII, Espagne, II, 1701–1722 (Paris, 1898), 344–383, esp. 377 (for a discussion of these instructions, see Max Savelle, The Origins of American Diplomacy: The International History of Anglo-America, 1492–1763 [New York, 1967], 310); Dubois to regent, CP, Espagne, 295, 89r (author identification from Leland, Archives of the Ministry of Foreign Affairs, 348); July 21, 1721, CP, Espagne, 303, 115r–115v; "Memoires utils dans la conjoncture des préliminaires de la paix, d'un futur congrès, et d'une ambassade en Espagne, par raport au comerce de France en 1727," CP, Espagne, 352, 21r–21v; Maurepas, "Memoire concernant la prise de l'Isle Royale par les Anglois ... ," May 13, 1747, CP, Espagne, 494, 211v, 213v–214r. See also Silhouette to Jean-Jacques Amelot de Chaillou, Aug. 3, 1741, CP, Angleterre, 412, 227r–227v.

17. "Mémoire pour la Louisiane," MD, Amérique, 2, 198v.

Implications of the Utrecht Settlement for
French Maritime Exploration

Such western visions should have and did attract French explorers, but as we saw in the previous chapter, the terms of the Utrecht settlement made French reconnaissance difficult. As diplomatic agreements channeled French American ambitions toward the North American West, they also closed routes to the region. When the War of the Spanish Succession ended, French knowledge of the lands and waters beyond the Mississippi, Great Lakes, and Hudson Bay was limited. Exploration would have to precede exploitation. By excluding French shipping from the Pacific, the Utrecht settlement precluded one exploratory approach. Indeed, the previous chapter's recitation of the sometimes tedious diplomatic calculations and maneuvers of the eighteenth century's first two decades makes comprehensible the otherwise puzzling absence of pre-1763 French efforts to reconnoiter western North America from the sea. Following one scholarly estimate, 181 ships left France for the South Sea in the early decades of the long eighteenth century, 148 reached it, and 107 made it back to France. Strikingly, records indicate that, in one extension of these voyages, French ships circumnavigated the globe at least ten times in this period. But it was only on July 3, 1786, that French naval explorer Jean-François de Galaup, comte de La Pérouse, sailed into Lituya Bay (between modern Juneau and Anchorage) aboard his ship the *Boussole,* bringing the French flag into company with the British, Spanish, and Russian pennants flying between Oregon and Alaska.[18]

La Pérouse's voyage serves as a reminder that eighteenth-century French ships were capable not only of reaching the North American west coast but also of attempting continental exploration from it. The French search for a Northwest Passage need not have been simply an east-to-west land venture. Until reconnaissance revealed that Lituya Bay was "a cul-de-sac closed off by two immense glaciers," La Pérouse contemplated the possibility that it might form the Pacific entrance to the continental interior: "Port des Français" was, briefly, "the channel by which we planned to enter into the heart of America; we thought that it might lead to a great river running between the mountains, which had its source in one of the great lakes in northern Canada." Given the difficulties French land explorers encountered in their efforts to cross the continent and to

18. Estimates of the number of French ships vary. These figures come from Carlos Daniel Malamud Rikles, *Cádiz y Saint Malo en el comercio colonial peruano (1698–1725)* (Cadiz, 1986), 62, 147; John Dunmore, *French Explorers in the Pacific,* I, *The Eighteenth Century* (Oxford, 1965), 19; E. W. Dahlgren, "Voyages français à destination de la mer du Sud avant Bougainville (1695–1749)," *Nouvelles archives des missions scientifiques et littéraires,* XIV (1907), 423–554, esp. 436–438.

find western rivers leading to the Pacific, a maritime approach such as La Pérouse's could have made a valuable contribution to French knowledge of western geography. No such approach was made, however, until after the French
North American dominions that a Northwest Passage would have connected to
the Pacific had already been ceded to Britain and Spain.[19]

For if French ships had been found sailing after 1713 into the North Pacific in
search of the opening of a Northwest Passage, the same diplomatic complications following French South Sea merchant voyages could have ensued. Spain
or Britain could have accused France of reneging on the Utrecht agreements,
and such accusations could very well have produced the kind of crisis in Franco-
Spanish and Anglo-French relations French policy sought to forestall. The diplomatic importance of the early-eighteenth-century Pacific discouraged overtly
French maritime exploration of the North American West. Pre-1763 French geographers and officials hoping to learn about the region would have to find vantage points other than the docks of French ships.

19. John Dunmore, ed. and trans., *The Journal of Jean-François de Galaup de la Pérouse, 1785–1788*,
2 vols. (London, 1994–1995), I, 105, 109–110.

6

IMPERIAL COMPARISONS

In 1730, a report "Touching upon the Discovery of the Western Sea" appeared as an attachment to a letter from the governor of New France, the marquis de Beauharnois (1726–1747). The report's author was the French explorer Pierre Gaultier de Varennes et de La Vérendrye. While serving as commander of French fur-trading posts north of Lake Superior in 1728 and 1729, La Vérendrye had been seeking information about western lands and waters from Indians at the French forts. His interlocutors included a former slave of the Assiniboines; Auchagah (also spelled Ochagach), "a savage [Sauvage] of my post"; "a Monsoni chief"; and the Cree leaders Pako, Lefoye, Petit Jour, Tacchigis, and La Marteblanche. In addition to oral testimony, La Vérendrye received sketches of western territory from Tacchigis, from Auchagah, and from La Marteblanche and two other chiefs from his nation. The most enticing notion La Vérendrye obtained from these drawings and conversations was that of a "rivière de l'Ouest," "a great river which flows straight towards the setting sun, and which widens continually as it descends."[1]

1. A copy of La Vérendrye's report can be found in "Continuation of the Report of the Sieur de La Vérendrye Touching upon the Discovery of the Western Sea," in Lawrence J. Burpee, ed., *Journals and Letters of Pierre Gaultier de Varennes de La Vérendrye and His Sons: With Correspondence between the Governors of Canada and the French Court, Touching the Search for the Western Sea*, trans. [W. D. LeSueur] (Toronto, 1927), 43–63. Unless otherwise noted, I am using the translations of the La Vérendrye documents included in this volume. For a comprehensive and detailed account of the activities of La Vérendrye and his family, especially strong concerning the commercial aspects of French western exploration, see Antoine Champagne, *Les La Vérendrye et le poste de l'ouest* (Quebec, 1968). For an earlier report of a great western river similar in some respects to La Vérendrye's, see Antoine Denis Raudot, "Of the La Sapiniere Indians and the Assenipouals," 1710, in Raudot, "Memoir concerning the Different Indian Nations of North America," in W. Vernon Kinietz, *The Indians of the Western Great Lakes, 1615–1760* (Ann Arbor, Mich., 1965), 376–377. Good general works on French western exploration include Philippe Bonnichon, *Des cannibales aux castors: Les découvertes françaises de l'Amérique, 1503–1788* (Paris, 1994); William Brandon, *Quivira: Europeans in the Region of the Santa Fe Trail, 1540–1820* (Athens, Oh., 1990); Jean Delanglez, "A Mirage: The Sea of the West," *Revue d'histoire de l'Amérique française*, I (1947–1948), 346–381, 541–568; W. J. Eccles, "French Exploration in North America," in John Logan Allen, ed., *North American Exploration*, II, *A Con-*

La Vérendrye thought he was receiving information concerning the long-sought water route to the Pacific. What his Indian informants had in mind is less clear. Geographer G. Malcolm Lewis, the leading scholar of Amerindian cartography, has suggested that La Vérendrye, a victim of wishful thinking and thorny problems of cultural and linguistic translation, mistook descriptions and representations of the Nelson River, which flows into Hudson Bay, for a River of the West running to the South Sea. It is also possible that La Vérendrye misinterpreted and muddled accounts of rivers such as the Winnipeg, the Saskatchewan, the Missouri, and the Mississippi. In any event, La Vérendrye felt he had the information needed to justify and direct a search for the Sea and River of the West, and he and his sons moved beyond Lake Superior during the next two decades looking for them. They established a chain of posts on the lakes and rivers west and northwest of Lake Superior, visited Mandan villages, tromped around the western parts of what are now the Dakotas, and, in the case of La Vérendrye's son Jean-Baptiste, died at the hands of a Sioux war party on a Lake of the Woods island.[2]

Neither the La Vérendryes' efforts nor their Indian guides' assistance sufficed to carry the French explorers past the regions of the Black Hills and the Forks of the Saskatchewan. Nor, before the end of the French Empire in North America, do other French investigators seem to have made it past New Mexico in the south or Rocky Mountain foothills in the north. When La Vérendrye died in 1749, he remained hopeful about the river and sea he had imagined and unsuccessful in his efforts to find them. Geographers in Paris had still to content themselves with unverified reports and unsatisfying speculations.

La Vérendrye's efforts to peer into the West's obscurities point to a question left open by previous chapters. Discussion of Spanish American silver, European trade with Asia, and the Pacific's conjectural geography may make French fascination with the idea of a Northwest Passage more comprehensible, but it leaves lingering French uncertainty about the existence of such a passage unexplained.

This persistent geographic confusion is puzzling. North America's European invaders had been interested in the possibility of a passage to India from the

---

tinent Defined (Lincoln, Neb., 1997), 149–202; Eccles, "La Mer de L'Ouest: Outpost of Empire," in Eccles, *Essays on New France* (Toronto, 1987); Marthe Emmanuel, "Le passage du Nord et la 'mer de l'ouest' sous le régime français: Réalités et chimères," *Revue d'histoire de l'Amérique française*, XIII (1959), 344–373. Burpee's survey, *The Search for the Western Sea: The Story of the Exploration of North-Western America* (Toronto, 1935), has been superseded by later works.

2. G. Malcolm Lewis, "La grande rivière et fleuve de l'Ouest / The Realities and Reasons behind a Major Mistake in the 18th-Century Geography of North America," *Cartographica*, XXVIII (1991), 54–87. For a less pointed interpretation of the circumstances complicating the interpretation of Indian information, see Champagne, *Les La Vérendrye et le poste de l'ouest*, 90–99.

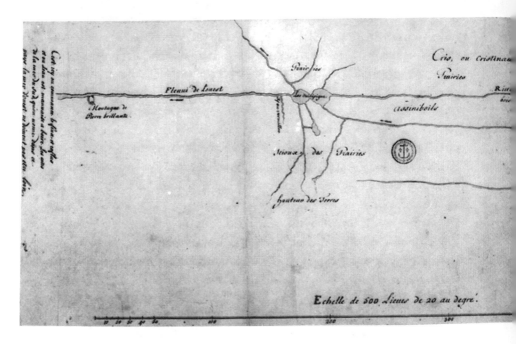

MAP 22. (above) Ochagach, *Cours des rivieres et fleuve, courant a l'ouest du nord du lac Superieur.* 1728–1729. Library and Archives Canada, e010771324

MAP 23. (right) Pierre Gaultier de Varennes et de La Vérendrye, *Carte des rivieres et fleuves, courant a l'ouest du nord du lac Superieur.* 1728. Library and Archives Canada, e010771323

earliest days of their encounter with the New World, and by the time La Vérendrye drafted his report, French explorers had been looking for it for more than two centuries. In fall 1534, while visiting the Iroquois village of Hochelaga (the site of modern Montreal), Jacques Cartier heard reports "that along the mountains to the north, there is a large river, which comes from the west." In 1615 and 1616, Samuel de Champlain reached the eastern shores of Lake Huron while seeking what he hoped would be a practicable riparian passage to China.[3]

It was not just that Europeans had long sought a resolution of the question of a Northwest Passage, but that a definitive answer to the question was possible. A twenty-first-century audience knows that no such practicable passage existed—

3. The Cartier quotation and the original French appear in H. P. Biggar, ed. and trans., *The Voyages of Jacques Cartier* (Ottawa, 1924), 170–172, 200–202. For Champlain, see Biggar, ed., *The Works of Samuel de Champlain,* 6 vols. (Toronto, 1922–1936), II, *1608–1613,* trans. John Squair, comp. J. Home Cameron, 326–327, 330–331, 345; ibid., III, *1615–1618,* ed. and trans. H. H. Langton and W. F. Ganong, comp. Cameron, 45–62, 95–105, 118–120.

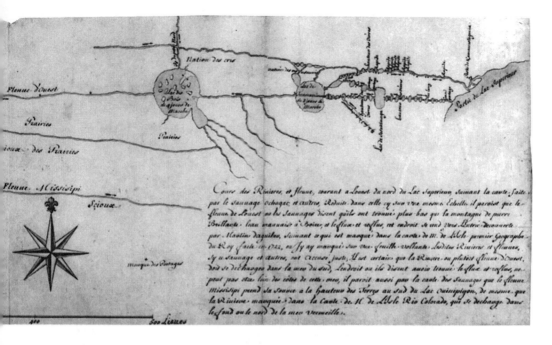
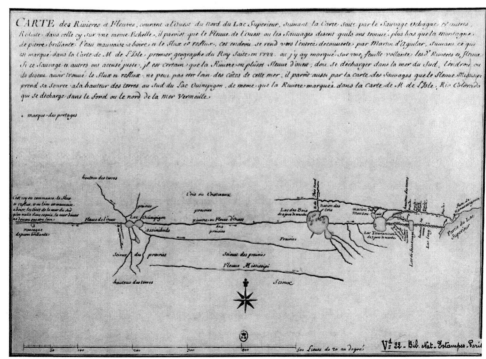

MAP 24. Jacques-Nicolas Bellin, *Carte de l'Amerique septentrionale pour servir à l'histoire de la Nouvelle France*. 1743. Geography and Map Division, Library of Congress, g3300 ct001126

though, with global warming and melting northern ice, one may soon come into being. An early modern ship could pass the straits of Magellan, Gibraltar, or Malacca, but no amount of effort or imagination would take it through the chimerical Strait of Anian. North America's western mountains as they were, and the Northwest Passage as it was envisaged, could not coexist. Later eighteenth- and early-nineteenth-century explorers would demonstrate this. Why, then, did eighteenth-century French explorers and scholars find the West's physical and human geography so difficult to grasp?

One way to begin answering this question is by using the excellent scholarly literature on French cartography to look at French efforts in other parts of the eighteenth-century world—in particular, areas where geographic investigators accomplished more of their objectives. The confused and conjectural quality of the French understanding of western American geography contrasts with the remarkable achievements of French cartography elsewhere. In France itself, the late seventeenth and eighteenth centuries witnessed a painstaking, systematic, and comprehensive survey of the kingdom that left it the most completely, elegantly, and accurately mapped area of Europe. French cartographic techniques became a model for other empires, and this gave French mapmakers entrée to the distant territories and invaluable archives of the Russian and Chinese empires. French cartographers assisted, observed, and participated in massive Russian and Chinese imperial surveying efforts. As a result, eighteenth-century French maps of enormous swaths of Eurasia became more detailed, complete, and correct than had been possible a century before.

On the face of it, it would seem that French explorers should have been able to avail themselves of comparable opportunities in western North America. Eighteenth-century French western expeditions were relatively numerous, and, where French expeditions themselves were insufficient, the French in the West would seem to have been in an excellent position to profit from Amerindian geographic understanding. Though early-eighteenth-century French explorers failed to dip their blistered toes in the waters of rivers flowing to the Pacific, they did advance far enough to encounter an array of western Indian nations. Many of these Indians descended from peoples who had inhabited the American West for centuries. They sent goods along trade routes extending thousands of miles, and in commerce, hunting, and war, numerous Indian nations covered considerable distances. Where goods and people were moving, ideas about geography could presumably have moved, too. When sharing tobacco and exchanging commodities, Frenchmen could reasonably have hoped to pick up geographic information as well, hearing tales and perusing sketches of lands they had not yet seen for themselves. Nonetheless, French explorers and geographers continued to understate, and probably to underestimate, the magnitude of the Rockies, to imagine inland seas where sagebrush grew, to expect Spanish, Chinese, and Japanese settlements where Chinook, Makah, and Nootka villages stood.

We could, of course, simply ascribe these differing outcomes to the many and manifest differences between North America and Eurasia, happily return to a narrow focus on one continent, and resolutely ignore the objects in our peripheral vision.

Three considerations suggest we should instead place eighteenth-century French investigation of the North American West in the larger context of French reconnaissance efforts elsewhere. First, in the period before Europeans had determined and defined — for themselves, at least — the geographic and cultural relationship between eastern Asia and western North America, French investigations in Russia and China were also exploration of America. French cartogra-

he Russian Empire and the occasional Jesuit missionary in the Chinese
ried to use participation in Russian North Pacific voyages and access to
records to learn about the Far West from the Far East.

nd, in accounting for the shortcomings of eighteenth-century French
ι reconnaissance, it helps to have some sense of the larger enterprise of
which it was a part. In particular, if we can ascertain the conditions favoring
French investigations in some areas, we can better evaluate the circumstances
hindering them elsewhere. Put differently, the point of the comparison is, not
to posit artificial similarities between France, Russia, and China on the one
hand and western North America on the other, but rather to determine which
and how differences mattered for French investigators and, indirectly, western
Indian peoples.

Finally, bringing French cartography of France, Russia, and China into the
story shows how western European approaches to western American geography fit into the larger story of early modern imperial expansion and dominion. Ideas suggested in Chapters 1 and 2 by Spain, Mexico, and Peru can be
checked and developed in Chapters 6, 7, and 8 by France, Russia, and China. In
the background are issues involving not just particular empires and their geographic horizons but empires and horizons in general. A bounded historical inquiry cannot resolve such issues, but it can at least begin to frame some questions. With multiple cases of imperial reconnaissance in view, we can obtain the
beginnings of a much broader appreciation of the relation between empire and
understanding than would be possible were attention directed to French western reconnaissance in isolation.

Achievements in France, Russia, and China demonstrated the formidable
capacities of eighteenth-century French cartographers and geographers. Outside of western North America, French surveyors and cartographers exhibited
the ability to generate comprehensive, precise, verifiable, and understandable
representations of vast stretches of territory, even when operating in distant
and difficult areas among people with markedly different cultures. The results of
these Eurasian efforts for understanding of the American West, however, proved
less solid than promised.

France

If French explorers, missionaries, and traders coming out of Canada and Louisiana were well situated for western exploration, why, then, were they so uncertain
and, in the eyes of someone familiar with modern maps of North America, so

confused about basic western geographic features? One possible line of investigation concerns French officials and savants' approach to geographic understanding. Twenty-first-century readers, having grown up enjoying the fruits of unrelenting technological innovation, having perhaps attended a modern research university, and having wasted their eyes and whiled away their hours watching *Star Trek* reruns, may take the notion of ever-broadening horizons of understanding for granted. Early modern Europe was, safe to say, different. Western Europeans had remained remarkably ignorant of large regions south of the Mediterranean and east of Muscovy for millennia. Early modern governments and scholars, despite their achievements since 1492, lacked the kind of financial and institutional resources that would later build atomic bombs and launch spacecraft. Their ships were marvels of the age, but the age was pre-industrial, and their feet differed little from their ancestors': Napoleon, it has been said, moved no faster than Julius Caesar. As historians such as Fernand Braudel have demonstrated, the pace of change in many areas of early modern life could be glacially slow, and continuity characterized many features of daily existence. Thinking in terms of the expectation of exploration or the assumption of technical and scientific advance runs the danger of anachronism, and we need to consider the possibility that a simple lack of French interest in or commitment to geographical and cartographical investigation prevented the French Empire from taking advantage of the opportunities its North American position afforded.

The French government's approach to mapping France itself suggests otherwise. From the 1660s, France witnessed a surge of interest in cartographic improvement as Louis XIV supported efforts to redraw maps of his kingdom. What ultimately made French cartographic efforts so remarkable, however, was not Louis's initial interest but the fact that French exertions turned out to be more assertive, more comprehensive, and more sustained than those of rival nations such as England.

These French cartographic projects began as one part of Louis and Jean-Baptiste Colbert's larger effort to rationalize French administration and more efficiently exploit French resources. The first step came in 1663–1664, when Colbert directed royal agents throughout France to submit well-drawn maps of their jurisdictions, to prepare new maps where necessary, and to indicate areas that remained poorly covered. Colbert also founded the Académie royale des sciences in 1666 and included cartography as one of its major functions. Despite these bright beginnings, initial results were disappointing. The challenges of surveying and mapping a country as large as France in accordance with observations based on modern scientific techniques such as triangulation proved formidable, and few satisfactory new maps had been drawn by the time of Colbert's death in 1683, or even Louis's in 1715.[4]

French cartographers found more immediate success on France's maritime periphery, verifying and redrawing charts of the French Mediterranean and Atlantic coastlines between 1670 and 1693. Even this venture proved hazardous in unexpected ways. New measurements of latitude and longitude, revised in accordance with tables of planetary motion recently constructed by the Italian-born French astronomer Jean-Dominique Cassini (1625–1712), demonstrated that the French Atlantic coastline lay farther east than had been previously thought. While Louis XIV was draining his kingdom's wealth and manpower in his efforts to extend its borders to the north and east, his cartographers were contracting the limits of France in the west. Their new calculations reduced the realm by 6,271 square leagues.[5]

The systematic mapping of France was not complete by the end of Colbert or Louis XIV's lives, but government-directed cartographic efforts continued after their deaths. Where the British Admiralty established a Hydrographic Office in 1795, the French navy set up its Dépôt des cartes et plans de la marine in 1720.

4. Colbert, Louis's contrôleur général des finances and secretary of state for naval affairs, is not to be confused with his nephew, the aforementioned Jean-Baptiste Colbert, marquis de Torcy. For Colbert's directive, see the "Extrait de l'instruction de Colbert pour les messieurs les maîtres des requêtes, commissaires départis dans les provinces," September 1663, rpt. in Nelson-Martin Dawson, *L'atelier Delisle: L'Amérique du Nord sur la table à dessin* (Sillery, Quebec, 2000), 51:

> Il est nécessaire que lesdits sieurs recherchent les cartes qui ont esté faites de chacune province ou généralité, en vérifiant avec soin si elles sont bonnes; et, au cas qu'elles ne soyent pas exactement faites ou mesme qu'elles ne soyent pas assez amples, s'ils trouvent quelque personne habile et intelligente, capable de les réformer, dans la mesme province ou dans les circonvoisines, Sa Majesté veut qu'ils les employent à y travailler incessamment et sans discontinuation; et, au cas qu'ils ne trouvent aucune personne capable de ce travail, ils feront faire des mémoires fort exacts sur les anciens, tant pour les réformer que pour les rendre plus amples.

This chapter's account of French cartography relies on the impressive monographic edifice of Josef W. Konvitz, *Cartography in France, 1660–1848: Science, Engineering, and Statecraft* (Chicago, 1987), 1–2; Monique Pelletier, "Cartography and Power in France during the Seventeenth and Eighteenth Centuries," *Cartographica*, XXXV (1998), 41–53, esp. 44; Pelletier, *Les cartes des Cassini: La science au service de l'État et des régions* (Paris, 2002), 39–41, 45, 99; Numa Broc, *La géographie des philosophes: Géographes et voyageurs français au XVIIIe siècle* (Paris, [1975]), 16–18, 21–22; and Dawson, *L'atelier Delisle*, 48–53. Dutch doctor Gemma Frisius (1508–1556) was the first to describe geodetic triangulation. Triangulation consists of precisely measuring a baseline, extending lines from its terminal points to form a triangle, and then using trigonometry to determine the lengths of the new lines as a function of the angles. See Konvitz, *Cartography in France*, 2, 6–7.

5. Broc, *La géographie des philosophes*, 17; Dawson, *L'atelier Delisle*, 15; Konvitz, *Cartography in France*, 4–8, 76; Pelletier, *Les cartes des Cassini*, 60–61.

The Dépôt compiled maps, charts, and explanatory texts such as logs, letters, and reports, and it recorded the sources of observations of location and distance and checked these results against reliable data. Moreover, while systematic surveying of France itself was often interrupted or delayed by the vagaries of war and royal finances, these hindrances proved temporary and surmountable. The effort to extend a north-south network of triangles along the length of France, begun in 1683, was interrupted by the wars of the late seventeenth and early eighteenth centuries and completed in 1718. Work in France ceased thereafter but recommenced in the 1730s, when France's comptroller general of finances, Philibert Orry, was endeavoring to coordinate, control, and standardize public works projects such as a network of roads and canals. Orry viewed precise maps as a necessary tool of planning and management. In 1733, he ordered resumption of the triangulation of France, and a first map of the entire kingdom based on precise scientific measurements appeared in 1744. Still more detailed surveying continued thereafter, and when the British government established the Trigonometric Survey in 1791, it was three years after the French government had completed a second, more finely grained map of France. In little more than a century, the French government had accomplished a thorough and comprehensive remapping of a major European country using the latest scientific surveying and cartographic techniques. It was not quite the *Guide Michelin*, but it was a step in that direction.[6]

The mapping of France was a noteworthy achievement, but one can fairly ask whether it really discloses much about the French approach to the geography of regions outside the hexagon. The rationale for surveying France itself was more immediate, the difficulties of doing so — even if one cartographer bemoaned the quality of cheese in Auvergne — less imposing than was the case for the lands and waters beyond. One could imagine French scientific and geographic interest and efforts focusing narrowly on the home country.[7]

Such was not the case. French officials and savants cast their gaze far beyond the rivers, mountains, and seas forming France's apparent physical limits. In the second half of the seventeenth century, with much royal encouragement, French emissaries and cartographers sought to enhance French understanding of the spaces and peoples of the Great Lakes region. In the 1730s and 1740s, navi-

6. "Remarques de M. Bellin, ingenieur de la marine, sur les cartes et les plans . . . ," in P[ierre-François-Xavier] de Charlevoix, *Histoire et description generale de la Nouvelle France, avec le journal historique d'un voyage fait par ordre du roi dans l'Amérique septentrionnale*, 3 vols. (Paris, 1744), III, i–ix, esp. ii; Konvitz, *Cartography in France*, 6–9, 16, 25, 73–75; Pelletier, "Cartography and Power," *Cartographica*, XXXV (1998), 47–49; Pelletier, *Les cartes des Cassini*, 7, 55–56, 79–80; Broc, *La géographie des philosophes*, 22, 42–43.

7. Konvitz, *Cartography in France*, 14.

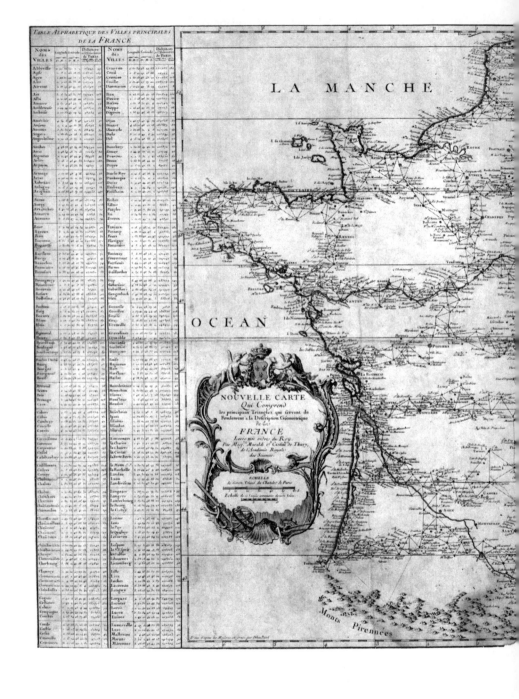

TABLE ALPHABETIQUE DES VILLES PRINCIPALES
DE LA FRANCE

LA MANCHE

OCEAN

NOUVELLE CARTE
Qui Comprend
les principaux Triangles qui servent de
Fondement à la Description Géometrique
de la
FRANCE
Levée par ordre du Roy
Par Mess.rs Maraldi et Cassini de Thury,
de l'Academie Royale
des Sciences.

ECHELLE
de Grosse Toises du Chatelet de Paris

Monts Pirennées

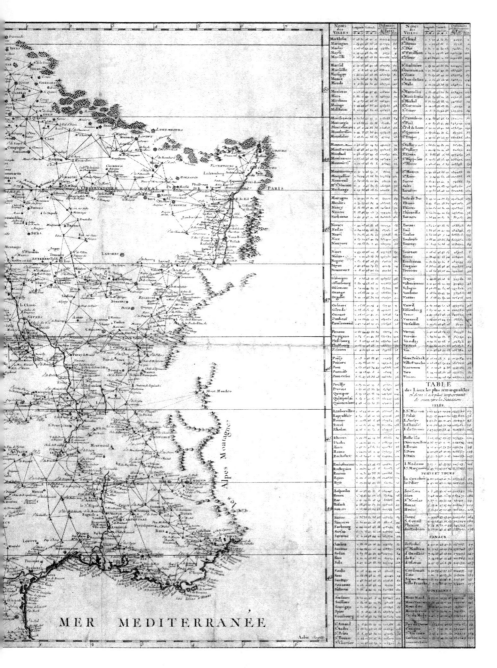

MAP 25. Jean-Dominique Maraldi and César-François Cassini,
*Nouvelle carte qui comprend les principaux triangles qui servent
de fondement à la description géométrique de la France.* [1744?].
Geography and Map Division, Library of Congress, g5830 ct001183

gation's demands and perils induced French hydrographers to use newly precise observations from both astronomers and navigators to revise their charts not only of the Mediterranean but also of the Gulf of Mexico. More strikingly, from 1671 to 1744, French scientists journeyed to sites as far afield as the Caribbean and Cape Verde Islands (1682), Siam (1685), Cayenne and Saint-Domingue (1699–1700), Lapland (1734–1737), and Peru (1707–1711, 1735–1744). Their tasks included not simply the acquisition of more precise latitudinal and longitudinal data to facilitate navigation but also the gathering of far-flung locational, chronometric, and astronomical measurements needed to resolve the hotly debated question of the globe's exact shape. The issue was of both general scientific interest and immediate cartographic importance, for the exact distance between lines of longitude, and therefore the exact position of sites in France and elsewhere relative to them, depended on whether the earth was an oblate or prolate spheroid. (The newly taken measurements demonstrated, in accordance with the views of Newton and others, that the earth is an oblate spheroid: it flattens toward the poles.) At the same time that French cartographers were intensely concerned with ascertaining the exact distance between hilltops in Burgundy, French scientific and geodetic interests extended to other parts of the world.[8]

In France itself, cartographers and officials evinced a robust inquisitiveness and the capacity to conduct sustained and large-scale endeavors. Outside France, they would exhibit the ability not just to gather data at specific points from Peru to Lapland but also to obtain a geographic overview of a region larger than France or the North American West.

Russia

In 1725, two members of early-eighteenth-century France's most accomplished and renowned family of cartographers, Joseph-Nicholas Delisle and Louis Delisle de La Croyère, set out for Saint Petersburg. Both, as their names suggest, were brothers of the celebrated Guillaume Delisle. Joseph-Nicholas was to establish an observatory in Saint Petersburg and join the newly established Russian Academy of Sciences. Trained by Cassini, Joseph-Nicholas would, during

8. Gilles Havard, "La domestication intellectuelle des Grands Lacs par les Français dans la second moitié du XVIIe siècle," in Charlotte de Castelnau-L'Estoile and François Regourd, eds., *Connaissances et pouvoirs: Les espaces impériaux (XVIe–XVIIIe siècles), France, Espagne, Portugal* (Pessac, France, 2005), 63–82; Broc, *La géographie des philosophes*, 17–22; E. W. Dahlgren, *Les relations commerciales et maritimes entre la France et les côtes de l'Océan pacifique (commencement du XVIIIe siècle)*, I, *Le commerce de la mer du Sud jusqu'à la paix d'Utrecht* (Paris, 1909), 552–556; Konvitz, *Cartography in France*, 4–5, 10–13, 76; James E. McClellan III, *Colonialism and Science: Saint Domingue in the Old Regime* (Baltimore, 1992), 118–120.

two decades spent in Russia, play a major role in the development of Russian cartography. His brother Louis would train Russian scientists and accompany Vitus Bering's second North Pacific expedition.

The origins of this French journey, and the roots of the French cartographic assimilation of Russia and Siberia that would arise from it, lay in the growth of the early modern Russian Empire. The extraordinary speed of Russian expansion across Eurasia contrasts with the often halting efforts of Spain, Britain, and France in western North America. Between 1581 (when Ermak Timofeev and his "band of cossacks" crossed the Urals, defeated the forces of Kuchum Khan, and began "the subjugation of the Siberian lands") and 1639 (when Ivan Moskvitin reached the Sea of Okhotsk and gained Muscovy a toehold on the South Sea's shores), Russian officials, explorers, adventurers, and trappers moved more than four thousand miles east of Moscow. By 1732, Russian sailors had crossed the North Pacific and sighted Alaska.[9]

The impetus for this headlong drive across Siberia in the seventeenth century, and across the North Pacific in the eighteenth, derived from economic imperatives and ecological limitations, state interests and private motivations. Like the western European rulers they increasingly sought to emulate, seventeenth- and eighteenth-century Russian czars needed precious metals to finance their ambitions and wars. Having not yet developed a Russian Potosí in the Urals, Russian rulers had to acquire bullion by taxing trade with other nations. At the same time, Russia possessed plenty of subjects willing to take risks and undergo hardships in pursuit of gain. Muscovy's proximity to Siberia provided an opportunity. As was the case with Canada, the frigid climate making Siberia redoubtable also made the "soft gold" of its fur-bearing animals a profitable commodity in European or Asian markets. In the late sixteenth and the seventeenth centuries, sometimes on their own initiative, sometimes at officials' urging, a diverse cast of characters pushed east to acquire pelts themselves or to extract them from

9. Quotations from "The Conquest of Siberia by Ermak Timofeev and His Band of Cossacks, as Reported in the *Stroganov Chronicle* (Excerpts)," in Basil Dmytryshyn, E. A. P. Crownhart-Vaughan, and Thomas Vaughan, eds. and trans., *To Siberia and Russian America: Three Centuries of Russian Eastward Expansion; A Documentary Record*, 3 vols. ([Portland, Ore.], 1985–1989), I, 14–23, esp. 14, 20, and see also xxix, xxxix–xli, xlvii; W. Bruce Lincoln, *The Conquest of a Continent: Siberia and the Russians* (New York, 1994), 63; Gerhard Friedrich Müller, *Bering's Voyages: The Reports from Russia*, ed. and trans. Carol Urness (Fairbanks, Alaska, 1986), 6. On the sighting of Alaska, see "A Statement from the Cossack Ilia Skurikhin concerning the Voyage of the Sv. Gavriil to the Shores of Bolshaia Zemlia [America] in 1732," Apr. 10, 1741, and "The Report of the Geodesist Mikhail Spiridonovich Gvozdev to Martyn Petrovich Spanberg concerning His Voyage of Exploration to the Coast of North America in 1732," Sept. 1, 1743, in Dmytryshyn, Crownhart-Vaughan, and Vaughan, eds. and trans., *To Siberia and Russian America*, II, 132–134, 161–167; James R. Gibson, "The Exploration of the Pacific Coast," in Allen, ed., *North American Exploration*, II, 333.

brutalized locals. Ecological restraint was as lacking as humanitarian scruples, and, as nearer regions' furs ran out, the quest for new supplies pushed Russians across Eurasia's northern tier to the Pacific. In the course of this fur rush, the seventeenth-century Russian Empire demonstrated that an early modern Eurasian state, when operating primarily on its own landmass and in areas distant from other empires' spheres of activity (Russian expansion slowed or stopped when it ran up against Chinese imperial claims) and when unconstrained by concern for the lives of its agents or objects of conquest, could extend its dominion over thousands of square miles and innumerable indigenous peoples.[10]

For those personally involved in the Russian conquest and settlement of Siberia, the essential tasks were to surmount distance and climate's challenges, to kill or cow resisting natives, and to trap and extort the furs making these efforts worthwhile. As the Russian Empire matured, its rulers and officials felt the need to think more broadly and to gain a better understanding of the lands, waters, and peoples they hoped to dominate and exploit. The reasons were, in part, practical. To profit from the resources of Siberia and whatever lay beyond it, Russian officials needed to know where and what those resources were, how to reach them, and how to establish that they lay within Russian imperial boundaries. Modern scholarship, especially a series of articles by Mark Bassin, has suggested that ideological considerations also played a role. Beginning with Peter the Great's acceptance of western Europe as a model for development, Russian rulers sought a respectable place among Europe's great states. But unlike nations such as Spain, France, and England, Russia began the eighteenth century without an American empire and without the wealth and prestige such overseas possessions could provide. In the aftermath of Russia's victory in the Great Northern War (1700–1721), Peter sought a way to emulate the western imperial model even as he had already begun trying to reproduce its urban, scientific, and economic achievements. Russia would not only continue to move east toward the Americas themselves, but it would imagine Siberia as the equivalent of the European New World dominions. The appeal of this idea was evident in the names Russians used: Bassin mentions "the common practice of referring to" Siberia as "'our Peru' or 'our Mexico,' a 'Russian Brazil,' or even 'our East India.'" In the 1730s, Russian geographer and historian Vasilii N. Tatischev would contribute

---

10. This chapter's view of Russian expansion derives especially from Mark Bassin, "Expansion and Colonialism on the Eastern Frontier: Views of Siberia and the Far East in Pre-Petrine Russia," *Journal of Historical Geography*, XIV (1988), 3–21, esp. 7–8, 11. See also Dmytryshyn, Crownhart-Vaughan, and Vaughan, eds. and trans., *To Siberia and Russian America*, I, xlii, II, xxxii; Gibson, "Exploration of the Pacific Coast," in Allen, ed., *North American Exploration*, II, 340; Lincoln, *Conquest of a Continent*, 46, 58–59, 87, 101.

to this process by arguing that the Ural Mountains, not the Don River, formed Europe's eastern boundary. Just as the western European empires possessed a European core and overseas colonial possessions, Russia would possess a metropolitan European center west of the Urals and a set of extra-European territories and subjects east of them.[11]

Ruling and defining Siberia required information. Its populations would have to be categorized, its wealth inventoried, its extent measured. To the cossacks, *streltsy* (musketeers), and the *promyshlenniki* (private traders, hunters, and trappers) so prominent in sixteenth- and seventeenth-century Russian expansion across Siberia, the eighteenth century added the *geodezist* and the scientist. Russian cartographic projects provide an illuminating example of this development. Systematic eighteenth-century Russian mapping efforts can be traced back to Peter the Great's concerns. Like Louis XIV, Peter found himself dissatisfied with the completeness and reliability of the maps of his realm. He also found himself inspired by the maps he observed on European trips. On a 1717 visit to France, he examined charts bearing the imprint of fifty-four years of concentrated French cartographic effort. With a problem and a remedy in mind, Peter initiated the first Russian state survey (c. 1717–1752). This project would involve adaptations of the most recent western European techniques of precise instrumental measurement and traverses from astronomically fixed points as well as data-gathering expeditions to the Russian Empire's far corners. It would ultimately cover more than half of European Russia and a significant part of Siberia. In the words of one scholar, the "ambition of both Peter and several of his associates was to commission a comprehensive cartographic and written geographical survey of the Russian state," a breathtaking notion, of course, if one considers the difficulties involved in surveying even a much smaller and more developed European nation such as France.[12]

---

11. The quotation is from Mark Bassin, "Inventing Siberia: Visions of the Russian East in the Early Nineteenth Century," *AHR*, XCVI (1991), 763–794, esp. 770 (see also 767–768). Also useful are Bassin, "Russia between Europe and Asia: The Ideological Construction of Geographical Space," *Slavic Review*, L (1991), 1–17, esp. 5–8; Gibson, "Exploration of the Pacific Coast," in Allen, ed., *North American Exploration*, II, 332.

12. Quotation is from Denis J. B. Shaw, "Geographical Practice and Its Significance in Peter the Great's Russia," *JHG*, XXII (1996), 160–176, esp. 167. See also Dawson, *L'atelier Delisle*, 61–65; L. A. Goldenberg and A. V. Postnikov, "Development of Mapping Methods in Russia in the Eighteenth Century," *Imago mundi*, XXXVII (1985), 63–80, esp. 63–69; Larry Wolff, *Inventing Eastern Europe: The Map of Civilization on the Mind of the Enlightenment* (Stanford, Calif., 1994), 144–147; S Ye Fel, "The Role of Petrine Surveyors in the Development of Russian Cartography during the 18th Century," in James R. Gibson, ed. and trans., *Essays on the History of Russian Cartography, 16th to 19th Centuries* (Toronto, 1975), 27–42; "An Account by the Cossack Piatidesiatnik, Vladimir Atlasov, concerning His Expedition to Kamchatka in 1697," Feb. 10, 1701, "A Petition to Tsar Peter

Remarkably, Peter was looking beyond Europe and Asia to North Pacific waters and North American lands. As early as the late seventeenth century, the breakneck pace of Russian exploitation had begun to deplete the easily accessible sources of Siberian furs. By 1744, an observer could say of one part of Siberia, "At first there were plenty of furbearing animals there, but now there are no sables and not many foxes in those Iakut lands, from the shores of the [Arctic] ocean all the way south to the great Lena River." With a visionary mind stretching even farther than his famously tall and well-traveled body, Peter saw that the North Pacific and western North America might provide new fur stocks to replace Eurasia's dwindling supplies. He also saw that the European powers already involved in North America might oppose Russian efforts to impinge upon their American claims. The solution was to sponsor scientific expeditions that could establish the rising Russian state's intellectual credentials while, especially when incompletely or misleadingly described, simultaneously providing a cover for the gathering of strategically useful and politically sensitive information about the peoples and regions stretching from the Urals to North America.[13]

Peter died in 1725, before he could bring his plans to fruition, but work continued on the scientific, cartographic, and exploratory projects he had initiated. Before his death, Peter had directed Bering to "discover where it ['Kamchatka or some other place'] is joined to America." In 1728, Bering sailed through the strait that now bears his name and observed Saint Lawrence Island, but he "did not see land [the mainland of North America]." Bering would have another opportunity during the Great Northern Expedition of 1733–1742, a venture "the like of which there" had "never been before." Compared to the American exploratory

Alekseevich from Servitors Who Killed the Prikashchicks of the Kamchatka Ostrogs: A Report of Expeditions to the Kuril Islands and an Account of Native Resistance," Sept. 26, 1711, and "An Ukaz from Tsar Peter Alekseevich to Navigators and to Iakutsk Servitors concerning Establishing a Route to Kamchatka and the Kuril Islands," July 3, 1714, all in Dmytryshyn, Crownhart-Vaughan, and Vaughan, eds. and trans., *To Siberia and Russian America*, II, xxxv, 3–12, 43–46, 49–53.

13. Quotation is from Heinrich von Füch, "An Eyewitness Account of Hardships Suffered by Natives in Northeastern Siberia during Bering's Great Kamchatka Expedition, 1735–1744 . . . ," Feb. 28, 1744, in Dmytryshyn, Crownhart-Vaughan, and Vaughan, eds. and trans., *To Siberia and Russian America*, II, 168–169, esp. 170. On diminishing fur supplies, see Bassin, "Expansion and Colonialism on the Eastern Frontier," 11; Lincoln, *Conquest of a Continent*, 101. On exploration and international relations, see Glynn Barratt, *Russia in Pacific Waters, 1715–1825: A Survey of the Origins of Russia's Naval Presence in the North and South Pacific* (Vancouver, B.C., 1981), 40, 46–47; Gibson, "Exploration of the Pacific Coast," in Allen, ed., *North American Exploration*, II, 332–334; Lincoln, *Conquest of a Continent*, 101; Dmytryshyn, Crownhart-Vaughan, and Vaughan, eds. and trans., *To Siberia and Russian America*, II, xxxv. Dmytryshyn, Crownhart-Vaughan, and Vaughan point out also (xxxi–xxxiii) that more extensive government involvement in expeditions became necessary because the challenges of North Pacific navigation exceeded the capabilities of unaided private explorers in a way that Siberia, with its many navigable rivers and native guides, had not.

efforts of contemporary Spain, France, and Britain, the scale of this project was remarkable. It involved directly as many as one thousand people, with roughly another two thousand men supporting the venture by performing tasks such as transporting supplies and equipment across Siberia. The goals of the expedition were as grand as its personnel were numerous. In the words of Ivan Kirilov, senior secretary of the Russian Administrative Senate and one of the key planners, they were

> (1) to find out for certain whether it is possible to pass from the Arctic Ocean to the Kamchatka or Southern Ocean sea . . . (2) to reach from Kamchatka the very shores of America . . . (3) to go from Kamchatka to Japan . . . (4) . . . to search for new lands and islands not yet conquered and to bring them under subjection; (5) to search for metals and minerals; (6) to make various astronomical observations both on land and sea and to find accurate longitude and latitude; (7) to write a history of the old and the new, as well as natural history, and other matters.
>
> The benefit to be expected is that from the eastern side Russia will extend its possessions as far as California and Mexico.[14]

These efforts involved more than Russians. Russia's eastward expansion had always been the enterprise of a variety of ethnic groups. German and Lithuanian prisoners captured in Russia's wars might later find themselves in the czar's service beyond the Urals. The eighteenth-century Russian Empire needed savants as well as soldiers, more than its own population and institutions could provide; so foreigners figured prominently in Russia's intellectual and exploratory undertakings. Vitus Bering was Danish. The Gerhard Friedrich Müller who was quoted in Chapter 3 and who suffered with the others on Bering's second expedition was German. French scientist Joseph Nicholas Delisle wrote in 1729 of Swedish officers taken in Peter's 1709 victory over Charles XII at Poltava who were then

---

14. Bering expedition quotations from "Instructions from Empress Catherine Alekseevna to Captain Vitus Bering for the First Kamchatka Expedition," Feb. 25, 1725, and "A Report from Captain Vitus Bering to Empress Anna Ivanovna concerning the First Kamchatka Expedition," Mar. 1, 1730, in Dmytryshyn, Crownhart-Vaughan, and Vaughan, eds. and trans., To Siberia and Russian America, II, 69, 79–86, esp. 86. Earlier, in 1719, Peter had directed "Ivan Evreinov and Fedor Luzhin to Explore Kamchatka and the North Pacific Ocean to Determine whether Northeast Asia Is Connected to Northwest America." See his official orders ibid., 65. For a strong biography of Bering, see Orcutt Frost, Bering: The Russian Discovery of America (New Haven, Conn., 2003). The Kirilov memorandum, circa 1733, is reprinted in Raymond H. Fisher, Bering's Voyages: Whither and Why (Seattle, 1977), 184–187 (see also 120–132). Also pertinent are the memoirs and instructions in To Siberia and Russian America, II, 87–131; Gibson, "Exploration of the Pacific Coast," in Allen, ed., North American Exploration, II, 334; Lincoln, Conquest of a Continent, 107–108, 116.

sent into exile beyond the Urals. One of these exiles, Philipp Johann von Strahlenberg, produced a beautiful map of Central Asia.[15]

Of most immediate importance for an understanding of French geographic understanding of Asia and North America was the participation of Joseph-Nicholas Delisle and Louis Delisle de La Croyère in Russia's cartographic and exploratory efforts. Peter the Great had met Guillaume Delisle in France in 1717, and the great scholar had encouraged Peter to investigate the question of Asia's geographic relation to America. Peter had hoped Guillaume would journey to Russia, but it was other Delisles who ended up making the trip. They were no mere tourists. Joseph-Nicholas helped direct the effort to compile a general map of the Russian dominions. He trained surveyors and cartographers, assisted with expedition planning, drew maps based on the data these expeditions were generating, and irritated Russian surveyors with his requests for "new astronomical observations to verify or correct Russian mapping." On Bering's second expedition, Joseph-Nicholas's brother Louis was "responsible for all the astronomical, physical and other observations" and was charged "to advise and assist" Bering "on all matters pertaining to Your Imperial Majesty's interests and on any needs Bering himself may have." Louis died of scurvy on the expedition, in part because of efforts wasted trying to locate islands appearing on one of his brother's North Pacific maps. It was Joseph-Nicholas's charts that sent Müller and his comrades on a fruitless search for Gama Land.[16]

While working for Russia's rulers, Joseph-Nicholas was also reporting to French officials. Specifically, he kept minister of marine Maurepas apprised of Russian discoveries. Delisle was, in fact, only allowed to accept the position outside of France on the condition that he use it to acquire information of value to his home country. So Delisle gathered all the Russian maps and reports he could and sent copies of them, or memoirs and maps based on them, to France. Scholarly estimates have him copying and sending hundreds of Russian maps, accompanied by boxes of Russian travel journals, reports, and books. The results of Russian mapping efforts included, therefore, not only the production of

15. J.-N. Delisle to the Abbé Bignon, May 25, 1729, in H. Omont, "Lettres de J.-N. Delisle au comte de Maurepas et à l'abbé Bignon sur ses travaux géographiques en Russie (1726–1730)," in *Bulletin de la section de géographie*, XXXII (1917), 148–154, esp. 152; Peter C. Perdue, *China Marches West: The Qing Conquest of Central Eurasia* (Cambridge, Mass., 2005), 450–453.

16. Quotation about Delisle's requests in Müller, *Bering's Voyages*, ed. and trans. Urness, 44. See also Albert Isnard, "Joseph-Nicolas Delisle: Sa biographie et sa collection de cartes géographiques à la Bibliothèque nationale," *Bulletin de la section de géographie*, XXX (1915), 39–41, 62–63, 79–80. Louis's tasks from "Instructions from Empress Anna Ivanovna to Vitus Bering, for His Second Kamchatka Expedition, as Prepared by the Admiralty College and the Senate," Dec. 28, 1732, in Dmytryshyn, Crownhart-Vaughan, and Vaughan, eds. and trans., *To Siberia and Russian America*, II, 108–125, esp. 122–123.

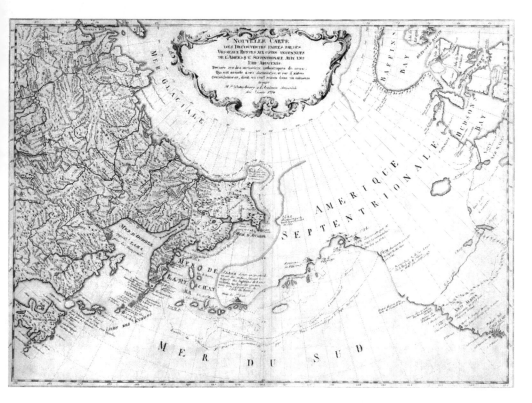

MAP 26. Gerhard Friedrich Müller, *Nouvelle carte des decouvertes faites par des vaisseaux russes aux côtes inconnues de l'Amerique septentrionale avec les pais adjacents.* 1754.
Courtesy of the John Carter Brown Library at Brown University

the 1734 *Atlas Vserossiiskoi imperii* and the 1745 *Atlas rossiiskoi*, works offering the first comprehensive view of Russia's Eurasian domains, but also the possession in France of new and revised maps and reports of Russia and Siberia. France had no claims to Siberia, neither France nor its colonies bordered on Siberia, and Siberia was larger and in many respects more forbidding to explorers than western North America. Nevertheless, through the involvement of French personnel in Russia's cartographic and exploratory efforts, eighteenth-century French geographers were able to rapidly enhance their cognizance of this expanse of Asian territory.[17]

17. On Delisle's information-gathering, see the letters reprinted in Omont, "Lettres de J.-N. Delisle au comte de Maurepas," *Bulletin de la section de géographie*, III (1915), 131–164; Müller, *Bering's Voyages*, ed. and trans. Urness, 20, 44–45, 157n. II97; Isnard, "Joseph-Nicolas Delisle," *Bulletin de la section de géographie*, XXX (1915), 44, 83–161; Broc, *La géographie des philosophes*, 30, 159–160; Marie-Anne Chabin, "Moscovie ou Russie? Regard de Joseph-Nicolas Delisle et des savants

They were trying to do the same for the North Pacific and North America. Delisle spoke to Bering after the Dane's first expedition and sent Maurepas a description of it in 1730. In fact, French involvement with Russian exploration was one way French officials could obtain the fruits of maritime Pacific investigation without violating treaty provisions by dispatching French ships into the Spanish Lake. More broadly, as the Russian Empire was trying to find its Peru east of the Urals, French investigators were trying to use the Russian Empire for the North Pacific and western North America as the Spanish had used the Inca Empire for western South America. They were trying to have another empire "discover" space to them.[18]

The fullest expression of the French effort to gain from Russian exploration came after Delisle's 1747 return from the Saint Petersburg Academy of Sciences to France—a move not, perhaps, entirely unrelated to Delisle's continuing to transmit information regarding Russian exploration even after the Russian government had moved toward a policy of secrecy concerning its North Pacific activities. In France, Delisle collaborated with cartographer Phillippe Buache (Buache was Guillaume Delisle's son-in-law and held the titles of *premier géographe du roi* and *géographe adjoint* to the Académie royale des sciences) on a new interpretation of northwestern American geography. In devising their notions, Delisle and Buache made use of Delisle's familiarity with the recent Russian reconnaissance, accounts of 1740s British exploration, theories passed down from the illustrious Guillaume Delisle, and what was later shown to be an apocryphal account of the 1640 expedition of a Spanish Admiral Bartholomew Fonte into a maze of islands, channels, lakes, and rivers in the North Pacific and northwest North America. The Fonte document claimed that, in the course of his explorations of northwestern American waterways, Fonte came upon a "Ship . . . of New England, from a Town called Boston." Logically, this meeting could only have occurred if some kind of water passage connected the Atlantic and Pacific Oceans from which the ships came. Delisle and Buache felt that they had identified consistencies between the account of Fonte's voyage and the reports of Bering's discoveries ["... la relation de l'Amiral de Fonte. Nous l'avons com-

français sur les États de Pierre le Grand," *Dix-huitième siècle: Revue annuelle,* XXVIII (1996), 43–56, esp. 48–50; Lincoln, *Conquest of a Continent,* 116; Shaw, "Geographical Practice and Its Significance in Peter the Great's Russia," *JHG,* XXII (1996), 160–176, esp. 165, 168–171; Dawson, *L'atelier Delisle,* 61–68.

18. See "Lettres de J.-N. Delisle au comte de Maurepas," *Bulletin de la section de géographie,* III (1915), 131–164; Müller, *Bering's Voyages,* ed. and trans. Urness, 20, 44–45, 157n. II97; Isnard, "Joseph-Nicolas Delisle," *Bulletin de la section de géographie,* XXX (1915), 44, 83–161; Broc, *La géographie des philosophes,* 30, 159–160; Chabin, "Moscovie ou Russie?" *Dix-huitième siècle,* XXVIII (1996), 48–50; Dawson, *L'atelier Delisle,* 61–68.

parée avec la route de mon frere et les autres connoissances que j'avois tirées de la Russie, et nous y avons trouvé un si grande conformité que cela nous a surpris"].[19]

In April 1750, they presented their ideas to a meeting of the Académie des sciences. Delisle's essay posited a navigable route from Hudson Bay to the Pacific, and Buache and Delisle's maps offered visions of it and the surrounding lands, some of eighteenth-century cartography's more arresting and fanciful images of North America. Some French geographers, impressed by the presenters' familial connections and official status and by the specious appeal of their evidence, initially accepted Delisle and Buache's views. But many others greeted their hypotheses with skepticism or even hostility. Delisle and Buache's claims were not based on the kind of precise, rigorous, verifiable, and extensive observations available for parts of Russia and France. Fonte's voyage had never happened, British ships had not sailed west of Hudson Bay, and Bering's expedition had not even landed on the North American mainland, much less surveyed it. Buache and Delisle's views of the Northwest would disintegrate in succeeding years under the weight of careful examination and new exploration. Their images of the West were appealing in some respects, but they were not durable. Russian North Pacific voyages had not provided a stable foundation for French North American policy.[20]

China

Though the results for western North America and the North Pacific were decidedly mixed, French participation in Russian mapping efforts shows how quickly French geographers could comprehend a huge territory under the sway of an at least semi-European empire. The French Jesuits' experience in eighteenth-century China furnishes the most striking instance of French scholars' abilities

19. Müller, *Bering's Voyages,* ed. and trans. Urness, 45; Numa Broc, "Un géographe dans son siècle: Philippe Buache (1700–1773)," *Dix-huitième siècle,* III (1971), 223–235, esp. 224, 227–228, 232; French quotation from Joseph-Nicolas de L'Isle, *Explication de la carte des nouvelles découvertes au Nord de la mer du Sud* (Paris, 1752), 10; quotation of Fonte from "The Voyage of Bartholomew de Fonte (1708)," in Glyn[dwr] Williams, *Voyages of Delusion: The Northwest Passage in the Age of Reason* (London, 2002), 417–422, esp. 421.

20. Good discussions of the controversy ignited by Delisle, Buache, and the Fonte letter can be found ibid., 133–136, 248–268; Lucie Lagarde, "Le passage du Nord-Ouest et la mer de l'Ouest dans la cartographie française du 18e siècle, contribution à l'etude de l'oeuvre des Delisle et Buache," *Imago mundi,* XLI (1989), 19–43, esp. 31–38; L. Breitfuss, "Early Maps of North-Eastern Asia and of the Lands around the North Pacific: Controversy between G. F. Müller and N. Delisle," *Imago mundi,* III (1939), 87–99; Henry R. Wagner, "Apocryphal Voyages to the Northwest Coast of America," American Antiquarian Society, *Proceedings,* n.s., XLI (1931), 179–234, esp. 205–215.

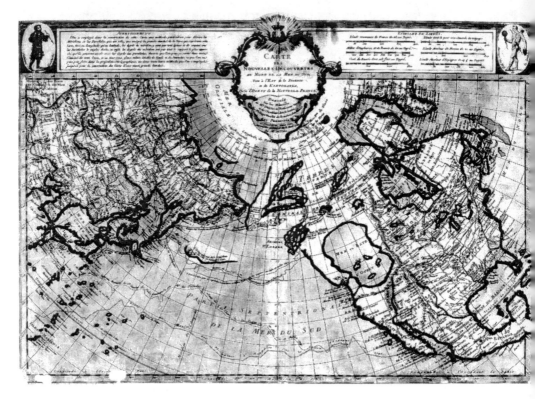

MAP 27. Joseph-Nicolas Delisle and Philippe Buache, *Carte des nouvelles découvertes au nord de la mer du Sud*. 1752. Library and Archives Canada, NMC 18579

to assimilate and represent material from and for an area inhabited by non-European peoples and without a European political overlay.[21]

Jesuits, mainly from Italy and Portugal, had been present in China since the early 1580s and had been involved in cartographic activities there from the start. As Theodore Foss has observed, the Jesuits, seeking converts in China to a reli-

21. The early-eighteenth-century Jesuit mapping of China provides a good example of a difficult and important subject rendered accessible to the nonspecialist by superb scholarship. I have found three secondary works especially useful. Theodore N. Foss offers a precise overview of the subject in "A Western Interpretation of China: Jesuit Cartography," in Charles E. Ronan and Bonnie B. C. Oh, eds., *East Meets West: The Jesuits in China, 1582–1773* (Chicago, 1988), 209–251. Perdue places the Jesuit project within the larger Chinese imperial context in *China Marches West*. Perhaps most important for this book, Laura Hostetler offers a highly suggestive interpretation of Chinese imperial mapping in *Qing Colonial Enterprise: Ethnography and Cartography in Early Modern China* (Chicago, 2001). Du Halde's preface to the first volume of his 1735 *Description géographique, historique, chronologique, politique, et physique de l'empire de la Chine et de la Tartarie chinoise*, 4 vols. (Paris, [1735]), provides a primary-source description of the Jesuit effort, especially useful when his Jesuito-centrism is balanced by the information in Foss, Perdue, and Hostetler.

gion based in Rome, struggling to surmount linguistic barriers, and feeling the need to explain their distant origins, saw the value of producing pictorial representations of the geographic relation between China and other parts of the world. At the same time, the Jesuits also wanted to know their way around the territories in which they were working and to satisfy European curiosity about the Far East; so they drafted maps of China for European eyes. By 1700, Jesuit surveyors had established the longitude and latitude of most of China's major urban centers, and Jesuit maps and other forms of information had already begun to exhibit and produce the kinds of cartographic changes evident in the case of France.[22]

The presence of the French Jesuits as a distinct group began in 1687 with the arrival of the *Mission française*. French cartographers' interest in China was one of the considerations leading Louis XIV to dispatch the mission. With an eye to the progress of French global cartographic efforts, Jean-Dominique Cassini hoped that mathematically and scientifically trained Jesuits could acquire Chinese astronomical and geographic data and transmit it back to France. Many of these missionaries were selected, in fact, specifically because of their considerable cartographic skills. As Foss has observed, the same French Jesuit schools teaching geography to the young Delisles also educated the missionaries sent to China.[23]

Although the French Jesuits carried cartographic interests and skills with them to China, it was the aspirations and concerns of the Kangxi emperor (1662–1722) that gave these outsiders the opportunity to conduct their famous survey. Scholars such as Laura Hostetler and Peter C. Perdue have emphasized the extent to which the common challenges faced by the expanding and consolidating French, Russian, and Chinese empires generated simultaneous and similar interests in the latest cartographic and surveying techniques. Like Russia in Eurasia and France in Europe and the Americas, China in the late seventeenth and the eighteenth centuries was adding large stretches of territory and vast numbers of people to its dominions. Like the rulers of Russia and France, those of China needed to know who and where their subjects were, how best to move armies and goods, where to fortify, and what to seize. Like their Russian and French counterparts, Chinese rulers wanted to tax their subjects efficiently and to establish their realm's borders securely. And like Louis XIV and Peter the Great, the Kangxi emperor saw that new maps drawn in accordance with the techniques developed in France could help him to govern. He also saw that,

22. Foss, "Western Interpretation of China," in Ronan and Oh, eds., *East Meets West,* 210–222.
23. Ibid., 219–222, 236.

with Russians lurking to the north, the Portuguese ensconced in Macao, and the Dutch only recently driven from Taiwan, maps clearly conveying Chinese territorial claims to European audiences could serve as valuable diplomatic tools.[24]

It was not that the Chinese government lacked access to indigenous cartographic skill. By 1700, the Chinese imperial use of maps for military and administrative ends seems to have extended back at least 1,850 years, and, though not based on astronomical or trigonometrical measures of distance, many Chinese maps drawn using traditional techniques were quite accurate in their representations of spatial relationships. Nonetheless, the Kangxi emperor wanted more. Existing Chinese maps might often have been quite accurate, but that does not mean that a ruler increasingly familiar with the surveys being undertaken in France deemed them reliably or sufficiently accurate. They were not drawn to the same scale, often required explanatory texts for interpretation, were less detailed and comprehensive in outlying and recently acquired regions, and could not be conveniently collated to meet "the Kangxi emperor's desire for 'a precise map which would unite all the parts of his empire in one glance.'" The emperor hoped the Jesuits might do for him what Cassini and his colleagues were trying to do for the kings of France. In 1700, the emperor asked for a map of the region around Beijing; in 1708, of a portion of the Great Wall. Satisfied with the value of the Jesuit techniques employed in these preliminary efforts, he then requested a comprehensive effort to survey and map the Chinese Empire using the most up-to-date western European methods.[25]

The cartographers and surveyors of the Mission française fanned out across the empire, applying the same techniques already in use in France and soon to be used in Russia. They employed triangulation extensively, checking their results against the locations of points fixed astronomically. As one might expect in light of the great size of the Chinese Empire, the short duration of the survey, and the small number of Jesuits involved (in 1717, the French mission numbered only twenty-eight), the French received a great deal of help. The role of the emperor himself was fundamental. He ordered local officials to cooperate, to open their records to the Jesuits and provide them with supplies. The Jesuits interrogated mandarins and local officials and pored over local maps and his-

24. Hostetler, *Qing Colonial Enterprise*, 4, 36, 70–74, 79–80; Foss, "Western Interpretation of China," in Ronan and Oh, eds., *East Meets West*, 222–223; Perdue, *China Marches West*, 443–449; Broc, *La géographie des philosophes*, 146. Castelnau-L'Estoile and Regourd offer a nice introduction to issues of empire and knowledge in *Connaissances et pouvoirs*.

25. The quotation is from Hostetler, *Qing Colonial Enterprise*, 4 (see also 70–74 and 79–80); and see Foss, "Western Interpretation of China," in Ronan and Oh, eds., *East Meets West*, 210–216, 222–224, 236–240; Perdue, *China Marches West*, 445–449; Du Halde, *Description géographique*, I, xxviii–xxx; Broc, *La géographie des philosophes*, 146.

tories. They were thus relying not only on techniques developed in Europe but also on the collective geographic understanding of China. Especially in the case of frontier zones of the Chinese Empire, the Jesuits also made use of Manchu, Mongol, and even Russian maps. The availability of a variety of sources of geographic information was crucial because the Jesuits did not actually visit all of the areas they mapped. For Tibet, for example, they had to rely on detailed accounts of the region and on the work of "Tartares" trained in mathematics, sent by the emperor, and instructed by the Jesuits in the necessary surveying techniques. In the case of Korea, the Jesuits made use of an existing map after checking its representation of the border region the Jesuits had been able to observe for themselves.[26]

It was evident in the 1720s and 1730s, and it remains apparent in retrospect, that the Jesuit mapping of China was a stupendous achievement. Father Jean-Baptiste Regis claimed justifiably that it was the greatest systematic geographic project ever undertaken. Taken together, the maps in the Jesuit atlas of China were wider in their geographic coverage, more systematic in the methods on which they were based, and more accurate than any maps of China yet produced. Drawn to the same scale using the same methods, the maps of individual regions could be combined to form a single view. Most important for a study of the development of French geographic thought, their circulation did not remain confined to China. In 1725, Louis XV received a copy of the atlas from his Jesuit confessor, and in the ten years that followed, cartographer Jean-Baptiste Bourguignon d'Anville edited forty-two maps of China for inclusion in Du Halde's 1735 *Description géographique, historique, chronologique, politique, et physique de l'empire de la Chine.* Despite the Chinese Empire's size, despite the cultural and linguistic challenges the French Jesuits had to overcome, the precision, detail, and extent of the geographic understanding of China available to French readers had increased enormously in the first four decades of the eighteenth century.[27]

26. Father Jean-Baptiste Regis, quoted in Du Halde, *Description géographique,* I, xxviii–xxx, xxxv–xl, esp. xxxv: "On a examiné les Cartes et les Histoires que chaque Ville garde dans ses Tribunaux; on a interrogé les Mandarins et leurs Officiers, aussi bien que les Chefs des Peuples dont on a parcouru les terres" (for "Tartares" quotation, see [xlvii]); [Antoine Gaubil] to Laurent Lange, May 15, 1732, Peking, in Gaubil, *Correspondance de Pékin, 1722–1759* (Geneva, 1970), 301–303, esp. 302; Foss, "Western Interpretation of China," in Ronan and Oh, eds., *East Meets West,* 224–233, 239; Hostetler, *Qing Colonial Enterprise,* 63, 65; Perdue, *China Marches West,* 449.

27. Regis's opinion quoted in Du Halde, *Description géographique,* I, xl: "Le plus grand Ouvrage de Géographie, qu'on ait jamais fait en suivant les régles de l'Art"; see also Broc, *La géographie des philosophes,* 147. On combining the maps for a single view, see Du Halde, *Description géographique,* I, xxlvii: "Toutes ces Cartes, tant de la Chine et de la Tartarie, que de la Corée et du Thibet, ont été mises non-seulement au même point, mais même sous une projection générale, comme si toutes les pieces n'en devoient composer qu'une seule, et effectivement on pourra les rassembler toutes,

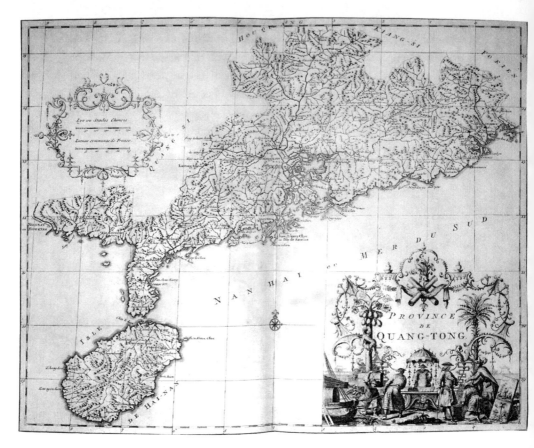

MAP 28. Jean-Baptiste Bourguignon d'Anville, "Province de Quang-tong,"
from *Nouvel atlas de la Chine, de la Tartarie chinoise, et du Thibet* (La Haye, 1737).
Geography and Map Division, Library of Congress, g7823g ct000750

Information gleaned from the Jesuit presence in China could even pertain to issues of North American exploration. Jesuit Father Antoine Gaubil, author of an important set of letters from Beijing (1722–1759), was specifically instructed by Louis XV and the duc d'Orléans to use missionary service in China to contribute to the progress of geographic science. Gaubil conducted a Beijing–Saint

et n'en faire qu'un seul morceau." On d'Anville, Du Halde, and Louis XV, see Du Halde, *Description géographique*, I, xlviii; Gaubil to Lange, Peking, May 15, 1732, in Gaubil, *Correspondance de Pékin*, 302; Foss, "Western Interpretation of China," in Ronan and Oh, eds., *East Meets West*, 234, 236, 240; Numa Broc, "Voyageurs français en Chine: Impressions et jugements," *Dix-huitième siècle*, XXII (1990), 39–49, esp. 45. Perdue rightly points out that the price of the cartographic standardization that made the Jesuit atlas so impressive was the exclusion of much of the culturally specific local detail and many of the territorial counterclaims that appeared on other maps of the Chinese Empire and its neighbors (*China Marches West*, 453).

Petersburg correspondence with Joseph-Nicholas Delisle, sought to acquire information about the geography of "Tartarie," and transmitted books back to Europe. Perhaps most interesting, in 1755, he used his familiarity with Chinese texts and history to refute the claims advanced by French scholar Joseph de Guignes—and entertained by Gaubil's correspondent Joseph-Nicholas Delisle—that Chinese Buddhist priests had journeyed to "Californie" (le "pays de Fou sang") in A.D. 458. Events did not turn out that way, but if, in fact, venerable Chinese documents had evinced a knowledge of mysterious North Pacific waters and North American lands, Gaubil could have reported this information to Delisle and France. And Gaubil, like Delisle in Russia, could have done so without the services of the kind of French Pacific naval expeditions precluded by the Utrecht settlement. China might have been for France, even more than the Russian Academy of Sciences in Saint Petersburg, a window on the West.[28]

The cases of France, Russia, and China show what eighteenth-century French geographers and cartographers could accomplish where circumstances were favorable. They could create new, accurate, comprehensive, and comprehensible maps covering large areas of difficult terrain. They could work productively with counterparts and trainees from different and distant empires and cultures. They could impress French, Russian, and Chinese eyes. A French minister sitting in a Parisian bureau could see large parts of Europe and Asia in 1745 in ways quite impossible in 1700. But western North America, though it had turned out to be close to Asia, had not turned out to be Asia.

28. Gaubil, *Correspondance de Pékin*, 9, 373–374, 486, 824–826, 830.

7

COMMUNICATION AND
INTERPRETATION

Eighteenth-century French surveyors triangulated their way across France and China. A French geographer trod the forests of Siberia and sailed the waters of the North Pacific. His brother collected in Saint Petersburg and dispatched to Paris maps of an empire spanning the world's largest continent. French cartographers produced maps of stunning clarity and precision revealing not only the heart of Gaul but also the reaches of "Tartary." In North America, in contrast, travel-weary explorers found their westward progress checked short of the South Sea; and, in France, bewigged cartographers pondered western rivers and inland seas whose existence they suspected and imagined but could not confirm. French achievements in one hemisphere call for an explanation of French disappointments in the other. What about the American West hindered the reconnoitering and representational efforts succeeding so brilliantly elsewhere?

One way to respond to this question is by analyzing explorers' reports from the long eighteenth century, using those from early decades to establish the factors hindering French exploratory endeavors between 1713 and 1763 and examining those from later years for corroborating evidence or contrary indications. Inferences from these primary sources can be supplemented by and checked against recent scholarly works covering pertinent subjects such as Indian cartography, the western Indian slave trade, and the spread of smallpox.

Examination of such materials suggests that a basic difficulty for French western reconnaissance arose from the unreliability of the human sources of geographic information, whether French explorers eager to find the region's rumored treasures or Indians transmitting or withholding information in keeping with their own purposes. Even when such sources could be credited, they were hard to understand. Contributing to communicative difficulties between Indians and explorers were both the need to navigate the West's many languages and the very different ways in which Indian and European peoples of the region conceptualized and represented western geography.[1]

Intentions, Deceptions, and Desires

A first issue is the credibility of the French explorers and western Indians providing geographic information. Explorers, like the broader category of travelers, enjoyed a reputation for telling tales partaking more of fantasy than reality, like the stories of Lilliputs and El Dorados satirized in *Gulliver's Travels* (1726) and *Candide* (1759). Over the course of the eighteenth century, explorers' accounts and the maps and conjectures resting upon them would be subjected to increasing and unfavorable scholarly and official scrutiny. The results would be manifold, influencing French territorial choices at the end of the Seven Years' War and inducing later eighteenth-century explorers such as James Cook and George Vancouver to invest their formidable acumen and energies in demonstrating the fantasticality of some of their predecessors' accounts. Already, in the first half of the eighteenth century, crucial characteristics of western exploration, explorers, and Indians made appraising the available descriptions of the region difficult.

A basic problem for those trying to assess reports of western exploration arose from the attitude many French explorers and geographers brought to their investigations. As historians of exploration such as Bernard DeVoto and Numa Broc have long emphasized, and as the Spanish scouts discussed in earlier chapters showed, wishful thinking often led explorers and geographers to hear what they wanted from their sources and to believe that western realities would conform to European desires and western lands would easily yield their secrets. Eighteenth-century French explorers and the officials and savants who evaluated their prospects often betrayed a confidence in the ease and products of discovery that hard-earned experience would fail to justify. La Vérendrye provided a good example when he claimed in his 1730 report that his new Indian information enabled him to speak with certainty about western waterways. He assured Governor Beauharnois that, if given "instructions to go and establish a fort at Lake Winnipeg," he would "have the honour in the second year thereafter to give" the governor "positive information respecting the Sea" of the West. Reflection upon the experiences of even the most accomplished French explorers such as Cartier, Champlain, and the sieur de La Salle should have suggested that

1. For an excellent overview of French imperial communications more generally, see Kenneth J. Banks, *Chasing Empire across the Sea: Communications and the State in the French Atlantic, 1713–1763* (Montreal, [2002]).

such easy confidence was premature, but visions of what lay beyond the horizon could enchant nonetheless. Over the next two decades, western exploration's challenges would cost La Vérendrye one son and many *"certitudes."* [2]

While La Vérendrye stood on the optimistic end of the spectrum of French explorers, other, more discerning figures shared some of his sentiments. Father Pierre-François-Xavier de Charlevoix, who had himself traveled extensively in France's North American territories in the early 1720s and gathered much information about the West, doubted the reliability of La Vérendrye's new Indian information about the Western Sea. Nevertheless, Charlevoix favored a search for the rumored American Mediterranean. It was possible, Charlevoix accepted, that the Western Sea was "so distant and the road thither so impracticable, that the discovery would be of no advantage to us. On the other hand," he asserted, "it may also be comparatively near and easy to reach . . . and . . . besides, in our search for it that may happen which has often happened in like circumstances, namely, that, in searching for what we are not destined to find, we may find what we were not looking for and what would be quite as advantageous to us as the object of our search." Enticed by the possibilities of the West, both the Pollyannish and the shrewd found reason to investigate beyond the horizon. [3]

This is not surprising. Although previous explorers' experience should have suggested caution, one might say the same about the track record of most gamblers. A French scout who discovered the West's imagined bounties, or who at least pointed the way to them, would have justified his efforts and suffering and might acquire material rewards and lasting renown as compensation for his travails. France — ideally, a grateful and generous France — would benefit, too. As discussed in Chapter 5, the eighteenth-century French Empire possessed two colonies on the North American mainland, and western geographic features would go a long way toward determining the value of these overseas territories. As those speculating about the West so often pointed out, the region must contain some kind of river draining into the sea. If that river were navigable, and if it led to wealthy Spanish or East Asian colonies on the Pacific or some inland extension of it, western exploration could transform the marginal colonies of Louisiana and New France into fountains of riches for the French Empire. The

2. Bernard DeVoto, *The Course of Empire* (Boston, 1952), 51–79; Numa Broc, *La géographie des philosophes: Géographes et voyageurs français au XVIIIe siècle* (Paris, [1975]), 153; "Continuation of the Report of the Sieur de La Vérendrye Touching upon the Discovery of the Western Sea," in Lawrence J. Burpee, ed., *Journals and Letters of Pierre Gaultier de Varennes de La Vérendrye and His Sons: With Correspondence between the Governors of Canada and the French Court, Touching the Search for the Western Sea*, trans. [W. D. LeSueur] (Toronto, 1927), 46–63, esp. 47, 63.

3. Charlevoix quoted in Pierre-François-Xavier de Charlevoix to comte de Maurepas, Aug. 8, 1737, in Burpee, ed., *Journals and Letters of La Vérendrye*, trans. [LeSueur], 73–81, esp. 75, 80.

often-realized danger was that the appeal of these kinds of hopes for the West could distort interpretations of evidence pertaining to the region; that French desire would lead explorers, geographers, and officials to deceive themselves into understating the difficulties and exaggerating the remunerations of western investigations.[4]

Self-deception constituted only one pitfall of the exploratory enterprise. The possibility also existed that French explorers might not just delude themselves but also deliberately mislead French officials and the gullible readers of tall tales about far-off lands. The same considerations inducing an explorer or author to unconsciously interpret evidence favorably could lead the less scrupulous to do so consciously and creatively. Such was thought to have been the case with the reputedly exaggerated accounts of Louis Hennepin and the allegedly apocryphal western voyage of the baron de Lahontan. The fear of being duped seems to have left certain ministerial eyebrows permanently raised. Maurepas, who oversaw La Vérendrye's activities, was notoriously suspicious that he was more interested in fur trade profits than western exploration. Still, La Vérendrye and his sons' accomplishments were considerable, and it may simply be that Maurepas under-estimated western exploration's difficulties. Nonetheless, the point raised by his distrust is an important one. Because the evidence for the eighteenth-century West is so scanty, an investigation of activities in the region has to make use of explorers' journals. The authors of these documents had more on their mind than leaving a transparent record of their activities for posterity, and the pos-sibility that they were strategically omitting, restating, exaggerating, and even fabricating cannot be dismissed out of hand.[5]

Finally, French officials were and later historians are further frustrated by the

4. On the need for a western river, see marquis de Beauharnois to Maurepas, Oct. 15, 1730, ibid., 63–66, esp. 65.

5. Lahontan's account of a western journey has long been viewed with suspicion, and I initially shared this sentiment. A June 11, 2006, talk by Peter Wood at the Omohundro Institute of Early American History and Culture Conference in Quebec City persuaded me to take Lahontan's report more seriously. Wood argued that Lahontan's famous Letter XVI describes, in fact, an actual 1688 journey up the Missouri River. Réal Ouellet and Alain Beaulieu, the editors of the best edition of Lahontan's writings, argue for a trip up the Minnesota River instead. I incline toward the Missouri argument, but the matter remains unresolved. Because, moreover, of lingering doubts about the reliability of specific details in Lahontan, I employ him as a supplemental source rather than as the foundation for arguments. See [Louis Armand de Lom d'Arce, baron de] Lahontan, *Oeuvres com-plètes*, ed. Réal Ouellet and Alain Beaulieu, 2 vols. (Montreal, 1990), I, 392–423. For a recent case for the evidentiary value of Hennepin, see Catherine Broué, "En filigrane des récits du Père Louis Hen-nepin 'trous noirs' de l'exportation louisianaise, 1679–1681," *Revue d'histoire de l'Amérique française*, LIII (2000), 339–366. For Maurepas's suspicions, see Maurepas to marquis de Beauharnois, Apr. 22, 1737, Maurepas to comte de La Galissonnière, Mar. 1, 1748, and "Report of La Jonquière," all in Burpee, ed., *Journals and Letters of La Vérendrye*, trans. [LeSueur], 269–270, 471–472, 481–482.

more intractable difficulty of the illegal and therefore silent trader. La Vérendrye might or might not have been more interested in dealing in furs and slaves than advancing French exploration, but he was at least seeking official approval for his presence in the West and transmitting reports of his activities to the French government. Far to the west of New Orleans and Montreal, other French traders were likely operating without any official sanction, and indeed in violation of governmental edicts. It was not in the interest of these enterprising and illicit traders or of the often conniving agents of French imperial authority in North America to call attention to their unauthorized conduct. Those illegally ascending the Saskatchewan or trudging toward New Mexico were not sending reports detailing their itineraries to the minister of the navy. The best evidence of these unauthorized activities sometimes comes from non-French sources, like the reports of Spanish officials in New Mexico. For the most part, though, evidence of illegal conduct is rare, and we can only speculate about the geographic information gleaned by illicit traders and withheld from metropolitan officials.

The various possibilities for deception and omission were not unique to French exploratory efforts in the North American West, but they figured more prominently there than in French cartographic efforts in the Russian and Chinese empires. This was in part because the stakes were different. As we have seen, France certainly had interests in China and Russia. The French Jesuits would very much like to have converted the Chinese Empire to Christianity, and French diplomats recognized the utility of a Russian military alliance. Where the success of French imperial policy regarding North America might depend on the character of western geography, however, the achievement of French objectives in Russia and China hinged on factors other than the results of trigonometric surveys of Tobolsk or Yunnan. The incentives to misinterpret or misrepresent the results of geographic investigation were fewer. In contrast, when the results of Russian exploration or the contents of Chinese records seemed to touch on the geography of northwestern North America and thus to relate more directly to French imperial interests, Joseph-Nicholas Delisle would find himself tempted to use the tentative indications of such geographic evidence as the basis for the most fanciful speculations.

More significant, the nature of the cartographic projects in France, Russia, and China militated against the dangers of witting and unwitting deception present in North America. Surveyors tended to work in teams, as parts of a hierarchical organization such as the French mission or the Russian government. They kept an eye on each other as well as on the landscape. Their methodical and repeated taking, recording, and verifying of their own and each other's measurements of position and distance left less room for personal interpretation than did the more improvisational efforts of French explorers to ascertain the course of that big river they seemed to be hearing about. More broadly, the eighteenth century's systematic surveys, with their striving for a kind of objective, verifiable representation, were trying to move beyond the kind of well-established but imprecise local knowledge that had led seventeenth-century cartographers to misplace the French coastline.

French explorers' rough western reconnaissance fell short of these new standards. Joseph-Nicholas Delisle drove his Russian colleagues to distraction with his demands that Russian maps be drawn to the same scale and with his complaints that Russian surveyors were showing too few of their calculations to allow verification of their data. The French explorers working in North America had not yet reached the point where these kinds of criticisms were even applicable. Most were writing of rough distance and direction from one point to another rather than providing latitude and longitude measurements. French scout and trader Louis Juchereau de Saint-Denis declared in 1717 that he had not recorded the latitude and longitude of places he had seen during his travels between Mexico and the Mississippi River because he did not know how to do so. Explorers who were endeavoring to provide such data might still be trying to figure out their equipment. In the journal of their 1742–1743 assay of parts of what are now the Dakotas and Wyoming, La Vérendrye's sons, Louis-Joseph and François, claimed they were unable to take latitudes at all because the ring of their astrolabe was broken. When the La Vérendryes were able to take the latitude of a Mandan village in December 1738, they appear to have been off by roughly a degree.[6]

The possibilities of deception and error sprang from more than French explorers' personalities and capacities. These scouts constituted only one of the sources of information about the North American West and were frequently not the origin of the ideas they were trying to convey. They were often deriving their

6. Albert Isnard, "Joseph-Nicolas Delisle: Sa biographie et sa collection de cartes géographiques à la Bibliothèque nationale," in *Bulletin de la section de géographie*, XXX (1915), 31–164, esp. 62; Gerhard Friedrich Müller, *Bering's Voyages: The Reports from Russia*, ed. and trans. Carol Urness (Fairbanks, Alaska, 1986), 44; Charmion Clair Shelby, [ed. and trans.], "St. Denis's Declaration concerning Texas in 1717," *SWHQ*, XXVI (1923), 165–183, esp. 172, 175; on errors in measurement, see Burpee's judgment in "Journal in the Form of a Letter Covering the Period from the 20th of July 1738, When I Left Michilimackinac, to May, 1739, Sent to the Marquis de Beauharnois," *Journals and Letters of La Vérendrye*, trans. [LeSueur], 290–361, esp. 346. For the broken astrolabe, see "Journal of the Expedition of the Chevalier de La Vérendrye and One of His Brothers to Reach the Western Sea, Addressed to M. the Marquis de Beauharnois, 1742–43," ibid., 406–432, esp. 427–428: "J'aurais fort souhaité de prendre hauteur à cet endroit [the site of modern Pierre, South Dakota]; mais notre astrolabe étoit, depuis le commencement de notre voyage, hors d'état de servir, l'anneau en étant cassé."

notions of western geography from the region's indigenous inhabitants. This multiplied the possibilities for misdirection. Just as French informants could delude their audiences, western Indians might have been deliberately misleading their French interlocutors, intentionally obstructing their efforts to move west or complaisantly confirming their preconceptions. Part of what makes this plausible is that it appears to have been a feature of the interaction between native peoples and European explorers from their first encounters. Already, on his first voyage, if his heavily edited logbook is to be believed, Columbus suspected that Indian captives were telling him tales of gold in hopes that the information would lead to their release. Eighteenth-century French explorers rarely acted like Columbus and his conquistador successors, but that does not preclude their being treated like them.

Lending credence to this possibility are the glimpses French sources offer into the agendas of western Indian nations. In 1724, in what is now Kansas, Étienne Véniard de Bourgmont conferred with "the head chief of the Padoucas [Plains Apaches]." The chief showed Bourgmont "about 800 warriors" and claimed that he led four times that many. When he then criticized the stinginess of Spanish trade policy, declared, "I am the emperor of all the Padoucas, and they go neither to war nor to the Spaniards without my permission," and added, "If you should ever need 2,000 warriors, you have only to ask," he was implicitly informing Bourgmont where real power on the plains lay and putting him on notice that Padouca commercial wants would have to be met: those same 2,000 warriors could be directed against the French.[7]

Mandan conduct fourteen years later would show that the fear of this kind of Indian force on the plains could easily serve the cause of trickery. La Vérendrye's visit to the Mandan villages (near modern Bismarck) in 1738 depended on the assistance of Assiniboine companions willing to show him the route. His journal reports that Mandan concern about the quantity of provisions being consumed by the Assiniboines led the Mandans to spread a false report "that the Sioux were not far away." The Assiniboines fled rather than remained to fight. Meanwhile, "a Mandan chief made a sign" to La Vérendrye, which the Frenchman interpreted as meaning that he should "wait" and that "the report about the Sioux was only to get the Assiniboin to go." If a Mandan chief could use fear of the Sioux to scare off Assiniboines, he, or another Indian leader—concerned, perhaps, about French traders' providing arms to enemy nations, treading on sacred ground, or depriving a well-situated people of its role as commercial middleman—could

7. "Journal of the Voyage of Monsieur de Bourgmont . . . to the Padoucas," in Frank Norall, *Bourgmont, Explorer of the Missouri, 1698–1725* (Lincoln, Neb., 1988), 125–162, esp. 155–159.

use the same tactic to frighten French explorers away from the routes to the West. None of this means that we can assume deliberate deception on the part of western Indians, but we must allow for that possibility. British sources suggest that Indian complaisance might have constituted another, less Machiavellian source of deception. James Isham and Andrew Graham, who served as factors in the Hudson's Bay Company's York and Churchill posts between the 1730s and 1770s, suspected their Indian interlocutors of telling them what they wanted to hear "to please the Factor" and gain importance as the bringers of "good news."[8]

French and Indian intentions and wishes constituted one difficulty for French investigators of western geography and continue to pose problems for historians trying to understand eighteenth-century events. The question of the extent to which sources could be believed recurred and recurs. Nonetheless, imperfect sources were what was available and, gullibly or grudgingly, eighteenth-century officials and geographers had to work with them. Even when the sources of information were deemed credible, however, challenges of interpreting them arose from the different ways French explorers and western Indians conceptualized and conveyed that information.

Language and Interpretation, Conceptions and Representations

The experience of asking directions is a reminder that even under the best of circumstances — when two people speak the same language and have been educated in similar ways of representing and navigating space — communicating geographic information is difficult. The best of circumstances did not prevail in the eighteenth-century North American West.

For French as for Spanish scouts, linguistic diversity was one problem. Among the Indian nations on the eastern side of the Rockies, Crees and Black-

8. La Vérendrye quotations in "Journal in the Form of a Letter," in Burpee, ed., *Journals and Letters of La Vérendrye*, trans. [LeSueur], 290–361, esp. 333. On Indian commercial motives for hindering European movement, see John C. Ewers, "The Indian Trade of the Upper Missouri before Lewis and Clark," in Ewers, *Indian Life on the Upper Missouri* (Norman, Okla., 1968), 14–33, esp. 30. Final quotation from Glyndwr Williams, ed., *Andrew Graham's Observations on Hudson's Bay, 1767–91* (London, 1969), 153. See also E. E. Rich, ed., *James Isham's Observations on Hudsons Bay, 1743; and Notes and Observations on a Book Entitled "A Voyage to Hudsons Bay in the Dobbs Galley, 1749"* (Toronto, 1949), 92. For later traders' suspicions that Indians were exaggerating the dangers of territories to come, see "François-Antoine Larocque's 'Yellowstone Journal,'" and "Charles McKenzie's Narrative," in W. Raymond Wood and Thomas D. Thiessen, eds., *Early Fur Trade on the Northern Plains: Canadian Traders among the Mandan and Hidatsa Indians, 1738–1818; The Narratives of John Macdonell, David Thompson, François-Antoine Larocque, and Charles McKenzie* (Norman, Okla., 1985), 156–220, 221–296, esp. 165, 168, 274, 276. Later events recounted in their journals suggest that Larocque and McKenzie's suspicions might have been unjust.

feet spoke Algonquian languages; Missouris, Mandans, Dakotas, Hidatsas, Kansas, and Iowas spoke Siouan languages; Shoshones and Comanches, a Uto-Aztecan language; Pawnees and Arikaras, Caddoan languages; Apaches, an Eyak-Athabaskan language.

Naturally, Frenchmen in the West frequently complained about their linguistic difficulties. Before a Sioux war band killed him in 1736, Jesuit missionary Father Jean-Pierre Aulneau was endeavoring to learn the Cree language but had to concede that he was not yet "very skillful at it." Two of the Frenchmen who accompanied La Vérendrye on his 1738 visit to the Mandan villages mentioned to him that when they asked their interlocutors "about one thing they replied about something else through failure to comprehend." For La Vérendrye himself during the same expedition, it was a disaster when his interpreter, "a Cree by nationality, who spoke good Assiniboin," "decamped . . . in order to follow an Assiniboin woman of whom he was enamoured." La Vérendrye's son spoke Cree, and several Mandans spoke Assiniboine, so La Vérendrye had been able to talk to his son in French, who could speak to the interpreter in Cree, who could speak to the Mandans in Assiniboine. Without the interpreter, La Vérendrye and his French companions were "reduced to trying to make ourselves understood by signs and gestures." Frenchmen and Indians would repeatedly have difficulty understanding what was being said to them.[9]

The more radical difficulty in the North American West arose not simply from the many and varied sounds and signs Indians and Frenchmen used to express themselves but from the different, culturally influenced ways in which the European and native American peoples of the region conceptualized, articulated, and represented physical and human geography. Evidence of these differ-

9. "Letter from Reverend Father Aulneau, of the Society of Jesus, to Reverend Father Bonïn," Fort Saint Charles, April 30, 1736, translated text in Reuben Gold Thwaites, ed., *The Jesuit Relations and Allied Documents: Travels and Explorations of the Jesuit Missionaries in New France, 1610–1791,* LXVIII, *Lower Canada, Crees, Louisiana, 1720–1736* (Cleveland, Ohio, 1900), 287–305, esp. 299 (French text on 298). For the quotation from La Vérendrye's companions, see "Journal in the Form of a Letter," in Burpee, ed., *Journals and Letters of La Vérendrye,* trans. [LeSueur], 290–361, esp. 345; for La Vérendrye himself, see ibid., 334. The utility of sign language on the northern plains should not be casually dismissed. Meriwether Lewis would later refer to it as "the common language of all the Aborigines of North America, . . . understood by all of them . . . [appearing] sufficiently copious to convey with a degree of certainty the outlines of what they wish to communicate." "The strong parts of the ideas are seldom mistaken," he stated, though he also acknowledged the language was "imperfect and liable to error," even if "much less so than would be expected." See Gary E. Moulton, ed., *The Journals of the Lewis and Clark Expedition,* V, *July 28–November 1, 1805* (Lincoln, Neb., 1988), 88, 196–197; "Charles Mckenzie's Narrative," in Wood and Thiessen, eds., *Early Fur Trade,* 274. Still, whatever Lewis's feelings, his French predecessors do not appear to have been satisfied with sign language, and their misunderstandings concerning rivers, inland seas, distance, direction, salinity, relief, and ethnicity suggest that its liability to misinterpretation was significant.

ing modes of forming and expressing geographic ideas is limited, and most of it was produced by often perplexed Europeans. This constrains efforts to comprehend the ways in which indigenous peoples understood western geography. One response to this difficulty is to make incomprehension the subject of inquiry: to use the confusions of European authors and evidence to illuminate the underlying communicative problems in the eighteenth-century West. A useful way to begin is by considering the different ways of categorizing the physical and human features of the region, the absence of agreed-upon bases for comparison that often generated these different classificatory schemes, and the resulting imprecision of understanding bedeviling the exchange of ideas.

Mountains provide a good starting point, especially in light of the West's topography and the eighteenth century's reliance on transport by boat, foot, and hoof. Just as a navigable river or inland sea could facilitate transcontinental movement, mountain ranges could check it. French explorers hoped for the former but were also picking up references to the latter. In 1721, Charlevoix wrote that a Missouri woman had confirmed what he had already learned from the Sioux: "that the Missouri rises from very high and bare mountains [Montagnes Pelées, fort hautes], behind which there is another large river, which probably rises from thence also and runs to the westward." His report raised the question of what, precisely, was meant by "very high mountains" and how an explorer or scholar should envisage them. To one familiar with the Rockies, the image of successive lines of jagged, snowcapped, heaven-scratching peaks might spring to mind. Eighteenth-century French geographers had others points of comparison. Jean-Baptiste Bénard de La Harpe, in describing terrain he encountered on his 1719 trip through what is now western Arkansas and southeast Oklahoma, wrote of "mountains that were very difficult because of the number of large overturned rocks that we encountered and because of the heights and descents that it was necessary to pass over." In their 1742–1743 journal, Louis-Joseph and François La Vérendrye described what were "probably the Black Hills" as "mountains . . . well wooded . . . and . . . very high." The Ozarks and the Black Hills constitute major terrain features, and they could loom large in the mind of an eighteenth-century French explorer, but they fall far short of the magnitude of the great mountain chains farther west.[10]

10. Charlevoix quotation from Pierre-François-Xavier de Charlevoix, *Journal of a Voyage to North America*, ed. Louise Phelps Kellogg, 2 vols. (1761; rpt. Chicago, 1923), II, 209; French text in P[ierre-François-Xavier] de Charlevoix, *Journal d'un voyage fait par ordre du roi dans l'Amerique septentrionale* (Paris, 1744), 396. Similarly, La Vérendrye's son Louis-Joseph reported that Cree Indians told him (probably in 1739) that the Saskatchewan River "came from very far, from a height of land where there were lofty mountains [montagnes fort hautes]; that they knew of a great lake on the other side of the mountains, the water of which was undrinkable" ("Abridged Memoran-

On the other hand, the Rockies themselves could register less than one might expect. Although many scholars find it likely that Hudson's Bay Company employee Anthony Henday approached the Rockies in 1754, his journal does not mention them. The 1750–1753 "Brief Report or Journal of the Expedition of Jacques Legardeur de Saint-Pierre" (Saint-Pierre was a military officer put in charge of the search for the Western Sea in 1750) speaks of Saint-Pierre's subordinate Joseph Boucher de Niverville ascending the Saskatchewan "as far as the Rock [sic] Mountains," but it remains unclear whether or not this trip actually took place and to what, exactly, the journal was referring. On a map drawn in 1801 by the Blackfoot chief Ac ko mok ki, the Rockies are represented by two long, parallel lines, symbols perfectly understandable to someone personally familiar with the mountains or able to hear Ac ko mok ki's explanation; but, for someone in less favorable circumstances for comprehension, less evocative than, say, a set of sharp triangles, twenty long and five deep. In part because of the ambiguity of representations such as Ac ko mok ki's, more than eighty years after Charlevoix's letter, Lewis and Clark would be surprised by the rigors of the

dum respecting the Map Which Represents the Establishments and Discoveries Made by the Sieur de La Vérendrye and His Sons" [1749], in Burpee, ed., *Journals and Letters of La Vérendrye*, trans. [LeSueur], 483–488, esp. 487). See also Jean-Baptiste Bénard de La Harpe, *The Historical Journal of the Establishment of the French in Louisiana*, ed. Glenn R. Conrad, trans. Joan Cain and Virginia Koenig (Lafayette, La., 1971), 174–175.

   Translated La Harpe quotation from Ralph A. Smith, ed. and trans., "Account of the Journey of Bénard de La Harpe: Discovery Made by Him of Several Nations Situated in the West," *SWHQ*, LXII (1958–1959), 75–86, 246–259, 371–385, 525–541, esp. 381–382. The French text given in La Harpe, "Relation du voyage de Bénard de La Harpe: Découverte faite par lui de plusieurs nations situées a l'Ouest," in Pierre Margry, ed., *Découvertes et établissements des Français dans l'ouest et dans le sud de l'Amérique septentrionale (1614–1754): Mémoires et documents originaux*, VI, *Exploration des affluents du Mississipi et découverte des montagnes Rocheuses (1679–1754)* (Paris, 1886), 243–306, esp. 282, reads, "assez difficiles à cause de la quantité de grosses pierres renversées que nous rencontrasmes, et des hauteurs et descentes qu'il falloit passer." The "assez" Smith translates as "very" might also be given as "rather" or "quite." In the context of the document, I find his "very" reasonable. A few pages after the passage in question, La Harpe describes mountains of the same region as "montagnes très difficiles à passer à cause de l'espaisseur de bois et des rochers bouleversés que l'on y rencontre, joint à cela que ce vallon va tousjours en montant (284)" [Smith: "mountains very difficult to pass, because of the denseness of the forest and of the overthrown rocks that are found there, in addition to the fact that this dale goes always upward" (383)]. Margry's texts are unreliable. I give them here as a supplement to other material and with the understanding that they must be treated with caution. For a more comprehensive account of French and Indian interactions in the Arkansas Valley, see Kathleen DuVal, *The Native Ground: Indians and Colonists in the Heart of the Continent* (Philadelphia, 2006). For the La Vérendryes on the Black Hills, see "Journal of the Expedition of the Chevalier de La Vérendrye and One of His Brothers," in Burpee, ed., *Journals and Letters of La Vérendrye*, trans. [LeSueur], 406–432, esp. 420: "Montagnes . . . bien boisés . . . et . . . fort hautes" (Black Hills identification from William H. Goetzmann and Glyndwr Williams, *The Atlas of North American Exploration: From the Norse Voyages to the Race to the Pole* [New York, 1992], 96–97). DeVoto mentions the issue of interpreting mountain height in *Course of Empire*, 578.

Rockies. Speaking generally, in the eighteenth-century West, in the absence of systematic, technical, and detailed descriptions of mountain characteristics such as relief, depth, and extension, considerable room for misunderstanding existed, and the dominant feature of western topography could elude European understanding after three centuries of European experience with the continent.[11]

Where mountains constituted a potential barrier to westward exploration, salt water formed one of its principal objectives. French geographers remained famously unsure of where North American land gave way to Pacific water and famously interested in the possibility that western territories might encompass an inland arm of the South Sea. This interest made reports of salt water inviting, but the nature of such reports made them misleading. Like descriptions of mountains, references to salty water created a problem of classification and assessment because of the lack of a clear standard of comparison. This question arose often. In 1738, Mandans told La Vérendrye that, lower down the Missouri, the "river . . . was very wide so that you could not see the land on the other side. The water was not drinkable." In the winter of 1738–1739, an Assiniboine spoke of journeying to a county where "you can't see the other side of the river; the water is salt; it is a mountainous country, with wide spaces between the mountains consisting of fine land." To a French explorer eager to reach a Western Sea or the Pacific itself, a fair inference would be that these references to a large, brackish body of water pointed to what he was seeking.[12]

11. On Henday, see Goetzmann and Williams, *Atlas of North American Exploration*, 110–111; Barbara Belyea, "Tracing Henday's Route," in Anthony Henday, *A Year Inland: The Journal of a Hudson's Bay Company Winterer*, ed. Belyea (Waterloo, Ontario, 2000), 325–342, esp. 328–337. Belyea is skeptical of the claim that Henday came within sight of the Rockies. The Saint-Pierre quotation is from "Brief Report or Journal of the Expedition of Jacques Legardeur de Saint-Pierre, Knight of the Royal and Military Order of Saint Louis, Captain of a Company of the Detached *Troupes de la Marine* in Canada, Assigned to Search for the Western Sea," in Joseph L. Peyser, ed. and trans., *Jacques Legardeur de Saint-Pierre: Officer, Gentleman, Entrepreneur* (East Lansing, Mich., 1996), 180–191, esp. 183 (see also 184, 189). For good discussions of Saint-Pierre's journal, see W. J. Eccles, "French Exploration in North America, 1700–1800," in John Logan Allen, ed., *North American Exploration*, II, *A Continent Defined* (Lincoln, Neb., 1997), 149–202, esp. 189–195; Antoine Champagne, *Les La Vérendrye et le poste de l'ouest* (Quebec, 1968), 407–420, 527–535; and Goetzmann and Williams, *Atlas of North American Exploration*, 105. Eccles (195) and Goetzman and Williams (105) quote Saint-Pierre's journal as referring to the "Montagnes des Roches." The copy of Saint-Pierre's journal in Margry, ed., *Découvertes et établissements des Français*, VI, 643, 650, and Champagne's quotation from the original (528) have "la montagne de Roche." Peyser and Champagne have done the most thorough work on Saint-Pierre, and I follow their translation and quotation. See also D. W. Moodie and Barry Kaye, "The Ac Ko Mok Ki Map," *The Beaver*, CCCVII (Spring 1977), 4–15. The vocabulary of western Indian languages might also have created problems of interpretation. Belyea notes, for example, that the Cree language does not make "a distinction between hill and mountain" (Belyea, "Tracing Henday's Route," in Henday, *A Year Inland*, ed. Belyea, 335).

12. "Journal in the Form of a Letter," in Burpee, ed., *Journals and Letters of La Vérendrye*, trans. [LeSueur], 336: "La rivière qui se trouvoit fort large, ne voyant point la terre dun bord a lautre, l'eau

Pays des Aricara & des
Padouka Blancs

Province de Quivire

Pays des Padouka Noirs

Prov. de la Louisianne

Nouveau   Prov de Lastek
Mexique

Californie

Mer de Californie

Illinois

R. du Mississipi

Golfe du Mexique

MAP 29. Louisiana and the
West. Drawn by Jim DeGrand,
after Bénard de La Harpe,
*Carte nouvelle de la partie de
l'ouest de la Louisiane.* 1723–1725.
For a full-color photographic
reproduction of the original
map at the Library of Congress,
see Derek Hayes, *America
Discovered: A Historical Atlas
of North American Exploration*
(Vancouver, 2004), 88–89

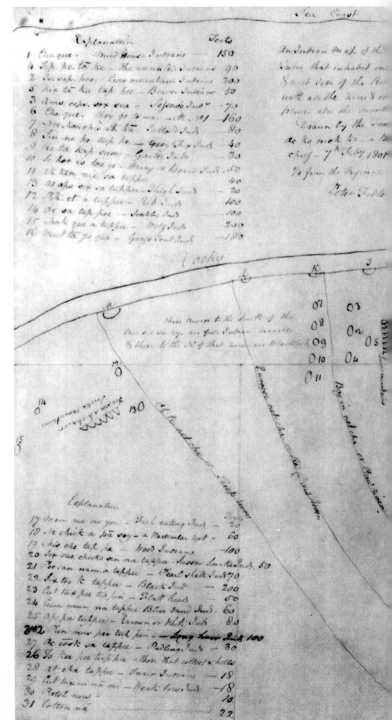

MAP 30. Ac ko mok ki map
of the West. c. 1801–1802.
Provincial Archives of
Manitoba, HBCA G. 1/25

Mountains

Explanation

A Devils head — or Smock ...
B King ...
C Road ...
D Hap ...
E Rain bow ...
G the Bell ...
H the Rattle ...
S the Heart ...
I Iron mountain ...
K Bad mountain ...
L Bull ...
M the Red ...

From B to M 33 Days walk ...

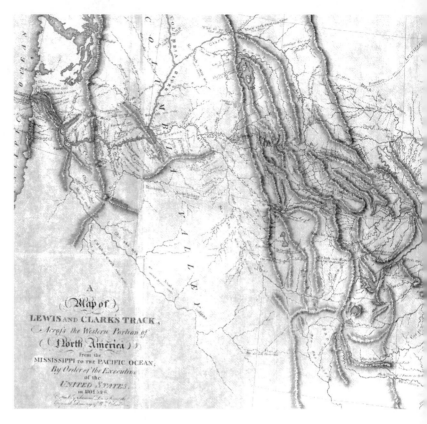

MAP 31. Samuel Lewis, after William Clark, *A Map of Lewis and Clark's Track*. 1810. Special Collections, University of Virginia Library

A fair inference, but not necessarily a correct one. Historians have long suspected that tales of the Great Salt Lake confused eager explorers. An even simpler possibility arises from the less extreme characteristics of other bodies of nonoceanic water. Rivers like the Rio Grande and the Pecos are large, and the volume of water flowing in them varies with the time of year. Moreover, seasonal evaporation, use of river water for irrigation, and local saline springs make the river systems of which they are a part susceptible to brackishness. Spanish explorers and authors noted the salinity of the Gila and Colorado rivers and sometimes referred to the Pecos River as the "Río Salado." Closer to La Vérendrye's sources of information, Charles Nolan Lamarque and La Vérendrye's son Louis-Joseph told him in 1739 that the water in the part of the Missouri flowing past other Mandan villages they had scouted was "not of the best quality for drinking, being rather brackish [un peu salée]. From the last mountain we have

---

mauvaise à boire"; 355: "On ne voye point l'autre cauté de la rivière leaux est salée c'est un païs de montagne, grande espace entre les montagnes de beau terein."

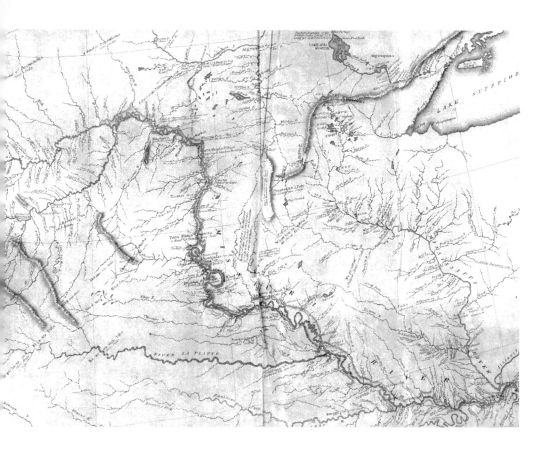

always found most of the marshes and pools saline and sulphurous [salées, ou soufrée]." If the standard of comparison was the ocean, a river might be considered fresh. If the standard was another river, it might fall into the salty category. Indian informants could have been reporting quite accurately what they experienced, only to have their ocean-obsessed French listeners draw the wrong conclusion.[13]

13. On salinity in the Rio Grande Basin, see United States Geological Survey, "Monitoring the Water Quality of the Nation's Large Rivers," http://water.usgs.gov/nasqan/docs/riogrndfact/riogrndfactsheet.html (accessed June 23, 2009). For a more general discussion of the confusion engendered by different conceptions of salinity, see G. Malcolm Lewis, "Native North Americans' Cosmological Ideas and Geographical Awareness: Their Representation and Influence on Early European Exploration and Geographical Knowledge," in Allen, ed., *North American Exploration*, I, *A New World Disclosed* (Lincoln, Neb., 1997), 71–126, esp. 103–104. On the salinity of rivers like the Gila, Colorado, and Pecos, see "Anza's Complete Diary, 1774," and "Díaz's Diary from Tubac to San Gabriel, 1774," in Herbert Eugene Bolton, ed. and trans., *Anza's California Expeditions*, II, *Opening a Land Route to California: Diaries of Anza, Díaz, Garcés, and Palóu* (Berkeley, Calif., 1930), 1–130, 245–290, esp. 45, 264; S. Lyman Tyler and H. Darrel Taylor, eds. and trans., "The Report of Fray Alonso de Posada in Relation to Quivira and Teguayo," *NMHR*, XXXIII (1958), 285–314, esp. 292;

The South Sea was one goal of western exploration; the kinds of precious metals the Spanish had found in mountains farther south, another. Like the description of mountains and waters, the classification of metals constituted a fertile source of confusion. In a manner reminiscent of Spanish scouts in the Southwest, La Vérendrye, when writing of nations in the region around the Mandans in 1738, and Cook, when describing the Indians he met at Nootka Sound forty years later, indicated that their interlocutors used the same word for "all metals," or at least "all white metal." For Europeans desperate to find silver and inclined to place it with gold in the category of precious metals rather than with iron in the category of white metals, these different categorizations could easily lead to befuddlement. French fur trader Nicolas Jérémie, in his 1720 account of Hudson Bay, claimed that Indian captives with whom he had spoken there talked of a nation farther west that lived next to "bearded men . . . not dressed like them" and using "white kettles." Jérémie showed his interlocutors "a silver cup, and they told me it was the same kind of thing as the others had spoken about, and they also told me that these others cultivate the land with tools of this white metal." If an Indian considered iron and silver to be similar metals, then the iron composing agricultural implements might very well have been "the same kind of thing" as the silver used for a cup, and this a reference to the Spanish in poor New Mexico to the south. To Frenchmen raised on tales of Potosí, however, it could signify an unknown and wealthy Spanish colony to the west.[14]

More intriguing than misunderstandings about metals and mountains were confusions concerning people, in particular agricultural town dwellers such as the Mandans of the upper Missouri and the Spanish and Indian nations of the

Fray Alonso de Posadas, "Informe á S. M. sobre las tierras de Nuevo Méjico, Quivira y Teguayo," in Cesáreo Fernández Duro, *Don Diego de Peñalosa y su descubrimiento del reino de Quivira* (Madrid, 1882), 53–67, esp. 56; "Itinerary of Juan Domínguez de Mendoza, 1684," in Bolton, ed. and trans., *Spanish Exploration in the Southwest, 1542–1706* (New York, 1916), 320–344, esp. 329–331, 342–343. On the brackish Missouri, see "Journal in the Form of a Letter," in Burpee, ed., *Journals and Letters of La Vérendrye*, trans. [LeSueur], 345.

14. "Journal in the Form of a Letter," in Burpee, ed., *Journals and Letters of La Vérendrye*, trans. [LeSueur], 336: "Le mot de fer parmi toutes les nations dicy est toutes sorte des mesteaux sapelle fer"; second quoted phrase in J. C. Beaglehole, ed., *The Journals of Captain James Cook on His Voyages of Discovery*, III, *The Voyage of the "Resolution" and "Discovery," 1776–1780* (Cambridge, 1967), 321; Nicolas Jérémie, *Twenty Years of York Factory, 1694–1714: Jérémie's Account of Hudson Strait and Bay*, trans. R. Douglas and J. N. Wallace (1720; rpt. Ottawa, 1926), 33; the original French, from Jérémie, "Relation du Detroit et de la baie de Hudson," in Jean Frédéric Bernard, *Recueil d'arrests et autres pièces pour l'établissement de la Compagnie d'Occident* (Amsterdam, 1720), 1–39, esp. 26–27, reads, "Ils disent que ces hommes portant barbe, ne sont point habillez comme eux, et qu'ils se servent de chaudieres blanches. Je leur montrai une tasse d'argent, et ils me dirent que c'étoit de cela même dont les autres leur avoient parlé. Ils disent aussi que ces gens-lá cultivent la terre avec des outils de ce metal blanc."

Southwest. As appears to have been the case with Spanish investigators also, different ideas about the appropriate categories into which these populations should be placed seems to have led to considerable misunderstanding by French explorers of what they were hearing from western Indians. La Vérendrye and the Mandans provide a good example. In 1734, La Vérendrye heard from a group of Crees and Assiniboines of a people called "Ouachipouennes or Caserniers," whom

> they took . . . for French; their forts and houses were much like ours,
> . . . their houses are large . . . having cellars where they keep their Indian
> corn in large wicker-work baskets. . . . Men and women, work in the
> fields.
>
> These Caserniers . . . are . . . white. . . . Their hair is light in colour,
> chestnut and red; a few have black hair. They have beards which they cut
> or pull out, some, however, allowing them to grow. They are engaging
> and affable with strangers who come to see them. . . . They are clothed in
> leather or in dressed skins skilfully worked and of different colours. . . .
>
> This tribe is very industrious. They sow quantities of corn, beans,
> peas, oats, and other grains, which they trade with the neighbouring
> savages, who come to their settlements to get them. The women do not
> work as hard as our Indian women, but take charge of domestic affairs
> and keep things neat and clean; and when work is pressing they render
> help in the fields.
>
> . . . He [the Assiniboine interpreter] believes that they are Frenchmen
> like us.
>
> . . . For kettles they use pots of sandstone or earthen vessels worked
> on the outside in compartments and with flowers.[15]

15. "Report in Journal Form of All That Took Place at Fort St. Charles from May 27, 1733, to July 12 of the Following Year, 1734, to be Transmitted to the Marquis de Beauharnois . . . ," in Burpee, ed., *Journals and Letters of La Vérendrye*, trans. [LeSueur], 153–155, 159–160. The "much" in the first line could also be translated as "more or less." The French text reads, "Ils les prenoient pour des françois, que leurs forts et leurs maisons êtoient à peu près comme les Nôtres . . . leurs maisons sont grandes . . . les maisons qui ont des caves, c'est la ou ils conservent les bleds d'inde dans de grands paniers d'ozier . . . ils travaillent tous à la terre hommes et femmes"; "ces Caserniers sont . . . blancs . . . leurs cheveux sont blonds, chatins et rouges, peu les ont bien noirs, Ils ont de la barbe qu'ils coupent ou arrachent Et quelques uns la laissent croître; ils sont caressans et affables aux Etrangers qui viennent les voir . . . ils sont habillés de Cuir ou de peaux passés bien travaillés et de differentes couleurs"; "Cette Nation est fort laborieuse, Elle sême quantité de bled, fêves, pois, avoine et autres grains qls commercent avec les sauvages voisins qui viennent les chercher chez Eux, les femmes ne travaillent pas tant que nos sauvagesses, mais elles sont chargées du menage qu'Elles tiennent propre et aux ouvrages pressans elles aident et travaillent aux Champs"; "il croit que ce sont des

La Vérendrye's own description of the Mandans he met in 1738 would resemble in many respects what he reported hearing:

> Of all the tribes they [the Mandans] are the most skilful in dressing leather, and they work very delicately in hair and feathers. . . .
>
> . . . All the streets, squares and cabins are uniform in appearance. . . . They keep the streets and open spaces very clean; the ramparts are smooth and wide. . . . If all their forts are similar you may say that they are impregnable to savages. Their fortification, indeed, has nothing savage about it. . . .
>
> This tribe is of mixed blood, white and black. The women are rather handsome, particularly the light-coloured ones; they have an abundance of fair hair. The whole tribe, men and women, is very industrious. Their dwellings are large and spacious, divided into several apartments by wide planks. Nothing is lying about; all their belongings are placed in large bags hung on posts. . . .
>
> . . . Their fort is very well provided with cellars, where they store all they have in the way of grains, meat, fat, dressed skins and bearskins. They have a great stock of these things, which form the money of the country. . . . They do very fine wicker work, both flat and in basket form. They use earthen vessels, which they make like many other tribes, for cooking their victuals. . . .
>
> . . . The men are . . . for the most part, good-looking, fine physiognomies, and affable. The women generally have not a savage cast of features.[16]

With their sedentary agriculture, solid fortifications, permanent houses, and, to many eighteenth-century eyes, European aspect, the Mandans might, to the As-

---

françois comme Nous"; "Ils se servent pour chaudieres de pots de grais ou de Terre ouvragés en dehors en compartimens et fleurs."

16. "Journal in the Form of a Letter," in Burpee, ed., *Journals and Letters of La Vérendrye*, trans. [LeSueur], 332, 339–343: "Ce sont gens qui passent mieux le cuire de toute les nations et travaille bien delicatment en poilles et plumes"; "toutes les rües places et cabannes se ressemble, . . . il tienne les rues et place fort nette, les rempard bien unie et Large . . . si tous leurs fort sont pareille on les peut dire imprenable a des sauvages, leurs fortification n'est point du sauvage; cette nation est d'un sang melée blanc et noir les femmes sont assés belles surtout les blanches beaucoup de cheveu blon et blanc, c'ést une nation fort laborieuse, hommes et femmes, leurs cabanne sont grande espacieuse séparé en plusieurs apartemens par des madriers fort large rien ne traine, tout leurs equipage est dans de grands sac suspendue a des poteaux"; "leurs fort est rempli de cave ou ils ser tout ce quil ont comme grains, viande, graisse, robe passée, peaux d'ours, ils sont bien muni, cést la monoie du pays . . . il travaille en osier fort proprement plat et corbeille, il se serve de pots de terre quils font comme bien d'autres nations, pour faire cuire leur manger"; "les hommes . . . pour la plus grande partie assés beaux du visage, belle fisionomie fort afable la plupart des femmes n'ont point la fisionomie sauvage."

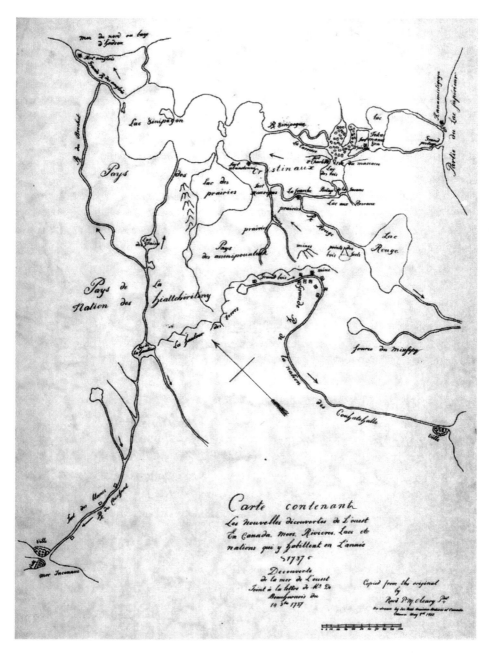

MAP 32. Pierre Gaultier de Varennes et de La Vérendrye, *Carte contenant les nouvelles découvertes de l'ouest en Canada.* 1737. Library and Archives Canada, NMC 24561

siniboines and others, have seemed more reminiscent of the French than of wandering Plains Indian hunters.

In a manner that shows how slippery human classification could be, La Vérendrye felt otherwise: "I was greatly surprised, as I expected to see people quite different from the other savages according to the stories that had been told us. They do not differ from the Assiniboin, being naked except for a garment of buffalo skin carelessly worn without any breechcloth. I knew then that there was a large discount to be taken off all that had been told me." Despite the many similarities between what he was told he would see and what he said he saw, La Vérendrye declared that he felt misled. He did not take the Mandans for French. Instead, with a reaction reminiscent of Montaigne's essay on cannibals, La Vérendrye kept his eye focused on the central issue: no matter the Mandans' crops, comportment, cleanliness, or crafts — they weren't wearing pants! Category was in the eye of the beholder, and communications could be confused and expectations disappointed as a result.[17]

Just as misunderstanding could arise over the respective classifications of the Mandans and the French, similar confusion could spring from the categorization of the Spanish and Mandans. La Vérendrye had heard that the Mandans were "whites." He would only go so far as to posit that they were partially so. Perhaps, though, he had misunderstood, and it was not the Mandans who were supposed to be white but the Spanish. In 1738, Mandans told La Vérendrye of a people down the Missouri, *"blanc"* like the French. They "worked in iron," went "always on horseback," had "iron armour," and "fought with lances and sabres, which they handled with great skill. You never saw a woman in their fields; their fort and houses were of stone." This could certainly be a description of the Spanish in New Mexico, perhaps based on testimony from Plains Indians who raided and traded between that province and the upper Missouri. Other witnesses echoed the account. In 1740, La Vérendrye reported on information he had received from two men left in the Mandan villages to learn the language. They spoke of an Indian who had "been brought up from childhood among whites." He told them that these whites were like the French: "They had beards . . . prayed to the great Master of Life in books which they described as made of leaves of Indian corn . . . sang holding their books in great houses where they assembled for prayer." They had houses "made of bricks," "their towns and forts" were "surrounded by good walls with wide ditches filled with water," and they used "powder, can-

17. Ibid., 319–320. For a different interpretation of La Vérendrye's reaction, see Barbara Belyea, "The Sea of Dreams: La Vérendrye's Mapping of Desire," *Australian-Canadian Studies,* XII (1994), 15–27, esp. 19.

nons, guns, axes, and knives." They produced "all kinds of grain, ploughing their land with horses and oxen." They marched "in column," and "their cuirasses" were "made of iron net." This again sounds reminiscent of the Spanish in New Mexico, and it is entirely possible that an Indian held since childhood as a slave could speak in detail about them after his escape or manumission. More interesting, in December 1738, when La Vérendrye "reproached" his Assiniboine informants for misleading him about the Mandans, "They replied that they did not mean the Mandan when they spoke of a nation like us, that they meant the nation that dwells down the river and that works in iron." As one Assiniboine put it more pointedly, "You did not rightly understand what was said to you." Perhaps La Vérendrye had conflated what were intended as distinct descriptions of the Mandans and the Spanish. Perhaps, quite reasonably, his informants had seen Mandan, French, and Spanish settlements as similar in many respects and had not emphasized the differences among them that so concerned La Vérendrye. In any case, it was not always easy for a French explorer to discern which "blancs" he was hearing about.[18]

It is possible that a similar confusion arose regarding the identities of the Indian nations living along the northwest coast. One hint of this comes from the records of a French trading expedition from the Illinois country. Led by the brothers Pierre and Paul Mallet, the party succeeded in reaching Santa Fe in 1739, then remained in New Mexico for more than nine months. An abstract of the Mallets' journal has them reporting a tradition among the Indians of New Mexico of a land three months to the west containing white men who wore silk and lived in large towns by the sea. This may represent no more than a traveler's tale springing from credulity, incomprehension, or the desire of traders or officials to believe in the presence of Asian settlements in North America. Before dismissing it out of hand, however, it is worth recalling that Manila galleons

18. For the first 1738 report, see "Journal in the Form of a Letter," in Burpee, ed., *Journals and Letters of La Vérendrye*, trans. [LeSueur], 336–337. The French text reads "travaillait le fer"; "il ne marchoit que a cheval"; "estant couvert de fer"; "se batoit avec des lances et sabre dont il estoit bien adroit l'on ne voyait jamais de femme dans les champs, leurs fort et maisons estoit de pierre." For the 1740 report, see "Extract from Journal of La Vérendrye," ibid., 367–372. The French text reads "été élevé chez les Blancs dès son enfance"; "ils avoient de la barbe, et prioient le grand maître de la vie dans des livres, en leur dépeignant qu'il étoient faits avec des feuilles de bled d'inde, qu'ils chantoient en tenant leurs livres dans de grandes maisons où ils s'assembloient pour la Prière"; "maisons faites de briques"; "leurs villes et forts sont entourés de bonnes murailles avec de grands fossés remplis d'eau"; "ils ont l'usage de la poudre, canons, fusils, haches et couteaux ... ils font toutes sortes de grains à la charue, tirée par chevaux et boeufs"; "ils marchent en corps de troupes, et que leurs cuirasses sont de fer maillé." For the final 1738 quotations, see 354–355. La Vérendrye had failed to elicit descriptions of a people resembling the Spanish from the Assiniboine in 1734, but that does not mean that different Assiniboine speaking in 1738 might not have been mentioning them.

laden with Chinese silk wrecked from time to time on Pacific shores. More intriguing, European explorers and traders visiting the northwest coast later in the century often wrote of peoples there having fair skin, weaving beautiful cloth from cedar bark, and living in big, coastal villages. It is easy to imagine that interior Indian accounts of the distant, wealthy, and artistically sophisticated peoples of the Pacific slope led eager European listeners to conclude that some kind of Asian or European presence existed on North America's western shores.[19]

Such a possibility may help to explain one of Charlevoix's conjectures about the West's inhabitants. He wrote in 1721 of having "met at la Baye [Green Bay] some Sioux, whom I closely questioned about the Regions which are West and North-West of Canada." Charlevoix evinced some skepticism about Indian testimony, conceding "that one cannot always accept to the letter all that the Savages say." Nonetheless, he averred that he had "every reason to believe, in comparing what they have related to me with what I have heard from several other sources, that there are on this Continent Spanish or other European Colonies — far more to the North than those of which we have knowledge in New Mexico or California." Though the presence of a handful of enterprising cossacks or shipwrecked Spaniards — or even windblown Ainu traders or Japanese fishermen — cannot be ruled out, actual European colonies beyond New Mexico, Baja California, or Kamchatka had not yet formed. What had was a set of northwest

19. For the Mallets, see "Extrait du journal de voyage des frères Mallet à Santa Fé," May 29, 1739–June 24, 1749, in Donald J. Blakeslee, *Along Ancient Trails: The Mallet Expedition of 1739* (Boulder, Colo., 1995), 215–225, esp. 218: "Une terre qui suivant la tradition vraye ou fausse des Indiens du pays est a trois mois dans les terres du côté du Ouest ou ils disent quil y a des hommes Blancs retus de soye qui habitent de grandes villes sur le bord de la mer." For examples of European descriptions of Northwest Coast Indians, see "Diary of Fray Tomás de la Peña Kept during the Voyage of the 'Santiago' — Dated 28th August, 1774," and "Journal of Fray Juan Crespi Kept during the Same Voyage — Dated 5th October, 1774," in Donald C. Cutter, ed., *The California Coast: A Bilingual Edition of Documents from the Sutro Collection*, trans. Cutter and George Butler Griffin (Norman, Okla., 1969), 135–201, 203–278, esp. 176–181, 254–259; "Juan Pérez's Letter to Viceroy Bucareli," Aug. 31, 1774, "Juan Pérez's 'Diario,'" June 11–Aug. 28, 1774, and "Francisco Mourelle's Narrative of Pérez's Voyage in 1774," all in Herbert K. Beals, ed. and trans., *Juan Pérez on the Northwest Coast: Six Documents of His Expedition in 1774* (Portland, Ore., 1989), 50–55, 58–99, 103–117, esp. 54, 89–90, 113; Bruno de Hezeta, *For Honor and Country: The Diary of Bruno de Hezeta*, ed. and trans. Herbert K. Beals (Portland, Ore., 1985), 76; Miguel de la Campa, *A Journal of Explorations Northward along the Coast from Monterey in the Year 1775*, ed. [and trans.] John Galvin (San Francisco, 1964), 43; Beaglehole, ed., *Journals of Captain James Cook*, III, 303, 311–313, 1097–1100, 1326, 1404–1405, 1411–1412; John Dunmore, ed. and trans., *The Journal of Jean-François de Galaup de la Pérouse, 1785–1788*, 2 vols. (London, 1994), I, 138; John Meares, *Voyages Made in the Years 1788 and 1789, from China to the North West Coast of America* (London, 1790), 230, 250–252; John Boit, "Boit's Log of the Columbia, 1790–1792," Massachusetts Historical Society, *Proceedings*, LIII (1919–1920), 217–275, esp. 247–248; J. Neilson Barry, "Columbia River Exploration, 1792," *Oregon Historical Quarterly*, XXXIII (1932), 31–42, 143–155, esp. 37, 41, 143.

coast peoples characterized by large and dense populations, social stratification, oceangoing canoes, and art and architecture that impressed then and captivates today. From Green Bay, it might have been possible to discern the presence of such cultures without recognizing their indigenous character.[20]

In some cases, European investigators were trying to determine the identity of the "whites" they were hearing about; in others, of those designated as "black." Jonathan Carver (who traveled west of the Great Lakes in 1766 and 1767) wrote that the Winnebagos called the Spanish "the Black People," and Hudson's Bay and later North West Company surveyor David Thompson claimed that the same denomination for the Spanish was in use among the Pieagans (Blackfeet) in 1788. On the other hand, the same Indian who told La Vérendrye's men he had been raised among Spanish-sounding *"Blancs"* whose "towns" were "near the great Lake the water of which rises and falls and is not good to drink," in a "country . . . of very high mountains," also was reported to have said that there were "black men there who have beards and who work in iron; before reaching the whites you have to pass through the country of the blacks." It is unclear who these "blacks" were. It seems unlikely that the informant was referring to Spanish peoples in New Mexico as black and Frenchmen in Louisiana as white, for example, because there is no indication that the whites among whom he was allegedly raised spoke or taught him French. La Vérendrye's men "did not understand" the "language of the white men" he was speaking. For historians as for explorers, tales of distant lands frustrate simple interpretation.[21]

20. Translated Charlevoix text from "Charlevoix Visits Wisconsin: His Description of the Tribes," in Reuben Gold Thwaites, ed., State Historical Society of Wisconsin, *Collections*, XVI, *The French Regime in Wisconsin, I, 1634–1727* (Madison, 1908), 417–418. French text in Charlevoix, *Journal d'un voyage*, 300–301. See also Antoine Denis Raudot, "Of the La Sapiniere Indians and the Assenipouals," 1710, in Raudot, "Memoir concerning the Different Indian Nations of North America," in W. Vernon Kinietz, *The Indians of the Western Great Lakes, 1615–1760* (Ann Arbor, Mich., 1965), 376–377. On Northwest Coast cultures, see Philip Drucker, *Cultures of the North Pacific Coast* (San Francisco, 1965); Erna Gunther, *Indian Life on the Northwest Coast of North America: As Seen by the Early Explorers and Fur Traders during the Last Decades of the Eighteenth Century* (Chicago, 1972); and Kenneth M. Ames and Herbert D. G. Maschner, *Peoples of the Northwest Coast: Their Archaeology and Prehistory* (London, 1999). On Japanese boats' reaching North America, see Fumio Kakubayashi, "Japanese Drift Records and the Sharp Hypothesis," *Journal of the Polynesian Society*, XC (1981), 515–524; George I. Quimby, "Japanese Wrecks, Iron Tools, and Prehistoric Indians of the Northwest Coast," *Arctic Anthropology*, XXII (1985), 7–15; and, more generally, Katherine Plummer, *The Shogun's Reluctant Ambassadors: Japanese Sea Drifters in the North Pacific*, 3d ed. (Portland, Ore., 1991).

21. Jonathan Carver, *Three Years' Travels through the Interior Parts of North-America, for More Than Five Thousand Miles . . .* (Philadelphia, 1789), 18; Richard Glover, ed., *David Thompson's Narrative, 1784–1812* (Toronto, 1962), 50; "Extract from Journal of La Vérendrye," in Burpee, ed., *Journals and Letters of La Vérendrye*, trans. [LeSueur], 367–372; the French text concerning the "Blacks" reads, "Que les villes sont proches du grand Lac, où l'eau monte et baisse, et qu'elle n'est pas bonne

One tantalizing possibility is that La Vérendrye's men were conflating elements of an account of the Spanish and Pueblos in New Mexico and the Utes or Paiutes farther north. In Franciscan Father Escalante's diary of his and Father Domínguez's 1776 journey through the Southwest, he described a band of "Tirangapuis" encountered south of the Utah and Great Salt lakes near modern Mills, Utah. They were "more fully bearded" than the Timpanogotzis farther north around Utah Lake, and "in their features they" looked more like the "Spaniards than they" did "all the other Indians" previously "known in America." Five "were so fully bearded that they looked like Capuchin padres or Bethlemites." From the Timpanogotzis, Domínguez and Escalante heard reports of the "harmful and extremely salty" waters of the nearby Great Salt Lake, and Escalante and Domínguez themselves crossed "salt flats" west of their meeting place with the Tirangapuis. One could easily imagine that Utes or Paiutes who looked like Spaniards generated confusion in accounts of the Southwest. Escalante himself thought as much. The "iron" La Vérendrye's men heard about remains puzzling. Indians northwest of New Mexico could have acquired metal items through raiding or trading, but Escalante does not mention the Tirangapuis' or Timpanogotzis' having metal in 1776. La Vérendrye's men might have misheard, or their Indian informant might have been speaking on the basis of hearsay — or he might simply have been talking about a different group of Indians.[22]

à boire, que cet endroit est un pais de montagnes fort hautes, qu'il y a des hommes noirs qui ont de la barbe, et qui travaillent le fer, qu'avant d'ariver chés les Blancs, il faut passer chés les Noirs."

22. Ted J. Warner, ed., *The Domínguez-Escalante Journal: Their Expedition through Colorado, Utah, Arizona, and New Mexico in 1776*, trans. Angelico Chavez (Provo, Utah, 1976), 60–66, 165–168. The Spanish text, in the order of the above quotations, reads "mucho más cerrados de barba"; "En la fisiognomía se parecen a los españoles más que a todos los demás indios hasta ahora conocidos en esta América"; "tan crecida la barba que parecían padres capuchinos o betlemitas"; "nocivas o extremadamente saladas"; "salinas." For similar references to southwestern Indians resembling Spaniards, see Francisco Garcés, *A Record of Travels in Arizona and California, 1775–1776*, ed. and trans. John Galvin (San Francisco, 1965), 19, 49, 66, 68, 73; Garcés, *Diario de exploraciones en Arizona y California en los años de 1775 y 1776* (Mexico City, 1968), 29, 55, 68–69, 70, 74.

One possible example of Utes' or Paiutes' being mistaken for Spaniards comes from Charmion Clair Shelby, [ed. and trans.], "St. Denis's Declaration concerning Texas in 1717," *SWHQ*, XXVI (1923), 181. Saint-Denis testified that he had "heard the Indians say that to the northwest of the Tejas there are bearded white men who do not go to the east nor to these parts to trade, as it is said that they have their dealings with people of the sea; the Indians say that in the land of these bearded men there is a mountain from which is seen the sea." The fair and bearded men could be Utes, the sea Lake Utah or Great Salt Lake. Another, more distant possibility is that Saint-Denis was hearing about Northwest Coast Indians, also often described by later Europeans as light-skinned and bearded. I think the former interpretation more probable because it involves less travel, but it's best to keep an open mind.

On Escalante's sensing confused ethnicities, see "Carta del Padre Fray Silvestre Velez de Escalante, escrita en 2 de abril de 1778 años," in *Documentos para servir a la historia del Nuevo México, 1538–1778* (Madrid, 1962), 305–324, esp. 323: "Por relacion de indios infieles mal entendidas, se per-

Another, in some respects simpler, explanation may arise from the complex composition of New Mexico's population. New Mexico was a Spanish colony. Some of its inhabitants traced their ancestry to the Iberian peninsula, spoke Spanish, and worshipped in Catholic churches. Many others descended from the Pueblo populations the Spanish had conquered. One group, the *genízaros*, consisted of Indian captives from the regions around New Mexico who had been captured or redeemed by the Spanish, had adopted aspects of Spanish culture while serving in households or on ranches, and had struck out on their own after paying their debts to those who held them. Historians have noted, moreover, that the peripheral settlements of New Mexico—in the mountains especially—often attracted a variety of marginal figures from the colony and surrounding regions whose mingling created a kind of "mixed society" exhibiting features associated with both Iberian and Indian cultures. Still farther afield, Spanish military expeditions venturing onto the plains and under the eyes of Indian observers there sometimes included more Indian allies than Iberian soldiers. The celebrated Segesser hide paintings, probably done during the 1720s, portray what appear to be some of these Indian allies or auxiliaries wearing Spanish armor and using Spanish weapons. In 1694, Governor Diego de Vargas of New Mexico reported that Ute Indians who had attacked a Spanish party claimed to have done so by accident, because the Pueblo enemies of the Utes had been availing "themselves of the ruse and stratagem of coming on horseback, some with suits of armor, dressed in leather jackets and hats they make of rawhide like those the Spaniards wear. They also bring firearms they took from the Spaniards and even their bugle." Part of the reason why descriptions of what was probably New Mexico seem to have blended the features of many populations and cultures was that the province being described did so also.[23]

suadieron muchos à que de la otra parte del rio Colorado . . . habitaba una nacion parecida á la española, la que usaba barba larga, armamento como el antiguo nuestro, de peto, morrion y espaldar. Y éstos sin la mayor duda son los yuttas barbones . . . los cuales viven en rancherías y no en pueblos, son muy pobres, no usan mas armas que las flechas y algunas lanzas de pedernal, ni tienen otro peto, morrion, ni espaldar que el que sacaron del vientre de sus madres."

23. For good discussions of genízaros, see the "Declaration of Fray Miguel de Menchero, Santa Bárbara, May 10, 1744," in Charles Wilson Hackett, ed., *Historical Documents Relating to New Mexico, Nueva Vizcaya, and Approaches Thereto, to 1773*, comp. Adolph F. Bandelier and Fanny R. Bandelier, 3 vols. (Washington, D.C., 1923–1937), III, 395–413, esp. 401–402; Russell M. Magnaghi, "Plains Indians in New Mexico: The Genízaro Experience," *Great Plains Quarterly*, X (1990), 86–95; James F. Brooks, *Captives and Cousins: Slavery, Kinship, and Community in the Southwest Borderlands* (Chapel Hill, N.C., 2002), 121–143. By 1733, groups of genízaros were requesting grants of land that would enable them to form new communities. By 1740, colonial authorities were acceding to these requests, hoping that genízaro settlements on the edges of the colony would serve as a first obstacle to raids by other, less acculturated Indians. It is conceivable that these peripheral settlements would have seemed distinct from the Spanish to an observer such as the one with whom La Vérendrye's

Differentiating among European and Indian, and perhaps Asian and Amer-indian, peoples proved difficult enough for French explorers. In some cases, French authors might even have encountered difficulties distinguishing humans and animals. In 1758, French author Antoine-Simon Le Page du Pratz wrote of the transcontinental journey of Moncacht-apé, an elderly Yazoo Indian with whom Du Pratz claimed he had spoken while living in Louisiana between 1718 and 1734. In Du Pratz's account, Moncacht-apé had, decades earlier, ascended the Missouri, crossed the mountains, and descended "the Beautiful River" to "the Great Water." The Beautiful River of the tale bears more than a passing re-semblance to the Snake and Columbia rivers later traveled by Lewis and Clark. Moncacht-apé told Du Pratz that he descended the watercourse for some eigh-teen days, and "I should have liked to push on further, following always the Beautiful River, for I did not become fatigued in the pirogue, but it was neces-sary for me to yield to reasons opposing this plan. They [his 'Otter' Indian com-panions] told me that the heat was already great, that the grass was high and the serpents dangerous in this season, and that I might be bitten in going on the chase." Serpents do indeed slither in the Columbia River region, but the claim that they were once so numerous or venomous as to inhibit hunting sounds a bit

---

men spoke, while simultaneously displaying characteristics associated with Spanish culture such as ironwork. The problem is chronology. La Vérendrye was writing of a description of the "blacks" through whom one reportedly had to pass on the way to New Mexico in 1740, and it is not clear that distinct communities of genízaros had formed on the fringes of the colony before that year. For peripheral New Mexico settlements, see Brooks, *Captives and Cousins*, 117. On Spanish military expeditions on the plains, see "The Diary of Juan de Ulibarri to El Cuartelejo, 1706," "Hurtado's Re-view of Forces and Equipment, Picuríes Pueblo," and "Diary of the Campaign of Juan Páez Hurtado against the Faraon Apache, 1715," in Alfred Barnaby Thomas, ed. and trans., *After Coronado: Spanish Exploration Northeast of New Mexico, 1696–1727; Documents from the Archives of Spain, Mexico, and New Mexico* (Norman, Okla., 1935), 59–77, 89–98, esp. 60, 77, 93–94; Robert Ryal Miller, ed. and trans., "New Mexico in Mid-Eighteenth Century: A Report Based on Governor Vélez Capuchín's Inspection," *SWHQ*, LXXIX (1975), 166–181, esp. 174–175. For the Segesser paintings, see Gottfried Hotz, *Indian Skin Paintings from the American Southwest: Two Representations of Border Conflicts be-tween Mexico and the Missouri in the Early Eighteenth Century*, trans. Johannes Malthaner (Norman, Okla., [1970]). For Vargas's report, see John L. Kessell, Rick Hendricks, and Meredith D. Dodge, eds., *Blood on the Boulders: The Journals of Don Diego de Vargas, New Mexico, 1694–97*, 2 vols. (Albu-querque, N.M., 1998), I, 306–309. Juan María Rivera's accounts of his 1765 journeys into the Ute and Paiute country beyond New Mexico furnish other instances of the difficulty of ethnic classi-fication. He was hearing about people who sounded like Spaniards, with beards, fair complexions, helmets, and a Spanish-like language, for example, but it is difficult to identify the exact people to whom these reports refer. Illicit Spanish and genízaro traders? Westering Frenchmen? Hopis, Utes, or Paiutes dressed or bearded like Spaniards? Spaniards far down the Colorado? See Joseph P. Sánchez, "Translation of Incomplete and Untitled Copy of Juan María Rivera's Original Diary of the First Expedition, 23 July 1765," and "Translation of Juan María Rivera's Second Diary, October 1765" in Sánchez, *Explorers, Traders, and Slavers: Forging the Old Spanish Trail, 1678–1850* (Salt Lake City, Utah, 1997), 137–147, 149–157, esp. 145, 155.

peculiar. Interestingly, though, Lewis and Clark's journals of their exploration of the river in 1805 allude frequently to hostilities between peoples living north of the Columbia and the "Snake" (perhaps Northern Paiute) Indians living south of it. That Moncacht-apé's serpents were people, their bites metaphorical, and Du Pratz and Moncacht-apé's account a classic case of miscommunication suddenly becomes plausible. Du Pratz might have misunderstood Moncacht-apé. In turn, Moncacht-apé, who mentions often in Du Pratz's text the process and difficulties of learning the languages of the people he encountered, might have misinterpreted their words.[24]

Here, as elsewhere, it is perhaps best to take the opacity of the sources as the point. A shifting combination of translation difficulties, the cultural determinants of classification, the complexity of the communities being described, the credulousness of French explorers, and the indulgence of their Indian informants could leave the identity and characteristics of western peoples a puzzle.

Descriptions of mountains, metal, salinity, and ethnicity often remained obscure because of differing habits of classification and bases for comparison. Other geographically crucial information might simply have resisted translation and communication because of differing cultural understandings. Direction, on maps and for travel, appears to have been one such case. It seems often not to have translated well between French and western Indian cultures and perhaps not among western Indian cultures either. This is a counterintuitive suggestion in many respects. One would think that, if there was anything two distinct cultures could agree upon, it would be how to describe the direction of the setting sun. Indian myths, legends, and religious practices, moreover, demonstrate a clear sense of cardinal direction, and numerous sources attest to the capacity of various western Indians to maintain a consistent direction when traveling. La Harpe noted, "It is remarkable that the savages do not make any mistake when they show the part of the world where the nation dwells of which they have knowledge, and that, taking the bearing of the places with the compass, one is

24. The English text of Du Pratz comes from Gordon Sayre's translation, available online at http://darkwing.uoregon.edu/gsayre/LPDP.html, III, chap. 7, 110–111 (accessed June 17, 2007). The original French from [Antoine-Simon] Le Page du Pratz, *Histoire de la Louisiane: Contenant la découverte de ce vaste pays; sa description géographique; un voyage dans les terres; l'histoire naturelle, les moeurs coûtumes et religion des naturels, avec leurs origines . . .* , 3 vols. (Paris, 1758), III, 110–111, reads, "J'aurois bien désiré pousser plus loin en suivant toujours la Belle Riviere; car je ne fatiguois point dans la Pirogue; mail il fallut me rendre aux raisons que l'on m'opposa. On me dit que les chaleurs étoient déja grandes, les herbes hautes et les Serpens dangereux dans cette saison; que je pourrois en être mordu en allant à la chasse." See also Sayre, "A Native American Scoops Lewis and Clark: The Voyage of Moncacht-apé," *Common-Place*, V, no. 4 (July 2005), http://www.common-place.org/vol-05/no-04/sayre/index.shtml (accessed June 24, 2009); Moulton, ed., *Journals of the Lewis and Clark Expedition*, V, 318–320, 325, 339, 342–343, 349, 351–352.

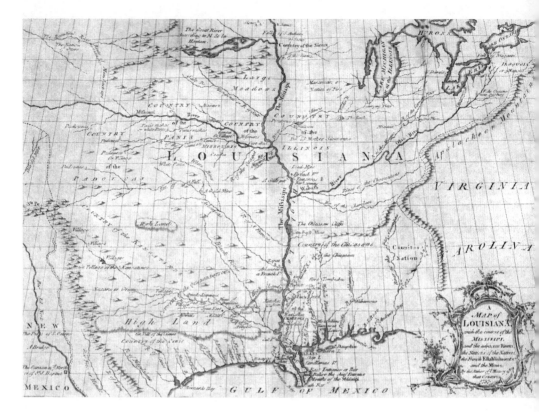

MAP 33. "A Map of Louisiana, with the Course of the Missisipi, and the Adjacent Rivers, the Nations of the Natives, the French Establishments, and the Mines," in Antoine-Simon Le Page du Pratz, *The History of Louisiana, or of the Western Parts of Virginia and Carolina: Containing a Description of the Countries That Lye on Both Sides of the River Missisipi....* (London, 1763). Special Collections, University of Virginia Library

certain of their situation." Similarly, British exploration promoter Arthur Dobbs claimed in 1744 that the half-French, half-Ojibwa fur trader Joseph La France had told him that Indians returning from a hunt "go home in a direct Line, never missing their Way, by Observations they make of the Course they take upon their going out, and so judge upon what Point their Huts are, and can thus direct themselves upon any Point of the Compass."[25]

25. Smith, ed. and trans., "Account of the Journey of Bénard de La Harpe," *SWHQ*, LXII (1958–1959), 253; French text in La Harpe, "Relation du voyage de Bénard de la Harpe," in Margry, ed., *Découvertes et établissements des Français*, VI, 263. For La France, see Arthur Dobbs, *An Account of the Countries Adjoining to Hudson's Bay, in the North-West Part of America* (London, 1744), 41–42. Andrew Graham, who served in a variety of capacities at the Hudson's Bay Company's York and Churchill posts between 1749 and 1774 and was familiar with the Crees and Assiniboines who visited them, described techniques that made this direction-finding possible. "Sometimes the sun

When one moves from the level of individual Indian travelers to that of Indian cartographic representations of western landscapes, however, much documentation suggests a different way of imagining and picturing cardinal direction. As mentioned above, Malcolm Lewis has argued that La Vérendrye's Indian-derived maps of the West appear compatible with the topography of the region west of Lake Superior and Hudson Bay, if one takes Lake Winnipeg and the river running west from it on La Vérendrye's map and rotates them so that the river in question now flows north from Lake Winnipeg toward Hudson Bay. What appeared to La Vérendrye, and what might have been conceptualized by his informants, as a straight-line journey would contain a right-angle turn on a modern map. Lewis's reasoning and documentation are commanding, and his contention is persuasive, but for his argument to hold up, La Vérendrye and his Indian informants had to somehow miss a common understanding of the difference between north and west.

Evidence from outside the scope of Lewis's investigation supports this possibility. Saint-Pierre's 1750–1753 journal expressed the concern that a push west toward the Pacific might end up instead on the shores of Hudson Bay. To avoid this, the journal proposed a move toward the sources of the Missouri and suggested that La Vérendrye's failure to grasp the importance of this strategy vitiated his exploratory efforts. Saint-Pierre's journal implies an awareness on his part that the kind of misdirection Lewis finds in the case of La Vérendrye might have occurred or at least could occur. Historians William Goetzmann and Glyndwr Williams mention another set of maps exhibiting the kinds of characteristics that might have confused La Vérendrye. The Chipewyan sketches upon which Hudson's Bay Company employees James Knight and Moses Norton based their c. 1716 and 1767–1768 maps of the west coast of Hudson Bay show the shoreline as a straight north-south line, giving "no indication of the great bulges of the Melville and Boothia peninsulas" nor the "great right-angle turn above Hudson Bay which gives the Arctic coastline its east-west orientation."[26]

---

is their guide," he observed. When the sun failed, the orientation of snowdrifts might serve the purpose, because prevailing winds tended to pile snow in the same direction. More reliable than accumulated snow was the growth of trees: the thickness of bark and the distribution of branches varied according to the way the wind most commonly and forcefully blew. The result was that "every tree the travelling American passes serves as a compass to direct his course, and never fails of guiding him to the desired place" (Williams, ed., *Graham's Observations on Hudson's Bay,* 171). See also Rich, ed., *Isham's Observations on Hudsons Bay,* 102–103.

26. Peyser's translated text of Saint-Pierre's "Brief Report or Journal" on page 183 of *Jacques Legardeur de Saint-Pierre* reads, ". . . being very determined to press my explorations forward. I only feared ending up toward Hudson Bay, which I strongly intended to avoid by traveling to the West to find the sources of the Missouri River, in the hope that they would lead me to several rivers which would flow in the region I am seeking to enter, without which I felt that it would be impossible

The root of these varying representations may lie in differences between two notions of straightness: one derived from the abstractions of Euclidean geometry and a system of Cartesian coordinates superimposed upon the features of the earth, and one drawn from the practical experience of moving in accordance with the most practical routes and the most recognizable landmarks. Barbara Belyea, the author of a set of provocative and valuable essays about eighteenth-century European and Indian western cartography, has observed that, in contrast with "grid maps," which "operate by locating positions along axes of latitude and longitude," "Amerindian maps rely not on fixed positions in space but on a pattern of interconnected lines. Spacing and directions of north / south / east / west are simplified, even ignored, since the key to reading the map is not to locate points in space but to trace a continuous path from one geographic feature to another." This is perhaps most easily imagined in the case of Indians whose usual movements followed watercourses and shorelines. Hudson's Bay Company factor James Isham claimed that at least some of the "natives" of "these Norther'n parts" who visited the company's forts were rarely "out of Lakes, Rivers, and creeks" and consequently, though conversant with the four cardinal directions, were less concerned with other points of the compass because these subdirections were not needed by travelers who could follow the contours of bodies of water to find their way. (Perhaps the closest modern analogy would be the oft-used example of a metro or subway map, on which the precise compass direction of the lines is less important than the relation of lines and stations to each other.) Speaking broadly, the notions of cardinal direction that were so central to early modern European ways of representing and moving across the world might simply have possessed a different significance for many

to penetrate farther due to the difficulty of transporting the food and supplies essential to such an undertaking. Which made me realize that the plans of [the late] M. de la Vérendrye were not very solid, it not being possible to succeed by any route other than the Missouri" (brackets in Peyser indicating material "present in the original but omitted in Margry"). Saint-Pierre, "Rétablissement d'un poste chez les Sioux," in Margry, ed., *Découvertes et établissements des Français*, VI, 642, gives, "Bien résolu de pousser bien avant mes découvertes, je n'avois à craindre que d'aboutir du côté de la baie d'Hudson, ce que je me proposois grandement d'éviter, en me jetant à l'Ouest pour trouver les sources de la rivière du Missoury, dans l'espoir qu'elles me conduiroient à quelques rivières, qui auroient leur cours dans la partie où je cherche à pénétrer, sans quoy je sentois bien qu'il seroit impossible de pénétrer plus avant, par la difficulté du transport des munitions et vivres indispensables pour une pareille entreprise. Ce qui me fit connoître que les projets de M. de la Verendrye n'étoient pas bien solides, n'étant pas possible de réussir par d'autre voye que celle du Missoury" (see page 644 also). For Knight and Norton's maps of Hudson Bay, see Goetzmann and Williams, *Atlas of North American Exploration*, 102–103, 108–109. See also Williams's fuller discussion of the issue in *Voyages of Delusion: The Northwest Passage in the Age of Reason* (London, 2002), 13–16, 218–220, 224–227.

western Indian cultures, thoroughly confusing attempts by outsiders to grasp western geography.[27]

Not only do available Indian maps seem to indicate a conception of direction different than that appearing on conventional European maps from the same period, but other French investigators seem to have come to errant conclusions like La Vérendrye's, often about the locations of western peoples. Indian prisoners told Nicolas Jérémie that "they were at war with another nation in the west, much further away than their own country. Those other men say that they have for neighbours men who are bearded and who build stone forts and live in stone houses, a custom which native tribes do not follow. . . . From the way which they describe the grain raised by these people, it must be maize." This could be the Mandans, or, as mentioned above, given the "white kettles" and "tools of this white metal" that the "bearded men" were reported to possess, it could be a description of the more distant Spanish or Indian peoples of the Southwest. In either case, the identification depends on substituting "southwest" for "west." It is also possible that Charlevoix's 1721 hypothesis about European colonies north of those known in Baja California and New Mexico might have stemmed, not from rumors of the distant Northwest Coast Indians, but from a misunderstanding of the direction to Santa Fe. On the whole, and as surprising as it initially appears, examples drawn from accounts such as those of Jérémie, Saint-Pierre, and perhaps Charlevoix sustain the suggestion of Lewis and others that cardinal direction resisted easy translation.[28]

Thus opportunities abounded for French explorers and scholars to misunderstand or fail to understand western geography. The many differences among Indian and European cultures and languages and the consequent difficulties of

27. Barbara Belyea, "Inland Journeys, Native Maps," *Cartographica*, XXXIII (1996), 1–16, esp. 6. Also interesting are Belyea's "Amerindian Maps: The Explorer as Translator," *JHG*, XVIII (1992), 267–277, and her "Mapping the Marias: The Interface of Native and Scientific Cartographies," *Great Plains Quarterly*, XVII (1997), 165–184. For a useful theoretical treatment of this subject that builds on the work of scholars such as Belyea and Lewis, see Tim Ingold, "To Journey along a Way of Life: Maps, Wayfinding, and Navigation," in Ingold, *Perception of the Environment: Essays on Livelihood, Dwelling, and Skill* (London, 2000), 219–242. For Isham, see Rich, ed., *Isham's Observations on Hudsons Bay*, 65.

28. Translated text from Jérémie, *Twenty Years of York Factory*, trans. Douglas and Wallace, 32–33. Douglas and Wallace suggest that the Mandans were the people described and make the same point about direction. The original French in Jérémie, "Relation du Detroit et de la baie de Hudson," in Bernard, *Recueil d'arrests et autres pièces*, 26–27, reads, "Ils m'ont dit avoir guerre avec une autre Nation beaucoup plus éloignée qu'eux dans l'Ouest. Ceux-là disent avoir pour voisins, des hommes barbus qui se fortifient avec de la pierre, et se logent de même; usage que les Sauvages n'ont point. . . . De la maniere qu'ils dépeignent le grain que ces gens cultivent, il faut que ce soit du Maïs."

interpretation comprised one facet of the problem. A different way to view the circumstances of this French befuddlement is to emphasize, not the presence of cultural and linguistic differences, but rather the absence of means to overcome them. The importance of such means becomes more apparent when one views French efforts to comprehend western American geography in comparison with French cartographic projects in France, Russia, and China.

Linguistic diversity, for instance, was not unique to the American West, but the difficulties generated for French investigators were much greater there than elsewhere. The Russian and Chinese empires were far from linguistically uniform, and France itself retained speakers of Breton and a bewildering array of French dialects often quite incomprehensible to learned Parisians. These Old World linguistic difficulties were more easily surmounted because of the political context in which surveying occurred. French Jesuits and geographers did not need to speak the many languages and dialects in China and Siberia. In the case of Russia, this was in part because, by the eighteenth century, enough residents of Russia spoke languages such as German, Latin, or French to allow for communication with western Europeans. More basically, Russian imperial agents were doing most of the surveying fieldwork in Russia and Siberia and thus most of the interacting with the Russian Empire's diverse populations. In China, French Jesuits who needed an understanding of imperial and local languages and dialects benefited from the superb academic educations that characterized their order as well as the opportunity for and imperative of years of scholarly study in China. Moreover, because they received the cooperation of Chinese and Manchu officials in their surveying efforts, the Jesuits had translators of local tongues and dialects available. In the Russian and Chinese empires, centralized state structures without equivalent in the American West were already in place and were already coping with polyglot populations and foreign-tongued outsiders.

Moving from the specific matter of linguistic variety to the more general issue of cultural difference, much distinguished the cultural backgrounds of French surveyors and cartographers from those of the peoples of Russia, China, and even France. As Laura Hostetler has pointed out, however, part of what was so significant about the scientific cartographic techniques being developed and employed in eighteenth- and late-seventeenth-century France was that they could be applied in places like the Chinese and Russian empires to produce maps whose utility transcended cultural difference. Louis XIV, Peter the Great, and the Kangxi emperor, whatever the differences in their upbringings, world-views, and personal habits, could all perceive the value for administration, diplomacy, and warfare of the kinds of precise, combinable, uniform, and—with a little training—easily grasped maps that French and French-trained cartographers were producing.[29]

Western American conditions precluded emulation of the survey techniques applied in Russia and China. This was in part because of the evident differences among the societies of China, European Russia, and regions beyond the Mississippi and Hudson Bay. There is nothing novel about noting that the scientific academies springing up in seventeenth- and eighteenth-century France and Russia and the venerable corpus of written mathematical "classics" studied in early modern China had no counterparts among the Indian nations of the eighteenth-century North American West; nor is there anything innovative in observing that these nations possessed an intimate and functional knowledge of the region that was the envy of the blundering European explorers seeking their help. What is significant, though, is that in the absence of the kinds of scientific and technical institutions and training available in China, for example, the transmission of Amerindian geographic expertise to those from different cultural backgrounds was difficult.[30]

Nor were people from the North American West acquiring education abroad or in their home regions that might have facilitated the exchange of geographic information. Peter the Great had compensated for Russia's educational wants by sending others—and himself—to western Europe's capitals. In contrast, although Bourgmont succeeded in conducting four chiefs from the Illinois, Missouri, Osage, and Oto nations on a two-month visit to Paris in 1725, there was no Indian exodus to Europe for technical training comparable to that of Peter's Russians. Nor was there a migration of European instructors to the American West. A contrasting example from an earlier century and a different empire calls attention to what was missing. In postconquest sixteenth-century Mesoamerica, Spanish mendicant friars added education in European languages, writing, and representational techniques to the highly sophisticated training received by indigenous cartographers. Over the course of the century, as cartographic historian Barbara E. Mundy has noted, Mesoamerican cartographers adapted and altered the conventions of indigenous mapmaking—substituting alphabetical for hieroglyphic place-names, for example—to make their maps more intelligible to Spanish audiences and therefore more useful in settings such as Spanish courts adjudicating land claims. But eighteenth-century western Indians were not receiving in their own territories the kind of rigorous scientific instruction

29. Laura Hostetler, *Qing Colonial Enterprise: Ethnography and Cartography in Early Modern China* (Chicago, 2001), 1–4, 21–22, 73–74.

30. On Chinese mathematics, see Joseph Needham and Wang Ling, *Science and Civilisation in China*, III, *Mathematics and the Sciences of the Heavens and the Earth* (Cambridge, 1959).

Chinese and Manchu surveyors were obtaining from the Jesuits, Russian cartographers were getting from the Delisles, and Mesoamerican mapmakers had received from Spanish missionaries. The kinds and numbers of teachers present in France and the Russian and Chinese empires had not yet made their way to the North American Far West. French explorers like La Vérendrye were not finding the kinds of people they were looking for, and French investigators were not making them. In contrast with Russia and China, they do not seem to have been, either on their own initiative or at the behest of an indigenous leader, turning North America's "Tartares" into scientifically trained cartographers.[31]

This is not to say that western Indians were not producing maps of the region. Frenchmen were, however, finding these images difficult to interpret. Unlike the maps coming out of Russia and China, comprehension of western Indian maps often hinged on the comprehension of Indian cultures and languages. Modern scholarship has emphasized the extent to which understanding of Indian maps depended on grasping the oral explanations accompanying them. As Mark Warhus has put it:

> Unlike western society, maps were not created as permanent documents in native American traditions. The features of geography were part of a much larger interconnected mental map that existed in the oral traditions. The world was perceived and experienced through one's history, traditions, and kin, in relationships with the animal and natural resources that one depended upon, and in union with the spirits, ancestors, and religious forces with whom one shared existence. . . . This indigenous knowledge was passed down in songs, stories, and rituals, and the understanding of the landscape it imparted was as sophisticated as that of any western map.
>
> . . . Unlike western cartography, where the "map" becomes one's picture of the landscape, native American maps are always secondary to the oral "picture" or experience of the landscape.[32]

31. Norall, *Bourgmont*, 81–87. On adaptations of Mesoamerican cartography, see two works by Barbara E. Mundy: "Mesoamerican Cartography," in David Woodward and G. Malcolm Lewis, eds., *Cartography in the Traditional African, American, Arctic, Australian, and Pacific Societies* (Chicago, 1998), 183–256, esp. 197–199, 213, 225, 241–245; and *The Mapping of New Spain: Indigenous Cartography and the Maps of the Relaciones Geográficas* (Chicago, 1996), 68–84; and see David Buisseret, "Meso-American and Spanish Cartography: An Unusual Example of Syncretic Development," in Dennis Reinhartz and Gerald D. Saxon, eds., *The Mapping of the Entradas into the Greater Southwest* (Norman, Okla., 1998), 30–55.

32. Mark Warhus, *Another America: Native American Maps and the History of Our Land* (New York, 1997), 3, 8.

Because the form and content of Indian maps were so closely tied to the traditions, beliefs, and practices of the cultures producing them, they were much easier for Europeans to understand when they were accompanied by oral explanations from representatives of the nations from which they came. A good example of the importance of these ancillary oral elucidations can be seen in Simon Fraser's journal of his 1808 exploration of the British Columbian River now bearing his name. Fraser was hearing from nations such as the Chilcotins and the Atnahs about the difficulty of the river below. He received, moreover, from a slave among the Atnahs, a "sketch" from which Fraser "could plainly see a confirmation of the badness of the navigation" to come. Fraser then had a group of Indians who spent the night at his camp draw him another "chart." On this occasion, however, Fraser's companion John Stuart "got [the map] explained, and took them [the Indians] down in writing, by which the road appears more practicable than by the information they gave us before." Oral explanation and visual representation together created a different impression than they had separately.[33]

It follows that the Europeans most likely to most fully understand Indian maps were those most thoroughly conversant with particular Indian cultures. It is probably no coincidence that Francisco Garcés, the Spanish North American missionary perhaps most confident in his understanding of native maps, was also notable among his countrymen for his ability to communicate with the Indians he encountered, willingness to spend months alone among Indian peoples, and practice of doing as the local Romans did. As one of his contemporaries, Fray Pedro Font, put it, "Father Garcés is so well fitted to get along with the Indians and to go among them that he appears to be but an Indian himself. . . . He sits with them in the circle, or at night around the fire, . . . and there he will sit musing two or three hours or more, . . . talking with them with much serenity and deliberation." Cultural awareness picked up during those long fireside jaws might have helped Garcés understand better the visual representations of the regions through which he was passing in the mid-1770s. Those less adaptable and durable often fared less well.[34]

33. "Journal of a Voyage from the Rocky Mountains to the Pacific Ocean Performed in the Year 1808," and "Second Journal of Simon Fraser from May 30th to June 10th 1808," in W. Kaye Lamb, ed., *The Letters and Journals of Simon Fraser, 1806–1808* ([Toronto], 1960), 61–128, 131–161, esp. 64–69, 132–140.

34. Garcés, *Record of Travels in Arizona and California,* ed. and trans. Galvin, 64, 89; Garcés, *Diario de exploraciones en Arizona y California,* 67, 89; Herbert Bolton, ed. and trans., *Anza's California Expeditions,* IV, *Font's Complete Diary of the Second Anza Expedition* (Berkeley, Calif., 1930), 121; Pedro Font, "Diario extendido de Font de 1776," or "Expanded Diary of Pedro Font," Dec. 8, 1775, at Web de Anza, http://anza.uoregon.edu/archives.html (accessed June 17, 2009).

ndeed, the connections among cartographic representation, verbal elucida-
ı, and cultural context appear to have hampered more generally the trans-
ssion of geographic ideas between European and Indian cultures and among
...tive American nations themselves. Scholars of Indian cartography have observed one result and indication of this: the area covered by most Indian maps
rarely extended beyond the bounds of a particular "linguistic region." Where
Jesuit surveys made it possible for the Kangxi emperor to bring together a set of
local maps to form a single picture of his dominions, only smaller, less compatible pieces of an overall picture of the Indian West were available.[35]

ONE REASON FRENCH explorers and the geographers and officials who relied
upon them found it difficult to comprehend western geography was that communicating ideas about the region across space, language, and culture was extremely difficult in its own right and relatively more difficult than was the case
for French cartographers operating in France, Russia, and China.

But more than communication was involved. There was also the question of
getting into the areas French investigators wanted to communicate about. Crucial for understanding the limited progress of French geographic understanding
of the mysterious lands beyond the Mississippi and Great Lakes were questions
of access to western territories, not simply for French explorers seeking a way
across the region but also for the Indian nations inhabiting the various parts of
it. Even when western explorers and Indians' testimony could be credited and
comprehended, it was often based only on the reports of distant others rather
than on personal acquaintance with the areas about which French geographers
wanted information.

35. G. Malcolm Lewis, "Metrics, Geometries, Signs, and Language: Sources of Cartographic
Miscommunication between Native and Euro-American Cultures in North America," *Cartographica*, XXX (1993), 98–106, esp. 103: "The areal extent of most early maps was usually confined to
one linguistic region."

8

RESTRICTED PATHWAYS

The previous chapter's discussion of difficulties arising from linguistic, conceptual, and cultural differences gives some sense of how the communication of geographic information could be impeded even when French investigators were speaking with Indians familiar with a region. Often, however, it appears that western Indians were not personally acquainted with the territories French explorers were asking about. Aggravating communicative difficulties, and further limiting the range of French geographic comprehension, were the hostilities among western nations restricting the safe movement of both Frenchmen and Indians.

These difficulties stand out more sharply when viewed against the backdrop of French cartographers' experiences in France and in the Russian and Chinese empires. In these areas, French investigators received royal or imperial sponsors' support, and they operated within political structures rendering that assistance effectual. We have already seen Louis Delisle de La Croyère traversing Siberia and the North Pacific with the Great Northern Expedition and French Jesuit cartographers crisscrossing China in the service of the Kangxi emperor. We have also seen that events in the American West unfolded differently. Here, lacking the state patronage they enjoyed elsewhere, a primary problem confronting French explorers and cartographers was their inability to cast their eyes on or deploy their instruments in large parts of the region they sought to comprehend.

Eighteenth-century French North American explorers, like their predecessors since Columbus's landfall and like their celebrated successors such as Alexander Mackenzie and Lewis and Clark, relied heavily on Indian guides and counselors. This was in part a matter of acquiring food and water and of knowing which road to travel and which stream to ascend. More critical were issues of human geography. French explorers who hoped to succeed or survive needed to know which Indian nations were likely to prove helpful or hostile. They needed companions who could translate unfamiliar languages and attest to the good intentions and commercial and political utility of these French emissaries.

But unlike China, where the emperor's commands and local officials' efforts opened provincial roads and civic archives to Jesuit surveyors, in eighteenth-century North America, Indian guides' warnings repeatedly closed western routes. Such was the case in the southern plains region. In his account of a 1719 journey up the Red River and into what is now southeastern Oklahoma, Bénard de La Harpe reported that he "had much difficulty persuading" his "Nassonite [Caddoan]" guide to lead him to

> some metallic stones in the mountains, . . . of which rocks the Spaniards
> were making much of importance . . . because of the fear that he had of
> meeting on this route some enemy party. . . . The first two days our guide
> was brave, but the third some tracks of men . . . which he recognized to
> be those of the Anahons [Osages], disconcerted him entirely. It was nec-
> essary to use threats in order to make him advance for some leagues. We
> came down into vast prairies then, in the sight of mountains, from where
> we perceived the fire of the enemies. Then it was not possible to reassure
> our guide; nothing could engage him to lead us further; it was necessary
> to decide to turn back the same evening.[1]

French explorers often could not go on when their pathfinders knew the hazards of local warfare and deemed avoiding them more important than the progress of French geographical research.[2]

1. Translated text and ethnic identifications from Ralph A. Smith, ed. and trans., "Account of the Journey of Bénard de La Harpe: Discovery Made by Him of Several Nations Situated in the West," *SWHQ*, LXII (1958–1959), 371. French text given in Bénard de La Harpe, "Relation du voyage de Bénard de La Harpe: Découverte faite par lui de plusieurs nations situées a l'Ouest," in Pierre Margry, ed., *Découvertes et établissements des Français dans l'ouest et dans le sud de l'Amérique septentrionale (1614–1754): Mémoires et documents originaux*, VI, *Exploration des affluents du Mississipi et découverte des montagnes Rocheuses (1679–1754)* (Paris, 1886), 243–306, esp. 272.

2. The observations of Spanish missionaries venturing in the late seventeenth and early eighteenth centuries into the Trinity and Neches river region of what is now East Texas bolster this suggestion that the plains constituted dangerous terrain for Caddoan peoples. In 1716, Fray Francisco Hidalgo noted that, when the Caddoan Assinais moved north and west to hunt buffalo, they had "their enemies in sight. Here there are extensive plains where every year the Assinai have wars with these Indians in order to secure meat and because of the ancient hostility between them." Similarly, in 1691, Fray Francisco Casañas Jesús María wrote that the "Tejas" or "Asinai" Indians around the Santísimo Nombre de María Mission hunted buffalo in bands for "fear of other Indians, their enemies"; and in 1722, Fray Isidro Félix de Espinosa observed that the Indians from the "Texas Country" went "well armed" to the hunt, "because at this time if they fall in with the Apaches the two murder each other unmercifully." See Mattie Austin Hatcher, trans., "Descriptions of the Tejas or Asinai Indians, 1691–1722," *SWHQ*, XXX (1926), 206–218, esp. 211, XXXI (1927), 50–62, 150–180, esp. 55, 157; for the original Spanish, see Isidro Félix de Espinosa, *Crónica de los Colegios de propaganda fide de la Nueva España*, ed. Lino G. Canedo, 2d ed. (Washington, D.C., 1964), 692. On Apache hostilities, see also Charmion Clair Shelby, [ed. and trans.], "St. Denis's Declaration concerning Texas in 1717," *SWHQ*, XXVI (1923), 165–183, esp. 176–180; Jean-Baptiste Bénard de La

In later decades, as French trading networks and alliance systems expanded west, routes across the southern plains seem to have become occasionally passable but remained generally insecure for French travelers. The Mallet brothers pressed on to Santa Fe in 1739 despite warnings from an Arikara slave about his Comanche captors' hostile intentions and after repelling a Comanche assault. In 1752, as mentioned in earlier chapters, French traders Jean Chapuis and Louis Feuilli succeeded in reaching New Mexico's Pecos Mission, demonstrating the practicability of traversing the southern plains and of cooperating with many of the Indian nations along the route. That getting to New Mexico from the Illinois country was possible, however, did not mean that it was risk free. Chapuis and Feuilli reported placating with goods Comanches initially reluctant to let them proceed. When asked by Spanish interrogators about the possibility of transporting merchandise to the colony in the future, Chapuis replied that it was feasible, that, "with caravans of horses which they bought from the Pawnees and Comanches, they could bring the goods to this capital of Santa Fe." He mentioned also that, "because of the risk they would run with the Comanche tribe, in whom they do not have total confidence, they would escort their traders with fifty or sixty armed men."[3]

Farther north, in the region between Lake Manitoba and the Black Hills, French explorers found Indian wars an apparently insurmountable barrier to westward movement. La Vérendrye wrote in the journal of his 1738 stay in the Mandan villages that, when speaking to the Mandans about whether "it took a long time to go to the country where the whites, the men who rode horses, were," the Mandans told him that "since they [the Mandans] had been at war with the Panana they did not venture to go very far. The roads were blocked so

Harpe, *The Historical Journal of the Establishment of the French in Louisiana,* ed. Glenn R. Conrad, trans. Joan Cain and Virginia Koenig (Lafayette, La., 1971), 104.

3. For the Mallets, see Donald J. Blakeslee, *Along Ancient Trails: The Mallet Expedition of 1739* (Niwot, Colo., 1995), 39–40, 131, 217–219; Don Gaspar de Mendoza to Fray Pedro Navarrete, in José Antonio Pichardo, *Pichardo's Treatise on the Limits of Louisiana and Texas: An Argumentative Historical Treatise with Reference to the Verification of the True Limits of the Provinces of Louisiana and Texas . . . ,* ed. and trans. Charles Wilson Hackett, trans. Charmion Clair Shelby, 4 vols. (Austin, Tex., 1931), III, 349. Pierre Mallet and his companions on a subsequent 1750–1751 journey to New Mexico told Spanish interrogators that a Comanche band had despoiled them of their goods and papers. See "Pedro Malec, Juan Bautista Boisex, and José de Cadalso to the Governor and Captain-General, Tomás Vélez Cachupín to Count of Revilla Gigedo, Feb. 25, 1751," Mendoza to Navarrete, "Testimony of Juan Bautista Boyer, Jun. 28 and Jun. 30, 1751," ibid., esp. 335, 339, 350, 354, 357. For Chapuis and Feuilli, see "Declaration of Juan [Jean] Chapuis, Frenchman," in Alfred Barnaby Thomas, [ed. and trans.], *The Plains Indians and New Mexico, 1751–1778: A Collection of Documents Illustrative of the History of the Eastern Frontier of New Mexico* (Glendale, Calif., 1940), 103–106; Tomás Vélez Cachupín, Aug. 9, 1752, in Pichardo, *Pichardo's Treatise,* ed. and trans. Hackett, trans. Shelby, III, 366–368.

far as they were concerned." Farther west, La Vérendrye's two sons reported that, during their 1742–1743 journey through the Dakotas and Wyoming, their "Gens de l'Arc" companions refused to continue their trip into the Black Hills because of their "*terreur*" of the "Gens du Serpent."[4]

In a similar fashion, sources bearing on the Saskatchewan River region west of Hudson Bay and Lake Winnipeg speak of violence among Indian nations and suggest its inhibiting effects on French exploration. The 1750–1753 "Brief Report or Journal of the Expedition of Jacques Legardeur de Saint-Pierre" claimed, on the basis of what Saint-Pierre had seen himself and what he had heard from the Indians with whom he had conferred, that it was not "possible to penetrate any farther than" he had "because of the war that all the nations of that continent [the region of the Saskatchewan River] wage on each other." Because the reliability and even the authenticity of Saint-Pierre's journal have often been called into question, it is important to have some kind of corroborating evidence for this claim. Anthony Henday and Henry Kelsey's descriptions of their respective 1754–1755 and 1690–1692 Saskatchewan River valley treks provide it. Henday's journal does not make clear why he stopped his westward progress, but it does lend credence to Saint-Pierre's description of northern plains dangers. The "Archithinues" Henday encountered camped "in open plains" so that "they should not be surprized by the Enemy." They, and their neighbors, seem to have had good reason to worry. On one occasion, Henday "spoke with 4 Indian Men who told us that the far distant Archithinue Natives had killed 30 of the nigh ones and 7 of our Indians"; on another, "two Asinepoet Natives came to

4. "Journal in the Form of a Letter Covering the Period from the 20th of July 1738, When I Left Michilimackinac, to May, 1739, Sent to the Marquis de Beauharnois," and "Journal of the Expedition of the Chevalier de La Vérendrye and One of His Brothers to Reach the Western Sea, Addressed to M. the Marquis de Beauharnois, 1742–43," in Lawrence J. Burpee, ed., *Journals and Letters of Pierre Gaultier de Varennes de La Vérendrye and His Sons: With Correspondence between the Governors of Canada and the French Court, Touching the Search for the Western Sea*, trans. [W. D. LeSueur] (Toronto, 1927), 290–361, 406–432, esp. 337, 414–425. The French text in Burpee for the first quotation reads, "Je demandés sil metoit bien du temps a aler ou estoit les blanc, gens de cheval, on me repondit . . . depuis qu'ils avoit guerre avec les *panana* ils nausoit entreprendre d'aller bien loing, Les chemins estoit bouchés pour eux." For an earlier example of potential hostilities looming over an explorer's trip, see [Louis Armand de Lom d'Arce, baron de] Lahontan, *Oeuvres complètes*, ed. Réal Ouellet and Alain Beaulieu, 2 vols. (Montreal, 1990), I, 393. Regarding the identity of the various Indian groups denominated "Snakes," Colin G. Calloway has written in *One Vast Winter Count: The Native American West before Lewis and Clark* (Lincoln, Neb., 2003), 293, that "the term Snake was widely used, referring sometimes to Shoshones and Comanches together, sometimes to Shoshones, Bannocks, and Paiutes, sometimes to any tribes living on the eastern slope of the central Rockies, and perhaps also to the Kiowas when they lived in the Black Hills." On northern plains warfare generally, see John C. Ewers, "Intertribal Warfare as the Precursor of Indian-White Warfare on the Northern Great Plains," in Ewers, *Plains Indian History and Culture: Essays on Continuity and Change* (Norman, Okla., 1997), 166–179.

us and informed us the Archithinue Natives had killed and scalped 6 Indians." Kelsey's glimpse into the late seventeenth century sustains the observations of Saint-Pierre and Henday by suggesting that the conditions they described were not novel. The clarity of Kelsey's account suffers from his decision to write in awkward verse rather than limpid prose, but the bellicose propensities and corresponding vulnerabilities of the Cree, Assiniboine, and (most likely) Blackfoot Atsina bands he encountered stand out nonetheless. His description of one incident gives a sense of his style and content: "I had no sooner from those Natives turnd my back / Some of the home Indians came upon their track; And for old grudges and their minds to fill / Came up with them Six tents of wch, they kill'd."[5]

5. Saint-Pierre quotation from "Brief Report or Journal of the Expedition of Jacques Legardeur de Saint-Pierre, Knight of the Royal and Military Order of Saint Louis, Captain of a Company of the Detached *Troupes de la Marine* in Canada, Assigned to Search for the Western Sea," in Joseph L. Peyser, ed. and trans., *Jacques Legardeur de Saint-Pierre: Officer, Gentleman, Entrepreneur* (East Lansing, Mich., 1996), 180–191, esp. 190 (see 184 also); Jacques Le Gardeur de Saint-Pierre, "Rétablissement d'un poste chez les Sioux: Le sieur Marin, père, envoyé, a son tour, chercher a la hauteur des sources du Mississippi, une rivière débouchant dans la mer de l'Ouest, legardeur de Saint-Pierre . . . ," in Margry, ed., *Découvertes et établissements des Français*, VI, 651. For questions about Saint-Pierre as a source, see W. J. Eccles, "French Exploration in North America, 1700–1800," in John Logan Allen, ed., *North American Exploration*, II, *A Continent Defined* (Lincoln, Neb., 1997), 149–202, esp. 189–195. For Anthony Henday, see his *Year Inland: The Journal of a Hudson's Bay Company Winterer*, ed. Barbara Belyea (Waterloo, Ontario, 2000), 43–198, esp. 89–90, 105–107, 151. Because the four surviving copies of Henday's journal differ significantly, his account presents a knotty interpretive problem. Barbara Belyea has done scholars a great service by publishing an edition of Henday with all four versions. The four copies of the journal agree on the substance of the second and third quotations in my text above, with one copy offering an additional detail for the third quotation. The fuller three of the four versions agree on the substance of the first quotation; the shorter version offers nothing to contradict it. For all three quotations, the different versions vary slightly with regard to wording, spelling, and capitalization. One could certainly argue that I employ this problematic text too aggressively in my attempt to illuminate western conditions. I think it better to use Henday carefully than not to use him at all. Two decades later, Matthew Cocking said of the same region, "The Natives in general are afraid of the Snake Indians and say they are nigh at hand." See Lawrence J. Burpee, ed., "An Adventurer from Hudson Bay: Journal of Matthew Cocking, from York Factory to the Blackfeet Country, 1772–1773," Royal Society of Canada, *Transactions*, 3d Ser., II (1908), 89–119, esp. 106. See also the references to Assiniboine fights with Snakes, Blackfeet, and Crees in Alexander Henry, *Travels and Adventures in Canada and the Indian Territories: Between the Years 1760 and 1776*, ed. James Bain (Toronto, 1901), 293, 303–304, 318, 329. For Kelsey, see "Kelsey's Introduction to the *Journal* of 1691," and "The *Journal* of 1691," in [Henry Kelsey], *The Kelsey Papers* (Ottawa, Canada, 1929), 1–4, 5–19, esp. 2–3, 9–10, 13–18. Additional examples of hostilities among the Crees, Assiniboines, and Blackfeet west of Hudson Bay appear in these letters: "From Thomas McCliesh," Aug. 8, 1728, Aug. 1, 1729, "From Richard Norton," Aug. 17, 1738, and "From Richard Norton and Others," Aug. 16, 1739, Aug. 9, 1740, all in K. G. Davies, ed., *Letters from Hudson Bay, 1703–40* (London, 1965), 133–138, 141–144, 249–259, 292–298, 318–322, esp. 135, 141, 249, 292, 318. Note also the mention of "Earchethinues" warring with "Sinnepoets and other Indians" in E. E. Rich, ed., *James Isham's Observations on Hudsons Bay, 1743; and Notes and Observations on a Book Entitled "A Voyage to Hudsons Bay in the Dobbs Galley, 1749"* (Toronto, 1949), 113,

Despite such dangers, European explorers ventured west during the first six decades of the eighteenth century. Some, like Henday, probably came quite close to the Rockies. Others, like the Mallets, made it over the first range into a valley beyond. In general, however, from the Red River and the southern plains to the Saskatchewan River and the North, European explorers were hearing warnings of hostile Indian intentions from their guides and were finding their way partially or entirely impeded by fear. This alone might seem to explain why French geographers found western North America so much more difficult to handle than the Russian or Chinese empires.

Upon closer examination, the explanation is less clear. French Jesuits had access to much of the Chinese Empire, but not all of it. As mentioned above, for sensitive regions such as Tibet, they had to rely on the work of Chinese or Manchu surveyors trained in western techniques. Nor could French cartographers see for themselves all the corners of the Russian Empire. La Croyère could join in Russian exploration of the distant North Pacific, but it is hard to imagine any mortal touching upon the whole of Russia and Siberia, even if Joseph-Nicholas claimed that his brother had tried. Joseph-Nicholas Delisle's scholarly responsibilities seem to have kept him generally confined to Saint Petersburg, and he had to rely on receiving and compiling the work of Russians, often surveyors trained by himself or his brother. In the case of neither China nor Russia do the French cartographers seem to have been mistaken in their reliance on the work of others. Some corrections to the initial maps of Tibet were necessary, and the extent to which Russian maps conformed to the exacting requirements of the Delisles varied, but even allowing for the inevitable imperfections, these new maps represented a huge advance in terms of consistency, accuracy, and intelligibility over what French geographers had before the eighteenth century. If eighteenth-century French geographers could use local informants to enhance their understanding of parts of the Chinese and Russian empires never visited by Frenchmen, perhaps they could do something similar with Indian informants who had journeyed to regions of North America unseen by Canadians or Jesuits.[6]

---

and the reports of Sioux preparing for war against Crees and Assiniboines in La Harpe, *Historical Journal of the Establishment of the French in Louisiana,* ed. Conrad, trans. Cain and Koenig, 37.

6. On La Croyère's efforts, see Joseph-Nicolas de L'Isle, *Explication de la carte des nouvelles découvertes au Nord de la mer du Sud* (Paris, 1752), 1–2: "Mon Frere De la Croyere, . . . avoit aussi entrepris, depuis quelques années, de parcourir de même tout le reste de la Russie et de la Sibérie, jusqu'aux dernieres extrémitez de l'Orient." On the use of non-French observers, see Jean Baptiste Du Halde, *Description géographique, historique, chronologique, politique, et physique de l'empire de la Chine et de la Tartarie chinoise,* 4 vols. (Paris, 1735), I, xxlvii; Theodore N. Foss, "A Western Inter-

The same Indian hostilities preventing French explorers from making their way west, however, seem also to have been curtailing the trips of western Indians. Much evidence indicates that many western Indians were making what might be called middle-distance, or regional, journeys: from the southern to the northern plains; back and forth across the Rockies; from the Mississippi's banks or Hudson Bay's shores to the slopes of the Rockies; from west of the Rockies to the Dalles. Such trips, and the personal ties and cultural arrangements sustaining them, seem to have been quite common. They would have given individual Indians and Indian nations ample opportunity to gain a personal acquaintance with many square miles of territory, to transport and acquire a variety of goods, to encounter and form alliances with Indians from different regions and of different ethnicities. On the other hand, the available evidence suggests that what might be designated long-distance, or semicontinental, journeys were quite rare in the North American West during the long eighteenth century. A long-distance journey would be something like a trip from Lake Superior, Hudson Bay, or the Mississippi to the Pacific; from Vancouver Island to New Mexico; or from the Saskatchewan River valley to southern California.[7]

It must be conceded that the evidence available for this argument leaves much to be desired. There simply are not enough extant documents, maps, and artifacts for certainty. What is presented here, then, is a hypothesis compatible with the available evidence and offering an explanation for observed events rather than a definitive statement proved by unambiguous documentation. It must also be stated that what held true for the eighteenth-century West, or for the peoples whom Europeans were encountering, need not have been the case some years before or a few hundred miles beyond the region of Euro-Indian interaction. To give one example, when Mandans told La Vérendrye in 1738 that war with the "Panana" had closed the roads to the horse-riding whites, they also indicated that they had formerly made the journey to this distant people. If Europeans, with their weapons, viruses, national rivalries, and coveted trade

pretation of China: Jesuit Cartography," in Charles E. Ronan and Bonnie B. C. Oh, eds., *East Meets West: The Jesuits in China, 1582–1773* (Chicago, 1988), 209–251, esp. 230–231, 234, 240; L. A. Goldenberg and A. V. Postnikov, "Development of Mapping Methods in Russia in the Eighteenth Century," *Imago mundi*, XXXVII (1985), 63–80, esp. 63–74; Albert Isnard, "Joseph-Nicolas Delisle: Sa biographie et sa collection de cartes géographiques à la Bibliothèque nationale," in *Bulletin de la section de géographie*, XXX (1915), 47, 62, 79–80.

7. For a nice discussion of plains trade, including the peacemaking mechanisms rendering it possible despite intertribal hostilities, see W. Raymond Wood, "Plains Trade in Prehistoric and Protohistoric Intertribal Relations," in Wood and Margot Liberty, eds., *Anthropology on the Great Plains* (Lincoln, Neb., 1980), 98–109.

goods, were exciting western hostilities, then regions that had less contact with them might have experienced more stable conditions.[8]

Why, then, should it be considered plausible that Indian movement was restricted in this fashion? The most basic reason is that the documents keep saying so. The trepidations of La Harpe's Nassonite guide and the younger La Vérendryes' Bow companions have already been alluded to. Other sources refer to similar incidents when Indians were moving without French companions. Hearing of a river flowing west, Nicolas Jérémie tried to send Indian scouts out from Hudson Bay to find it: "I did all I could, while I was at Fort Bourbon, to send the natives in that direction, so as to learn if there were not some sea into which this river discharged." But as far as he could discern, war blocked any such effort. The Indians he hoped would go west for him were "at war with a nation which bars them from this road." The nation impeding progress seems to have been in similar straits: "I have questioned prisoners of this nation whom our natives had brought expressly to show me. They told me they were at war with another nation in the west, much further away than their own country." Fear of violence and consequent hesitations to travel far appear repeatedly in different documents referring to different territories in different decades.[9]

Sources also refer explicitly to the considerable, impressive, but ultimately bounded range of western Indians' geographic cognizance. In his introduction to his *Journey from Prince of Wales's Fort in Hudson's Bay to the Northern Ocean in the Years 1769, 1770, 1771, and 1772*, for example, Hudson's Bay Company explorer Samuel Hearne wrote that he had "seen several Indians who" had "been so far West as to cross the top of that immense chain of mountains which run from North to South of the continent of America." This, in and of itself, is a remarkable statement, given the distance separating the lands west of Hudson Bay from the Rockies and the difficulties involved in traversing the flatlands and high mountains of Canada. The westward extent of Indian movement and familiarity was therefore striking, but Hearne did not find it to be infinite. He

8. "Journal in the Form of a Letter," in Burpee, ed., *Journals of La Vérendrye*, trans. [LeSueur], 337.

9. Translated text from Nicolas Jérémie, *Twenty Years of York Factory, 1694–1714: Jérémie's Account of Hudson Strait and Bay*, trans. R. Douglas and J. N. Wallace (1720; rpt. Ottawa, 1926), 33; the original French in Jérémie, "Relation du Detroit et de la baie de Hudson," in Jean Frédéric Bernard, *Recueil d'arrests et autres pièces pour l'établissement de la Compagnie d'Occident* (Amsterdam, 1720), 1–39, esp. 26, reads, "J'ai fait tout mon possible pendant que je suis resté au Fort Bourbon, pour envoyer des Sauvages de ce côté-là, sçavoir s'il n'y auroit point quelque Mer dans laquelle se déchargeât cette Riviere; mais ils ont guerre contre une Nation qui leur barre ce passage. J'ai interrogé des prisonniers de cette Nation, que nos Sauvages avoient amenez exprés pour me les faire voir. Ils m'ont dit avoir guerre avec une autre Nation beaucoup plus éloignée qu'eux dans l'Ouest."

observed that, when he was at his "greatest Western distance, upward of five hundred miles from Prince of Wales's Fort [Churchill], the natives, my guides, well knew that many tribes of Indians lay to the West of us, and they knew no end to the land in that direction." He also stated that he had not "met with any Indians, either Northern or Southern, that ever had seen the sea to the Westward." Indians Hearne or Jérémie had not met might, of course, have journeyed from Lake Winnipeg to the South Sea, but these authors' testimony at least suggests that this was not commonly the case.[10]

On his great voyages down the Mackenzie River to the Arctic Ocean in 1789, and across the Rockies to the Pacific in 1793, Alexander Mackenzie sought and failed to find evidence of such long Indian treks to the Pacific. Mackenzie suspected that his Indian informants were telling him less than they knew about paths to the west—and their desire to avoid being dragooned into service by the relentless Mackenzie would certainly explain why. Nonetheless, the testimony he heard in his 1789 conversations with Indians between Fort Chipewyan and the Arctic Ocean and in his 1793 talks with the Sekanis just east of the Rockies corroborate the notion that northern Indians living shy of the Rockies knew of many nations to the west, had in some cases traversed the first range of the

10. Samuel Hearne, *A Journey from Prince of Wales's Fort in Hudson's Bay to the Northern Ocean in the Years 1769, 1770, 1771, and 1772*, ed. J. B. Tyrrell (Toronto, 1911), 55. Arthur J. Ray offers an insightful discussion of the early-eighteenth-century western Indian peoples traveling to Hudson Bay in his *Indians in the Fur Trade: Their Role as Trappers, Hunters, and Middlemen in the Lands Southwest of Hudson Bay, 1660–1870* (Toronto, 1974), 53–61.

There are indications that, far to the northwest, beyond the range of pre–Seven Years' War French exploration and therefore the topic of this book, some Indians were moving between the shores of the Pacific and of Great Slave Lake and Lake Athabasca. Fur trader Alexander Henry, Sr., claimed that in 1776 he met "Chepewyans, or Rocky Mountain Indians" from the Lake Athabasca region who "made war on the people beyond the mountains, toward the Pacific Ocean, to which their warriors had frequently been near enough to see it." One 1789 source reported that the redoubtable fur trader Peter Pond had "met with two Indians who came, as they said up a River from the Northern Pacific Ocean, all the way to the Slave Lake." This is possible, though the mountainous terrain of western Canada and Alaska would have made such trips noteworthy accomplishments. Aspects of Henry's and Pond's accounts, moreover, make their reliability less than certain. Pond had vigorously sought for his North West Company an exclusive right to the far western fur trade, and it is conceivable that he exaggerated the ease of western commerce to make the proposal more palatable. Future investigations would demonstrate, moreover, that geographic hypotheses purportedly drawn from Pond were incorrect: the Rockies do not end at latitude 62½°; nor does a river extend from Great Slave Lake to the Pacific. See Henry, *Travels and Adventures in Canada*, ed. Bain, 331–332; "To the Honorable Henry Hamilton Esq. Lieutenant Governor and Commander in Chief in and over the Province of Quebec and Frontiers Thereof in America etc. etc.: The Memorial of Peter Pond on Behalf of the North West Company in Which He Is a Partner," Quebec, Apr. 18, 1775, and ". . . An Extract of a Letter from Isaac Ogden, Esq., at Quebec, to David Ogden, Esq., of London, Dated Quebec, 7th November, 1789," in Henry R. Wagner, *Peter Pond: Fur Trader and Explorer* ([New Haven], Conn., 1955), esp. 82–84, 86–95.

mountains on raiding and trading missions, had heard reports of the Pacific and the Indian and later European peoples along its shores, but had rarely seen the ocean themselves. Mackenzie met Indians who had heard of a river or rivers flowing west to a "White-man's Lake" and a "White Mens Fort," but they indicated that their information came from the reports of others rather than from personal experience. The Sekanis "all pleaded ignorance, and uniformly declared, that they knew nothing of the country beyond the first mountain." On the Pacific side of these mountains, George Vancouver, after three years' exhaustive coastal surveying from the Columbia River to the Bering Strait, remarked, "In all the parts of the continent on which we landed, we nowhere found any roads or paths through the woods, indicating the Indians on the coast having any intercourse with the natives of the interior part of the country, nor were there any articles of the Canadian or Hudson's bay traders found amongst the people with whom we met on any part of the continent or external shores of this extensive country." Vancouver underestimated the extent of contact between coast and interior, as demonstrated by Mackenzie's journey along one of the paths Vancouver missed, but what is striking is, not that Vancouver failed to recognize the evidence of tramontane trade, but that such trade was sufficiently inconspicuous that an observer as exacting as he could overlook it.[11]

Lewis and Clark's experiences farther south paint a similar picture of impressive but finite trade and geographic familiarity. Once again, Indians living east of the Rockies evinced knowledge of lands as far as the mountains or just over the first range. In January 1805, the "Big White Chef of the Lower Mandan Village" gave Clark a "Scetch of the Countrey" as far as the Rocky Mountains, but not as far as the Pacific. Lewis found that the geographic familiarity of the Hidatsas

11. "Journal of a Voyage from Fort Chipewyan to the Arctic Ocean in 1789," "Journal of a Voyage from Fort Chipewyan to the Pacific Ocean in 1793," in W. Kaye Lamb, ed., *The Journals and Letters of Sir Alexander Mackenzie* (Cambridge, 1970), 161–234, 235–418, esp. 211–215, 218, 223, 261, 278, 287– 288, 318–319. See also Glyndwr Williams, ed., *Andrew Graham's Observations on Hudson's Bay, 1767– 91* (London, 1969), 256. For Vancouver, see George Vancouver, *A Voyage of Discovery to the North Pacific Ocean and Round the World, 1791–1795*, ed. W. Kaye Lamb, 4 vols. (1752; rpt. London, 1984), IV, 1556. Interestingly, Spanish scientist José Mariano Moziño, who visited Nootka Sound between May and September 1792, claimed that Vancouver indicated that the Indians there acquired their iron and copper through trade on the continent with other Indians at a place fewer than four hundred miles east of a port whose name Moziño could not recall: "Cuando los vio el capitán Cook por la primera vez, encontró que ya tenían conocimiento del hierro y cobre, y parece indisputable que adquirieron estos metales, traficando en el continente con otras naciones que van a hacer sus *cambios*, y que según las observaciones del comandante Vancouver, no dista más de cuatrocientas millas al este de un puerto en que él estuvo fondeado dentro del estrecho, y cuyo nombre no tengo ahora presente." See Fernando Monge and Margarita del Olmo, eds., *Las "Noticias de Nutka" de José Mariano Moziño* (Aranjuez, Spain, 1998), 153. Perhaps Moziño misunderstood Vancouver; perhaps Vancouver changed his mind about Indian trade.

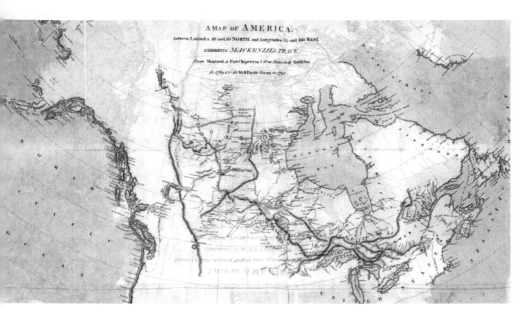

MAP 34. "A Map of America, between Latitudes 40 and 70 North, and Longitudes 45 and 180 West," in Alexander Mackenzie, *Voyages from Montreal on the River St. Laurence, through the Continent of North America to the Frozen and Pacific Oceans, in the Years 1789 and 1793* (London, 1801). Special Collections, University of Virginia Library

(who lived just north of the Mandans) extended farther, beyond the Rockies, but not farther than one could see from the Continental Divide.[12]

More subtly, one indication of the apparently bounded geographic horizon of the Indians living between the Rockies and Hudson Bay, Lake Superior, and the Mississippi was the way in which the detail and plausibility of information drawn from Indian sources diminished with distance and the manner in which accounts drawn from Indians seem to be conflating information regarding distinct peoples. Charlevoix provides a good example of both phenomena. The description he received from Iowa informants of the "*Omans* [Mandans], who have white skins and fair hair, especially the women," and who are "continually at war with the Panis and other more remote Indians towards the west," accords nicely with La Vérendrye's firsthand assessments from two decades later. On the

12. Gary E. Moulton, ed., *The Journals of the Lewis and Clark Expedition*, III, *August 25, 1804–April 6, 1805* (Lincoln, Neb., 1987), 269, 368. See also "John Macdonell's 'The Red River,'" "François-Antoine Larocque's 'Missouri Journal,'" in W. Raymond Wood and Thomas D. Thiessen, eds., *Early Fur Trade on the Northern Plains: Canadian Traders among the Mandan and Hidatsa Indians, 1738–1818; The Narratives of John Macdonell, David Thompson, François-Antoine Larocque, and Charles McKenzie* (Norman, Okla., 1985), 77–92, 129–155, esp. 85, 151.

other hand, when the Iowas spoke on the basis of information they had received from the Mandans about people living on the shores of "a great lake very far from their country," they gave an account that might very well have been blending features of two separate peoples. Those "resembling the French, with buttons on their cloaths, living in cities" could be the Spanish. Those "using horses in hunting the Buffalo, and cloathed with the skins of that animal; but without any arms except the bow and arrow" sound like one of the Indian nations of the southern plains. When Iowas were describing peoples they had not seen, it would be easy to imagine distinct nations as one.[13]

A comparable conflation may occur in Saint-Pierre's journal. It avers that Saint-Pierre heard from "an old Indian of the Kinongé8ilini" that "only a short time ago, a settlement had been established which is very far from where they live, to which they go to trade." The goods brought were "almost identical" to those coming from Canada. The Indian declared categorically that the traders were not English — he thought instead that they were "French" — but they were "not men as completely white" as Saint-Pierre and his French companions. Saint-Pierre estimated the direction to them as "West-Northwest." He said that he himself had seen "the horses and the saddles" obtained at this trading site. In what seemed to Saint-Pierre to be a confirmation of this report, Niverville told him that he had heard at "the settlement that he had built at the Rock Mountains" that an Indian war party "had encountered a tribe loaded with beaver which was going, by a river which comes out of the Rock Mountains to trade with Frenchmen who had their main settlement on an island at very little distance from the mainland." In exchange for beaver, those with whom the Indians traded gave "them knives, some lances, but no firearms; that they also sell them horses with saddles, which protect them from arrows when they go to war." The direction to the "settlement" was "East to West-South-West." One possibility is

<hr />

13. Pierre-François-Xavier de Charlevoix, *Journal of a Voyage to North America*, ed. Louise Phelps Kellogg, 2 vols. (1761; rpt. Chicago, 1923), II, 210; French text in P[ierre-François-Xavier] de Charlevoix, *Histoire et description generale de la Nouvelle France . . .* , III, *Journal d'un voyage fait par ordre du roi dans l'Amerique septentrionnale* (Paris, 1744), 397. Admittedly, given the confusion mentioned above about the alleged Frenchness of the Mandans and the circulation of goods on the plains, instead of referring to two peoples, the description Charlevoix heard might describe Comanches or Apaches who had acquired Spanish horses and articles of Spanish clothing and lived in large tent encampments but had not yet obtained firearms. See also Jean-Bernard Bossu, *Travels in the Interior of North America, 1751–1762*, ed. and trans. Seymour Feiler (Norman, Okla., 1962), 105: "When I questioned the nomadic Sioux, they told me that they had heard from other Indians that to the west of their country there were people who wore clothing and sailed their canoes on a large saltwater lake. These people lived in large villages built of white stone, and were ruled by a despotic chief who commanded formidable armies." The canoes and saltwater lake could very well indicate Puget Sound or the Inside Passage, the villages of white stone the communities of the Southwest.

that these were simply accounts of Plains Indians trading with the Spanish in New Mexico, that the Spanish seemed so similar to the French as to merit the same designation, and that Saint-Pierre's "West-Northwest" was another case of direction becoming muddled in translation. Many Indian nations traded with New Mexico, and Samuel Hearne would later offer the substantiating observation, "It is . . . well known to the intelligent and well-informed part of the [Hudson Bay] Company's servants, that an extensive and numerous tribe of Indians, called E-arch-e-thinnews, whose country lies far West of any of the Company's or Canadian settlements, must have traffic with the Spaniards on the West side of the Continent; because some of the Indians who formerly traded to York Fort, when at war with those people, frequently found saddles, bridles, muskets, and many other articles, in their possession, which were undoubtedly of Spanish manufactory." It is also possible that those trading on the coast for beaver, or perhaps more accurately otter, could be poorly documented Russians to the west and northwest. Tales of that distant commerce might have crossed the Rockies and been blended by people lacking personal familiarity with either Spanish or Russians into an account of a hybrid people.[14]

In the cases above, available evidence has indicated the rarity of long-distance journeys. Other testimony, however, suggests the opposite. Two extant accounts, for example, purport to describe transcontinental Indian journeys. The first is Moncacht-apé's. After hearing the aforementioned warning about serpents, Moncacht-apé eventually descended the "Beautiful River" to the "Great Water"; purportedly heard about and fought short, stocky, heavily clad, firearm-bearing, bearded white men who raided the coastal Indians for slaves and dyewood; continued up the coast to a land of long days and short nights; and, finally, returned. In the other account, Arthur Dobbs recounted a tale he claimed to have received from Joseph La France of a Cree Indian "who had, about 15 Years ago, gone at the Head of 30 Warriors to make war against the *Attimospiquais, Tete Plat,* or *Plascotez de Chiens,* a Nation living Northward on the Western Ocean of *America.*" The vivid story includes an epic journey, a raid on a coastal town, and even a sighting of whales, "a great many large black

---

14. Saint-Pierre quotations from "Brief Report or Journal," in Peyser, ed. and trans., *Jacques Legardeur de Saint-Pierre,* 184–185 (brackets in original indicate text present in original but absent in Margry) (see also Saint-Pierre, "Rétablissement d'un poste chez les Sioux," in Margry, ed., *Découvertes et établissements des Français,* VI, 644, 650–651); Hearne, *Journey from Prince of Wales's Fort,* ed. Tyrrell, 55–56. Interestingly, Isham's 1743 account of his conversation with an "Earchethinue" slave indicated that this particular group of Indians did not "traffick" with the "Spaniards" (Rich, ed., *Isham's Observations on Hudsons Bay,* 113). This lends credence to the claim of the "old Indian of the Kinongé8ilini," as reported by Saint-Pierre in 1753, that the Kinongé8ilini's trading for what may have been Spanish goods had begun "only a short time ago."

Fish spouting up Water in the Sea." Both tales are fascinating; neither lends itself to easy confirmation. My own suspicion is that a germ of truth lay at their core, though later raconteurs might have embellished the tales or combined elements from different sources to make a single epic. It is possible, for example, that the tale of Moncacht-apé springs from a western journey made by an actual person and that the apparently incongruous and far-fetched elements in it spring from Du Pratz's mistaken translations and interpretations. It is also possible that the story seems improbable simply because it casts a flickering light on a place and time unfamiliar to historians dependent on written sources. La France's account asks less of the reader. A Cree raid on head-flattening Indians somewhere in the Northwest is by no means outside the range of possibilities: Cree bands did raid beyond the first ranges of the Rockies, and Mackenzie found a woman on the Fraser River who spoke their language, having learned it after being captured and taken across the mountains by them. More broadly, many cultures produce Marco Polos and Odysseuses, and occasional trips could have produced fabulous stories grounded in mundane reality.[15]

Significant for the purposes of the discussion here of western Indian travel and geographic familiarity, however, is that, even in these two stories positing trips to the Pacific, the duration or difficulty of the journey is highlighted. Moncacht-apé "was five years making his journey to the West." The journey of La France's hero lasted more than two years, and after he and his thirty warriors raided the coastal town, "they all died one after another, except this old Man, of Fatigue and Famine, leaving him alone to travel to his own Country, which took him up about a Year's Time." The impression these unverifiable tales create is that long, perhaps even transcontinental journeys had occurred and perhaps could occur—but were not to be undertaken lightly.[16]

Later, more easily validated documentation corroborates this impression.

15. For Moncacht-apé, see Antoine-Simon Le Page du Pratz, *Histoire de la Louisiane: Contenant la découverte de ce vaste pays; sa description géographique; un voyage dans les terres; l'histoire naturelle, les moeurs coûtumes et religion des naturels, avec leurs origines* . . . (Paris, 1758), III, 102–140; for Gordon M. Sayre's translation, with same pagination, see http://darkwing.uoregon.edu/gsayre/LPDP.html, III, chap. 5, 102–140 (accessed July 15, 2009) (see also Sayre, "A Native American Scoops Lewis and Clark: The Voyage of Moncacht-apé," *Common-Place*, V, no. 3 [2005], http://www.common-place.org/vol-05/no-04/sayre/index.shtml [accessed June 24, 2009]); Arthur Dobbs, *An Account of the Countries Adjoining to Hudson's Bay, in the North-West Part of America* . . . (London, 1744), 44–45; "Journal of a Voyage from Fort Chipewyan to the Pacific Ocean in 1793," in Lamb, ed., *Journals and Letters of Sir Alexander Mackenzie*, 318–319.

16. Translated text from Sayre, History of Louisiana/L'Histoire de la Louisiane (1758), at http://darkwing.uoregon.edu/gsayre/LPDP.html, III, chap. 8, 129. The original French from du Pratz, *Histoire de la Louisiane*, III, 129, reads, "Il fut cinq ans à faire ce voyage de l'Ouest." La France quotation from Dobbs, *Account of the Countries Adjoining to Hudson's Bay*, 44–45.

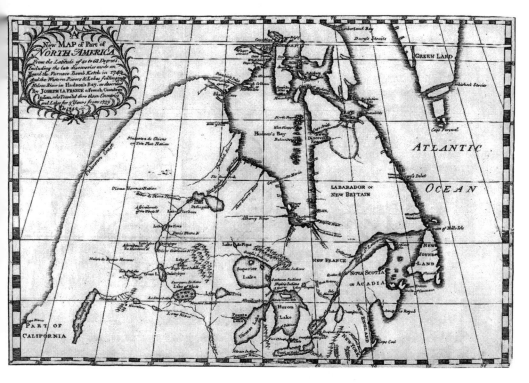

MAP 35. Joseph La France, *A New Map of Part of North America*. 1744.
Library and Archives Canada, NMC 6619

Evidence from late-eighteenth- and early-nineteenth-century explorers and traders establishes that Nez Perces and Blackfeet, for example, moved back and forth across the Rockies to hunt or raid, potentially linking the geographic ideas that circulated on either side of this mountainous barrier. The same documents demonstrating the reality of these trips, however, also highlight the hazards involved. David Thompson, who "spent a winter (1787–88?)" at a Piegan camp, averred, "The Peeagan Indians, and their tribes of Blood and Blackfeet, being next to the Mountains often send out parties under a young Chief to steal Horses from their enemies to the south and west side of the Mountains, known as the Snake, the Saleesh and the Kootanae Indians." This resembles, in a less spectacular fashion, the raiding world La France sketched. Thompson also declared that this kind of expedition was "allowed to be honourable, especially as it," like the raid described by La France, was "attended with danger" and required "great caution and activity." Lewis and Clark encountered Nez Perces on the western side of the mountains who had crossed the Rockies, had met Hidatsas, and had names for tributaries of the Missouri as far east as the Knife River upon which the Mandan villages of the first decade of the nineteenth century stood. Repeatedly, though, a reference to traversing the Rockies to the plains was followed by a description of the dangers to be met there. "In the Spring," they "cross the mountains to the Missouri to get Buffalow robes and meet etc. at which . . . time they frequent meet with their enemies and lose their horses and maney of ther people." They "would fondly go" to "the Plains of the Missouri," "provided" the "black foot Indians and the Minnetares of Fort de Prarie . . . would not kill them." Conditions in the late eighteenth and early nineteenth centuries need not have prevailed decades earlier, but they are at least consistent with the perilous pictures generated by earlier sources.[17]

Like the tales of travelers such as Du Pratz's Moncacht-apé and La France's old Cree man, the long-distance movement of goods through the West appears at first glance to constitute evidence of the long-distance movement of people. Upon scrutiny, the implications of such trade become inconclusive. The use of metals on the northwest coast offers one example. When Spanish and British mariners such as Juan Pérez and James Cook reached the northwest coast in 1774 and 1778, they found that iron, brass, and copper were being employed in various ways. They observed, for example, women wearing "bracelets of iron and copper." The question arose as to where these various metals came from. In the case of iron, Cook saw trade as one possible source: "One cannot be surprised at finding iron with all the Nations in America sence they have been so many years in a manner surrounded by Europeans and other Nations who make use of iron,

17. Richard Glover, ed., *David Thompson's Narrative, 1784–1812: A New Edition with Added Material* (Toronto, 1962), 267. Similarly, in 1776, Alexander Henry, Sr., found a slave woman among the Assiniboines who had been "taken . . . on the other side of the western mountains." The chief who did so "at the same time . . . had lost a brother and a son, in battle" (Henry, *Travels and Adventures in Canada*, ed. Bain, 312–313 [see also 278, 303–304]). Other sources mentioning long-distance raids and their dangers are "John Macdonell's 'The Red River,'" and "Charles McKenzie's Narrative," in Wood and Thiessen, eds., *Early Fur Trade*, 77–92, 221–296, esp. 92, 240. The 1801 Ac ko mok ki map discussed earlier is often adduced as evidence of extensive Blackfoot geographic expertise, and it is true that Blackfoot horizons reached far. The 1801 chart includes not only an impressive diagram of the plains and the Saskatchewan and Missouri rivers but also lines representing the Pacific coast and what seem to be the Snake and Columbia rivers. The map is nonetheless compatible with the argument advanced in this chapter. The letter accompanying it from Hudson's Bay Company surveyor Peter Fidler indicates that the chart's western rivers derived from Kootenay rather than Blackfoot information and that that some of Ac ko mok ki's notions about distant peoples were secondhand. The degree of detail on the map, moreover, diminishes west of the mountains: the peoples indicated appear to be those from the western slopes of the Rockies rather than the Pacific coast, and the lines seemingly representing the Snake and Columbia rivers fail to join each other before meeting the sea. See D. W. Moodie and Barry Kaye, "The Ac Ko Mok Ki Map," *The Beaver*, CCCVII (Spring 1977), 4–15. Lewis and Clark quotations in Moulton, ed., *Journals of the Lewis and Clark Expedition*, V, *July 28–November 1, 1805* (Lincoln, Neb., 1988), 259; ibid., VII, *March 23–June 9, 1806* (Lincoln, Neb., 1991), 248, 342–343.

and who knows how far these Indian Nations may extend their traffick one with a nother." Cook might have been correct, and modern scholars have suggested that "intertribal trade across the continent probably accounted for some" of the iron present on the northwest coast. The question has remained open, however, and Northwest Coast Indians might also have obtained iron from exchange networks extending east into Siberia or from storm-tossed Japanese fishing boats. A combination of sources seems most likely. In the case of copper, deposits from the Lake Superior and Coppermine River (west of Hudson Bay) regions could have been sources of the metal appearing on the Pacific coast, but it is worth noting that deposits of the material can be found in the more accessible Copper River region in Alaska and that Cook and his successors discerned a trade in the metal between Nootka Sound Indians and peoples dwelling farther north along the coast. Pieces of copper appearing in the region might have originated within it. The presence of metals in the Pacific Northwest might or might not indicate the existence of active transcontinental commercial patterns.[18]

More generally, the long eighteenth century was the period when trading and raiding was bringing firearms west from Hudson Bay, Canada, and Louisiana and horses north from Mexico and New Mexico. As many scholars have noted, long before the appearance on the plains of horses and guns, venerable trade networks had carried shells inland from the Pacific coast, obsidian east and west from the Rockies, agricultural products to animal hunters, and animal parts to village farmers. The Dalles on the Columbia constituted one of North America's major trading sites, and representatives from northwest coast and Columbia Plateau peoples met there and exchanged commodities that might have originated in California or the Great Lakes region. With the exception of

18. First quotation is my translation of the phrase *"brazaletes de hierro y cobre"* from the "Journal of Fray Juan Crespi Kept during the Same Voyage—Dated 5th October, 1774," in Donald C. Cutter, ed., *The California Coast: A Bilingual Edition of Documents from the Sutro Collection,* trans. Cutter and George Butler Griffin (Norman, Okla., 1969), 203–278, esp. 236. On iron, see J. C. Beaglehole, ed., *The Journals of Captain James Cook on His Voyages of Discovery,* III, *The Voyage of the Resolution and Discovery, 1776–1780,* 4 vols. (Cambridge, 1967), 321–322; Bill Holm, "Art," in William C. Sturtevant, ed., *Handbook of North American Indians,* VII, *Northwest Coast,* ed. Wayne Suttles (Washington, 1990), 602–632, esp. 603; Philip Drucker, *Cultures of the North Pacific Coast* (San Francisco, 1965), 23, 110; Erna Gunther, *Indian Life on the Northwest Coast of North America: As Seen by the Early Explorers and Fur Traders during the Last Decades of the Eighteenth Century* (Chicago, 1972), 249–251; George Quimby, "Japanese Wrecks, Iron Tools, and Prehistoric Indians of the Northwest Coast," *Arctic Anthropology,* XXII (1985), 7–15. On copper, see Beaglehole, ed., *Journals of Captain James Cook,* III, 322; John Meares, *Voyages Made in the Years 1788 and 1789 from China to the North West Coast of America* (London, 1790), lvii–lix, 247; Wallace M. Olson and John F. Thilenius, eds., *The Alaska Travel Journal of Archibald Menzies, 1793–1794* (Fairbanks, Alaska, 1993), 127; Drucker, *Cultures of the North Pacific Coast,* 65–66, 117; Gunther, *Indian Life on the Northwest Coast of North America,* 249–251.

horses, these goods were not carrying themselves, and long-distance movement of goods could therefore indicate the long-distance movement of people. This is possible, but far from certain. It is not clear from the evidence of trade whether the goods moved like news from Marathon or like the baton in a relay race. Items might have changed hands many times rather than moved on the back of one group of traders all the way across the continent.[19]

One way to illuminate the murky significance of long-distance exchange is to try to relate it to the long-distance movement of something other than trade goods. Disease offers one possibility. At first glance, recent histories of disease seem, like the evidence of trade, to indicate the magnitude of human movement across the continent. Elizabeth A. Fenn's superb book on smallpox treats what appears to have been a continental epidemic of the malady between 1775 and 1782. During these years, smallpox "ravaged the greater part of North America, from Mexico to Massachusetts, from Pensacola to Puget Sound," and Fenn suggests that the range of the disease's devastation "showed that a vast web of human contact spanned the continent well before Meriwether Lewis and William Clark made their famous journey to the Pacific in 1804–06." This is entirely plausible and would seem to militate against the hypothesis that few people were moving long distances across the West.[20]

Upon closer examination, however, the extent and intensity of the smallpox epidemic of 1775–1782 may sustain rather than undermine the hypothesis in question. Smallpox is particularly lethal among populations without prior exposure to the disease. Especially high mortality rates during a smallpox epidemic, therefore, although not proof that a population has been isolated from the illness, may at least indicate that possibility. Fenn suggests that areas such as western Canada had not been exposed to smallpox on a large scale before the 1770s, and they suffered appalling losses when they were. Fenn also shows the ways in which developments such as the spread of horses and the fur trade to the northern portions of the Far West might have increased the contact between those areas and the larger world to the point where a smallpox epidemic afflicting one part of the continent could now extend its devastation to all. Fenn's examples suggest that part of what might have made the outbreak so lethal in large parts of the Far West was that many of their inhabitants had not been personally exposed to the disease before, since many of the longer and denser strands forming the

19. For an overview of western Indian trade, see Calloway, *One Vast Winter Count*, 47–49, 93–115, 239–312.

20. Quotations from Elizabeth A. Fenn, *Pox Americana: The Great Smallpox Epidemic of 1775–82* (New York, 2001), 3, 6 (see also 28, 137–138, 172–173, 192–193, 222–223, 227, 238, 248, 252–253, 271–273); Calloway, *One Vast Winter Count*, 415–426.

transcontinental web discussed above were only spun by expanding commerce in the years before the 1770s. What had been largely regional trade networks had become more completely integrated into much larger continental and even global patterns of exchange. The intensity of the effects of smallpox in the West in the 1770s and 1780s may indicate the limitations of human movement earlier in the century.[21]

Moreover, just as evidence that may initially seem a sign of frequent long-distance journeys often turns out to suggest the contrary, phenomena fostering movement in some ways probably inhibited it more generally. The spread of horses on the plains and across the Rockies in the eighteenth century made it possible for Comanches, Shoshones, Blackfeet, and Nez Perces to cover greater distances at greater speeds than had been conceivable in the long millennia between the extinction of horses native to North America and the arrival of conquistadors so alien to it. Thompson spoke of a 1787 Piegan battle with the Spanish 1,500 miles away from the Blackfoot home territories. The same horses enhancing the mobility of nations like the Blackfeet, however, increased the vulnerability of many other peoples. The Nez Perce had to think long and hard before venturing onto the plains the Blackfeet roamed, and the travel-inhibiting terror of the "Snakes" mentioned above arose in part because people like the Shoshones were among the first Plains nations to acquire horses and use them in war.[22]

A similar conclusion can be drawn regarding the effects of the trade in Indian captives taken in raids or wars. Sacagawea is the most famous Indian captive to play a role in European and Anglo-American western exploration — her presence assured western Indians of the Lewis and Clark Expedition's peaceful inclinations, and her meeting with her brother Shoshone chief Cameahwait smoothed relations between the explorers and the Indian nation whose cooperation was necessary for crossing the Rockies. She was not alone. Captive Indians were appearing in western exploration records long before 1804. The Snakes who seem to have inspired so much fear among the peoples Louis-Joseph and François

21. For analogous eighteenth-century cases in Asia, see Peter C. Perdue, *China Marches West: The Qing Conquest of Central Eurasia* (Cambridge, Mass., 2005), 45–49, 91–92. I should point out that my interpretation of the implications of Fenn's examples may differ from hers.

22. Glover, ed., *David Thompson's Narrative*, 269–270. Similarly, fur trader Peter Pond, in his description of a 1773 visit to a Sauk village in what is now Wisconsin, claimed that the village men often raided on and beyond the Missouri, sometimes approaching Santa Fe: "The Men often join War parties with other Nations and Go aganst the Indans on the Miseure and west of that. Sometimes thay Go Near St. Fee in New Mexico and Bring with them Spanish Horseis." See "The Narrative of Peter Pond," in Charles M. Gates, ed., *Five Fur Traders of the Northwest: Being the Narrative of Peter Pond and the Diaries of John Macdonell, Archibald N. McLeod, Hugh Faries, and Thomas Conner* (Minneapolis, Minn., 1933), 9–60, esp. 41.

La Vérendrye met in 1742–1743 were rumored to have destroyed seventeen villages in 1741 and to have sold the villages' young women for "horses and merchandise." More generally, the numbers of slaves traded were substantial. In a 1757 memoir, French officer Louis-Antoine de Bougainville wrote that, in the trade with the Crees and Assiniboines during "an ordinary year" at the French "post of the Sea of the West" on the Saskatchewan River, there arrived not only beaver, otter, and lynx furs but also "more than fifty to sixty Indian slaves or panis of *Jatihilinine*, a nation situated on the Missouri, who play in America the role of the negroes in Europe." Perhaps more striking were the implications of Indian slave trafficking farther east. Brett Rushforth has estimated that, in "Montreal's commercial district around Rue Saint-Paul and the Place du Marché," "fully half of all colonists who owned a home in 1725 also owned an Indian slave." Many of the slaves in New France came from regions east of the Mississippi and the Great Lakes, but many others came from farther west. Because slaves were often captured in Indian raids on peoples in distant regions, they constituted a potential source of information about unfamiliar territory, and an aspiring French explorer could presumably have heard much about lands beyond the Mississippi and Great Lakes by striking up a conversation in the streets of Montreal or New Orleans. To give one example, in 1758, in a description of Missouri River Indian peoples, Louisiana governor Louis Billouart de Kerlérec (1752–1763) mentioned Arikara slaves' saying that the sources of the Missouri were unknown to old men of their nation. On the other hand, the circumstances in which captives were taken from their natal regions might not have facilitated the clearest memory of their journey, and captives who did recall routes home might have been less than enthusiastic about describing them to potential enemies. The kinds of communicative difficulties discussed earlier may also help explain how curious residents of slaveholding communties could have failed to understand more about distant regions.[23]

23. For the La Vérendryes, see "Journal of the Expedition of the Chevalier de La Vérendrye and One of His Brothers," in Burpee, ed., *Journals and Letters of La Vérendrye*, trans. [LeSueur], 412–417: "chevaux et quelques marchandises." These slaves were allegedly sold "on the coast [à la mer]," where there were "whites [Blancs]" or "French [Les François]" with "a large number of slaves [quantité d'esclaves]." Allowing for the apparent confusion about salt water and ethnicity mentioned above, these whites could have been the Spanish in New Mexico. See also Henry, *Travels and Adventures in Canada*, ed. Bain, 278, 312–313; "Slave Woman, The," in Davies, ed., *Letters from Hudson Bay*, 410–413; Glyn[dwr] Williams, *Voyages of Delusion: The Northwest Passage in the Age of Reason* (London, 2002), 8–10; "1757: Memoir of Bougainville," in Reuben Gold Thwaites, ed., State Historical Society of Wisconsin, *Collections*, XVIII, *The French Regime in Wisconsin—1743–1760, The British Regime in Wisconsin—1760–1800, The Mackinac Register of Marriages—1725–1821* (Madison, Wis., 1908), 187. Variations of "Panis" appear to be a generic term for "slave" rather than a clear indication that captives came from a distinct Pawnee ethnic group. See Brett Rushforth, "'A Little

La Vérendrye took his efforts far beyond urban chats and hoped Indian slaves would assist his explorations west of Lake Superior. He wrote in his journal for 1737 that, "to facilitate my discovery," he wanted "to withdraw from the hands of the Cree as many slaves as possible belonging to" the Assiniboine and "Pikaraminioüach" nations, "so as to form friendly relations with them, and have these slaves to accompany me on my journey and act as interpreters." The Crees were valuable sources of information, but La Vérendrye felt that their slaves offered something more. He was intrigued, for example, by reports of "white men" who lived down the River of the West, had "walled towns and forts" and provided the Pikaraminioüach with "axes, knives and cloth," but who evinced "no knowledge of fire-arms or of prayer." The Crees themselves had "no knowledge of these men except through the slaves they have made after having crossed the height of land." It is difficult to ascertain who these "white men" might be. Most likely, this account conflated the Spanish and the Mandans living across the height of land separating the Missouri and Hudson Bay–Great Lakes watersheds. La Vérendrye's informants might also have been referring in some way to the Northwest Coast Indians who dwelled beyond the east-west Continental Divide and who possessed cloth, fortified villages, and a certain amount of iron and copper when European sailors encountered them later in the century. Whatever the referent might have been, Indian slaves offered hints to European explorers about what lay beyond the horizon.[24]

Flesh We Offer You': The Origins of Indian Slavery in New France," *WMQ* , 3d Ser., LX (2003), 777, 788–789; "Memoir on Indians by Kerlérec," Dec. 12, 1758, in Dunbar Rowland, A. G. Sanders, and Patricia Kay Galloway, eds., *Mississippi Provincial Archives: French Dominion*, V, 1749–1763 (Baton Rouge, La., 1984), 203–227, esp. 207. Idea regarding slaves' difficulty in remembering routes home derives from a telephone conversation with Rushforth.

24. "Report of the Sieur de La Vérendrye, Lieutenant of the Troops and Commandant of the Posts of the West: Presented to Monsieur the Marquis de Beauharnois . . . to Be Sent to the Court," in Burpee, ed., *Journals and Letters of La Vérendrye*, trans. [LeSueur], 213–262, esp. 248–249:

> Le haut de la riviere du Couchant est habité de Sauvages errans comme les Assiniboils, nommés Pikaraminioüach fort nombreux sans armes a feu, mais ils ont haches, couteaux, etoffes etc. comme nous, qu'ils tirent du bas de la riviere habitée par des blancs qui ont des villes mûrées et des forts, ces blancs ne connoissent point les armes a feu ny la priere, il peut y avoir trois cent lieuës de la hauteur des terres a la mer, Les Cris n'ont connoissance de ces hommes blancs que par les esclaves qu'ils ont fait après avoir passé la hauteur des terres; cette nation porte ses tentes ou lôges comme les Assiniboils.
>
> Pour faciliter ma découverte . . . mon dessein seroit . . . de retirer d'entre les mains des Cris ce que je pourois d'esclaves de ces deux nations, pour lier amitié avec elles, m'accompagner dans le voyage, et me servir d'interprettes.

For an earlier example of the use of what might have been slaves, in this case *"Essanapés,"* as interpreters and guides, see Lahontan, *Oeuvres complètes*, ed. Ouellet and Beaulieu, I, 400, 411, 413. For an example of British acquisition of geographic information from an "Earchethinue" slave, see Rich, ed., *Isham's Observations on Hudsons Bay*, 113.

This acquisition of slaves required raids on distant peoples, and slaves were, of course, coerced into making journeys taking them far from their homes. In this sense, Indian slavery increased the extent of western Indian movement. It is to be suspected, however, that in a larger sense slave raids curtailed mobility by increasing insecurity. Jonathan Carver's observations, based on his own travels around and beyond the Great Lakes, have provided historians with one indication of this. Carver opined that the French purchase of Indian slaves, "instead of being the means of preventing cruelty and bloodshed, . . . only caused the dissensions between the Indian nations to be carried on with a greater degree of violence, and with unremitted ardor. The prize they fought for being no longer revenge or fame, but the acquirement of spirituous liquors, for which their captives were to be exchanged, and of which almost every nation is immoderately fond, they sought for their enemies with unwonted alacrity, and were constantly on the watch to surprise and carry them off." In such conditions, movement away from the protection of a friendly village or reliable allies could only be hazardous. Farther west, and in later decades, similar indications of fearful immobility appear. Explorers such as Mackenzie and Lewis and Clark encountered nations such as the Sekanis and Shoshones who had lost members to captivity (as well as horses to theft or lives to combat) and who, to protect life, liberty, and property, had resorted to precarious and penurious refuge in the mountains. The Utes that Fathers Escalante and Domínguez encountered evinced a great fear of raids by Comanches from the east, perhaps in part because Comanche bands were taking Utes captive in the second half of the eighteenth century. People who knew routes through the Rockies were often hiding in them.[25]

The character of Indian cartography renders these kinds of movement re-

25. Jonathan Carver, *Three Years' Travels through the Interior Parts of North-America, for More Than Five Thousand Miles*... (Philadelphia, 1789), 177; Moulton, ed., *Journals of the Lewis and Clark Expedition*, V, 83, 91, 109, 121–123; Lamb, ed., *Journals and Letters of Sir Alexander Mackenzie*, 287, 318; Ted J. Warner, ed., *The Domínguez-Escalante Journal: Their Expedition through Colorado, Utah, Arizona, and New Mexico in 1776*, trans. Angelico Chavez (Provo, Utah, 1976), 31–33, 38–39, 50, 54, 56, 58, 60, 61; Joseph P. Sánchez, "Translation of Juan María Antonio Rivera's Second Diary, October 1765," in Sánchez, *Explorers, Traders, and Slavers: Forging the Old Spanish Trail, 1678–1850* (Salt Lake City, Utah, 1997), 149–157, esp. 152–153, 156; James F. Brooks, *Captives and Cousins: Slavery, Kinship, and Community in the Southwest Borderlands* (Chapel Hill, N.C., 2002), 50, 64. Earlier in the century, Spanish sources report Apaches from the plains and sierras northeast of Taos abandoning their hunting grounds and rancherías and seeking refuge among Spanish or Navahos because of Comanche and Ute attacks on their settlements and seizure of their women and children. "Diary of the Campaign of Governor Antonio de Valverde against the Ute and Comanche Indians, 1719," "Decree for Council of War, Santa Fe, Nov. 8, 1723," "Bustamante to Casa Fuerte, Santa Fé," Jan. 10, May 30, 1724, in Alfred Barnaby Thomas, ed. and trans., *After Coronado: Spanish Exploration Northeast of New Mexico, 1696–1727; Documents from the Archives of Spain, Mexico, and New Mexico* (Norman, Okla., 1935), 112–118, 194, 201, 208.

strictions especially important for a study of geographic ideas' circulation. Because Indian maps seem to have been impermanent, culturally specific, and dependent on oral explanation, it is not clear how geographic expertise could be maintained if a group was excluded from an area once frequented, particularly if experienced members of the community died along with their knowledge. Nor is it certain how well groups confined to a particular area could have acquired precise and usable information from their neighbors. If western Indians were often speaking of areas they had heard about but not seen, the difficulty French investigators had in using Indian geographic information is more easily explicable.

THE UNRELIABILITY OF EUROPEAN and Indian sources of geographic information, the plethora of tongues spoken by the many European and Indian peoples of the West, and the varied ways of imagining and describing western landscapes created tremendous interpretive difficulties for French explorers and geographers investigating North America's mysterious western territories. Even when French explorers or western Indians had seen the regions geographers wanted to apprehend, their testimony resisted easy belief or understanding.

In the many cases where unsettled western conditions denied explorers and Indians personal familiarity with the geography of large parts of the region, comprehension was still more elusive. The contrast between circumstances favoring French cartographic enterprises in Eurasia and conditions inhibiting French reconnaissance in the North American West is telling. The political conditions facilitating movement and information-gathering in France, Russia, and China found no equivalent in the eighteenth-century West. No imperial overlord established or maintained stability and security. No one power could conduct French investigators to distant regions or send agents who could safely return from them. Examining 1713–1763 French western exploration in isolation, or in comparison to contemporary Spanish or British efforts, its dynamism and reach stand out. In the context of the larger French geographic enterprise, it becomes clear just how dependent French investigators were on vigorous state support.[26]

The anarchic international system pushing Spain, France, Britain, and ultimately Russia into competition for western North American dominion seems to have had its counterpart in a mutable international climate among the in-

26. For a comparable argument regarding the late-seventeenth- and early-eighteenth-century French maritime data-gathering mentioned in Chapter 6, see Nicholas Dew, "Vers la ligne: Circulating Measurements around the French Atlantic," in James Delbourgo and Nicholas Dew, eds., Science and Empire in the Atlantic World (New York, 2008), 53–72, esp. 58–59. Dew emphasizes the French scientific community's reliance on commercial transport, with all its vagaries.

digenous peoples of the eighteenth-century North American West. And just as European imperial rivalries often constrained European exploration by rendering large parts of the continent unwelcome to subjects of the wrong crown, native American hostilities seem frequently to have impeded the movements of Indian peoples and European explorers through the West.

French explorers like La Vérendrye were often trying to communicate using languages they understood imperfectly and discussing geographic features distinct cultures conceptualized differently, all in an effort to apprehend regions neither party had visited. Western terrain, it is true, humbled explorers then as it awes travelers now. But imperial and cartographic achievements in France, Russia, and China had shown—and explorers' accomplishments in the American West would soon demonstrate—that determination, repetition, and luck could overcome natural obstacles. Ultimately, it was a combination of the context in which humans related to one another and their contrasting fashions of representing the natural world that checked French western efforts. Had Frenchmen been able to crisscross the West as they had China, they could have rendered western geography as clear to their countrymen as that of China had become. Had western Indians been able to travel the continent as freely as French cartographers journeyed across France, and apt to represent what they had seen in a fashion as comprehensible to French geographers as French maps were to Russian and Chinese rulers, France could have grasped western North America without touching it.

Beyond illuminating French difficulties with western American geography, this discussion of eighteenth-century French investigations of France, Russia, China, and North America also recalls the earlier discussion of western conditions' contributing to lingering Spanish geographic uncertainty. The results of the two examinations generally support one another. In both cases, we find the same pattern of communicative difficulties, disruptive effects of European arrival, and impressive but ultimately bounded western geographic horizons as well as the suggestion of indigenous mobility constrained by unsettled relations among diverse peoples. In both cases, the presence or absence of empire is a critical variable. It is not that expansionist powers like Spain, France, Britain, and Russia required the presence of an existing empire to rapidly acquire a rough geographic understanding of a large body of land. Navigable rivers, exceptional intrepidity, and sometimes relentless brutality had proved efficacious substitutes in the sixteenth and seventeenth centuries in areas outside the West, and European and Euro-American powers would finally explore and survey the far West itself in the later eighteenth and nineteenth centuries. But Spanish and French experiences suggest that a relationship of cooperation or conquest—and conquest actually involved a good bit of cooperation — with an empire that had already or was currently engaged in investigating, organizing, and delineating space could make terrestrial reconnaissance much easier. Knowledge production was harder and slower than imperial parasitism. Empires, even those from radically different cultures, could, in the language of political scientist and anthropologist James C. Scott, make spaces and people "legible" not just to themselves but also to each other.[27]

The trickier question concerns the geographic horizons of non-state indigenous Americans. We cannot assume that, because Europeans failed to acquire comprehensive geographic understanding from western Indians, these Indians failed to possess it. Conceptions of space were so different, communication so difficult, and Amerindian peoples often so distant that much could go on in western Indian heads that never made its way to European sources. We cannot be sure that the absence of large-scale and centralized political structures precluded the kind of long-range interactions and geographic understandings we cannot see. We need not, in short, follow Thomas Hobbes and infer that "without a common Power to keep them all in awe," western Indians possessed "no Knowledge of the face of the Earth." The native maps we can examine and journeys we can document show that many knew quite a lot. If we are going to postulate the grandest indigenous geographic horizons, however, we need to search for the kinds of mechanisms that might have enabled semicontinental geographic understanding without Leviathan.[28]

More immediately, we can say that French explorers were not getting the kind of information from western Indians necessary to dispel the mists obscuring the Far West. As the French and British empires approached their climactic North American confrontation, French explorers and geographers seeking the West continued to grope and stumble in darkness.

27. James C. Scott, *Seeing Like a State: How Certain Schemes to Improve the Human Condition Have Failed* (New Haven, Conn., 1998), 2–3.
28. Thomas Hobbes, *Leviathan*, ed. C. B. Macpherson (London, 1985), 185–186.

PART IV.

BRITISH PACIFIC VENTURES AND THE EARLY YEARS OF THE SEVEN YEARS' WAR

9

BRITISH DESIGNS ON
THE SPANISH EMPIRE, 1713–1748

La Vérendrye and his compatriots were not alone in their search for a water route to the west, nor was a great river the only form such a passage might take. In 1731—a year after La Vérendrye's report "Touching upon the Discovery of the Western Sea"—Arthur Dobbs, Ulster landowner and member of the Irish House of Commons, wrote a seventy-page memoir positing the existence of a Northwest Passage from Hudson Bay to the Pacific and encouraging the Hudson's Bay Company or the British Admiralty to dispatch ships to look for it. Dobbs used different kinds of evidence to support his case. He offered a convoluted interpretation of the height and direction of the bay's tides, seeing in them indications of an oceanic connection. He noted the presence of whales in the bay who did not appear to have reached it through a known channel. He mentioned the relative lack of ice in the northwest parts of the bay during certain times of the year, an apparent sign of open sea in that direction. He noted the presence in native burial sites of copper that he presumed came from trade with nations farther west. And he cited reports of voyages such as that of "Valerianos" (Juan de Fuca), who, it was claimed, had sailed in 1592 through a northwest coast inlet to an interior sea and a "Rich and pleasant Country."[1]

1. English merchant and Northwest Passage enthusiast Michael Lok claimed to have spoken with Greek pilot Juan de Fuca, "named properly Apostolos Valerianos," in Venice in 1596. According to Lok, Fuca reported that in 1592 he took two ships up the North American Pacific coast in search of the "Straits of Anian." Fuca "came to the Latitude of fortie seven degrees, and . . . there finding that the Land trended North and North-east, with a broad Inlet of Sea, betweene 47. and 48. degrees of Latitude: hee entred thereinto, sayling therein more than twentie dayes, and found that Land trending still sometime Northwest and North-east, and North, and also East and Southeastward, and very much broader Sea then [sic] was at the said entrance." Fuca claimed "that he went on Land in divers places, and that he saw some people on Land, clad in Beasts skins: and that the Land is very fruitfull, and rich of gold, Silver, Pearle, and other things." See Samuel Purchas, *Hakluytus Posthumus; or, Purchas His Pilgrimes: Contayning a History of the World in Sea Voyages and Lande Travells by Englishmen and Others*, XIV, 20 vols. (Glasgow, 1906), 415–421. Interestingly, scholars have found a Greek pilot working in New Spain between 1588 and 1594, and the northwest

Dobbs promoted his project with abandon, even when it met with less-than-universal enthusiasm. Skeptics such as the managers of the Hudson's Bay Company evinced little interest in expending funds on the search for a passage of dubious existence and often fatal appeal. Between 1576 and 1616, in their search for a route to Asia, English navigators Martin Frobisher, John Davis, Henry Hudson, and William Baffin investigated the northern straits and bays that would come to bear their names. In 1719, Hudson's Bay Company employee James Knight probed and died on the western shores of Hudson Bay. None of these explorers had found a route to China. Such inauspicious precedents notwithstanding, Dobbs's faith in the existence of a Northwest Passage proved as unyielding as it was unfounded: neither icebergs, nor Inuits, nor extant and impassable land barriers ever shook it. He never visited the Hudson Bay region himself, but Dobbs pored over published descriptions of it and happily depicted with his pen what he never saw with his eyes. The prevailing European ignorance of the vast regions west of Hudson Bay gave Dobbs's faith ample room for survival. Although he could not prove that a practicable strait existed, lack of reliable information about the region made it difficult to prove that a Northwest Passage, or anything else for that matter, did not exist somewhere in North America. The challenges of exploring the subarctic regions around Hudson Bay made it difficult for expeditions to investigate every promising locality. The faithful could always point to some unturned stone concealing geographic treasure.

Dobbs's persistent hectoring resulted in action. In 1741, he persuaded the Admiralty to send an expedition to search for a passage, and, in 1742, Captain Christopher Middleton explored Hudson Bay's western shores. In 1747, an expedition commanded by William Moor and privately financed by Dobbs and his associates returned to the bay in quest of the long-desired route to the Indies.[2]

Situated on North America's northern edge, more hospitable to polar bears than people, endowed with few easily exploitable natural resources, Hudson Bay has generally seemed to historians too isolated a region to exert much influence on events elsewhere. In the mid-eighteenth century, however, because of the

coast opens between latitudes 48° and 49° north into Puget Sound and the passage between Vancouver Island and the mainland. The strait received Fuca's name in 1787. It remains unclear whether Lok or Fuca invented the voyage described or embellished or misunderstood the achievements of a trip that actually took place. See Warren L. Cook, *Flood Tide of Empire: Spain and the Pacific Northwest, 1543–1819* (New Haven, Conn., 1973), 21–29; Glyn[dwr] Williams, *Voyages of Delusion: The Northwest Passage in the Age of Reason* (London, 2002), 132–133. Dobbs's memoir appears in William Barr and Glyndwr Williams, eds., *Voyages to Hudson Bay in Search of a Northwest Passage, 1741–1747,* 2 vols. (London, 1994–1995), I, 9–36, esp. 28.

2. The best account of these events is in Williams, *Voyages of Delusion,* 46–214. For a short overview, see William H. Goetzmann and Glyndwr Williams, *The Atlas of North American Exploration: From the Norse Voyages to the Race to the Pole* (New York, 1992), 42–43, 106–107.

PLATE 1. Arthur Dobbs. N.d. Oil on canvas.
Courtesy North Carolina Museum of History

bay's imagined connection to the Pacific, British exploration of this marginal area contributed to the origins of a conflict of global extent and consequences. During the years leading into the Seven Years' War, British expeditions to Hudson's Bay and, more broadly, British designs on the South Sea attracted French and Spanish interest and influenced French and Spanish policy.

Understanding why requires an investigation of the diplomatic consequences of British Hudson Bay reconnaissance and Pacific projects, but that investigation depends on addressing two preliminary questions. As with Spain and France, one issue is why British far western exploration was so limited and why, when it occurred, it involved ships in a forbidding northern bay rather than the kind of temperate land expeditions launched by Britain's North American competitors. A second question, one that preoccupied mid-eighteenth-century French observers as well, was how this Hudson Bay venture fit into the larger context of 1730s and 1740s British imperial expansionism. One answer will serve for both questions. For reasons rooted in geography, geopolitics, and the still-essentially-maritime character of the pre-1763 British Empire, Spanish America played for Britain the role the undiscovered North American West did for France.

Limited British Western Exploration

The French and Spaniards . . . possessed more imagination and spirit of adventure than the English, and were better fitted to be the explorers of a new continent even as late as 1751.

    This spirit it was which so early carried the French to the Great Lakes and the Mississippi on the north, and the Spaniard to the same river on the south. It was long before our frontiers reached their settlements in the west, and a *voyageur* or *coureur de bois* is still our conductor there. Prairie is a French word, as Sierra is a Spanish one. Augustine in Florida, and Santa Fé in New Mexico . . . both built by the Spaniards, are considered the oldest towns in the United States. Within the memory of the oldest man, the Anglo-Americans were confined between the Apalachian Mountains and the sea, "a space not two hundred miles broad," while the Mississippi was by treaty the eastern boundary of New France. . . . So far as inland discovery was concerned, the adventurous spirit of the English was that of sailors who land but for a day, and their enterprise the enterprise of traders.

    —HENRY DAVID THOREAU, *Cape Cod*, 1855

Thoreau exaggerated. South Carolina traders pushing toward the Mississippi were hardly inert, even if New Englanders confined to the Northeast were. But his basic point is valid. Before 1763, few Britons set foot in the western two-thirds of North America. Navigators like Francis Drake on the Pacific coast and pathfinders like Anthony Henday west of Hudson Bay were exceptional. Their successors arrived only after a century and a half of British Atlantic settlement and maritime activity. Spaniards like Coronado and Kino and Frenchmen like La Salle and La Vérendrye lacked serious British rivals.[3]

Accounting for the limits of Spanish and French western North American exploration was a complicated and lengthy business. In the case of the British Empire, the answer is relatively straightforward. Prospective British explorers generally lacked easy entrée to far western territories, and in Hudson Bay, the one area where British posts stood in an ideal location for launching western expeditions, the restrictive policies of the Hudson's Bay Company limited pre-1763 western reconnaissance.

Before the end of the Seven Years' War, the fundamental difficulty for British western reconnaissance was Anglo-America's limited access to the West's territories and inhabitants. This was most evident in the case of the British Atlantic seaboard colonies. The farthest outliers of these colonies before the Seven Years' War were the fur-trading posts at Fort Oswego on Lake Ontario (fortified in 1727) and Pickawillany on the Miami River (established in 1748), and a scattering of trans-Appalachian settlers. A handful of explorers and traders had gone farther west, but a host of obstacles limited the range of their wanderings.

It was not just the distance separating British settlements along the Atlantic coast from the trans-Mississippi interior; nor was it just the successive Appalachian ridges to be surmounted or bypassed, the English colonies' lack of great rivers like the Mississippi and the Saint Lawrence, or the orientation of northeastern corridors like the Hudson–Lake Champlain and Mohawk valleys toward New France and Iroquoia. It was also the awkward fact that, from Louisiana's 1699 foundation, a continuous belt of putatively French territory intervened between Anglo-America and the Far West. Although it is easy in retrospect to belittle the French Empire's ability to control the American interior territory it claimed, British voyagers often found the French presence a substantial impediment. In 1699, a ship from Charles Town ascended the Mississippi for one hundred miles, only to be repulsed by the recently arrived Iberville. In 1742, a British party led by John Howard and John Peter Salley descended the Ohio and much of the Mississippi, only to be captured by "a company of men, to the number of ninety, consisting of French men Negroes and Indians who took us prisoners and carried us to the town of New Orleans." In 1752, a Franco-Indian party destroyed Pickawillany. Although the Louisiana and Illinois country French were thinly spread, they traveled widely and frequented and dominated key routes and junctions.[4]

3. See Henry David Thoreau, *A Week on the Concord and Merrimack Rivers: Walden, or, Life in the Woods; The Maine Woods; Cape Cod* (New York, 1985), 1012–1013.

4. John Bartlet Brebner, *The Explorers of North America, 1492–1806* (Cleveland, Ohio, [1966]),

More important, as Howard and Salley could attest, although the French were few, their Indian associates were many. A Choctaw raiding party, whether acting in its own interest or at French behest, might prove harder for a British explorer to elude than a French post. Perhaps more fundamentally, as events in the decades after the Seven Years' War would remind Anglo-Americans, trans-Appalachian Indians possessed reasons quite apart from French encouragement to contest westward British advances.[5]

Topographic, French, and Indian obstacles hindered but did not extinguish pre–Seven Years' War British geographic interest and trans-Appalachian exploration. Individuals and institutions on both sides of the Atlantic were seeking to acquire information about the territories beyond the mountains. From its 1696 inception, England's Board of Trade served as a destination, repository, and occasional producer of maps of and reports about areas such as the French territories west of the Appalachians. When board member Martin Bladen was dispatched to France in 1719 to serve as a British negotiator, he was instructed while there "to get the best Information you can, concerning the Situation, Trade, strength, Laws and Government of the French Colonies in America." The fruit of this kind of interest was apparent in the board's 1721 "State of the British Plantations in America," a forty-page survey of the "situations, Governments, strengths and Trade" of Britain's continental colonies and their western French and Indian neighbors. The Board of Trade was, moreover, delighted to accept charts such as John Mitchell's 1755 "Map of the British and French Dominions in North America." Despite its interests, the board lacked the funding and personnel to conduct large-scale and systematic exploration, and its acquisition of geographic information was consequently more often passive than active. It conducted no operation on the scale of the great Russian and Chinese surveying efforts.[6]

267; "Journal of the Voyage of the Chevalier d'Iberville on the King's Ship the *Renommée,* in 1699, from Cape François to the Shore of the Mississippi and His Return," in Richebourg Gaillard McWilliams, ed. and trans., *Iberville's Gulf Journals* (University, Ala., 1981), 105–156, esp. 107–109; McWilliams, ed. and trans., *Fleur de Lys and Calumet: Being the Pénicaut Narrative of French Adventure in Louisiana* (Baton Rouge, La., 1953), 30, 160–162; Jean-Baptiste Bénard de La Harpe, *The Historical Journal of the Establishment of the French in Louisiana,* ed. Glenn R. Conrad, trans. Joan Cain and Virginia Koenig (Lafayette, La., 1971), 65; John Peter Salley, "A Brief Account of the Travels of Mr. John Peter Salley," in William M. Darlington, *Christopher Gist's Journal: With Historical, Geographical, and Ethnological Notes and Biographies of His Contemporaries* (Pittsburgh, 1893), 253–260, esp. 255; Fred Anderson, *Crucible of War: The Seven Years' War and the Fate of Empire in British North America, 1754–1766* (New York, 2001), 29.

5. On Choctaw raiding parties, see Daniel H. Usner, Jr., *Indians, Settlers, and Slaves in a Frontier Exchange Economy: The Lower Mississippi Valley before 1783* (Chapel Hill, N.C., 1992), 81–87.

6. For introductions to the Board of Trade, see Charles M. Andrews, *The Colonial Period of*

Imperial offices constituted, however, only one of the possible sources of exploratory initiative, and Anglo-American trans-Appalachian journeys could take place with or without imperial impulsion. Eighteenth-century traders, trappers, and adventurers such as Thomas Nairne, George Croghan, Conrad Weiser, and many anonymous others pushed west from the Atlantic seaboard, sometimes on their own initiative, sometimes prompted by organizations such as the Ohio Company or colonies such as Virginia and Pennsylvania. Such explorers were dynamic, so much so that they would ultimately prove uncontrollable, and they were becoming more numerous as the eighteenth century progressed. But because they often lacked rudimentary cartographic or compositional skills and were frequently pursuing local or personal agendas, their territorial coverage was erratic, their sketches often opaque. They could provide a hint of western possibilities, not a comprehensive and reliable picture of them.

The deficiencies of British cartography and British geographic familiarity became evident long before British thoughts reached the trans-Mississippi West. In the mid-eighteenth century, despite more than a century of experience with the continent, Britons and Anglo-Americans continued to display a surprising lack of familiarity with basic eastern American geographic features. It was only in 1752 that John Finley established the relation between the Cumberland Gap and the Kentucky lowlands. In 1755, following the advice of Lewis Evans, draftsman of the famous "General Map of the Middle British Colonies in America," General Edward Braddock followed the Virginia route to Fort Duquesne by way of Alexandria and Winchester rather than what turned out to be the manifestly easier Pennsylvania road from Carlisle. In 1764, speaking of the North American territories acquired during the Seven Years' War, the Board of Trade could still justly complain, "We find ourselves under the greatest difficulties arising from the want of exact surveys of these countries in America, many parts of which have never been surveyed at all and others so imperfectly that the Charts and

American History, IV, England's Commercial and Colonial Policy (New Haven, Conn., 1938), 309–310; Oliver Morton Dickerson, American Colonial Government, 1696–1765: A Study of the British Board of Trade in Its Relation to the American Colonies; Political, Industrial, Administrative (Cleveland, Ohio, 1912), 286; I[an] K. Steele, Politics of Colonial Policy: The Board of Trade in Colonial Administration, 1696–1720 (Oxford, 1968), 153–154, 165–167. For Bladen's instructions, see "Instructions for Martin Bladen Esqr," July 3, 1719, SP 78/166, 15r; "State of the British Plantations in America, in 1721," Sept. 8, 1721, in E. B. O'Callaghan, ed., Documents relative to the Colonial History of the State of New York: Procured in Holland, England and France, V, London Documents: XVII–XXIV; 1707–1733, comp. John Romeyn Brodhead, 15 vols. (Albany, N.Y., 1855), 591–630. On the limited cartographic capacities of the Board of Trade, see Andrews, Colonial Period of American History, IV, 299; G. R. Crone, Maps and Their Makers: An Introduction to the History of Cartography (London, [1953]), 145; William P. Cumming, British Maps of Colonial America (Chicago, 1974), 12.

Maps thereof are not to be depended upon." In more ways than one, eighteenth-century British America was a long way from France, Russia, or China.[7]

What was true for the British colonies along the Atlantic seaboard need not have applied to the British fur-trading posts on Hudson Bay. Here a great continental indentation took the British Empire and the Hudson's Bay Company to territories deep within the North American interior. For all Dobbs's rhetorical excesses and geographic credulity, he was right to see Hudson Bay as a jumping-off point for exploration. The obstacles of a harsh climate and Amerindian peoples who might wish to check British westward movements existed, but mountains were distant and a set of navigable rivers ran conveniently into the bay. Although the French Empire endeavored after the War of the Spanish Succession to establish inland fur-trading posts interrupting Hudson's Bay Company access to the interior Indian trade, these incipient efforts sufficed to intercept many furs that might otherwise have gone to the bay but not to entirely obstruct occasional British western forays. When Anthony Henday journeyed inland in 1754–1755, he encountered French trading posts and many Indians "in the French interest." Nonetheless, he succeeded in venturing hundreds of miles, probably to within a short distance of the Rockies. When French traders at Fort Le Pas tried to detain Henday, either he, his guide Attickashish, or both of them scoffed at the effort. Whether the French lacked the power to arrest a British trader or, more precisely, to interfere with a companion of Attickashish is not quite clear. In any case, Henday and Attickashish pushed on two days later. In the decades after the Seven Years' War, many more Hudson's Bay men would follow, surveying huge tracts of what is now Western Canada.[8]

A few outliers like Henday and his 1690–1692 predecessor Henry Kelsey aside, where the British Empire stood in the best position to move into the American Far West before the Seven Years' War, its subjects chose for the most part to confine themselves to the neighborhood of the Hudson's Bay Company's littoral posts. As Glyndwr Williams has noted in his treatments of the company, its directors generally preferred the modest but steady profits to be gained by receiving Indian hunters at coastal forts over the costs and dangers inherent in interior exploration. A company blessed, furthermore, with a charter granting it

7. On Finley and the Cumberland Gap, see Goetzmann and Williams, *Atlas of North American Exploration,* 82. On Evans and Braddock, see Lawrence Henry Gipson, *The British Empire before the American Revolution,* VI, *The Great War for the Empire: The Years of Defeat, 1754–1757* (New York, 1946), 73–75. Quotation from Keith R. Widder, "The Cartography of Dietrich Brehm and Thomas Hutchins and the Establishment of British Authority in the Western Great Lakes Region, 1760–1763," *Cartographica,* XXXVI (1999), 1–23, esp. 2.

8. Anthony Henday, *A Year Inland: The Journal of a Hudson's Bay Company Winterer,* ed. Barbara Belyea (Waterloo, Ontario, 2000), 43–198, esp. 54, 57–58.

exclusive rights to the bay's trade often deemed it wise to avoid allowing exploration to attract the attention of potential British competitors. Dobbs bitterly criticized the company for being asleep "at the edge of a frozen sea." Only beginning in 1754, when the aftermath of Dobbs's criticisms made it urgent, did the Hudson's Bay Company succeed in regularly pushing its employees to visit and winter in lands beyond Hudson Bay's shores. In contrast with late-seventeenth- and eighteenth-century Carolina, Virginia, and Pennsylvania traders pushing across the Appalachians in quest of Indian trade, before the 1750s, Hudson's Bay Company traders largely relied on Cree or Chipewyan Indians' coming to them.[9]

This gave the company access to geographic informants, but the selection was limited. For Indians from more distant nations, the trip to the bay was difficult. The Crees near it often proved less than cooperative with Indians from other nations who wished to pass through their territory. Moreover, the wet or forested areas near Hudson Bay could be tough going for Plains Indians familiar with other types of terrain and unaccustomed to using canoes. Henday asked an "Archithinue [Blackfoot] Leader" to allow "some of his young men" to journey to York Factory and received the reply that "it was far off, and they could not live without Buffalo flesh; and that they could not leave their horses etc: and many other obstacles, though all might be got over if they were acquainted with a Canoe, and could eat Fish, which they never do. The Chief further said . . . he was informed the Natives that frequented the Settlements, were oftentimes starved on their journey." In addition, once French traders like the La Vérendryes had advanced French posts to Lake Winnipeg and the Saskatchewan River, the question was not simply whether Indians could make the long journey to the bay but why they would prefer it to a shorter trip to the French forts. In 1728, York Fort governor Thomas McCliesh reported "forty canoes of Indians this summer, most of them clothed in French clothing that they traded with the French last summer." These Indians "upbraided" the British for the poor quality of their goods. Chief factor Richard Norton wrote from Prince of Wales Fort in 1736 that "our trade" had "suffered greatly from those French settlements" on the lakes west of the bay. Too few Britons were moving inland from the bay, and too few

9. Glyndwr Williams, *The British Search for the Northwest Passage in the Eighteenth Century* (London, 1962); Williams, *Voyages of Delusion*, esp. 178, 215. On the Hudson's Bay Company's move inland, see Richard I. Ruggles, "British Exploration of Rupert's Land," in John Logan Allen, ed., *North American Exploration*, II, *A Continent Defined* (Lincoln, Neb., 1997), 203–269, esp. 215–233. An exception to the general rule of Hudson's Bay Company employees' staying near their forts was William Stuart, who went inland in 1715–1716 with a party including a Chipewyan woman, Thanadelthur, who had escaped to York Fort from captivity among the Crees and who served as interpreter and informant. Williams has reconstructed this episode in *Voyages of Delusion*, 8–13.

erior Indians were journeying to it for the company to acquire the full range of Indian geographic information.[10]

COMPARED TO FRENCH VISIONS of western Louisiana and the lands beyond, pre–Seven Years' War British designs on the North American Far West were undeveloped. This was in part because Britain lacked convenient gateways to the temperate lands sloping west from the Mississippi, in part because of the discreet policies of the Hudson's Bay Company. It was in part also because British expansionist attentions were more often directed elsewhere. Britons might find it difficult or fruitless to get into or across the North American interior, but they could sail to the shores of Spanish America. Where post-Utrecht French projects for Spanish America were often routed through the development of Louisiana or licit trade at Cadiz, many Britons maintained more immediately aggressive designs on Spanish American territories well south of New Mexico or northern New Spain. This was especially evident in the lead-up to and course of the colorfully named War of Jenkins' Ear.

British War of Jenkins' Ear Projects

Although many wars are commercially predatory, few have been as open about it as the War of Jenkins' Ear. To some extent, the war can be traced to tensions aris-

10. In obstructing passage through their territory, the Crees were probably protecting their position as middlemen. Henday surmised that the Indians residing near York Factory acquired many furs from trade with inland nations such as the Blackfeet and would have to trap more furs for themselves if their suppliers began conveying their wares directly to the British posts. See Lawrence J. Burpee, ed., "York Factory to the Blackfeet Country: The Journal of Anthony Hendry [Henday], 1754–1755," Royal Society of Canada, *Transactions*, 3d Ser., I (1907–1908), 307–363, esp. 350–351; Henday, *Year Inland*, ed. Belyea, 180–182; Samuel Hearne, *A Journey from Prince of Wales's Fort in Hudson's Bay to the Northern Ocean in the Years 1769, 1770, 1771, and 1772*, ed. J. B. Tyrrell (Toronto, 1911), 199–203; [Henry Kelsey], "Kelsey's Introduction to the *Journal* of 1691," "Indian 'Belief and Superstitions,'" in [Henry Kelsey], *The Kelsey Papers* (Ottawa, Canada, 1929), 1–4, 15–24, esp. 2–3, 18; "From Richard Norton," 1725, Aug. 14, 1726, Aug. 17, 1738, "From Anthony Beale," July 26, 1729, "From Thomas McCliesh," Aug. 1, 1729, all in K. G. Davies, ed., *Letters from Hudson Bay, 1703–40* (London, 1965), 111–113, 116–118, 139–141, 141–144, 249–259, esp. 111–112, 117, 139, 142, 249. On the difficult terrain near Hudson Bay, see Burpee, ed., "York Factory to the Blackfeet Country," 307–363, esp. 338; Henday, *Year Inland*, ed. Belyea, 105–107; "From Richard Norton," Aug. 17, 1738, "From Richard Norton and Others," Aug. 16, 1739, both in Davies, ed., *Letters from Hudson Bay*, 249–259, 292–298, esp. 249, 292. On the effects of French forts, see "From Thomas McCliesh," Aug. 8, 1728, Aug. 17, 1732, "From Richard Norton and Others," Aug. 17, 1736, "From Richard Norton," July 26, 1737, all in Davies, ed., *Letters from Hudson Bay*, 133–138, 168–171, 210–216, 225–229, esp. 136, 168, 174, 211–212, 225–226; Burpee, ed., "York Factory to the Blackfeet Country," 327; Henday, *Year Inland*, ed. Belyea, 63; Ruggles, "British Exploration of Rupert's Land," in Allen, ed., *North American Exploration*, II, 217–230.

ing from the clash between unsatisfied British mercantile ambitions and restrictive Spanish commercial policies. Under the provisions of the Anglo-Spanish commercial agreement forming part of the Utrecht settlement, the British South Sea Company was granted the asiento, the exclusive concession to sell a fixed number of slaves in Spanish American ports. The agreement also allowed the firm to send one ship of limited tonnage per year to trade in the company of the official Spanish fleets when they docked in Portobello, Cartagena, and Veracruz. These privileges were to last for thirty years. British merchants hoped to use them as the thin end of a wedge opening Spanish imperial markets. They sold merchandise other than slaves, and in quantities well beyond that allotted for the single ship allowed. Especially aggressive British traders, often from Jamaica, claimed no legal pretext and simply acted as smugglers. Spanish officials feared that the silver of Spanish America would flow into British pockets rather than Spanish coffers, and they protested to the British government and South Sea Company. Spanish *guarda costas* (coast guard ships) tried the more immediate remedy of seizing British ships thought to be carrying illegal goods. Frustrated by the post-Utrecht restrictions on access to Spain's American markets and provoked by Spanish guarda costas' heavy-handed methods of enforcing them, a bellicose British merchant community and its parliamentary allies pushed the habitually pacific and cautious Walpole ministry into declaring war on Spain in 1739.

Epic British successes and spectacular British failures characterized the ensuing War of Jenkins' Ear (so-called because of the guarda costas' alleged severing of one of the eponymous captain's). Among the triumphs were Admiral Edward Vernon's 1739 capture of Portobello and Commodore George Anson's 1740–1744 circumnavigation. Anson's expedition was especially noteworthy. His flotilla rounded Cape Horn, sacked the Peruvian town of Paita, traversed the South Sea, seized the Manila galleon off the Philippines, and returned around Africa to England with a king's ransom in Spanish silver, 1,313,843 pesos and 35,682 ounces of virgin silver. Less happily for the British Empire, its forces under General Thomas Wentworth tried to seize Cartagena in 1741, only to be repelled by determined Spanish defenders and implacable tropical diseases. A British attack on Santiago de Cuba was equally fruitless. Admiralty plans for coordinated attacks on Panama City by Anson from the Pacific and Vernon from the Caribbean — with the objective of seizing the best available route between the Pacific and the Atlantic — misfired.[11]

11. The value of the silver Anson seized amounted to about £400,000. By way of comparison, the total value of the "800,000 pounds of tea, 400 chests of china, 7000 pieces of wrought silk, 60 chests of raw silk, and £20,000 in bullion" two British East India Company ships brought back

These military operations have furnished material for gripping memoirs, novels, and grand narratives, but the more prosaic concern here is with what the documents surrounding such ventures reveal about British attitudes toward American empire. A war launched for the express purpose of gaining commercial and territorial advantages from the Spanish Empire produced not only thousands of deaths but also a great deal of paper, as British planners and officers considered which of their many objectives were most desirable and attainable. To help in the conduct of his 1740 naval expedition, for instance, Britain's Major General Lord Charles Cathcart was provided with copies of a host of policy proposals, many of which are treated below. These papers considered, modified, and expanded upon the themes of the South Sea Company documents mentioned in Chapter 4's discussion of the Treaty of Utrecht. In their optimism about the possibilities afforded by Spanish America, they recall French writings about the mysterious North American West, writings with which they can be profitably compared.

One focus of the late 1730s and early 1740s British imperial documents was the conquest of key Spanish American territories. A popular target was Darien, the eastern portion of modern Panama, lying between the Atlantic Gulf of Darien and the Pacific Gulf of San Miguel. Unlike Hudson Bay, Darien, of course, really did offer a practicable route between the Pacific and Atlantic oceans. It was from Darien that Balboa had first espied the South Sea and across Darien that Peruvian silver passed on its journey from Panama to Portobello. The pirates on their way to and from easy pickings in the Pacific had traversed Darien repeatedly in the 1670s and 1680s. Scots trying to muscle their way to commercial prominence by securing the shortest route between the oceans had attempted a Darien colony in 1698 and 1699. They had succeeded only in squandering Scotland's capital and perhaps in demonstrating that union with England offered a shorter road to prosperity than did the junction of the Americas.[12]

---

from China in the same year was about £240,000. See Glyndwr Williams, *The Great South Sea: English Voyages and Encounters, 1570–1750* (New Haven, Conn., 1997), 243–246. On the failed British attack on Santiago de Cuba, see "Santiago, and the Freeing of Spanish America, 1741," *AHR*, IV (1898–1899), 323–327, esp. 323. On Anson and Vernon, see Manuel Luengo Muñoz, "El Darién en la política internacional del siglo XVIII," *Estudios americanos*, XVIII (1959), 139–156, esp. 145–147, 150; Lucio Mijares Pérez, "Programa político para América del marqués de la Ensenada," *Revista de historia de América*, LXXXI (1976), 82–130, esp. 96, 101–102; Richard Pares, *War and Trade in the West Indies, 1739–1763* (Oxford, 1936), 104–105; George Anson, *A Voyage round the World in the Years, MDCCXL, I, II, III, IV*, comp. Richard Walter and Benjamin Robins, ed. Glyndwr Williams (London, 1974), 195–196.

12. Peter T. Bradley, *The Lure of Peru: Maritime Intrusion into the South Sea, 1598–1701* (Hampshire, U.K., 1989), 184; Guillermo Céspedes del Castillo, "La defensa militar del istmo de Panama a fines del siglo XVII y comienzos del XVIII," *Anuario de estudios americanos*, IX (1952), 238–246;

PLATE 2. Jan Wanderlaar, *Admiral Lord Anson.* N.d. Oil on canvas.
Courtesy, Trustees of the Goodwood Collection, The Bridgeman Art Library

The earlier Scottish failure seems to have done little to dissuade the late 1730s and early 1740s partisans of British conquest. New proposals to seize the region came from a variety of sources, including member of Parliament and leading Board of Trade member Martin Bladen. In 1739, Bladen proposed a Darien settlement to the Lords of the Admiralty and to secretary of state William Stanhope, Lord Harrington (1739–1742). A Darien establishment would place Britain within range of Isthmian gold-mining districts and between Cartagena and Portobello. It would therefore "be of great service to distress the Spaniards." Bladen argued further that Darien was "very convenient for Opening a Passage to the South Sea, by Rivers . . . from whence a Comunication might likewise be had to the East Indies" — it was not just Dobbs thinking about routes to the Pacific. Bladen's scheme found receptive ears. First Lord of the Admiralty Charles Wager took it under consideration, and the proposal was included among the papers given to Lord Cathcart to assist his West Indies expedition.[13]

Perhaps the most important advocate of British Darien designs was Jamaica governor Edward Trelawny (1739–1752). Trelawny avowed his opposition to indiscriminate acquisition of large West Indian territories, but he argued nonetheless in a February 1740 letter to General Wentworth that it was "absolutely necessary to have such as may be able with the help of our men of war to command the trade there. Panama, as the Isthmus of Darien would go with it, may

---

Ignacio Gallup-Díaz, "The Spanish Attempt to Tribalize the Darién, 1735–50," *Ethnohistory*, XLIX (2002), 281–317; Peter Gerhard, *Pirates on the West Coast of New Spain, 1575–1742* (Glendale, Calif., 1960), 147–148; Manuel Luengo Muñoz, "El Darién en la política internacional del siglo XVIII," *Estudios americanos*, XVIII (1959), 139–156; Pares, *War and Trade*, 1; Geoffrey J. Walker, *Spanish Politics and Imperial Trade, 1700–1789* (Bloomington, Ind., 1979), 21.

13. Martin Bladen to Lord Harrington, June 12, 1739, 21r–23v, "Reasons for Making a Lodgement on the Ihstmus [sic] of Darien, Where the Scotch Had Formerly a Settlement," Paper Delivered by Bladen to the Lord of the Council, Sept. 10, 1739, 31r–32v, and Charles Wager, "Abstract of Several Schemes under Consideration," October 1739, 61v, all in Add. Mss. 32,694; "Minute," Whitehall, Sept. 10, 1739, Add. Mss., 19,030, 438r–439r. For information on Bladen, see Pares, *War and Trade*, 78–79; and Steele, *Politics of Colonial Policy*, 152, 165–167. James Knight, a figure associated with the sugar-planting interest, offered a similar proposal for a Darien settlement, which, he claimed, could "be effected without any great difficulty or Expence by the Crown." Noting the swift isthmus crossings by English buccaneers like Henry Morgan and William Dampier, and the isthmus's location in the midst of Spain's most important American possessions, Knight highlighted "how Easy will it be when a Settlement is made there to obstruct and Restrain the Trade of Panama, Cartagena, and Porto bello; And at any time to make a descent into the S Seas. it will be in our power to prevent great part of the Riches of Peru, and Lima, from being Exported to old Spain, or to make them our Own; to Cut of [sic] all Communication between those places, and prevent their being Supplied with necessary's for their Support, which will reduce them to the greatest Extremity." See James Knight to the duke of Newcastle, Stoke Newington, Nov. 20, 1739, Add. Mss. 32,694, 9r–10v (another copy of this proposal is in Add. Mss., 22,677, 25r–28v); Knight to Newcastle, Stoke Newington, Apr. 3, 1740, Add. Mss. 22,677, 36r–37r; William Beckford to Knight, Spanish Town, Feb. 10, 1741, Add. Mss. 12,431, 119r.

be reckoned I think the most convenient of any for that purpose, consequently that ought to be the first object of our views after the destruction of the [Spanish] Fleet." Like so many others, Trelawny saw the desirability of having "a Port in the South Seas, in the narrowest part of the isthmus, so as not only to command the trade of Peru, but to have the shortest land carriage that may be between the North and South Seas." Trelawny's was an influential voice. Noting the difficulty of directing New World campaigns from London, the official instructions for the West Indies commanders Admiral Vernon, Lord Cathcart, and General Wentworth explicitly stated that they, their seconds-in-command, and Trelawny were to determine in council the appropriate operations to conduct against the Spanish. Trelawny convinced Wentworth and a majority of this council of war to launch a Panama expedition (Vernon's initial opposition and half-hearted execution impeded its success).[14]

One reason for interest in Darien was the belief, reminiscent of British views concerning southern South America and French visions of the North American West, that the area not only lay near known Spanish mines but also abounded itself in undeveloped mineral deposits. Rumors and reports to this effect were manifold. Sir Charles Wager's October 1739 "Abstract of Several Schemes under Consideration" mentioned Darien Indians' knowing "(as it is said) of many Gold Mines." An anonymous April 1740 "Account of the Havana and the Other Principal Places Belonging to the Spaniards, in the West Indies" contended that Darien "by all Reports is the richest Spott of Ground for Gold in the World"; the Indians there had "industriously conceal'd the richest mines from the Spaniards." Authors extended these sorts of assumptions into territories beyond Darien as well. Captain William Lea recommended conducting British operations "on the South Side of the Gulph of Honduras, which are very near the Gold and Silver Mines where is produced the greatest part of the Plate which is sent to be coined at the Mint at Guatimala," and opening "the Trade to the City of Guatimala, and all the Country back to the South Seas, which produces in great abundance, Gold Silver, Cochineal, the best Indigo, Cocoa, Balsam of Peru, and a great variety of useful Drugs, with all sorts of Dying woods." British authors

14. Edward Trelawny to Newcastle, Jamaica, Jan. 15, 1740, 28r–29r, "Extract of M Trelawny's Letter to General Wentworth dated the 5th of Febry 1740," 99r, both in CO 137/57, part 1; "Instructions for Our Trusty and Welbeloved Edward Vernon Esqr., Vice Admiral of the Blue Squadron of Our Fleet, and Commander in Chief of Our Ships Employed, or to Be Employed in the West Indies: Given at Our Court at Herrnhausen the Tenth Day of July O.S. 1740," Add. Mss., 32,694, 242v–245r; Newcastle to Thomas Wentworth, Whitehall, Oct. 15, 1741, Add. Mss. 32,698, 161v–163r. See also Trelawny to Newcastle, Jamaica, Jan. 15, 1740, CO 137/57, 26r. On Trelawny's convincing others to attack Panama and the failure of the expedition, see Trelawny to Newcastle, Jamaica, Jan. 31, Feb. 9, 1741, CO 137/57, part 1, 126r–129v; Pares, *War and Trade*, 93–97.

were portraying Spanish America as a treasure trove awaiting British exploitation.[15]

Darien was merely one among many Spanish American targets considered by British planners. Havana, all of Cuba, and even Mexico constituted others. William Wood, a former customs official and frequent advocate for Bristol and London-based West Indian merchants, wrote to Thomas Pelham Holmes, duke of Newcastle, in September 1741, arguing "that the sending 5000 Effective Men at this Time will, with the Troops already in America, and upwards of 1000 Able-bodied Negroes, be sufficient to Take the Havanna, and by the following these Troops, in a Month or thereabouts, and so every two Months after, with 1000 recruits or more, to gain Great Britain the Possession of the whole Island of Cuba." Wood saw Havana as one in a series of British conquests culminating in Mexico: "Whenever We shall be in possession of the Havanna, . . . it will be a Proper Time to think of Executing a Scheme or Proposal which may have been made to take La Vera Cruz and march to Mexico." Others shared Wood's sentiments. Trelawny spoke of taking and keeping Havana or Veracruz, in order to "command the trade of Mexico." The "Account of the Havana and the Other Principal Places Belonging to the Spaniards, in the West Indies," written by an informant who claimed to have been held prisoner in Mexico City, put forward that British operations against Mexico would be easy and successful. Taking such suggestions to their most grandiose conclusion, William Lea recommended in March 1740 taking not only Guatemala but also "the whole Kingdom of Mexico," so as to "enlarge the British Empire in America: quite round the Bay of Mexico, 'till it join'd with Carolina."[16]

British conquest schemes focused on the Caribbean and Gulf of Mexico but extended occasionally to and around southern South America. Numerous authors expressed interest in establishing a British presence at Buenos Aires.

15. Wager, "Abstract of Several Schemes under Consideration," 62r–62v, "An Account of the Havana and the Other Principal Places Belonging to the Spaniards, in the West Indies," Apr. 14, 1740, 87r, both in Add. Mss. 32,694. See also Knight to Newcastle, Stoke Newington, Nov. 20, 1739, 10v, Bladen to Harrington, June 12, 1739, 23r, "Reasons for Making a Lodgement on the Ihstmus [sic] of Darien," 31r–31v, and "Considerations How the British Nation Could Annoy That of Spain in America," Add. Mss. 32,694, 68r. For Lea, see William Lea, London, Mar. 3, 1740, Add. Mss. 32,698, 145v–146r (copy sent to Thomas Wentworth and Edward Vernon, Oct. 15, 1741).

16. William Wood to Newcastle, Whitehall, Sept. 10, 1741, Add. Mss. 32,698, 26r (on Wood, see Pares, War and Trade, 78); Trelawny to Newcastle, Jamaica, Jan. 15, 1740, CO 135/57, part 1, 29r; "An Account of the Havana," Add. Mss. 32,694, 81r: "The taking of Mexico [City] is practicable, is not difficult; and with an Army of 8 or 10000 Men fit for Service, nothing can hinder your March to that City; there is not any thing like a Place fortified, nor one great Gun in the whole inland Country." See also Knight to Newcastle, Stoke Newington, Apr. 3, 1740, Add. Mss. 22,677, 36v; Lea, London, Mar. 3, 1740, Add. Mss. 32,698, 147r.

Wager's June 1738 paper, "Places Where the Spaniards May Be Attack'd . . . in Europe and in the West Indies," indicated that Buenos Aires "may be taken, and kept, with a sufficient Land force." Wood declared that possession of Buenos Aires would facilitate British trade with surrounding areas and prevent Spanish navigation around Cape Horn and through the Straits of Magellan. An anonymously authored document Robert Walpole (leading British minister, 1721–1742) gave Anson recommended Valdivia be taken and "made an arsenal or magazine for the English Men of war and merchant Ships," after which the English "may open and Secure a trade in the whole South Sea even to California, and therby render it a certain drain to the Peruvian wealth." The author spoke further of plundering Lima and Callao, where the Peruvian capital's annual produce of five million silver dollars would be collected and which could "be taken in one night if proper directions are pursued." That such schemes were taken seriously is evident not only from the fact that Anson received them but also from official instructions to him that "You may, if You shall think it proper, leave one or Two of Our Ships in the South Sea for the Security of any of the Acquisitions, You may have been able to make."[17]

Still farther afield, the boldest British promoters and officials contemplated not just operations off the Philippines, like Anson's capture of the Spanish galleon, but assaults on the islands themselves. In a 1740 memorandum, Wager noted, "The Phillipin [sic] Islands on the Coast of China, belonging to the Spaniards are said to be very Rich, particularly Manilla, a City on the Island Luconia, where two Spanish Ships come yearly with Eight or Ten Millions of pieces of Eight from Acapulca in Mexico; It is thought that two Ships of 50 Guns . . . would be sufficient to go upon such an Expedition." What could be raided might also be held. Wager's October 1739 "Abstract of Several Schemes under Consideration" indicated that Manila "may be kept with a Garrison, if the King thinks fit, or given to the East India Company, for a Valuable Consideration." Such schemes illustrate the reach of British territorial ambitions. Planners at the highest level of government, plotters who could capture their attention, and officials empowered to act on the designs in question were all casting their eyes upon the Spanish Empire's core territories and farthest limits.[18]

17. Charles Wager, "Places Where the Spaniards May Be Attack'd (with a Proper Force) in Europe and in the West Indies," June 5, 1738, Add. Mss. 32,694, 52v (see also Wager, "Attempts That May Be Made, upon the Spanish Coast of Europe, or America," 1740, 57r–57v); Wood to Newcastle, Whitehall, Sept. 10, 1741, Add. Mss. 32,698, 27r–27v; "Mr. Walpole: Copy Given to Commodore Anson, June 25, 1740," 37r–38r, "Instructions for Our Trusty and Wellbeloved George Anson . . . Commander in Chief of Our Ships Designed to Be Sent into the South Seas in America, Given at Our Court at St. James's the Thirty First Day of January 1739/40," 8v (draft), both in SP 42/88.

18. Wager, "Attempts That May Be Made," 56v–57v (see also "Places Where the Spaniards May

British schemes did not confine themselves to contemplating how British forces acting alone could seize Spanish American territory or plunder its inhabitants. Like the authors of French projects for Louisiana, British writers devoted considerable attention to cooperation with Spanish America's creole or Indian inhabitants. Belief in the feasibility of such alliances rested on several assumptions. One was that even Spanish Americans of European descent chafed under the peninsular yoke, "a tyranny they have long groaned under, and which they are ready to shake off, whenever they shall have a proper opportunity," as Stephen DeVere put it in 1741. If this were true, prodding the residents of colonies like Chile and Peru toward independence seemed a likely means of arousing and exploiting latent pro-English sentiments. A draft of Anson's instructions spoke of "Reason to believe, from private Intelligence, That the Spaniards in the Kingdom of Peru, and especially in that Part of it, which is near Lima, have long had an Inclination to revolt from their Obedience to the King of Spain, (on Account of the great Oppressions and Tyrannies exercised by the Spanish Vice-Roys and Governors)." Anson was directed, "If You should find, that there is any Foundation for these Reports, by all possible Means, to encourage, and assist such a Design." Creole independence, it was hoped, would forward British trade: "And in case of any Revolution, or Revolt from the Obedience of the King of Spain, either amongst the Spaniards, or the Indians, in those Parts, and of any new Government being erected by Them, You are to insist upon the most advantageous Conditions for the Commerce of Our Subjects." The mention of Spaniards or Indians in the last sentence shows the flexibility, or perhaps inconsistency, of Anson's instructions and the ideas behind them. He was to be ready to ally with Spanish Americans against Spain or with unconquered American Indians against Spanish creoles. "It has been represented unto Us," a draft of his instructions stated, "That the Number of native Indians on the Coast of Chili, greatly exceeds That of the Spaniards, and That there is reason to believe, That the said Indians may not be averse to join with You against the Spaniards, in order to recover their Freedom."[19]

The same notions of restive creoles and Indians welcoming British intervention appeared in discussions of the Darien region, and the same commercial benefits were anticipated. Darien contained — along with pestilential, often pre-

<hr>

Be Attack'd," 52v), Wager, "Abstract of Several Schemes under Consideration," 60r, both in Add. Mss. 32,694.

19. Stephen DeVere [DeVereux], "Some Thoughts Relating to Our Conquests in America," June 6, 1741, in "Santiago, and the Freeing of Spanish America," *AHR*, IV (1899), 327; "Instructions for Our Trusty and Wellbeloved George Anson," SP 42/88, 4r–5v. See also "Mr. Walpole," SP 42/88, 37v; Wager, "Abstract of Several Schemes under Consideration," Add. Mss. 32,694, 60v–61r.

cipitous, and frequently impenetrable jungle—one of those pockets of Amer-indians unconquered by Spain and therefore of great interest to interlopers seeking native allies. Bladen, perhaps revealing more concerning the charac-ter of British ambitions than was seemly, noted that the natives of the Isth-mus of Panama had always assisted "Buccaneers" and "Pyrates" on their way to the South Sea. He wrote to Lord Harrington of Indian clans who had not yet acknowledged Spanish sovereignty and "who would willingly receive any other Nation amongst 'em, none more willingly than the English." Other writers echoed this contention. Referring more broadly to Indians and Spanish creoles, Trelawny spoke optimistically of shattering the Spanish Empire and its com-mercial system. He wrote to the duke of Newcastle, "If we consider the great disproportion in number of the Indians to the Spaniards, and that most of the creole Spaniards are discontented with their Governments, it cannot be in the nature of things but we must dismember the Spanish Monarchy at least, if we do not entirely destroy it in these parts, and lay a foundation for a most exten-sive and beneficial trade with the Inhabitants in spite of France and Spain." "So great and rich a part of the world," he continued, "should be in the hands of the natives, who would naturally break into so many independent governments, no one of which could arrogate to itself the commerce of the whole." As is evident in Trelawny's proposals, for British authors, sentiments favoring Indian and cre-ole independence and conditions favoring British profits coincided. This is most vividly illustrated by the "Account of the Havana and the Other Principal Places Belonging to the Spaniards, in the West Indies." In speaking of a British con-quest of Panama, the paper contended that "Millions of miserable People wou'd bless their Deliverers, their Hearts and their Mines wou'd be open to us." One could hardly ask for a better parody of twentieth-century interventionism.[20]

Spanish gold and silver mines were surely dearer to British imperial expan-sionists than creole and Indian hearts. Indeed, as was the case during the closing years of the War of the Spanish Succession, British War of Jenkins' Ear writings concerning Spanish America averred that British gains there could justify and

20. On Darien's Indian population, see Gallup-Díaz, "Spanish Attempt to Tribalize the Darién," *Ethnohistory*, XLIX (2002), 281–317. For Bladen's views, see "Reasons for Making a Lodgement on the Ihstmus [sic] of Darien," 31r, and Bladen to Harrington, June 12, 1739, 23r, both in Add. Mss. 32,694 (see also Knight to Newcastle, Stoke Newington, Nov. 20, 1739, 9v–11r, and "An Account of the Havana," 87r). For other authors expecting Indian aid, see Wager, "Abstract of Several Schemes under Consideration," 62r, and "Considerations How the British Nation Could Annoy That of Spain in America," 68r, both in Add. Mss. 32,694; Trelawny to Newcastle, Jamaica, Jan. 15, 1740, CO 137/57, part 1, 28r; Lea, London, Mar. 3, 1740, Add. Mss. 32,698, 147r. For Trelawny's optimistic views, see Trelawny to Newcastle, Jamaica, Jan. 15, 1740, CO 137/57, part 1, 26r–29v (see also Knight to Newcastle, Stoke Newington, Nov. 20, 1739, Add. Mss. 32,694, 12r); "An Account of the Havana," Apr. 14, 1740, Add. Mss. 32,694, 88r.

cover the cost of the current conflict. A June 1740 paper Walpole transmitted to Anson asked "whether the present war does not require an object proportioned to the expence of it and other inconveniency with which it must be attended." The author hoped in response to this question that, by launching a British Pacific expedition, "many national advantages may be gain'd and great riches acquired." In 1741, William Wood, proposing seizure of Cuba, argued that it was "the only country, unless Buenos Ayres, belonging to Spain in America, that is worth putting the Nation to any great Expence of either Men or Money to have the possession of, or that by Our Taking is capable of answering to the Nation the Great Expence of the Armament under Lord Cathcart, as well as the present Armament, and that by the keeping Possession of will bring Lasting Riches and Lasting Power to Great Britain." Hopes of mercantile advantage drew Britain into a costly Spanish conflict; wartime expenses then enhanced those commercial gains' importance.[21]

Accumulation of optimistic British plans for raiding the Spanish Empire does not mean, of course, that all Britons favored such actions. Walpole famously opposed the Spanish war altogether, and less celebrated figures critiqued projects discussed in the preceding pages. Dissension regarding specific approaches to the Spanish Empire was, however, often accompanied by acceptance of the more general idea of preying on it. Authors skeptical about British designs on particular Spanish imperial objectives often favored schemes aimed at other regions. William Wood, writing to Newcastle in 1741, thought the Isthmus of Panama and the Spanish Main's climate unhealthy for Britons, and he therefore discouraged operations there. He favored instead British establishments around nominally salubrious Buenos Aires or allegedly healthful Havana. Other figures opposed specific operations while still accepting part of the rationale for them. Admiral Vernon gainsayed attempting to settle Darien because he feared that reportedly improved Spanish-Indian relations would deprive British forces of native assistance. He shared, however, belief in the existence of Darien gold mines, and he implicitly acknowledged the recency of the rumored amelioration of Spanish-Indian hostility. Some writers saw weaknesses in the arguments for particular ventures but favored the projects anyway. Bladen conceded that, even if Britain succeeded in seizing Darien, diplomats might be obligated to re-

21. "Mr. Walpole," SP 42/88, 37r, 40r; Wood to Newcastle, Sept. 10, 1741, Add. Mss., 32,698, 26r. Wood computed the annual product of the island at more than £350,000 a year and contended that supplying it would bring great wealth to British manufacturers and traders and Anglo-America's inhabitants. William Lea favored conquering Guatemala and Mexico, "as Success in such an Attempt would be the best method to indemnify the Nation for the Depredations they have suffer'd, as well as for the Expence of the War" (Lea, London, Mar. 3, 1740, Add. Mss. 32,698, 147r–147v).

turn the territory in peace negotiations. He argued, however, that the benefits of possessing Darien for even a few years would justify the effort and cost of acquiring it. And some counselors repudiated the whole notion of British conquests, only to advance more radical projects. Stephen DeVere opposed seizing Spanish territories: the Spanish government would "never consent" to their loss, new British establishments would require expensive "garisons and Colonies," and even the mere "keeping of a cautionary town" would "give great offence to the naval powers of Europe." What DeVere recommended instead was freeing the American colonies from Spanish rule and opening their trade to all Europe. Routine thinking for the nineteenth century, ahead of its time in 1741.[22]

Criticisms, skepticism, and caution notwithstanding, pugnacious counsels carried Britain into a war of choice and launched expeditions against Portobello, Cartagena, Cuba, Peru, and the Philippines. It was audacity and a touch of megalomania rather than caution and skepticism that characterized many prominent British figures' thoughts about Spain's dominions, and an aggressive spirit guided British actions. One indication of this was the multiplicity of conquest objectives. Vernon's instructions directed him to "take into Consideration the Making an attempt upon some of the Places following, Viz. The Havana, La Vera Cruz, Mexico, Cartagena, and Panama." For the most fervid, American imperium beckoned. In Trelawny's imagining, "the French Fleet being destroyed, the Spaniards will be at our mercy, our fleet will be masters of these seas, and our troops may invade and seise at pleasure, and give laws to this new world."[23]

The British documents discussed above merit attention in their own right because they reveal some of the conceptions animating the empire's early-eighteenth-century Spanish policy. They become more interesting when viewed alongside French documents regarding the trans-Mississippi West. British authors describing Spanish America and French authors discussing western North America envisaged territories filled with precious metals. Both saw new settle-

22. Wood to Newcastle, Whitehall, Sept. 10, 1741, Add. Mss. 32,698, 26v; Vernon to Newcastle, Jan. 23, 1739/40, SP 42/85, 122r–123v; "Reasons for Making a Lodgement on the Ihstmus [sic] of Darien," Add. Mss. 32,694, 31r–32v; DeVere, "Some Thoughts Relating to Our Conquests in America," June 6, 1741, in "Santiago, and the Freeing of Spanish America," AHR, IV (1899), 325–327.

23. "Instructions for Our Trusty and Welbeloved Edward Vernon," Add. Mss. 32,694, 243r; Trelawny to Newcastle, Jamaica, Jan. 15, 1740, CO 137/57, part 1, 27r. Others echoed this inclination to seize American dominion. In a letter to Newcastle regarding a proposed Darien settlement, Knight spoke of its enabling "us to prescribe laws to the Kingdom of Peru and great part of Lima." In a later missive, he went further, suggesting that "Settlement of Darien . . . will Enable us with the assistance of Jamaica more Effactually to Curb the Insolance [sic] of the Spaniards, and Even to prescribe laws to them in America" (Knight to Newcastle, Stoke Newington, Nov. 20, 1739, 12v, "Mr Knight's Second Letter," Dec. 3, 1739, 18r–18v, both in Add. Mss. 32,694).

ments enjoying a lucrative trade with Spanish American populations. Both predicted friendly, profitable, and militarily vital relations with American Indian nations.

Differences existed also. British authors placed more emphasis on the seizure of key Spanish imperial points, whereas French authors — though not indifferent to the possibility of conquering Spanish provinces such as New or even old Mexico — put more weight on the development of western lands still unoccupied by Europeans. French authors showed less interest than their British counterparts in the possibility of Spanish creoles' breaking away from the Spanish Empire. French portrayals of Louisiana devoted more attention to agricultural possibilities than British authors did for any region other than perhaps that around Buenos Aires. British writers generally focused more on establishing trading outposts in Spanish lands.

Fundamentally, though, visionaries from both nations saw fortunes to be made in lands neglected or poorly managed by Spain, and both empires evinced a fascination with reaching the Pacific Basin's rich markets. For the British, South Sea access would come by way of Darien, southern South America, or even Hudson Bay; for the French, by way of a great western river and inland sea. For both empires, grandiose projects and confident spirits would finally provide means of capturing the New World wealth Spain had enjoyed for two centuries.

It was in this climate of opinion that French observers would interpret British exploration in Hudson Bay.

10

FRENCH REACTIONS TO THE BRITISH SEARCH FOR A NORTHWEST PASSAGE FROM HUDSON BAY AND THE ORIGINS OF THE SEVEN YEARS' WAR

Après avoir dominé la période de la succession d'Espagne, la question des rapports de l'Amérique espagnole avec l'Europe et les colonies européennes d'Amérique fut probablement la question primordiale de la politique internationale au milieu du XVIIIe siècle.
—PIERRE MURET, *La prépondérance anglaise (1715–1763)*, 1937

Dans l'histoire du dix-huitième siècle, peu d'événements ont donné lieu á autant de controverses et à plus de jugements contradictoires que les incidents diplomatiques que précédèrent la guerre de Sept Ans.
—RICHARD WADDINGTON, *Louis XV et le renversement des alliances*, 1896

During and immediately after the War of Jenkins' Ear and the larger War of the Austrian Succession it joined, British explorers, promoters, and officials sought ways to overcome the physical and diplomatic barriers to British Pacific navigation. French officials observed these British efforts and contemplated their implications. The question raised was how the French Empire would respond to the prospect of its leading rival's obtaining the South Sea access that the Utrecht settlement and North America's obstinate impermeability had so far denied to France.[1]

An earlier version of material in this chapter appeared in Paul Mapp, "French Reactions to the British Search for a Northwest Passage from Hudson Bay and the Origins of the Seven Years' War," *Terrae Incognitae*, XXXIII (2001), 13–32. My thanks to the journal for allowing me to reuse it.

1. Pierre Muret, *La prépondérance anglaise (1715–1763)* (Paris, 1937), 381 ("After having dominated the period of the Spanish Succession, the question of the relations of Spanish America with Europe and the European colonies of America was probably the primordial question of international politics in the middle of the eighteenth century"); Richard Waddington, *Louis XV et le renversement des alliances: Préliminaires de la guerre de Sept Ans, 1754–1756* (Paris, 1896), v ("In the history of the eighteenth century, few events have given rise to as many controversies and to more

One response was writing. French evaluations of British South Sea and North American probes appear in the letters and memoirs circulating among mid-eighteenth-century French ministers, ambassadors, and foreign ministry officials, and the availability of these papers allows analysis of the geostrategic considerations driving French diplomats between the two great mid-eighteenth-century European wars.

Presentiments of another French confrontation with Britain were prominent in these papers. In the years after the War of the Austrian Succession, French officials knew that another Anglo-French conflict was possible, suspected it was probable, and feared it was unavoidable. Territorial and commercial disputes in India, Acadia, the Caribbean, and the Ohio Valley remained unresolved, and tensions arising from these squabbles unallayed. Still counting the lives and fortunes lost in the last war and fearful of the disasters that a more momentous future war might bring, most members of the French government hoped to prevent or delay the outbreak of new hostilities. Indeed, French officials often disparaged the value of contentious North American areas like the Ohio Valley and questioned whether their possession justified the hazards of another struggle with the British Empire.[2]

Despite the desire to avoid war and doubts about whether disputed North American territories were worth fighting over, French statesmen responded aggressively between 1748 and 1755 to a variety of small-scale British actions in the two empires' North American marchlands. The French government had forts constructed in contested territories to repel British traders, land speculators, and soldiers. It sent expeditions into the Ohio Valley to establish a French claim to the region, and it destroyed or seized British forts challenging this assertion. It sent new army regiments to Canada, and its soldiers and their Indian allies defeated Washington at Fort Necessity in 1754 and Braddock on the Monongahela in 1755. British statesmen countered these French measures. The chain of increasingly bellicose actions and reactions leading to 1754 and 1755 Anglo-French Ohio Valley combat conduced also to 1755 and 1756 French decisions instigating a global Anglo-French war. This sequence of events raises the question of why French officials allowed seemingly minor frontier provocations in marginal

---

contradictory judgments than the diplomatic incidents that preceded the Seven Years' War"). In this chapter, translations from French are mine unless otherwise noted.

2. Lawrence Henry Gipson, *The British Empire before the American Revolution*, V, *Zones of International Friction: The Great Lakes Frontier, Canada, the West Indies, India, 1748–1754* (New York, 1942), 350; ibid., VI, *The Great War for the Empire: The Years of Defeat, 1754–1757* (New York, 1946), 100, 391; Patrice Louis-René Higonnet, "The Origins of the Seven Years' War," *JMH*, XL (1968), 57–90, esp. 57–58, 60, 77; Waddington, *Louis XV et le renversement des alliances*, 81, 97, 103–104, 112.

North American territories to embroil France in a global war the French government hoped to avoid. Examination of French reactions to the 1740s British quest for a Northwest Passage illuminates one part of this question; namely, the underlying geopolitical assumptions and concerns causing French statesmen to believe that 1748–1755 British North American aggression required a militant French response.[3]

French diplomats believed that British actions in North America's Anglo-French borderlands involved issues of more general importance. They suspected that murky North American wilderness incidents were related to a larger pattern of British activity, one involving British designs on the wealth of Spain's American empire. British Hudson Bay activities heightened French misgivings. French officials were skeptical about the existence of a usable Northwest Passage from the bay, but their ignorance of the region's geography left them apprehensive; and the very unlikeliness and temerity of British exploration in the bay's forbidding subarctic environment demonstrated to French officials the lengths to which the British would go to reach the Pacific's more inviting waters.

Hudson Bay exploration composed, moreover, one of a group of mid-

3. A superb narrative of these events can be found in Fred Anderson, *Crucible of War: The Seven Years' War and the Fate of Empire in British North America, 1754–1766* (New York, 2001). In this chapter, I respond to issues Anderson has raised in his book. Specifically, I assess the relationship between events in the Ohio Valley and in other parts of North America and the world. See also Gipson, *British Empire before the American Revolution*, IV, *Zones of International Friction: North America, South of the Great Lakes Region, 1748–1754* (New York, 1939), and see V and VI; Higonnet, "Origins of the Seven Years' War," *JMH*, XL (1968), 57–90; Max Savelle, *The Diplomatic History of the Canadian Boundary, 1749–1763* (New Haven, Conn., 1940); Savelle, *The Origins of American Diplomacy: The International History of Anglo-America, 1492–1763* (New York, 1967); Waddington, *Louis XV et la renversement des alliances*. Gipson, Savelle, and Anderson are good places to start for the British side of events. Julian S. Corbett, *England in the Seven Years' War: A Study in Combined Strategy*, 2 vols. (London, 1907), is the most insightful examination of British naval policy. Also useful are two articles: Steven G. Greiert, "The Board of Trade and Defense of the Ohio Valley, 1748–1753," *Western Pennsylvania Historical Magazine*, LXIV (1981), 1–32; and T. R. Clayton, "The Duke of Newcastle, the Earl of Halifax, and the American Origins of the Seven Years' War," *Historical Journal*, XXIV (1981), 571–603. Waddington provides a detailed narrative of French diplomacy. Savelle's *Origins of American Diplomacy* offers a reliable English-language overview of events; his *Diplomatic History of the Canadian Boundary* covers North American border disputes in great detail. Higonnet's is the most insightful analysis of the sequence of events leading to the war. He ascribed its outbreak to diplomatic misconceptions occurring within "an increasingly anarchic diplomatic system" ("Origins of the Seven Years' War," *JMH*, XL [1968], 89). Higonnet found that the misleading reports and bellicose actions of colonial officials with a "private motive" (65) for favoring imperial expansion, such as Governor Duquesne of Canada, drew the French and British governments into open conflict in North America. Diplomatic miscalculations in Europe then led to a series of escalating measures and countermeasures culminating in a full-blown imperial world war. I focus in this chapter on the geopolitical assumptions, fears, and ideas shaping the mental context that contributed to these miscalculations.

eighteenth-century British actions seemingly directed at the Pacific. Uneasiness about the British search for a Northwest Passage accentuated French concerns about these other British actions, and worry about these other British initiatives intensified French anxieties about the British hunt for a passage. Together, an array of British activities involving Hudson Bay, Central America, and Cape Horn convinced French officials of the reality of a British plan for Pacific maritime expansion. Existence of such a design indicated a British decision to disregard the provision of the 1713 Treaty of Utrecht declaring the Pacific the exclusive domain of Spanish shipping and a likely British intent to encroach upon Spain's silver-rich and lightly defended South Sea colonies.

British willingness to abandon one provision of the Utrecht settlement and to menace one part of the Spanish Empire suggested to French officials that British actions elsewhere in the Americas shared the same audacity and ambition. Incidents that might have appeared trivial by themselves seemed instead to be steps in a larger journey toward British imperial aggrandizement and French abasement. Such fears prodded French officials into responding assertively to the various instances of British North American aggression, contributing to the chain of increasingly provocative diplomatic and military actions that triggered the Seven Years' War.[4]

4. Scholars have left room for further discussion of the relation between British exploration in Hudson Bay, other British moves toward the Pacific, and the outbreak of the Seven Years' War. Savelle, in his 1940 *Diplomatic History of the Canadian Boundary*, 24–26, mentioned Dobbs's plans to expand the trade activities of the Hudson's Bay Company south and west of the bay, but not his promotion of the search for a Northwest Passage. He discussed French opposition to any Hudson's Bay Company expansion in the direction of territories claimed by New France, but not French concerns about British attempts to move west from the bay toward the Pacific. E. W. Dahlgren, in *Les relations commerciales et maritimes entre la France et les côtes de l'Océan pacifique (commencement du XVIIIe siècle)*, I, *Le commerce de la mer du Sud jusqu'à la paix d'Utrecht* (Paris, 1909), described French interest in the Pacific in the first decades of the eighteenth century, but he did not carry this work forward to the era of the Seven Years' War. John Dunmore did in *French Explorers in the Pacific*, I, *The Eighteenth Century* (Oxford, 1965), but he did not go into the diplomatic repercussions of interest in the Pacific in the 1750s. Dunmore noted the need for work in the French, British, and Spanish foreign ministry archives that would illuminate the political aspects of Pacific exploration (vi). Muret, in *La prépondérance anglaise* and "Le conflit anglo-espagnol dans l'Amérique centrale au XVIIIe siècle," *Revue d'histoire diplomatique*, LIV–LV (1940–1941), 129–148, exhibited the greatest appreciation for the significance of British encroachment on the Pacific Ocean and on the Spanish Empire, but he did not fully develop the implications of these ideas. Savelle, Dahlgren, and Muret lacked access to Glyndwr Williams's *British Search for the Northwest Passage in the Eighteenth Century* (London, 1962), and his more recent *Voyages of Delusion: The Northwest Passage in the Age of Reason* (London, 2002). Williams's monographs, with their thorough reconstruction of events and excellent accounts of European geographic thought, are the indispensable starting point for an appreciation of the diplomatic repercussions of British 1740s Hudson Bay exploration.

Anson's South Sea Expedition and Rumors of British Interest in a Patagonian Passage

After more than two decades of relative quiet, the 1739 outbreak of the Anglo-Spanish War of Jenkins' Ear gave French geopolitical thinkers revived cause for alarm about British South Sea ambitions. In particular, George Anson's 1740–1744 Pacific voyage troubled French diplomats because it seemed to indicate a renewed British intent to invade the Spanish Empire's Pacific preserve. The attentiveness with which French diplomats followed news of Anson's progress demonstrates the seriousness with which they viewed his South Sea foray, as does their nervous interest in various rumors about the goal of his mission, issues related to his voyage, and similar expeditions being prepared in Britain.[5]

French diplomats speculated widely about the true objectives of Anson's mission. Initially, they suspected an attack on Buenos Aires or the establishment of a British post somewhere in southern South America. Other rumors suggested that Anson's ships carried British South Sea Company merchants who would try to initiate trade with Chile. Some reports from Spain and England accurately described plans for Anson to attack Panama from the Pacific in conjunction with an assault by Admiral Vernon from the Caribbean, with the ultimate aim of seizing a route between the oceans. Foreign ministry memoirs recalled that British privateers had captured one of the galleons bound from Manila to Acapulco during the War of the Spanish Succession, and French speculations about Anson's goal came, correctly, to include the belief that he would try to capture a galleon himself.[6]

French diplomats worried also that Anson's mission portended the beginning

5. After Utrecht, French officials followed reports of British activities directed at the Pacific, but these had been too scattered and rudimentary to constitute a major threat to the French Empire. See marquis d'Huxelles to Pierre Lemoyne d'Iberville, March 1716, CP, Angleterre, 288, 196v–197r; "Extrait de la lettre escrite a Madrid le 21 Septemb. 1722," 24r–25v, "Memoire des observations faites sur une lettre ecrite de Madrid le 21 Septembre 1722," 25r, both in CP, Espagne, 326; "Memoires utils dans la conjoncture des préliminaires de la paix, d'un futur Congrés, et d'une ambassade en Espagne, par raport au comerce de France en 1727," CP, Espagne, 352, 18r–18v; La Boulaye, "Observations sur les revolutions qui sont arrivees au *commerce d'Espagne* dans le rojaume du Mexique," June 15, 1730, CP, Espagne, 374, 214r–215v; "Observations et remarques presentées a la Compie. du Sud pour l'engager a entreprendre de nouvelles decouvertes en Amerique," Aug. 3, 1732, CP, Angleterre, 378, 24r–25v.

6. Jean-Jacques Amelot de Chaillou to comte de La Marck, Oct. 17, 1740, 119r, La Marck to Amelot, Nov. 1, 1740, 237r–239v, both in CP, Espagne, 462; August 1741, CP, Angleterre, 412, 255v; Amelot to bishop of Rennes, June 18, 1741, 155r–156r, Rennes to Amelot, June 30, July 7, 1741, 282r, 310r–311r, all in CP, Espagne, 466; Rennes to Amelot, July 20, Aug. 7, Oct. 15, 1741, CP, Angleterre, 412, 208r–208v, 242v–243r, 313r–314v; "Memoire sur les interets de l'Espagne relatifs a la France et a l'Angleterre par rapport au commerce," Feb. 26, 1742, CP, Espagne, 473, 62r–63r, 66r–66v.

of a wave of British Pacific assaults. Confused reports about new expeditions circulated. In fall 1743, rumors spoke of one or possibly two expeditions in preparation, involving four to six ships, perhaps aimed at French ships returning from the Pacific (where they were taking advantage of a brief wartime relaxation of Spanish commercial regulations), perhaps at the establishment of a British post at Manila, perhaps at Buenos Aires. Some reports spoke of a privately organized and Admiralty-supported venture; others, of a British naval project under Commodore Charles Knowles. In 1745, two French naval officers returning to Paris from London captivity reported not only that the procession of forty-five (their estimate) carts carrying the silver and gold Anson had captured from the Spanish had had beneficial effects on British morale and finances but also that five more English warships had departed for the Pacific in search of prizes.[7]

One particularly tangled and peculiar set of French conjectures regarding the intent of Anson's mission merits more detailed treatment. These reflections concern a possible Patagonian route to the Pacific. Though involving the extremity of the Americas a world away from Hudson Bay, these speculations disclose the way French foreign ministry officials responded to reports of possible British interest in probing obscure areas for better routes to the South Sea.[8]

7. Directors of the French Compagnie des Indes to Rennes, Sept. 24, 1743, 320r, comte de Maurepas to Rennes, Oct. 13, 1743, 353r–354r, both in CP, Espagne, 474; M. d'Avringhen to Amelot, Oct. 14, Dec. 19, 1743, CP, Angleterre, 417, 304r–309r, 395r–396v. For discussions of Spanish imperial commerce during wartime, see John Lynch, *Bourbon Spain, 1700–1808* (Oxford, 1989), 153; Geoffrey J. Walker, *Spanish Politics and Imperial Trade, 1700–1789* (Bloomington, Ind., 1979), 211–215; Maurepas, "Mémoire: Les officiers du Vau; La marquise d'Antin pris à leur rétour du Perou et qui sont arrivés de Londres à Paris raportent," Dec. 19, 1745, CP, Espagne, 487, 391r–393r.

8. For the course of events concerning the Patagonian passage, this account relies especially on the following from CP, Angleterre, 411: "Relation des negotiations du Sr. . . . avec L'Angre, puis avec la Russie, sur le project de decouverte d'un passage, de l'Ocean meridional atlantique a la Mer pacifique," "Envoyé par M. de Bussy," Mar. 31, 1741, 270r–285r; François de Bussy to Amelot, Mar. 31, 1741, 286r–314r; "Instructions pour Mrs Jean Opie et Charles Cagnoni Sur la proposition que Je. . . . [ellipses in original] confie a leur prudence et habilité," included with Bussy's dispatch of Mar. 31, 1741, 315r–319r; and Antonio Béthencourt, "Proyecto de un establecimiento ruso en Brasil (1732–1737)," *Revista de Indias*, IX (1949), 651–668. Anson's mission offered yet another reminder of the arduousness of the known passages from the Atlantic to the Pacific. In his voyage around Cape Horn, one of Anson's ships wrecked in the *Golfo de Penas* (one of the survivors, Lord Byron's grandfather John Byron, wrote a harrowing tale of the resulting adventures). Two abandoned the attempt and returned to England. Scurvy and others of the "endless to enumerate . . . disasters of different kinds which befel" Anson's men and produced "the calamitous condition of the whole squadron" killed about two-thirds of those on the three ships that succeeded in reaching the island of Juan Fernández. Five Spanish ships attempted to follow Anson into the Pacific. One vanished, one was run aground by a mutinous crew, three returned to the River Plate with more than half of their sailors dead. By the time one of the Spanish ships finally reached Chile, Anson had already made it to China. See George Anson, *A Voyage round the World in the Years MDCCXL, I, II, III, IV*, comp. Richard Walter and Benjamin Robins, ed. Glyndwr Williams (London, 1974). See also Rennes to Amelot, Aug. 29, 1741, CP, Espagne, 466, 476v–477r; Rennes to Amelot, Oct. 29, Nov. 7, 1741, CP,

Initial reports of a possible British effort to locate and exploit a Southwest Passage came from the French minister plenipotentiary in London, the venal and resilient François de Bussy (1740–1743). The March 1740 instructions to Bussy had directed him to devote great *"attention"* and *"vigilance"* to "discovering the true state of the English enterprises in the seas of America." On October 3, 1740, Bussy reported rumors that Anson was bound neither for Buenos Aires nor Juan Fernández, nor even for a passage around Cape Horn. Instead, word had it that Anson was going to establish a post near Patagonia ["quil doit faire dans une des terres de la Mer du Sud prés des Patagons, un etablissement"]. Reports suggested that a gulf existed in this area through which one could "communicate with the two parts of the South Sea" ["un Golfe par lequel on peut communiquer avec les deux parties de la mer du Sud"] (for Bussy, the "South Sea" often included the Pacific and the South Atlantic). Bussy's source indicated this passage was entirely unknown to the Spanish and thus did not appear on maps of the area. Instead of dismissing all this as nonsense, the French secretary of state for foreign affairs, Jean-Jacques Amelot de Chaillou (1737–1744), instructed Bussy to continue seeking and sending reliable information about the matter.[9]

Bussy investigated as instructed and sent another report on March 31, 1741, which, when combined with Antonio Béthencourt's 1949 scholarly discussion of pertinent Spanish documents, offers some indication of events surrounding the proposed Patagonian passage expedition. In 1732, Antonio da Costa, a Portuguese slave trader with twelve years of South Atlantic experience, had his brother Juan, a merchant active in London, propose to Robert Walpole a project for the establishment of a British presence on a South American gulf linking the

---

Espagne, 467, 209r–210v, 245v; Rennes to Amelot, May 4, 1742, CP, Espagne, 471, 14r–14v; Étienne de Silhouette to Amelot, Aug. 7, 1741, and August 1741, 263–263v, both in CP, Angleterre, 412, 240v; Glyn[dwr] Williams, *The Prize of All the Oceans: The Triumph and Tragedy of Anson's Voyage round the World* (London, 1999), 39–108; John Bulkeley and John Byron, *The Loss of the Wager: The Narratives of John Bulkeley and the Hon. John Byron* (Woodbridge, Suffolk, U.K., 2004).

9. Bussy took the occasional payment from the British government, without much apparent damage to his French diplomatic career. He became a premier commis at the foreign ministry in 1749 and was entrusted with an important diplomatic mission to Hanover in May of 1755, with the late-1755 composition of the instructions for France's new ambassador to Prussia, and again with the job of French minister plenipotentiary in London in 1761. Bussy's instructions in "Instruction a moi-même," in Paul Vaucher, ed., *Recueil des instructions données aux ambassadeurs et ministres de France depuis les traités de Westphalie jusqu'à la Révolution française*, XXV, *Angleterre*, III, *1698–1791* (Paris, 1965), 313–322, esp. 314: "Le Sieur de Bussy . . . ne peut trop multiplier son attention sa vigilance et son activité pour découvrir le véritable état des entreprises des Anglais dans les mers de l'Amérique comme dans les autres mers où ils ont des armemens, pénétrer leurs vues et leurs desseins pour l'avenir, et être informé de leurs Ressources pour les exécuter." See also Bussy to Amelot, Oct. 3, 1740, 128r–129r, Amelot to Bussy, Nov. 22, 1740, 247r–247v, both in CP, Angleterre, 408. Note that French authors sometimes referred to both the east and west coasts of southern South America as Patagonia.

Atlantic and Pacific oceans. Juan da Costa seems to have spoken vaguely about the passage's exact location, affirming that it lay a considerable distance from territories occupied by Europeans. He claimed that he had brought this proposal to Britain because it seemed the commercial nation best suited to taking advantage of it. British discussions of Costa's proposal apparently stretched from 1732 to 1735. Bussy asserted that the Admiralty and its First Lord Charles Wager, various British businessmen and seamen, the "Bureau de Commerce," and the "Conseil d'Etat" all endorsed the proposal and that a well-funded company had formed to undertake the exploration. Walpole rejected the proposal, however, out of fear of arousing Spain's ire and jeopardizing British trade with the Spanish Empire.[10]

After hearing rumors about the destination of Anson's mission, Bussy contacted Juan da Costa for information about the old Southwest Passage project. Bussy suggested to Costa that France might be the ideal sponsor for the scheme, but that he needed more precise information about the passage's location. Costa claimed the gulf lay south of Cape San Antonio (Cape San Antonio is southeast of Buenos Aires), more than four hundred miles from any European settlement. A whirlpool disturbing waters forty leagues out to sea shielded the gulf's mouth

10. It seems strange that the Costas did not bring their project to the Portuguese government in Lisbon. One can only speculate about their true motives in all of this, but it may be worth mentioning that the last time Antonia da Costa was in Lisbon, in 1729, he had been brought there by the Inquisition. See Béthencourt, "Proyecto de un establecimiento ruso en Brasil," *Revista de Indias,* IX (1949), 651–668, esp. 655. The phase of the affair after Walpole's rejection, from 1735 to 1737, offers material for an entertaining novel about illusion, intrigue, and betrayal but also goes well beyond the scope of this chapter. It appears that the Russian empress Anna Ivanovna (1730–1740) heard of at least part of the Costa project, offered to support it, and prepared a Russian expedition to set up a colony in Brazil's Rio Grande de São Pedro region. Two of the Costas's associates, John Opie and Carlos Cagnoni, then tried to double-cross the brothers by conducting the Russian expedition without them, at which point Juan da Costa started selling information to the Spanish government in hopes that Spain would foil the plans of his treacherous agents. Spain threatened to sink Russian ships found in South American waters, and this, combined with Russia's need to direct all its forces toward a new war with the Ottoman Empire, caused Russia to abandon the expedition. Then an Italian, identified only as "un fripon d'Italien," intimate with Russia's representative in London, Prince Antiochus Cantemir, seems also to have obtained information about the planned Russian expedition that he sold to both Spain and Portugal. Bussy's contention that Portugal then sent a naval expedition to the Rio Grande de São Pedro region because of desire to forestall Spanish or Russian occupation of the area may or may not be accurate, but modern histories of Brazil confirm that, in 1737, Portugal established the fortified post of São Pedro do Rio Grande at the mouth of the river as part of a longer-term design to increase the Portuguese presence in the lands between the Portuguese city of Rio de Janeiro and the Portuguese colony of Sacramento, across the River Plate from Buenos Aires. See comte de Vaulgrenant to Amelot, Jan. 14, 1737, 82v–83r, Amelot to Vaulgrenant, Jan. 28, 1737, 88v, both in CP, Espagne, 440; Béthencourt, "Proyecto de un establecimiento ruso en Brazil," *Revista de Indias,* IX (1949); C. R. Boxer, *The Golden Age of Brazil, 1695–1750: Growing Pains of a Colonial Society* (Berkeley, Calif., 1962), 241; Frédéric Mauro, "Portugal and Brazil: Political and Economic Structures of Empire, 1580–1750," in Leslie Bethell, ed., *The Cambridge History of Latin America,* I, *Colonial Latin America* (Cambridge, 1984), 441–464, esp. 464–467.

and had caused Portuguese navigators to avoid the area. Large ships could go only twenty leagues into the gulf, but smaller boats could make the rest of the journey to the Pacific.

Bussy saw how far-fetched the whole notion of seeking a kind of Southwest Passage appeared. What lent credence to the idea for him was the apparent British interest in it. Bussy claimed to have coolly reflected on the whole business before forwarding his report to Amelot and the foreign ministry. "But the more I consider," he said, "the more I find the appearance of reality." He wrote of "the avidity with which this project has been seized upon by the people of this nation [Britain], the most experienced in the navigation and in the commerce of South America; the general approbation which the admiralty, the bureau of commerce, and the council of state of England have given it; the public praise in particular of Sir Wager who has navigated for so many years in these seas." He suggested that Walpole opposed the company's proposal only because he wanted the discovery's glory and profit to accrue to the king, and Bussy noted that Walpole had recently tried to resume discussions with Juan da Costa. For Bussy, "all these circumstances gathered together" appeared "to be equal to a demonstration in favor of the probability of the success of the project." He concluded that France should support Costa and attempt the discovery itself, with hopes of obtaining for Louis the "immense advantages" an establishment in the region would offer.[11]

Amelot expressed skepticism about Bussy's proposal, suggesting that this Southwest Passage might be similar to one of those invisible islands mariners so often reported and sought, but so rarely found. He also complained that Bussy's geographical descriptions were too vague to clarify the exact areas he was talking about. But Amelot also found the interest of reputable British figures sufficient cause for further French inquiries into the matter, and he requested that Bussy send Costa to France.[12]

11. The original French of the quotations reads: "Plus je la considere, plus J'y trouve d'aparence de realité"; "... l'avidité avec laquelle ce Projet a eté saisi par les gens de cette Nation les plus experimentéz dans la Navigation et dans le Commerce de L'Amerique Meridionale; l'aprobation generale qu'y ont donné L'Amirauté, le Bureau du Commerce, et le Conseil d'Etat d'Angleterre; l'aplaudissement en particulier du Cher Wager qui a Navigué tant d'années dans ces Mers"; "Touttes ces Circonstances rassemblées, Monseigneur, me paroissent valoir une demonstration en faveur de la probabilité du succés du Projet"; "... des avantages immenses que presente le projet d'Un Etablissement dans les Lieux decouverts." See the following from CP, Angleterre, 411: "Relation des negotiations du Sr . . . ," "Envoyé par M. de Bussy," Mar. 31, 1741, 270r–285r; Bussy to Amelot, Mar. 31, 1741, 286r–314r; "Instructions pour Mrs Jean Opie et Charles Cagnoni . . . ," included with Bussy's dispatch of Mar. 31, 1741, 315r–319r.
12. See letter to Bussy, Apr. 20, 1741, CP, Angleterre, 411, 293r–293v (this volume has two sets of pages numbered 293r–293v; this is the second group); Bussy to Amelot, June 2, 1741, CP, Angle-

Bussy and Amelot's reactions to this unlikely Patagonian scheme reveal a great deal about the way prominent French diplomats handled geographic information of uncertain validity. The incident shows that, before receiving reports of British Hudson Bay exploration, high French foreign ministry officials were open to the notion that new routes to the Pacific might be found in areas not yet fully explored by Europeans. They were susceptible to the suggestion that the British government was promoting the search for these routes, and, because of Britain's prestige as the foremost maritime power, they were receptive to the idea that official British interest in a rumored geographic feature constituted a good indication the feature might exist.

French Reactions to the British Search for a Northwest Passage

A Northwest Passage from Hudson Bay was one of these rumored geographic features, and, after the end of the War of the Austrian Succession in 1748, the French foreign office began to appreciate fully the implications of the British search for it.

Information regarding possible routes to the Pacific came from multiple sources. By 1745, the foreign ministry had begun receiving reports from agents in England about British interest in a passage. French geographers, animated by both the British expeditions Arthur Dobbs had inspired and the Russian North Pacific exploration discussed in Chapter 6, were also engaging in revivified discussions about the possible existence of a North American route to the South Sea. Pierre Louis Moreau de Maupertuis, a leading scientist of the day (at least until he became the target of Voltaire's caustic wit) and the leader of a 1736–1737 expedition to Lapland to measure the earth's exact shape, advocated, in his 1752 *Lettre sur le progrès des sciences,* exploration of the Pacific and of the globe's southern oceans. He considered the existence of a northern passage from the Atlantic to the Pacific an unresolved question ["ce passage, dont l'utilité n'est pas douteuse, mais dont la possibilité est encore indécise"]. Maupertuis was one of the inspirations for Charles de Brosses and his 1756 *Histoire des navigations aux terres australes.* In addition to postulating Pacific landmasses and recommending France establish South Sea outposts, Brosses echoed Maupertuis's call for additional Pacific exploration and his view that "the famous questions of the two passages of the northeast and northwest" ["les fameuses questions des deux passages du nord-est et du nord-ouest"] remained undecided. The works of figures

terre, 412, 122r–126r. It is not clear whether Juan da Costa ever reached France, and I have found no evidence that France actually sent any ships to Patagonia to seek the alleged gulf.

such as Guillaume Delisle, Phillippe Buache, Maupertuis, and Brosses furnish examples of the kinds of materials shaping the broader intellectual context in which French statesmen evaluated news of the British search for a passage to the Pacific.[13]

Other circumstances were shaping the political climate. For French diplomats after the War of the Austrian Succession, British interest in a new northern route to the Pacific represented more than an intriguing geographic problem to be viewed in analytical isolation and with detached scholarly interest. The period's intensifying Anglo-French imperial rivalry was pushing foreign office personnel to relate individual British actions to the larger pattern of British imperial policy, and incidents took on historical importance by virtue of their perceived connection to this larger configuration. Because a Northwest Passage would provide a new route to the Pacific, French diplomats tended to consider British interest in a passage in the context of other developments pertaining to the South Sea and the Spanish colonies on its shores. Three circumstances appeared especially relevant to the French.

The first was the 1749 appearance of a French edition of the authorized account of Anson's Pacific mission. Parts of the book amounted to a kind of blueprint for South Sea expansion, a public expression of the aggressive British sentiments discussed in the previous chapter. In addition to recounting the dramatic episodes of the voyage, Anson recommended that Britain search South America's southern coasts for havens for British ships rounding Cape Horn. He urged that Britain locate South Atlantic sites, perhaps on one of the Falkland Islands, that could provide refuge and supplies to British ships as the isle of Juan Fernández could in the South Pacific. Anson also painted a rosy picture of the prospects for British conquests among Spain's Pacific possessions. He claimed that Chile's Araucanian Indians hated the Spanish and would join the British in attacks on them. Western South American creoles would then be forced to separate themselves from the Spanish Empire and could easily be subjected to British domination.[14]

13. For reports from England, see, for example, Apr. 2, 1745, 297r, Apr. 3, 1745, "Sur le passage dans les mers du Nord par la Baye d'Hudson," 305r–306r, Apr. 16, 1745, 408r, all in CP, Angleterre, 419. See also Maurepas to marquis de Beauharnois and Gilles Hocquart, June 12, 1743, AC B76, 454r–454v; [Pierre Louis Moreau] de Maupertuis, *Lettre sur le progrès des sciences* (n.p., 1752), 7–26, 30–38 (quotation on 31); Charles de Brosses, *Histoire des navigations aux terres australes: Contenant ce que l'on sçait des moeurs et des productions des contrées découvertes jusqu'à ce jour ...* , 2 vols. (Paris, 1756), I, i–iv, x (quotation), 2–4, 13–18, 78–80, II, 259–304, 315–316, 362–369. Maupertuis felt that ice would make use of a Northwest or Northeast Passage difficult. He recommended investigation into the possibility of a route across the pole. Dunmore surveys this French intellectual interest in the South Sea in *French Explorers in the Pacific*, I, 44–49.

14. Actual authorship of Anson's volume has been variously ascribed to "Richard Walter, M.A.

Anson's belligerent plans were alarming in their own right, but his ascent to a position where he could transform his personal views into British policy made his book more significant to French observers. By the end of 1744, Anson had become a member of the Board of Admiralty and Rear Admiral of the Blue. He was elevated to the peerage in 1748 and became First Lord of the Admiralty in 1751, a position he retained until 1756 and regained from 1757 to 1762. As the duc de Noailles, a former ambassador to Spain and an influential advisor of Louis XV—despite being, in the eyes of some of his Spanish and British counterparts, "unsteady," "showish," and "not an able Man"—put it in February 1755, one had only to "open and read the relation of the voyage of Admiral Anson to know the ideas and projects of the person who has today perhaps the greatest influence on the affairs of the British Navy."[15]

The second circumstance involved abortive preparations in British ports for a series of Pacific naval expeditions. Moving toward execution of the kinds of plans outlined in Anson's book, the Admiralty prepared in 1749 "two Sloops" "to go in Search of" the mythical South Atlantic Pepys Island, "from thence to go to Falkland's Islands" to make "what Discoveries they can in those Parts." After refitting, the ships were then "to double Cape Horn, and to water at Juan Fernandez, from thence to proceed into the Trade Winds keeping between the Latitudes of 25 and 10 Sth. and to steer a traverse Course for at least 1000 Leagues or more if they have an opportunity of recruiting their Wood and Water." In 1751, another set of rumors arose suggesting that Anson and his naval colleagues proposed to send two frigates to establish a British settlement on Juan Fernández. Because of Spanish protests to be discussed more fully in the next chapter, these expeditions did not actually leave Britain's ports. Nonetheless, to the French foreign office, the rumored preparations for these Southern Hemispheric Pacific ventures were noteworthy.[16]

Chaplain of . . . [Anson's flagship] the *Centurion*" and to pamphleteer Benjamin Robins. Because French diplomats treated Anson as the author, and because factors such as "Anson's close scrutiny of the work in progress . . . point to the *Voyage* as being in everything except stylistic terms Anson's own interpretation of events," it is appropriate in this discussion to refer to the book as Anson's own (Williams, *Prize of All the Oceans*, 237–241). For the passages referred to, see Anson, *Voyage round the World*, comp. Walter and Robins, ed. Williams, 95–99, 255–264. The English edition appeared in May 1748.

15. "Il n'y a qu'à ouvrir et lire la relation du voyage de L'Amiral Anson pour connoitre les vues et les projets de la personne qui a peut être aujourdhuy le plus d'influence dans les affaires de la marine d'Angleterre." See duc de Noailles, "Memoire sur la conjoncture presente," February 1755, MD, Angleterre, 52, 109r; Jonathan R. Dull, *The French Navy and the Seven Years' War* (Lincoln, Neb., 2005), 23. For one opinion of Noailles, see Benjamin Keene to Thomas Robinson, Sept. 21, 1754, SP 94/147, 197r.

16. Quotations from earl of Sandwich to duke of Bedford, Apr. 14, 1749, Add. Mss. 43,423, 81r–82r.

The third circumstance concerned Central America, where the continued activities of some one to two thousand British log-cutters in Belize, and one to two hundred loggers and gatherers on the Mosquito Coast alarmed both French and Spanish officials. These British interlopers were harvesting a range of tropical products: mahogany, tortoise shell, sarsaparilla, and, most important, logwood. Logwood, *Haematoxylum campechianum L.,* of "the riparian lowlands of the Yucatán Peninsula and northern Belize," yields a dye capable of producing red, blue, black, and gray shades. Because cloth production constituted eighteenth-century Britain's most important manufacturing activity, because dyes produced from British materials were inferior to those generated from imports, and because officials and merchants wanted to avoid dependence on potentially hostile or grasping Spanish dye suppliers, the British government attached great importance to British extraction of Central American logwood.[17] Despite repeated Spanish rejection of such English claims, British logging and gathering camps on the Bay of Honduras and the Mosquito Coast remained in place after the Treaty of Aix-la-Chapelle. British officials sought in various ways to legitimate and entrench the British presence there. The government appointed a superintendent for the Mosquito Shore in 1749. Jamaica's Governor Trelawny sent troops to the shore in the early 1750s and claimed the region for Britain in 1752. British ministers demanded Spanish restoration of British settlers to their Belizean camps after a Spanish raid drove them away in 1754. British settlers, traders, and loggers also sought ways to render their presence more permanent. A fort stood alongside the Nicaraguan town of Black River by 1748, and loggers rebuilt their fortifications in Belize after Spanish attacks in 1754.[18]

17. Lawrence Henry Gipson, "British Diplomacy in the Light of Anglo-Spanish New World Issues, 1750–1757," *AHR,* LI (1946), 627–648, esp. 632–637; Richard Pares, *War and Trade in the West Indies, 1739–1763* (Oxford, 1936), 540–552; Jacob M. Price, "The Imperial Economy, 1700–1776," in W[illiam] Roger Louis, ed., *The Oxford History of the British Empire,* II, *The Eighteenth Century,* ed. P. J. Marshall and Alaine Low (Oxford, 1998), 78–104, esp. 82; Savelle, *Origins of American Diplomacy,* 423–429; Frank Griffith Dawson, "William Pitt's Settlement at Black River on the Mosquito Shore: A Challenge to Spain in Central America, 1732–87," *HAHR,* LXIII (1983), 677–706, esp. 677–690; Karl H. Offen, "British Logwood Extraction from the Mosquitia: The Origin of a Myth," *HAHR,* LXXX (2000), 113–135, esp. 120 (quotation on logwood trees).

18. Gipson, "British Diplomacy in the Light of Anglo-Spanish New World Issues," *AHR,* LI (1946), 632–637; Sylvia-Lyn Hilton, "Las Indias en la diplomacia española, 1739–1759" (Ph.D. diss., Universidad Complutense de Madrid, 1979; rpt. Madrid, 1980), 549–569; Muret, "Le conflit anglo-espagnol dans l'Amérique centrale," *Revue d'histoire diplomatique,* LIV–LV (1940–1941), 137; Vicente Palacio Atard, *Las embajadas de Abreu y Fuentes en Londres, 1754–1761* (Valladolid, Spain, 1950), 5–11; Pares, *War and Trade,* 540–552; Price, "Imperial Economy," in Louis, ed., *Oxford History of the British Empire,* II, 78–104, esp. 82; Savelle, *Origins of American Diplomacy,* 423–429; Dawson, "William Pitt's Settlement at Black River on the Mosquito Shore," *HAHR,* LXIII (1983), 677–690; Offen, "British Logwood Extraction from the Mosquitia," *HAHR,* LXXX (2000), 113–135; Bedford to Edward Trelawny, Oct. 5, 1749, 7r, Trelawny to Lords of Trade, July 17, 1751, 12r–13v, both in CO

Because all Central American territory lay within reach of the Pacific, British activities there seemed to the French foreign ministry to be potentially related to other British actions bearing on that ocean. French diplomats noted that British logging camps on the Mosquito Coast and the Bay of Honduras might serve as base camps for an advance to Central America's Pacific shores. The French minister of foreign affairs, the "moderate," "able," and, to some, "gullible" Louis-Philogène Brûlart, marquis de Puysieulx (1747–1751), mentioned in November 1750 that British dyewood harvesting in Campeche could lead in time to British Pacific establishments capable of destroying Spanish South Sea commerce. In April 1754 and July 1755, Noailles offered a similar warning that British Mosquito Coast settlements could facilitate British Pacific trade, and he observed that Anson's book provided evidence of British interest in acquiring a foothold on the South Sea. French officials spoke also of British desires for a beachhead on the Isthmus of Panama. In fall 1754, minister of foreign affairs Antoine-Louis Rouillé (1754–1757) — a reputedly rudderless figure who somehow kept floating into powerful ministries — and France's ambassador in Spain, the "impetuous," "disorderly," and ultimately ineffective Emmanuel Félicité de Durfort, the duc de Duras (1752–1755), discussed rumors that Britain was asking Spain for the cession of Darien. Rouillé and Duras worried that a British presence there would lead to British domination of Pacific navigation, to British control of the route by which Peruvian silver passed from the Pacific to the Atlantic, to more extensive British contraband trade with Spain's possessions, and, ultimately, to British seizure of more American territory. British Central American activities pointed to the conclusion that the British Empire was preparing to acquire, by conquest or commerce, the wealth of the Spanish Pacific.[19]

123/1; Trelawny to Bedford, Apr. 8, 1749, CO 137/58, 124r–124v; Trelawny to Bedford, Dec. 8, 1750, July 17, 1751, 60v, 114r–114v, Trelawny to earl of Holdernesse, May 25, 1752, 197r–197v, and "Copy of Governor Trelawny's Orders to Captn Robert Hodgson," May 25, 1752, 205r–206r, all in CO 137/59.

19. Puysieulx's letter refers to "des etablissemens qui y detruiroient le commerce Espagnol." See marquis de Puysieulx to Vaulgrenant, Nov. 23, 1750, CP, Espagne, 507, 165v–166r. The characterization of Puysieulx is from Dull, *French Navy*, 8, 22–23. For Noailles's views, see duc de Noailles to duc de Duras, Apr. 8, 1754, CP, Espagne, 511, 311v; Noailles to Antoine-Louis Rouillé, July 8, 1755, CP, Espagne, 518, 13r–13v. Rouillé had succeeded Maurepas at the naval ministry (1749–1754). For Rouillé and Duras, see Duras to Rouillé, Sept. 30, 1754, 163v–165v, Rouillé to Duras, Oct. 14, 1754, 236r–236v, both in CP, Espagne, 516. Around 1752, Duras had requested and received an unsigned memoir, the author of which warned that the British had long wished to control the Gulf of Darien on the Pacific side of the Isthmus of Panama, and that a British base in this region could disrupt Spain's Pacific and trans-Isthmian commerce. See "Description de l'Amerique meridional tant des côtes de la mer du Nord que d'une partie de celle du Sud . . . ," MD, Amérique, 2, 68r–69r. French worries had a basis in British realities. To give one example, in the late 1740s and early 1750s, Governor Trelawny of Jamaica was urging the British secretary of state for the Southern Department

To the French foreign office, these British activities in Hudson Bay, in Central America, around southern South America, and in English shipyards, government bureaus, and printing shops did not represent an entirely new threat of British Pacific expansion but rather a more mature and menacing stage of a long-developing danger. Descriptions in foreign office documents of postwar events involving the Pacific attested to this perspective. In a 1751 letter to Puysieulx, Rouillé mentioned the long-standing and long-familiar British plans for the acquisition of Pacific establishments ["les vües qu'on connoit depuis si longtems aux Anglois pour former des établissements dans la mer du Sud"]. Two years before, an unsigned 1749 memoir discussing English Pacific designs began by saying, "It has been known for a long time that the English have formed projects for establishments in the South Sea and in the lands alongside it." This author referred to incidents as far back as John Narborough's 1669–1671 voyage to Chile. French diplomats also referred to the more recent affairs discussed earlier in this chapter. In summer 1749, having heard of the rumored British preparations for another South Sea naval expedition, Puysieulx instructed the French *chargé d'affaires* in London, François-Michel Durand de Distroff, to ascertain whether the expedition would seek the Patagonian passage Bussy and Amelot had discussed eight years before. British Pacific interest in and of itself was not, to French diplomats, a novel element in international affairs.[20]

Certain aspects of this recent British interest, however, did appear both original and ominous. Where British efforts to reach the Pacific had formerly been halting and sporadic, they appeared to have become aggressive and recurring, and they were now coming from three directions simultaneously. The apparent relationship among these events was crucial. Each instance of British expansionist intent heightened French foreign office concern about the others.

---

and the Lords of Trade and Plantations to support his efforts "to cultivate a Friendship with the Darien Indians," which would lead to British conquest of the Isthmus of Panama. See Trelawny to Lords of Trade and Plantations, May 7, 1748, CO 137/25, 31r–32r; Trelawny to Bedford, May 7, 1748, CO 137/58, 98r–98v; Trelawny to Bedford, Apr. 14, 1750, CO 137/59, 22r–23v. On Duras, see Michel Antoine, *Louis XV* ([Paris], 1989), 670. On Rouillé, see Walter L. Dorn, *Competition for Empire, 1740–1763* (New York, 1963), 303; Dull, *French Navy*, 23.

20. Rouillé to Puysieulx, June 5, 1751, CP, Espagne, 508, 64r–65v. Rouillé went on to be minister of foreign affairs from 1754 to 1757. Because he was involved in foreign ministry deliberations before 1754, I treat him along with the foreign office in the late 1740s and early 1750s. See "Memoire sur les projets des Anglois pour des etablissemens dans la mer du Sud," Dec. 11, 1749, MD, Amérique, 9, 126r–127r: "Il y a long tems qu'il est connu, que les Anglois ont formé des projets pour des etablissemens dans la mer du Sud, et dans les Pays voisins" (another copy of this memoir appears in MD, Espagne, 82, 91r–95v). See also Puysieulx to François-Michel Durand de Distroff, July 6, 1749, 365r, Durand to Puysieulx, July 24, 1749, 377v–378v, both in CP, Angleterre, 426.

French foreign ministry personnel saw Anson's book as evidence of a general British commitment to Pacific expansion, and this apparent commitment made it more plausible that the British were seeking a Northwest Passage and hoping to use it in ways inimical to French interests. The 1749 memoir about British Pacific designs, after noting Britain's long-standing interest in establishing itself in the South Sea, went on to argue that the just-published account of Anson's expedition left no doubt that England would "work incessantly at the execution of these projects." The memoir then discussed plans such as the establishment of British outposts on the islands of Juan Fernández and Inchin off the coast of Chile, Quibo off the coast of Panama, and Tinian in the Marianas group. These examples, the author declared, gave an idea of what the English could do in the Pacific using the ordinary Cape Horn route.[21]

The author then closed with a section regarding a Northwest Passage from Hudson Bay. He contended that the British had by no means abandoned their design of reaching the Pacific from the north, appearing, on the contrary, more determined than ever to pursue this objective. A recently published account of the 1747 British Hudson Bay expedition further convinced the author of this resolve. Although the author did not name the account to which he was referring, timing, topic, and tone make the 1749 French translation of Henry Ellis's 1748 *Voyage to Hudson's-Bay, by the "Dobbs Galley" and "California," in the Years 1746 and 1747, for Discovering a North West Passage* a likely candidate. Ellis's book, in any case, exhibits the kind of optimism regarding the possible existence and discovery of a Northwest Passage that so alarmed French observers. Although the expedition "did not discover a North West Passage," Ellis wrote, "yet were we so far from discovering the Impossibility or even Improbability of it, that . . . we returned with clearer and fuller Proofs . . . that evidently shew such a Passage there may be." For the anonymous French memoirist, one could conclude from public expressions of such sentiments "that the English have high hopes of opening from the northwest a passage into the South Sea." Had the kind of British Hudson Bay investigation described by Ellis occurred by itself, or had French diplomats viewed it in isolation, the British quest for a Northwest Passage might have seemed more like a regional curiosity and less like an ominous sign. But when French diplomats placed Hudson Bay exploration alongside Anson's program

21. "Memoire sur les projets des Anglois pour des etablissemens dans la mer du Sud," Dec. 11, 1749, MD, Amérique, 9, 126r: "Mais la relation des voyages de l'Amiral Anson, qui a été publieé cette année ne permet pas de douter, que l'Angleterre ne travaille incessament a l'exécution de ces projets"; 134r: "Ces differentes descriptions ont paru necessaires pour donner une idée de ce que les Anglois peuvent faire dans la mer du Sud par la route ordinaire, qui est celle du Cap de hoorn."

for British Pacific expansion, the aggressive intent of the search for a passage seemed evident.[22]

Like Anson's book and his government position, British seaport activities seemed to nervous French officials to point to a British desire to find a Northwest Passage. The author of the 1749 memoir mentioned such maritime preparations among the reasons for his belief the British intended to find a northern route to the South Sea. The memoir indicated specifically that British expeditions would soon examine western Hudson Bay's Chesterfield Inlet and Repulse Bay; the Hudson's Bay Company intended to send five ships to seek a passage in the current year and two ships in the next. Similar reports came from a different source. Durand had indicated in July 1749 letters from London that the two frigates the Admiralty had been preparing for a South Sea expedition would not seek a Patagonian passage. Instead, after passing around South America, the frigates would proceed to the North Pacific. There they would seek the opening of a gulf that could "serve to establish a communication between the two seas [the North Pacific and North Atlantic]." The British had not despaired, he claimed, of finding the "famous passage" into the Pacific from the north. Despite the difficulties encountered by 1740s British explorers, it appeared to many in France that the British search for a Northwest Passage continued.[23]

As preparations in British ports and publication of Anson's book made the search for a Northwest Passage seem more threatening, the quest itself made British imperial ambitions look more vast and menacing. One aspect of this threat involved the apparent British desire to establish Pacific naval stations. Having already spoken of Anson's interest in naval bases in locations like Chilean islands, the author of the unsigned 1749 memoir noted that discovery of a Northwest Passage would lead to placing another British Pacific post on a

22. Ibid., 134r–134v: "Ils n'ont point abandonné le dessein de penetrer dans cette mer du côté Septentrion[al], au Contraire ils paroissent plus determinés qu'ils ne le furent jamais a suivre ce des[sein]"; "On voit que les Anglois ont de grandes esperances de s'ouvrir par le nord ouest un passage dans la mer du Sud." See also Henry Ellis, *Voyage to Hudson's-Bay, by the "Dobbs Galley" and "California," in the Years 1746 and 1747, for Discovering a North West Passage* (London, 1748), esp. 298 (see also xix, xxi, 300); Ellis, *Voyage de la baye de Hudson: Fait en 1746 et 1747, pour la découverte du passage de Nord-Ouest* (Paris, 1749), I, xlvi, xlix, II, 261–262, 264.

23. "Memoire sur les projets des Anglois pour des etablissemens dans la mer du Sud," Dec. 11, 1749, MD, Amérique, 9, 134v–135r; Durand to Puysieulx, May 26, 1749, 147v, June 8, 1749, 189v–190r, July 4, 1749, 287r–288v, July 24, 1749, 377v–378v ("On m'a ajouté qu'il y avoit un dessein de penetrer aussy avant qu'on pouroit dans la partie du Nord de la mer du Sud, afin d'y decouvrir quelque Golfe qui fut près de ceux qui sont connüs dans l'ocean Septentrional, et qui pût servir a etablir une communication entre les deux mers. On ne desesperoit point outre cela de pouvoir en tirer des inductions sur le fameux passage"), all in CP, Angleterre, 426, and see Puysieulx to Durand, July 6, 1749, 365r.

northern route to California. "In this manner," he averred, "they [the English] would render themselves masters of the entry into this sea [the Pacific] by its two opposite extremities." Rouillé used nearly identical language in his 1751 letter to Puysieulx, then went on to note that these British attempts to control Pacific entrances had "taken on a new intensity since the relation of the voyage of Admiral Anson, and that of the researches made for a passage to this ocean through Hudson Bay."[24]

French foreign ministry observers traced this British desire for South Sea mastery to interest in Spanish Pacific trade, and they connected British Hudson Bay exploration to designs on this commerce. In an unsigned memoir surveying American conditions, one foreign ministry scribe imputed the British searches for "a passage into this sea from Hudson Bay" to a British scheme to "carry away from the Spanish a portion of their commerce in the South Sea." The British might make use of Royal Navy ships based in British naval outposts to prevent

24. "Memoire sur les projets des Anglois pour des etablissemens dans la mer du Sud," Dec. 11, 1749, MD, Amérique, 9, 135r: "Ils ne manqueroient pas de faire un établissemt dans la mer du Sud vers Californie. De cette maniere ils se rendroient Maitres de l'entrée dans cette mer par les deux extremités opposées." Rouillé spoke of British attempts "to render themselves masters of its [the Pacific's] access by way of the two extremities [of the Americas]." See Rouillé to Puysieulx, June 5, 1751, CP, Espagne, 508, 64r–65v.

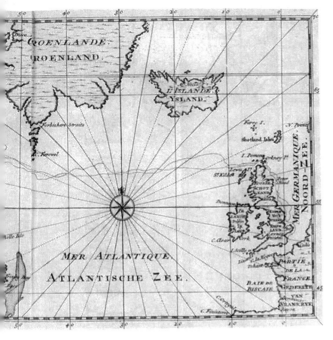

MAP 36. Nicolaes van Frankendaal, "Nouvelle carte des endroits, où l'on taché de découvrir en 1746 et 1747 un passage par le Nord-ouëst," in Henry Ellis, *Voyage à la baye de Hudson, fait en 1746. et 1747. . . .* (Leiden, 1750). Courtesy of the John Carter Brown Library at Brown University

Spanish or French vessels from entering the Pacific. Meanwhile, British ships sailing through Hudson Bay would carry European merchandise to Chile and Peru and American silver to the British isles. British trade would grow, while France remained excluded from half the globe.[25]

Did French officials really think a Northwest Passage existed for British explorers to find? They allowed that it might but evinced a healthy skepticism about the matter. The initial 1745 reports from London had mentioned Middleton's loss of crewmen to arctic cold and his observations that the passage's reality seemed unlikely, at least in a region whose climate would leave enough ice-free sailing time to allow its use. The unsigned 1749 memoir acknowledged that "the discovery of a Northwest Passage is something very uncertain" and conceded that, even if it were found, mariners would still encounter "great difficulties in making use of it."[26]

25. "Je crois que l'on peut attribuer au projet d'enlever aux Espagnols une portion de leur Commerce de la mer du Sud les recherches que les Anglois font d'un passage dans cette mer par la Baye d'hudson" ("Memoire sur l'Amérique," MD, France, 2008, 86v). This memoir is undated, but based on the presence of references to the published version of Anson's voyages and the absence of allusions to open Anglo-French hostilities in America, it appears to date from sometime between 1749 and 1754.

26. "Sur le passage dans les mers du Nord par la baye *d'Hudson*," Apr. 3, 1745, CP, Angleterre,

But French skepticism did not lead to total dismissal. Despite his doubts, the anonymous memoirist did not yet see the "impossibility" of finding such a usable passage, and he noted that the English appeared to expect the success of their exploratory enterprise. In the late 1740s and early 1750s, the combination of Britain's unique familiarity with the bay and prestige as the world's preeminent maritime nation lent a patina of plausibility to British exploratory objectives. Britain had enjoyed exclusive access to Hudson Bay since the Treaty of Utrecht and might have found reasons for believing in the existence of a Northwest Passage. Britons reported doing so. After returning to England from a 1739 voyage to Hudson Bay, Middleton had written to Dobbs, "This Year some of the Natives, who came down to trade at Churchill . . . informed" Churchill Factory governor Richard Norton that "they frequently traded with Europeans on the West Side of America, near the Latitude of Churchill by their Account, which seems to confirm that the two Seas must unite." Edward Thompson, surgeon on Middleton's 1741–1742 expedition, testified before a parliamentary committee in 1749 that he had "heard the Natives talk of a Sea to the Westward, which, by their Accounts, is not far distant; . . . [and] of a Streight which takes them five Days in crossing. . . . By their Account, this Streight has a Communication both with the Bay and the South Sea." (The Franco-Indian trade on and around Lake Winnipeg, a body of water variously denominated the "Little sea," the "Great Water," and the "Sea Lake," may help to explain reports such as these.)[27]

French diplomats confronted two issues. The first held implications for British naval capabilities: the British might find a new Pacific route facilitating British trade with western America and East Asia as well as British attacks on Spanish

419, 305r–306r; "Memoire sur les projets des Anglois pour des etablissemens dans la mer du Sud," Dec. 11, 1749, MD, Amérique, 9, 135r.

27. "Memoire sur les projets des Anglois pour des etablissemens dans la mer du Sud," Dec. 11, 1749, MD, Amérique, 9, 135r: "On ne dissimulera pas que la decouverte d'un Passage au Nord-oüest est quelque chose de trés incertain; et que quand même elle reussiroit, il se trouveroit de grandes difficultés a en faire usage; mais enfin on n'y voit pas jusqu'à present d'impossibilité; et si le projet avoit le succés, que les Anglois paroissent en attendre. . . ." Middleton quotation in William Barr and Glyndwr Williams, eds., *Voyages to Hudson Bay in Search of a Northwest Passage, 1741–1747*, 2 vols. (London, 1994–1995), I, 54 (Thompson quotation in II, 347–348) (see also Ellis, *Voyage to Hudson's Bay*, 304–305; E. E. Rich, ed., *James Isham's Observations on Hudsons Bay, 1743; and Notes and Observations on a Book Entitled "A Voyage to Hudsons Bay in the Dobbs Galley,"* 1749 [Toronto, 1949], 113; for references to Lake Winnipeg, see 68–69); "From James Knight," Sept. 17, 1716, "From Richard Norton and Thomas Bird," Aug. 3, 1723, in K. G. Davies, ed., *Letters from Hudson Bay, 1703–40* (London, 1965), 56–70, 83–87, esp. 57, 85; Anthony Henday, "A Copie of Orders and Instructions to Anthy Hendey upon a Journey in Land," June 26, 1754, in Henday, *A Year Inland: The Journal of a Hudson's Bay Company Winterer*, ed. Barbara Belyea (Waterloo, Ontario, 2000), 39–42, esp. 40; [Nicolas] Jérémie, "Relation du detroit et de la baie de Hudson," in Jean Frédéric Bernard, *Recueil d'arrests et autres pièces pour l'établissement de la Compagnie d'Occident* (Amsterdam, 1720), 1–39, esp. 25.

commerce and possessions. This possibility represented a threat to the sale of French goods in Spanish America. The second issue was a certainty elucidating British imperial intent. Whether or not they could find a practicable Northwest Passage, British explorers had unquestionably been looking for one. This seemed an undeniable indication that the British Empire wanted new means of reaching the Pacific Ocean and the Spanish American markets dotting its shores.

Quite apart from the question of a Northwest Passage's existence, British intent to reach the Pacific by means of it led to troubling conclusions about the evolution of British policy since the Treaty of Utrecht. When pondering the intensifying imperial rivalry of the post-1748 period, French officials often recalled the terms and principles of this earlier settlement. The December 1754 instructions to the duc de Mirepoix, France's ambassador extraordinary in London, indicated that the issues at stake in Anglo-French negotiations involved far more than the safety of French North American possessions. It was "a question also of the system established by the Peace of Utrecht to assure the sources of the commerce and power of all the sovereigns of Europe." To French diplomats in the years after the War of the Austrian Succession, the British Empire seemed to be engaged in the exact kind of Pacific expansion France had been compelled to renounce at the end of the War of the Spanish Succession. They exhibited growing fear that British imperialists might have abandoned entirely the peaceable spirit of the 1713 settlement and chosen, instead, a policy of imperial exaltation as threatening to France as Louis's quest for *la gloire* had been to Europe. If, despite long-standing and mutually accepted prohibitions, the British were trying to find a new Hudson Bay Pacific route, one could reasonably conclude they must be willing to disregard the Utrecht provision prohibiting British Pacific navigation. If the British intended to ignore one Utrecht stipulation, how could France rely on them to observe the others? And if the British government intended to flout the terms of an agreement as important as the Treaty of Utrecht, how could French statesmen count on its honoring any new compact to forego expansion and suspend hostilities in North America?[28]

It was not that Hudson Bay, or even the Pacific Ocean to which it might lead,

28. Memoir sent to duc de Mirepoix, Dec. 31, 1754, CP, Angleterre, 437, quotation and translation from "Private Memoir for Mirepoix," Dec. 31, 1754, in Theodore Calvin Pease, ed., *Anglo-French Boundary Disputes in the West, 1749–1763* (Springfield, Ill., 1936), 65–83, esp. 82–83. In a 1754 letter to ambassador Duras, Rouillé mentioned that acquisition by a non-Spanish power of a new right to sail in Spanish imperial waters or to trade with Spanish imperial territories would violate the Utrecht settlement (Rouillé to Duras, Oct. 14, 1754, CP, Espagne, 516, 235r). Puysieulx noted in a June 1749 letter to French ambassador Vaulgrenant in Spain, that if, for example, the Spanish were to allow the British to send ships to the Pacific, it would run contrary to the letter and spirit of the treaty (Puysieulx to Vaulgrenant, June 3, 1749, CP, Espagne, 503, 22r–23v).

had become the central issue of European international relations between 1748 and 1755. Hudson Bay remained closer to the North Pole than to London, Paris, or Madrid. But regions of little intrinsic weight can become significant in periods of acute international tension because of their relation to areas of greater importance. Hudson Bay derived its diplomatic importance in the interwar period not only from the possibility that Britain might find a new Pacific route there but also from what it indicated about the scale of British imperial ambition. Drake and Narborough's sixteenth- and seventeenth-century Pacific voyages had demonstrated English South Sea interest and seamanship, but those voyages had taken place long ago and led to nothing. The negotiations conducing to the Utrecht settlement had demonstrated British South Sea mercantile ambitions, but the agreement had stymied such aspirations. Anson's voyage revealed the reach of British naval power but could perhaps be viewed as a spectacular yet singular wartime achievement. British log-cutters refused to abandon their camps in Central America, but that could be ascribed merely to the value of the region's trees.

British Hudson Bay exploration during the 1740s could only be interpreted as evidence of a commitment to finding better routes to the Pacific, and the most likely objective of such interest was the wealth of Peru and Mexico. Such British exploration showed that Anson's Pacific voyage might be singular in execution, but it was typical in intent. It suggested that British actions at two extremes of the globe were parts of a unified scheme. It revealed just how far the British government was willing to go to satisfy its Pacific aspirations, neither northern ice, nor past failures, nor the unlikeliness of success sufficing to deter the attempt to find a new way to the South Sea. Then, the attacks on the Spanish Empire advocated in the published account of Anson's voyages, Anson's rise to Admiralty prominence, and the flurry of rumors concerning new British expeditions indicated the Treaty of Aix-la-Chapelle had not ended British Pacific designs. These incidents were like puzzle pieces: individually they meant little, but properly arranged they formed an ominous picture of British imperial expansionism.[29]

29. Much of this interpretation of French foreign office thinking rests on analysis of documents the French foreign office produced between 1748 and 1755. One might object to such a reading on the grounds that these letters and memoirs offer a less-than-reliable indication of the true sentiments of their authors. In particular, some of these memoirs were intended for Spanish eyes, and one might be tempted to regard them as diplomatic rhetoric designed to secure Spanish assistance for France rather than as sincere efforts to express the views of the French foreign ministry (for an example of this kind of argument, see Savelle, *Diplomatic History of the Canadian Boundary*, xiii). As will be discussed at greater length in succeeding chapters, French diplomats did transmit some of these memoirs to their Spanish counterparts, and they did hope that the arguments in them would persuade the Spanish government to renew its alliance with France. The rarity of diplomatic sincerity, however, does not render diplomatic frankness in documents such as these an impossibility.

Effects of British Pacific Encroachments on French
Interpretations of British Aggression Elsewhere

French assessments of the search for a Northwest Passage and other threatened Pacific incursions shaped evaluation of other instances of British aggression. Two aspects of this influence are significant here. The first concerns French perceptions of the scale of British maritime aspirations. After the War of the Austrian Succession, French officials believed they were observing not merely British attempts to obtain local maritime dominance in various areas but a comprehensive British attempt to obtain the "empire of the seas." French documents repeatedly allude to this British aspiration. The 1749 instructions for Ambassador Mirepoix stated that friendly Anglo-French relations would be impossible unless England abandoned its plan to "usurp the empire of the seas." Puysieulx referred in a 1751 letter to the comte de Vaulgrenant to England's project "to usurp the principal domination at sea and in America."[30]

The danger French officials envisioned was that the British Empire would use the resources gained from control of maritime commerce to dominate European affairs. Mariners trained on American trading voyages in times of peace

---

Rather than assume one way or the other, it is better to test the sincerity of the ideas expressed in the documents in question. If, for example, the French foreign office memoirs given to the Spanish government reflected what French diplomats wanted Spanish diplomats to think about an issue and not what French diplomats actually believed, there should be a substantial discrepancy between the style and substance of letters and memoirs intended solely for internal foreign ministry perusal and those being given to Spanish officials. For the topics discussed in this chapter, no such substantial discrepancy exists. Documents moving solely among French officials exhibited the same tone and content as those given by them to their Spanish counterparts. Both groups of documents evinced a concern about British designs on the Pacific and the Spanish commerce there. Both saw British activities around Hudson Bay, Central America, and southern South America as evidence of that design.

Furthermore, if French discussions of the British threat to the Pacific were essentially insincere diplomatic rhetoric, they should manifest some degree of incompatibility with other tendencies in French diplomatic thought and policy. In fact, they appear to be logical results of such tendencies rather than departures from them. This and earlier chapters have demonstrated a half century of French interest in Pacific Ocean trade and navigation. French post-1748 comments about British Pacific activities represent a more developed stage of this interest, not a newly conceived diplomatic stratagem.

30. "Mémoire pour servir d'instruction au sieur marquis de Mirepoix, chevalier des ordres du roi, lieutenant général de ses armées, commandant pour sa Majesté en provence et gouverneur de la ville de Brouage, allant en Angleterre," in Vaucher, ed., *Recueil des instructions données aux ambassadeurs et ministres de France*, XXV, 335–336, 343–344 ("Elle renonçât au dessein suivi d'usurper l'Empire des Mers"); Puysieulx to Vaulgrenant, Jan. 26, 1751, CP, Espagne, 507, 272v. See also [Nicolas-Louis?] Le Dran, "Reflexions sur les demêlés survenus entre'les cours de France et d'Angleterre," June 1755, MD, Angleterre, 41, 44r–44v: "Le sistême de la cour de Londres, est d'usurper l'empire de la mer et de rendre les Anglois maitres absolus de tout le commerce."

could be converted to ship-of-the-line sailors in times of war. Money made in colonial trade could be taxed or borrowed by the British government and used to build ships and outfit armies. In particular, silver obtained by selling British goods to Mexico and Peru could sustain British finances and buy the services of German princes and German mercenaries. Puysieulx warned in 1751 that England could render other European states dependent on the British Empire by using wealth gained from America to supply the wants of Europe's penurious princes. Noailles noted in 1755 that the Americas acted as the principal source of Europe's precious metals and constituted therefore a matter of grave consequence for European affairs. Universal monarchy might be an unattainable goal, but universal influence would not be if "one nation succeeded in rendering itself sole master of all the commerce of America" ["Quelque chimerique que soit le projet de la Monarchie universelle, celui d'une influence universelle par le moyen des richesses cesseroit d'etre une chimere, si une nation parvenoit à se rendre seule maitresse do tout le Commerce de l'Amerique"]. If the English became "masters of the sea" ["maitres de la Mer"], they would be "in a state to give the law to all of Europe" ["en etat de donner la loy à toute l'Europe"].[31]

31. Daniel A. Baugh presents the best treatment of the relation between commercial resources and European power in "Withdrawing from Europe: Anglo-French Maritime Geopolitics, 1750–1850," *International History Review*, XX (1998), 1–32, esp. 5–21. See also W[illiam] J. Eccles, "The Role of the American Colonies in Eighteenth-Century French Foreign Policy," in Eccles, *Essays on New France* (Toronto, 1987), 144–145, esp. 146; Gipson, *British Empire before the American Revolution*, V, 32–33; Max Savelle, "The American Balance of Power and European Diplomacy, 1713–78," in Richard B. Morris, ed., *The Era of the American Revolution* (New York, 1965), 140–169, esp. 158; Yves F. Zoltvany, "France's View of Canada — Asset or Liability?" in Kenneth A. MacKirdy, John S. Moir, and Yves. F. Zoltvany, eds., *Changing Perspectives in Canadian History* (Notre Dame, Ind., 1967), 1–15, esp. 2, 10. For Puysieulx and Noailles, see Puysieulx to Vaulgrenant, Mar. 2, 1751, CP, Espagne, 507, 329r; Noailles, "Memoire sur la conjoncture presente," February 1755, MD, Angleterre, 52, 104v, 108r. Noailles echoed Silhouette's 1747 argument that the Americas acted as the primary source of bullion for Europe and the key market for European goods, and that therefore, if the British were to entirely appropriate the commerce of the Americas, "they would be masters at sea by means of their fleets, and masters on land by means of their riches; and it is with the funds of America that they would draw the means of dictating the law to Europe" ["que l'augmentation de leurs Richesses et par consequent l'accroissement de leur puissance ne seroient plus contenus par aucunes bornes, s'ils pouvoient reussir a s'approprier le commerce de l'Amerique en entier; puisque c'est la source de l'or et de l'argent, le debouché le plus considerable comme le plus certain de toutes les manufactures." "Ils seroient maitres de la mer par leurs flottes, de la terre par leurs Richesses; et c'est du fonds de L'Amerique qu'ils tireroient les moyens de dicter la Loy a L'Europe"]. See "Observations sur les finances la navigation et le commerce d'Angleterre," October 1747, MD, Angleterre, 46, 86v–87r. For similar expressions of this sentiment, see Noailles to Duras, Dec. 31, 1753, CP, Espagne, 511, 244v–245r; [LeDran?], "Sur la politique des Anglois pour detruire l'equilibre des puissances en Amerique," 1755, MD, Angleterre, 41, 27r–27v: "C'est par la destruction de la liberté et de l'independance de l'Amerique qu'ils se proposent de parvenir au projet de dicter la loi a toute l'Europe" (author identification from Waldo G. Leland, John J. Meng, and Abel Doysié, *Guide to*

The center of gravity of these maritime efforts clearly lay in the Atlantic. The volume and value of British shipping involved in the North Atlantic fisheries and in the commerce among Europe, West Africa, and the Atlantic American and West Indian colonies was, of course, vastly greater than that involved in Pacific or Indian Ocean commerce. The threat of British Pacific forays, however, along with British operations in India, demonstrated that, although British ambition might originate and be concentrated in the Atlantic, it was not limited to that ocean. With its probes in and around Hudson Bay, Central America, and Cape Horn, the British Empire seemed to be trying to extend its Atlantic navigation into the Pacific Basin and toward the Spanish resources and markets waiting there. French concerns about this extension did not eclipse French apprehensions about the growth of British Atlantic power, but they did suggest that one ocean alone might be too little for the British Empire. The body of the British Empire might remain in the Atlantic, but its arms threatened to embrace the world.[32]

The second way in which the British search for a Northwest Passage and other indications of British Pacific interest shaped French interpretations of British conduct was by calling attention to the Spanish Empire as an objective. French discussions of the Ohio Valley show how this could work. French diplomats sometimes depreciated the region. A February 1755 letter from Rouillé to Duras concerning the objectives of British American policy illustrates this sentiment:

> But are our possessions in America the unique object of the jealousy, the ambition, and the cupidity of the English? . . . One need only cast one's eyes on a map to be left with no doubt about the designs of England. The territory of the Ohio which forms the subject of the current discussions does not approach in value the amount that the court of London is expending on armaments, and the nation would not pardon the ministry for engaging in a war of which all the advantage was limited to a portion of a barren and wild country, where it is not possible to establish a lucra-

Materials for American History in the Libraries and Archives of Paris, II, _Archives of the Ministry of Foreign Affairs_ [Washington, 1943], 934); comte de La Galissonière, "Memoire sur les colonies de la France dans l'Amerique septentrionale," December 1750, MD, Amérique, 2, 27r.

32. When seeking assistance from the Spanish court, French diplomats spoke of this British scheme to achieve mastery over the "universal commerce of Europe" and adduced as one of the "incontestable proofs of the reality of this design" the projects discussed in Anson's book that involved seizing from Spanish territory a route between the Atlantic to the Pacific. See Rouillé to Duras, Sept. 27, 1754, 88r–91v, and "Memoire," 108r–109r, both in CP, Espagne, 516.

tive commerce. The supposed rights to the Ohio are nothing but a mask artificially contrived to cover the true intended objective.[33]

Despite such disdain, French statesmen, aided by their British counterparts, allowed valley squabbles to become skirmishes, skirmishes to become battles, and battles to become war. Seeing the Ohio Valley in the wider geographic and strategic context of the Americas helps to resolve this anomaly. For French diplomats, the Ohio Valley, like Hudson Bay, derived its primary importance, not from what it was, but from what it led to. It provided a means by which British soldiers and settlers could sail, paddle, and float from west of the Appalachians to the Ohio's juncture with the Mississippi, down the Mississippi to the river's mouth on Gulf of Mexico, and from there to the possessions of Spain. Rouillé went on to say, "It is at the possessions of Spain that the English wish to arrive." But because France's colonies acted as a barrier between those of Britain and Spain, Britain could only achieve this goal if it became master of Louisiana first. The Ohio Valley lay on the route between Canada and Louisiana. British possession of it would cut French communications and allow Britain to attack either French colony as it pleased. With the French defeated, Mexico would lie open to British invasion.

Certainly, French diplomats could have imputed British actions regarding areas such as the Ohio Valley, Acadia, Saint Lucia, and Honduras to local concerns: activities in the Ohio Valley, to the greed of fur traders and land speculators; claims to the Saint John River, to long-standing New England fears of

---

33. "Mais Nos possessions en Amerique sont elles l'unique objet de la jalousie, de l'ambition et de la cupidité des Anglois? . . . Il n'y a quà jetter les yeux sur une carte géographique pour ne laisser aucun doute sur les vues de l'Angleterre. Le territoire de la belle riviere qui fait le sujet des discussions actuelles, ne vaut pas à beaucoup près la dépense que la Cour de Londres fait pour ses Armemens, et la Nation ne pardonneroit pas au Ministère de l'engager dans une guerre dont tout l'avantage se borneroit à une portion du pays stérile et désert, où il n'est pas possible d'établir un commerce intéressant. Les prétendus droits sur l'Ohio ne sont qu'un masque artificiellement imaginé pour couvrir le véritable but qu'on se propose" (Rouillé to Duras, Feb. 25, 1755, CP, Espagne, 517, 154r–155r). See also Rouillé to Mirepoix, Mar. 17, 1755, "Observations sur le contre projet des Anglois pour une convention preliminaire," "Jt à la lettre de Mr Rouillé," Feb. 13, 1755, all in CP, Angleterre, 438, 146v–147r (partially reproduced in Rouillé to Mirepoix, Mar. 17, 1755, "Observations on the English Counter-Project," Feb. 13, 1755, both in Pease, ed., Anglo-French Boundary Disputes in the West, 159–164, 164–177, esp. 163, 176); Rouillé to Duras, July 27, 1755, CP, Espagne, 518, 86v–87v. In the same month, Noailles argued that the trade of the Ohio Valley was at the moment practically nil, would never amount to much, and that "an object so mediocre is not worth perhaps, neither for the one nor the other nation, the cost of arming one squadron" ["Le commerce de cette riviere [Ohio] est presque nul et ne sera jamais considerable, tant par la nature des productions du païs qu'elle arose, que par sa distance de la mer. Un objet aussi mediocre ne vaut peut être ni pour l'une ni pour l'autre nation, les frais de l'armement d'une Escadre"]. See Noailles, "Memoire sur la conjoncture presente," February 1755, MD, Angleterre, 52, 109r.

Quebec; interest in Saint Lucia, to the security concerns of British Caribbean planters; presence in Central America, to the avarice of local log-cutters. These would have represented locally driven issues with potential but uncertain imperial significance.

Manifest British designs on the Spanish Pacific placed these individual incidents in a different strategic context. They appeared to be individual parts of a broader plan aimed at the Americas rather than isolated events directed at smaller areas. They were associated more with Anson's schemes for imperial expansion than with the petty machinations of minor colonial officials. Immediately after suggesting that British Ohio Valley encroachments aimed at Spanish possessions, Rouillé spoke of Anson's position as First Lord of the Admiralty and of the opportunity thus provided to him for the advancement of British designs on the Spanish Empire. In a 1754 letter, Duras spoke of the extension of English commerce "from the River Plate to Hudson Bay, precisely from one end of America to the other" and observed, "To possess finally all of America, the English will lack only an establishment on the shore of the South Sea . . . and the ownership of the Ohio River that we defend today against them."[34]

The magnitude of British ambitions and the specific objectives composing them affected French assessments of their ramifications. The stakes in the various Anglo-French disputes exceeded a single valley west of the Appalachians, or a single Caribbean island, or a single isthmus connecting Acadia to the mainland. If France allowed individual cases of British aggression to proceed unchecked, Britain might feel emboldened to attempt expansion elsewhere. If France allowed Britain to obtain territories controlling the movement of sailors or soldiers, Britain would be in a more favorable strategic position when open hostilities commenced. If France stood by while Britain gained the lion's share of American commerce in general and Spanish American commerce in particular, then Britain would be in a more favorable financial position when war began.

---

34. Rouillé to Duras, Feb. 25, 1755, CP, Espagne, 517, 154v–155r: "Comme il [Anson] est aujourd'huy á la tête de l'Amirauté en Angle., il est à partie de contribuer plus que personne par ses conseils et par son crédit, à l'exécution de ce vaste et dangéreux projet." Another memoir concerning British aggression in North America put it this way: "It suffices to read the relation of the voyage of Admiral Anson to know that their vast projects embrace all of Spanish America" ([LeDran?], "Sur la politique des Anglois," 1755, MD, Angleterre, 41, 25v). For Duras, see Duras to Rouillé, Sept. 30, 1754, CP, Espagne, 516, 166r–167r: "Depuis le Rio de la Plata jusqu'a la Baye d'Hudson, d'un bout à l'autre de l'Amerique exactement, on etend le commerce de l'Angleterre tant par elle même que par ses allies. Pour posseder en fin toute l'Amerique il ne manquera plus aux Anglois qu'un établissement sur le rivage de la mer du Sud deja indiqué dans les memoires de Milord Anson, et la proprieté de la belle riviere que nous deffendons aujourd'huy contre eux." See also "Differens entre Le Roi et le roi d'Ange. en Amere . . . et sur l'Interruption de la Common et celle de la Negoton," August 1754, MD, Amérique, 24, 181v.

The growth of British imperial power made it seem to French statesmen that French actions involving an immediate risk of war might be more prudent than those entailing the eventual certainty of defeat.[35]

After 1748, a variety of French figures animated by a wide range of concerns were making decisions about how to manage different aspects of intensifying Anglo-French imperial rivalry. Influencing these diverse figures and shaping these various concerns were an underlying logic and a pervasive anxiety. If the power of the British Empire was growing more quickly than that of France, then it would become more difficult with each passing year for France to prevent British achievement of a dominant position in European affairs. British Hudson Bay exploration and the wider array of British actions aimed at the Pacific accentuated French concern about this growing British imperial power. British moves toward the Pacific seemed evidence of an intent to dominate the wealthy colonies of the Spanish Empire and to acquire thereby mineral resources and commercial opportunities that would augment British imperial might. French officials interpreted other British American actions as components of this menacing British design and found it necessary to respond to them to prevent its attainment. This contributed to the sequence of French and British measures and countermeasures leading to the Seven Years' War.

ARTHUR DOBBS WAS APPOINTED governor of North Carolina in 1753 and remained in that office until his death in 1765. He retained his faith in the existence of a Northwest Passage to the end, writing in his final letter to Britain of his continuing hopes for the discovery of a route to and the opening of trade with the Pacific. His compatriots never found the passage Dobbs had coveted, but Dobbs's activities contributed to the Anglo-American acquisition of something more important: the better part of the North American continent in which the Northwest Passage had been thought to exist. Seven Years' War defeats forced the French government to cede Canada and those portions of Louisiana lying

35. As in the case of Central America, Cape Horn, the Ohio Valley, and a Northwest Passage, a European power controlling other areas associated with the outbreak of the Seven Years' War, such as the Caribbean island of Saint Lucia and the Saint John River in Acadia, could further or hinder movement of the sources and instruments of European imperial power: ships and boats, sailors, soldiers, and merchants. The Saint John River region in Acadia provided the only practicable route between New France and the Atlantic when winter ice closed the Saint Lawrence (Savelle, *Diplomatic History of the Canadian Boundary*, 27–28). Saint Lucia stood upwind of Martinique and possessed an excellent small harbor, and it could therefore serve as a base for French privateers preying on trade between Barbados and the British Leeward Islands (Pares, *War and Trade*, 199). Corbett, in *England in the Seven Years' War*, I, v, 27, although talking about slightly different topics from a somewhat different perspective, captures the essence of this implication of British policy.

east of the Mississippi to Britain. Spain took over the French claim to the trans-Mississippi West. Neither Spain nor Mexico could repel the westward advance of the United States and Canada. By contributing to the origins of the Seven Years' War, Dobbs helped bring about the removal of the French barrier to Anglo-American expansion, preparing the way for explorers such as Mackenzie and Lewis and Clark to reach the Pacific by land.[36]

36. Barr and Williams, eds., *Voyages to Hudson Bay,* II, 320.

11

SPANISH REACTIONS TO BRITISH PACIFIC ENCROACHMENTS, 1750–1757

Has heaven reserv'd, in pity to the poor,
No pathless waste, or undiscover'd shore;
No secret island in the boundless main?
No peaceful desart yet unclaim'd by Spain?
—SAMUEL JOHNSON, "London," 1738

Considerations of European interest in the Pacific, French efforts to comprehend American geography, and French reactions to British Hudson Bay exploration have pointed repeatedly to what—in terms of extent, populousness, wealth, and longevity—was the grandest polity of the early eighteenth-century Americas: the Spanish Empire. Because of its geographic position, lingering power, and manifest potential, rival mid-eighteenth-century French and British statesmen vied to bring Spanish policy in line with their own objectives.[1]

To contemporary French observers, and to later imperial and diplomatic historians, this policy has presented something of a riddle. In particular, Spanish relations with Britain were warmer and Spanish relations with France more distant than circumstances would seem to dictate. Between 1748 and 1756, while France and Britain tottered from tension to hostility to war, Spain stood largely apart from Anglo-French conflict, maintaining civil relations with both powers but following the lead of neither. In fact, not only did Spain avoid allying with France until 1761 but, in a century marked by six Anglo-Spanish wars and countless skirmishes, Anglo-Spanish relations between 1750 and 1757 were remarkably cordial.[2]

1. For the epigraph, see Samuel Johnson, "London: A Poem in Imitation of the Third Satire of Juvenal," in *The Yale Edition of the Works of Samuel Johnson,* VI, *Poems,* ed. E. L. McAdam, Jr., and George Milne (New Haven, Conn., 1964), 47–61, esp. 56.

2. Jean O. McLachlan, *Trade and Peace with Old Spain, 1667–1750: A Study of the Influence of Commerce on Anglo-Spanish Diplomacy in the First Half of the Eighteenth Century* (Cambridge,

A Spanish policy cultivating harmonious relations with Britain and leaving France to confront the British Empire without Spanish aid has seemed curious, for British threats to the Spanish Empire gave Spanish statesmen good reason to lash out at Britain. Some of these threats, such as Central American logging, the founding of Georgia, contraband trade with Spain's overseas possessions, and War of Jenkins' Ear attacks on Spanish imperial territory are well known to scholars. Spanish concern about British Pacific encroachments constitutes a less familiar subject.[3]

1940), 139–144; Lawrence Henry Gipson, "British Diplomacy in the Light of Anglo-Spanish New World Issues, 1750–1757," *AHR*, LI (1946), 627–648; Richard Pares, *War and Trade in the West Indies, 1739–1763* (Oxford, 1936), 556.

3. On familiar British threats to the Spanish Empire, see, for example, Herbert E. Bolton and Mary Ross, *The Debatable Land: A Sketch of the Anglo-Spanish Contest for the Georgia Country* (New York, 1968); Gipson, "British Diplomacy in the Light of Anglo-Spanish New World Issues," *AHR*, LI (1946); Sylvia-Lyn Hilton, "Las Indias en la diplomacia española, 1739–1759" (Ph.D. diss., Universidad Complutense de Madrid, 1979; rpt. Madrid, 1980), 452–479, 544–578; McLachlan, *Trade and Peace with Old Spain*, 139–144; Pares, *War and Trade*, 533–565; Trevor Richard Reese, *Colonial Georgia: A Study in British Imperial Policy in the Eighteenth Century* (Athens, Ga., 1963); Max Savelle, *The Origins of American Diplomacy: The International History of Anglo-America, 1492–1763* (New York, 1967), 419–429; J. Leitch Wright, Jr., *Anglo-Spanish Rivalry in North America* (Athens, Ga., 1971), 80–87, 101–106. Gipson, Pares, Wright, and McLachlan have discussed the reasons for outward 1750–1757 Anglo-Spanish amity, but they have not addressed the question of how it survived the British challenge to Spain's Pacific navigational monopoly. Scholars writing about the eighteenth-century British Empire and the Pacific have generally been more interested in, and have done more research on, the British side of affairs and the evolution of geographic thought and have devoted less attention to the details of French and Spanish diplomatic reactions to British actions.

On Spanish policies and policymakers of the 1750s and 1760s, see the following by Vicente Palacio Atard: *El tercer Pacto de familia* (Madrid, 1945); "El equilibrio de América en la diplomacia del siglo XVIII," *Estudios americanos*, I (1949), 461–479; *Las embajadas de Abreu y Fuentes en Londres, 1754–1761* (Valladolid, Spain, 1950); and "La neutralidad vigilante y constructiva de Fernando VI," *Hispania: Revista española de historia*, XXXVI (1976), 301–320. He has not generally discussed the importance of the Pacific Ocean and the Pacific coast of the Americas for Spanish diplomatic policy. More recently, Sylvia L. Hilton has written extensively and insightfully about the place of the American empire in eighteenth-century Spanish foreign policy. See her thesis, "Las Indias en la diplomacia española," and Iñigo Abbad y Lasierra, *Descripción de las costas de California*, ed. Sylvia L. Hilton (Madrid, 1981); and see Hilton, ed., *La alta California española* (Madrid, 1992). These works provide a strong introduction to Spanish complaints about threatened British incursions into the South Sea as well as to the topic of the next chapter, Spanish fears of French explorers' finding a North American river route to the Pacific. Hilton's writings encourage further efforts to integrate fully frontier provinces such as New Mexico into discussions of the South Sea, to employ French and British sources to establish the importance of the Pacific for European as well as Spanish diplomacy, and to establish an analytical framework within which Anglo-French threats to the Spanish Lake can be related to unresolved historical questions like that of why 1750s Spanish statesmen favored Anglo-friendly neutrality over Franco-Spanish alliance.

Pierre Muret, in *La prépondérance anglaise (1715–1763)* (Paris, 1937), and "Le conflit anglo-espagnol dans l'Amérique centrale au XVIIIe siécle," *Revue d'histoire diplomatique*, LIV–LV (1940–1941), 129–148, demonstrated an awareness of the importance of British Pacific encroachments, but he did not fully develop the implications of these ideas.

Spanish officials were no less attentive than their French counterparts to the apparent British South Sea designs treated in the previous chapter. Madrid exhibited more interest than Paris in reports of Commodore Anson's having sacked Peruvian Paita and captured the Manila galleon. Spanish ministers were as aware as French diplomats that British loggers clinging to Central America's Atlantic shores were within trudging distance of Pacific waters. Spanish geographers pondered, like their Gallic brethren, the attempts of British explorers to find a Northwest Passage from Hudson Bay. And, most materially for this discussion, Spanish ambassadors and ministers, even more than their French counterparts, pursued rumors of British expeditions to establish island bases near Cape Horn in 1749, 1750, and 1751.

Oddly enough, this collection of apparent British threats to Spain's Pacific prerogatives and American territories failed to subvert friendly Anglo-Spanish relations before 1757, as Britain's enthusiastic privateering and recalcitrant Honduran and Nicaraguan presence did after that year; and it was only late in the Seven Years' War that Spanish worries about rising British power finally pushed the Spanish government to join France in its campaigns against the British Empire threatening both Bourbon powers. British actions that should have produced Anglo-Spanish antipathy somehow allowed years of apparent Anglo-Spanish diplomatic amiability. This raises the question of why Spanish statesmen chose to remain friendly with the aggressive British Empire threatening the Spanish Pacific and distant from the French Bourbon supplicant offering to help defend Spanish territory.

The basis for an answer can be found in Spanish archives. The Archivo General de Simancas, Archivo Histórico Nacional in Madrid, and Archivo General de Indias in Seville retain correspondence from Spain's ministers of foreign affairs and of the Indies, from Spain's ambassadors in France and Britain, and from Spanish officials in the New World. In these letters, Paris and London rumors move from the ears of Spanish ambassadors to the hands of Spanish ministers. Isolated figures from the fringes of Spain's empire transmit their ominous observations and fearful speculations to Cadiz and Castile. Ministers compile and compare, judge and misjudge the reports coming in. They transmit their hopeful or fretful instructions to Spanish representatives at Europe's most splendid courts and Spanish officials in the empire's most distant outposts. The records of diplomacy and administration allow assessment of Spanish responses to the array of challenges facing the grand and precarious mid-eighteenth-century empire.[4]

4. Useful guides to the Spanish archives include Charles E. Chapman, *Catalogue of Materials in the Archivo General de Indias for the History of the Pacific Coast and the American Southwest* (Berkeley, Calif., 1919); Pedro González García, ed., *Discovering the Americas: The Archive of the Indies* (Paris,

Given the marginality of the early modern Pacific in much historical literature and its distance from the European core of Spain's embattled empire, it would be unsurprising to find in these Spanish papers indications that Spanish officials monitored apparent British South Sea designs without being especially concerned about them. Spanish Pacific pretensions had survived Drake's exploits, Dutch expeditions, piratical predations, French traders, and even the founding of a British South Sea Company. Perhaps Spanish ministers and diplomats serenely supposed that distance, scurvy, harsh weather, the defensive efforts of loyal Spanish-Pacific subjects, and the hostile reactions of Europe's other maritime nations would dash mid-eighteenth-century British South Sea ambitions as they had overcome earlier challenges to Spanish presumptions. Perhaps French and Spanish observers were coming to very different conclusions from the same evidence of British Pacific expansionism, and the French were alone in their anxiety.

Spanish papers suggest otherwise. They show Spanish officials not simply aware of British South Sea designs but apprehensive about them. Spanish diplomats and ambassadors monitored the progress of British Pacific projects, viewed these British designs as a violation of Spanish prerogatives and a threat to Spanish interests, and insisted upon the termination of schemes entailing the introduction of British ships into the Spanish Lake. Spanish statesmen saw, moreover, the significance of British actions and British intentions. They recognized the rising power of the British Empire and the danger it posed to the venerable but vulnerable American possessions of Spain.

Spain's "Vigilant Neutrality"

The significance of Spanish reactions to Britain's threatened South Sea incursions will appear most clearly when viewed in the context of traditional explanations of the unusually friendly tenor of 1750s Anglo-Spanish diplomacy.

The achievement initiating midcentury Anglo-Spanish amity was a 1750 Anglo-Spanish commercial treaty, and an immediately appealing hypothesis is that this agreement's resolution of trade issues that had poisoned Anglo-Spanish

1997); Miguel Gómez del Campillo, *Relaciones diplomatícas entre España y los Estados Unidos según los documentos del Archivo Histórico Nacional*, I, *Introducción y catálogo* (Madrid, 1944); Carmen Crespo Nogueira, *Archivo Histórico Nacional: Guía* (Madrid, 1989); Julian Paz and Ricardo Magdaleno, *Documentos relativos a Inglaterra (1254–1834)* (Madrid, 1947); José María de la Peña y Cámara, *Archivo General de Indias de Sevilla: Guía del visitante* (Valencia, Spain, 1958); Angel de la Plaza Bores, *Archivo General de Simancas: Guía del investigador* (Madrid, 1992); William R. Shepherd, *Guide to the Materials for the History of the United States in Spanish Archives: Simancas, the Archivo Historico Nacional, and Seville* (Washington, D.C., 1907).

relations after the Treaty of Utrecht accounts for the good feelings of the 1750s. The most important provision required Britain to give up its claim to four more years of the asiento in exchange for £100,000. This renunciation removed one channel through which British merchants had been smuggling foreign goods into the Spanish Empire and therefore mitigated one cause of Spanish complaint.[5]

One cause, but not all; for the commercial treaty left the more fundamental territorial, demographic, and economic bases of Anglo-Spanish antipathy untouched. Casualties during the War of Jenkins' Ear had not been high enough to kill off all the Britons hoping one day to flood the markets of Peru and Mexico, seize the ships connecting Spain and America, or plunder Spanish American cities like Cartagena, Portobello, and Havana. Georgia continued to threaten Spanish Florida, and, more generally, the populations of Britain's mainland colonies continued to expand, placing before prescient Spanish observers the likelihood that these Anglo-American communities would one day imperil all of Spain's North American possessions. South of Mexico, Spanish officials continued to fret about British logging and gathering camps on the Mosquito Coast and the Gulf of Honduras. It was not just that Spanish officials would have preferred to see their own subjects profit from providing dyewoods to Europe but also that the British camps lay close to Guatemalan and southern Mexican towns to which goods could be smuggled. Georgia stood at least on the edge of the Spanish Empire; the logging camps were close to its heart. This British Central American presence was so nettlesome, in fact, that, despite the general Spanish disposition to avoid antagonizing the British government, Spanish forces actually attacked the Gulf of Honduras settlements in 1754. These assaults temporarily drove the loggers from Belize to the Mosquito Coast, but the British interlopers soon returned and continued their cutting. One 1750 commercial agreement notwithstanding, British desires and opportunities to encroach upon Spanish territory and markets remained. Spanish officials inclined to suspicion and unease still had much to be suspicious and uneasy about.[6]

5. Gipson, "British Diplomacy in the Light of Anglo-Spanish New World Issues," *AHR*, LI (1946), 627–628, 647; Hilton, "Las Indias en la diplomacia española," 454–479; McLachlan, *Trade and Peace with Old Spain*, 139, 144; Didier Ozanam, "Les origines du troisième Pacte de famille (1761)," *Revue d'histoire diplomatique*, LXXV (1961), 307–340, esp. 314–315; Pares, *War and Trade*, 556, 558; Savelle, *Origins of American Diplomacy*, 419–429.

6. Vera Lee Brown, "The South Sea Company and Contraband Trade," *AHR*, XXXI (1926), 662–678; Walter L. Dorn, *Competition for Empire, 1740–1763* (New York, 1963), 122–127; John Lynch, *Bourbon Spain, 1700–1808* (Oxford, 1989), 150–152, 179; Abbad y Lasierra, *Descripción de las costas de California*, ed. Hilton, 26; Richard Lodge, ed., *The Private Correspondence of Sir Benjamin Keene, K.B.* (Cambridge, 1933), xvi; Derek McKay and H. M. Scott, *The Rise of the Great Powers, 1648–1815* (London, 1983), 159–162; McLachlan, *Trade and Peace with Old Spain*, 59–62, 73, 78–121; Pares, *War*

Because Anglo-Spanish relations improved despite these latent differences of imperial interest, and because the character of interactions among high British and Spanish officials constituted the most visible indication of Anglo-Spanish diplomatic warming, another line of reasoning emphasizes the personal qualities and inclinations of the individuals involved rather than the configuration of underlying strategic objectives. It points to the alleged Anglo- and Franco-philias and -phobias of Spanish officials and monarchs: the skill of British ambassador to Spain Benjamin Keene (1749-1757); the relative clumsiness of the French ambassadors the comte de Vaulgrenant (1749-1752) and the duc de Duras (1752-1755); and the personal relationships among figures such as Keene, Spanish foreign minister Don José de Carvajal y Lancaster (1746-1754), and Spanish ambassador (1748-1754) and then foreign minister (1754-1763) Richard Wall. Biography and ability figure prominently. Carvajal y Lancaster, it is re-marked, traced his ancestry back to John of Gaunt and seems to have enjoyed the prestige of this link to English nobility. Somewhat incongruously, the pro-tean Wall combined Irish ancestry, French birth, Spanish employment, and pro-British inclinations. His resultant ability to "put himself into as many Shapes and Humours as any one" might have contributed to his famously pleasant relations with his British counterparts. One of these, Benjamin Keene, also possessed an agreeable manner and, after many years of service in Iberia, had achieved flu-ency in both the Spanish language and Spanish affairs. Where a silver tongue proved insufficient, Keene employed hard money to extend his influence at the Spanish court.[7]

Factors such as Anglophilic sentiment and English skill, however, account for only a part of the story. Keene, for example, could not have exploited Spanish desires for peace in the 1750s had those desires not been especially strong. The same qualities underlying Wall's diplomatic ease limited his room for diplomatic maneuver. Because Wall's reputation was pro-English, he felt pressure to act in an anti-English fashion so as to avoid causing less-cosmopolitan Spaniards to question his loyalties. As George William Hervey, earl of Bristol, who replaced Keene as British ambassador to Spain in 1758, put it, "To obtain that Impartiality

and Trade, 14–28, 56–64, 540–550; Reese, Colonial Georgia, 53–73; Savelle, Origins of American Diplo-macy, 141–145, 151–152, 326–350; Geoffrey J. Walker, Spanish Politics and Imperial Trade, 1700–1789 (Bloomington, Ind., 1979), 71–74, 85–87, 181; Philip Woodfine, Britannia's Glories: The Walpole Min-istry and the 1739 War with Spain (Woodbridge, Suffolk, U.K., 1998); Wright, Anglo-Spanish Rivalry in North America, 101–106; J. H. Parry, Trade and Dominion: The European Overseas Empires in the Eighteenth Century (New York, 1971), 102–113; Gipson, "British Diplomacy in the Light of Anglo-Spanish New World Issues," AHR, LI (1946), 632–637; Hilton, "Las Indias en la diplomacia espa-ñola," 549–558.

7. Benjamin Keene to earl of Holdernesse, Nov. 6, 1751, SP 94/140, 221v–222r.

he aims at," Wall "is often hurried on to appear, and to conduct himself, as if he was in Interests diametrically opposite" to Britain. Nor was Wall's alleged Anglophilia consistent or durable. After his accession to the post of minister of foreign affairs and the commencement of the Seven Years' War, he became increasingly frustrated with British privateers' attacks on Spanish ships and with the British government's intransigent commitment to a continued presence in Honduras and on the Mosquito Coast. He became estranged from his British associates and came to favor joining France in its war against Britain. Wall's perception of Spain's strategic needs trumped his friendly sentiment toward England. Carvajal, on the other hand, despite his illustrious English forebears, never seems to have let regard for the country of his ancestors interfere with loyalty to the nation of his birth. He recognized that both France and Britain were pursuing their own interests and that it was a matter of accident rather than design when those interests happened to coincide with those of Spain.[8]

As Carvajal's case suggests, and as shrewd historians have contended, more persuasive than partisan loyalties and personal characteristics as an explanation for Spain's quest for good relations with Britain is the argument that Spain's ministers and king were pursuing the Spanish national interest, and they perceived outwardly friendly Anglo-Spanish relations as a means for advancing it. During the first half of the eighteenth century, Spain had increased its familiarity with fruitless wars, economic decline, and foreign threats to its overseas possessions. Acutely aware of the magnitude of the mid-eighteenth-century challenges confronting it, Ferdinand VI (1746–1759), minister of the Indies the marqués de La

8. Earl of Bristol to William Pitt, Aug. 31, 1761, SP 94/164, 75v–76r. See also Keene to duke of Bedford, Dec. 8, 1750, SP 94/138, 249r–249v; Keene to Pitt, Sept. 26, 1757, PRO 30/47/12; Gipson, "British Diplomacy in the Light of Anglo-Spanish New World Issues," *AHR*, LI (1946), 627–628, 638–641, 647; McLachlan, *Trade and Peace with Old Spain*, 139, 144; Pares, *War and Trade*, 556, 558, 562–565; Savelle, *Origins of American Diplomacy*, 419–429, 442; Didier Ozanam, ed., *La diplomacia de Fernando VI: Correspondencia reservada entre D. José de Carvajal y el duque de Huéscar, 1746–1749* (Madrid, 1975), 46–47; Ozanam, "Origines du troisième Pacte de famille," *Revue d'histoire diplomatique*, LXXV (1961), 314–315, 321–322; Vera Lee Brown, "The Spanish Court and Its Diplomatic Outlook after the Treaty of Paris, 1763," *Smith College Studies in History*, XV (1929–1930), 7–38, esp. 10; François Rousseau, *Règne de Charles III d'Espagne (1759–1788)*, 2 vols. (Paris, 1907), I, 17–18; Lynch, *Bourbon Spain*, 158–162, 183–185, 193; Palacio Atard, *El tercer Pacto de familia*, 116; Hilton, "Las Indias en la diplomacia española," 454–479, 557–569, 655, 661–662; Antonio Bermejo de la Rica, *La colonia del Sacramento: Su origen, desenvolvimiento y vicisitudes de su historia* (Toledo, [Spain], 1920), 42–44; José de Carvajal y Lancaster to Richard Wall, May 29, 1749, Estado 4503, AGS. Gipson provides the best example of a scholar emphasizing the importance of the personal qualities of mid-eighteenth-century diplomats. This may be owing to his reliance on British sources in which the perceptive and articulate Keene highlighted the personalities of the various officials and diplomats around the Spanish court. Spanish and Spanish-speaking historians reading what Spanish statesmen themselves wrote rather than what Keene wrote about them have often come to a different understanding of Spanish policy.

Ensenada (1743–1754), and foreign minister Carvajal agreed on the importance of protecting Spain's American possessions from rival European powers. They viewed overseas empire as the source of Spain's past and potential wealth and grandeur. At the same time that securing Spain's colonies was deemed essential, resorting to war was to be avoided, because its costs could derail Spanish attempts at economic revival, and its chances could render foreign threats to the empire immediate.

To simultaneously maintain peace and protect the empire from the depredations of European rivals, Spanish leaders sought a balance among the European powers sufficient to prevent any of them from dominating Spain or conquering Spanish territory. They wanted neutrality, but it was to be, as Spanish historian Vicente Palacio Atard has put it, a "vigilant neutrality," during which Spanish ministers would take care that no foreign power gained advantage at Spain's expense. This general policy overshadowed personal ministerial preferences. Ensenada, although ostensibly pro-French, suspected both French and British intentions. He particularly feared British naval might and felt that close ties between Spain and France would enable the Spanish government to intimidate Britain into good behavior. Carvajal, although reputedly pro-British, saw Britain as a threat to Spain's empire and simply felt that Anglo-friendly neutrality could, for the moment, preserve Spanish interests more effectively than a policy of alliance with France. Ensenada was "vain and presumptuous," "profuse . . . in his way of Living," gay, easy, and wanting in "application"—in short, so constructed as to cause British observers to deem him capable of passing for a Frenchman. Carvajal, in contrast, was "one of the most reserved" ministers a diplomat might meet, "mild, Complaisant . . . of a very dry Conversation, . . . a little Embarassed," and possessing "the simplicity and abstinence of an Old Roman"—to wit, for British partisans, presumptively Anglophile. But differing personalities need not imply contrasting loyalties. When it came to Spanish policy, Carvajal and Ensenada seem to have shared ends even if they sometimes disagreed about means. The perspicacious Keene, who penned parts of the contrasting descriptions above, saw that, though Carvajal and Ensenada might agree on little else, both hoped for a period of neutrality during which Spain could build its strength and after which the Spanish Empire could put its renewed power to use.[9]

9. The quoted descriptions of Ensenada and Carvajal come from Keene to Bedford, Feb. 25, 1749, SP 94/135, 58r–59r; Keene to Bedford, Feb. 20, 1751, SP 94/139, 68r–68v; Keene to Holdernesse, Nov. 6, 1751, SP 94/140, 222r; Bristol to Pitt, Aug. 31, 1761, SP 94/164, 79v–80r. See also Keene to duke of Newcastle, Aug. 13, 1750, SP 94/138, 169r–169v; marqués de La Ensenada to Francisco Pignatelli, May 31, 1749, Estado 6483, AHN; Don Joseph de Carvajal y Lancaster, "Pensamientos de don Joseph de Carvajal y Lancaster," in Bermejo de la Rica, Colonia del Sacramento, 249–259. The most recent and most perceptive scholarship supports this national interest account of Spanish

If Spanish ministers were pursuing policies calculated to advance Spanish interests, understanding their responses to British Pacific encroachments requires an assessment of how they viewed the relation between those incursions and their own imperial objectives. The kind of British movements toward the Pacific discussed in the previous chapter would certainly appear to have endangered Spanish imperial interests. They threatened to give Britain South Sea access. They seemed to presage British domination of parts of Spain's imperial economy or territory. This raises the question of why Spanish ministers during the years from 1750 to 1761 chose to reject a more intimate Franco-Spanish relationship that might balance the rising power and deter the impending conquests of the British Empire. To put it differently, the previous chapter showed how French concerns about British encroachments on the Spanish Lake and empire contributed to a more bellicose French policy toward Britain. Why did similar Spanish concerns about the same encroachments fail to push Spanish policy along the same track?

Spanish Concerns about British Pacific Encroachments

Developments during the War of Jenkins' Ear and during and after the War of the Austrian Succession gave the Spanish government fresh reasons to be concerned about Pacific intrusions. Anson's raid raised worries about the security of many parts of the Spanish Lake. His prowling off Acapulco in 1742, for instance, caused Spanish officials to think again about the unsecured upper and lower California coast, where pirate ships had found refuge before and where new enemies like Anson might locate shelter again. In the mid-1730s, before Anson's appearance off North America's Pacific coast, Jesuit Baja California missionaries Juan Francisco de Tompes and Clemente Guillén de Castro were already warning that the English or others might acquire a base on their poorly protected peninsula that would facilitate seizure of the Manila galleon. Already they were requesting Spanish fortifications and soldiers. Such fears of a foreign establishment, exacerbated by Anson's raid, contributed to the Council of the Indies' 1744 recommendation that the viceroy of New Spain maintain two warships in

---

policy in the 1750s. See M[arí]a Dolores Gómez Molleda, "El pensamiento de Carvajal y la política internacional Española del siglo XVIII," *Hispania: Revista española de historia*, XV (1955), 117–137, esp. 127–135; Hilton, *Las Indias en la diplomacia española*, 477, 554–556, 621–622, 625–628, 640, 658; Lynch, *Bourbon Spain*, 157–165; Lucio Mijares Pérez, "Programa político para América del marqués de la Ensenada," *Revista de historia de América*, LXXXI (1976), 82–130, esp. 109; Palacio Atard, "La neutralidad vigilante y constructiva de Fernando VI," *Hispania: Revista española de historia*, XXXVI (1976), 301–303; Pares, *War and Trade*, 523.

California waters and settle the Tres Marías Islands (southeast of the end of the Peninsula).[10]

Anson's assaults had menaced the Pacific during an Anglo-Spanish war, and Spanish officials might reasonably expect that such threats would diminish after the 1748 Treaty of Aix-la-Chapelle. Instead, the apparent British challenges to the Spanish Pacific monopoly continued, and Spanish concern about them subsisted. Reactions to the planned 1749 British expedition to the South Atlantic and Pacific that so alarmed the French foreign ministry provides a good example of this Spanish unease.[11]

Spanish officials had two ways of receiving information about this enterprise. The first was Spanish spies in British ports. Jorge Juan, for instance, after returning from the South American scientific expedition he and Antonio de Ulloa were involved in (mentioned in Chapter 3), was sent by Ensenada to England to gather information, equipment, and personnel useful for rebuilding Spain's navy. In 1749, Juan observed preparations involving two English sloops, the *Raven* and the *Porcupine,* intended for the South Atlantic and Pacific. He informed Wall and Ensenada. Wall reported the matter to Carvajal and also asked the duke of Bedford, the British secretary of state for the Southern Department, for an explanation. In addition to their own sources of information, Spanish officials were also receiving reports from interested French diplomats.[12]

10. "Guilléns 1737 Report on California," and "Tompes' 1735 Report: Need of a California Port of Call; Rebellion in the South," in Ernest J. Burrus, ed. and trans., *Jesuit Relations: Baja California, 1716–1762* (Los Angeles, Calif., 1984), 104–110, 161–181, esp. 106, 162. On the 1744 decision, see "Extracto del expediente que trata de la reduccion a curatos de las missiones de las provincias de Sinaloa, y Sonora . . . ," June 15, 1752, AGI, Guadalajara 137, 299–300; "Extractos de providencias para el descubrimto del Mar del Sur, y Californias . . . ," 1790, Estado 2848, carpeta 6, AHN; Charles Edward Chapman, *The Founding of Spanish California: The Northwestward Expansion of New Spain, 1687–1783* (New York, 1916), 30–32; Donald C. Cutter, "Plans for the Occupation of Upper California: A New Look at the 'Dark Age' from 1602 to 1769," *Journal of San Diego History,* XXIV (1978), 78–90, esp. 84–85; Abbad y Lasierra, *Descripción de las costas de California,* ed. Hilton, 26. These concerns proved timely, as other foreign visitors followed Anson to North America's west coast in the 1740s.

11. Alan Frost and Glyndwr Williams, "The Beginnings of Britain's Exploration of the Pacific Ocean in the Eighteenth Century," *The Mariner's Mirror,* LXXXIII (1997), 410–418; Hilton, "Las Indias en la diplomacia española," 460–462; and Glyndwr Williams, *The Great South Sea: English Voyages and Encounters, 1570–1750* (New Haven, Conn., 1997), 258–260, all discuss this affair.

12. On Juan's activities, see earl of Rochford, July 8, 1762, SP 94/167, 208r; Kenneth J. Andrien, "The *Noticias secretas de América* and the Construction of a Governing Ideology for the Spanish American Empire," *Colonial Latin American Review,* VII (1998), 175–192, esp. 177; María Pilar de San Pío, *Expediciones españolas del siglo XVIII: El paso del Noroeste* (Madrid, 1992), 34–35. On Juan and Wall's reports, see Wall to Carvajal, Apr. 17, 1749, Carvajal to Wall, May 10, 1749, both in Estado 6915, and Wall to Carvajal, May 5, 1749, Estado 4503, all in AGS. For the French reports, see "Extrait des nouvelles de Londres du 8 Juin 1749," Estado 6914, AGS.

Correspondence between Wall and Carvajal reveals the reasons underlying Spanish concern about this British expedition. Spanish officials recognized Anson's Admiralty influence, and they saw the proposed sally as an attempt by Anson and his supporters to execute the aggressive plans outlined in *A Voyage round the World*. This provided cause for worry. European ships had not thoroughly explored South Atlantic waters and coasts, and Spanish officials could not be sure of what might be found there. Carvajal examined maps of the waters between southern Africa and South America and observed that British reconnaissance of the area posed no obvious danger to Spain. But he also warned Wall that the British might possess specific information absent from general South Atlantic maps ["que no expliquen bien los mapas generales"] and that they might discover, or might have already found, something imperiling Spanish interests ["un descubrimiento que nos ofenda"]. Should British exploration lead to the establishment of bases on South Atlantic islands, Carvajal feared the British would then obstruct Spanish ships trying to reach the Pacific by way of Cape Horn or the Straits of Magellan ["impedir la navegacion al Mar del Sur tanto por el Cabo de Hornos, como por el estrecho de Magalones"].[13]

Carvajal and Wall also worried about diplomatic complications. They foresaw that France could not fail to hear about the planned British expedition and feared that the French government would then exploit the ensuing Anglo-Spanish tensions by demanding equal rights to Pacific and South Atlantic navigation. Carvajal recalled that Spain had found it necessary to send a naval force (the Martinet expedition) to the Peruvian and Chilean coasts to evict French ships ignoring the Peace of Utrecht's prohibition of their presence. Spanish officials wanted to forestall any French diplomatic inclination to demand restoration of direct French trade with western South America. Should such commerce be revived, the Spanish Empire would find its South Sea coasts swamped with foreign traders as they had been during the War of the Spanish Succession and its diplomats embroiled in an Anglo-French dispute over Pacific expansion, just as they had been during the negotiations leading to the Utrecht settlement.[14]

Other maritime powers might try the same approach. Spanish statesmen

13. The first quotation appears in Carvajal to Wall, May 10, 1749, Estado 4267, AHN; the second, in Wall to Carvajal, May 29, 1749, Estado 4503, AGS. Carvajal's geographic knowledge was limited: he was unsure whether the Falklands were in the Atlantic or the Pacific. See Carvajal to Wall, May 21, 22, 1749, Estado 4503, AGS (final quotation).

14. Wall to Carvajal, Apr. 17, 1749, Estado 6915, Wall to Carvajal, May 5, 1749, Carvajal to Wall, May 10, May 22, 1749, both in Estado 4503, and Wall to Carvajal, June 8, 1749, Estado 6914, all in AGS. See also E. W. Dahlgren, "Voyages français à destination de la mer du Sud avant Bougainville (1695–1749)," *Nouvelles archives des missions scientifiques et littéraires*, XIV (1907), 423–554, esp. 428–429.

knew that nations such as Britain and the Dutch Republic resented Spanish claims to exclusive navigational rights in American and Pacific waters. Such envious maritime powers might use a new squabble over the Pacific to press their case for more open oceans. In 1749, Spanish diplomats were engaged in negotiations with Britain regarding British trading privileges in the Spanish Empire. British efforts to enter Pacific waters might force Spain to invoke treaties granting its empire exclusive Pacific access, and this assertion could provoke a British counterdemand for free navigation ["puede esta disputa despertar el delicado punto de la libre navegazon"]. This was a can of worms Spanish diplomats preferred to keep closed.[15]

Spain could conceivably lose more than oceanic exclusivity. Wall recalled that France, in an attempt to lure the English away from the Grand Alliance against Louis XIV, had offered Chilean territory to England during the War of the Spanish Succession ["ofrecer a la Inglaterra . . . un establecimiento en el continente de la America meridional, en la costa de Chile"]. Should a new Anglo-French dispute over Pacific navigation arise, the parties involved might seek to end it by compensating their frustrated maritime ambitions with the offer of Spanish territory.[16]

These were serious concerns, and Spanish officials quickly transformed them into protests to British statesmen such as Keene, Bedford, First Lord of the Admiralty the earl of Sandwich, and the head of the ruling Whig ministry, Henry Pelham. Carvajal told Keene that British actions appeared designed to foster illicit commerce or foment hostilities against Spain. They evinced an ominous British belief in the transience of Anglo-Spanish concord ["estan resueltos los Ingleses a que dure mui poco la Paz"]. Wall complained to Bedford that it appeared the British were trying to acquire a permanent southern seas establishment at which they could maintain a squadron and from which they could attack Spanish possessions. Since these actions contravened earlier British treaty agreements, Wall claimed they would cause Spain to lose confidence ["perder toda la confianza"] in British trustworthiness. Both Wall and Carvajal warned that a British Pacific or South Atlantic expedition would jeopardize ongoing attempts to foster amicable Anglo-Spanish relations. They insisted that Spain alone had the right to enter the Pacific. The threat that Spanish force might sustain Spanish claims was tangible.[17]

15. Quotation in Wall to Carvajal, May 5, 1749, Estado 4503, AGS.
16. Ibid.
17. On Carvajal's warning to Keene, see Carvajal to Wall, May 22, 1749, Estado 4503, AGS; Keene to Bedford, May 21, 1749, SP 94/135, 265r–269v (another copy in Add. Mss. 43,423, 86r–88v). On Wall's complaint to Bedford, see Wall to Carvajal, Apr. 17, 1749, Estado 6915, AGS. On the de-

The British government heeded these protests. In April, concerned that sailing into the Pacific was too provocative, Bedford and the earl of Sandwich proposed confining the 1749 expedition to the South Atlantic. Because he thought that British establishments on South Atlantic islands could impede Spanish communications with the Pacific and could by themselves serve as bases for peacetime contraband and wartime attacks, Carvajal was unsatisfied. In June, worried about disrupting ongoing commercial negotiations with Spain, British ministers ordered the Admiralty to suspend the expedition entirely. Bedford, with a display of the haughty spirit that made him less than universally popular among British officials and, indeed, a frequent and determined opponent of other British statesmen like the duke of Newcastle and William Pitt, maintained nonetheless that the British king's "right to send out Ships for the discovery of unknown and unsettled Parts of the World must indubitably be allowed by every body." But Carvajal understood him to have conceded that treaties prohibited British Pacific navigation, and the Spanish minister was enthusiastic about the inference, inherent even in Bedford's assertion to the contrary, that the expedition's cancellation might be construed as implying a Spanish right to impede free navigation of American waters ["confesion del derecho de impedirles la libre navegacion en los Mares de la America"].[18]

Spanish diplomats had intercepted one danger, but they remained disquieted. Anson and many who thought as he did retained their positions, sentiment favoring British Pacific expansion was evident, and Britain possessed the ships necessary to launch a similar expedition in the future. Carvajal told Wall to continue watching the *Porcupine* and the *Raven*. In a September 1749 letter to Ensenada, Wall discussed Anson's belief in the possibility of opening a profitable trade with and providing dangerous weapons to Spain's South American Indian

---

terioration of Anglo-Spanish relations, see Wall to Carvajal, Apr. 24, 1749, Estado 6915, Carvajal to Wall, May 22, 1749, Estado 4503, both in AGS; Keene to Castres, May 29, 1749, in Lodge, ed., *Private Correspondence of Sir Benjamin Keene*, 128; Keene to Bedford, May 21, 1749, SP 94/135, 265r–269v; marqués de Tabuerniga to Newcastle, May 21, 1749, Add. Mss. 32,817, 27r–28r. On Spanish threats of force, see Wall to Carvajal, Apr. 17, 1749, Carvajal to Wall, May 10, 1749, both in Estado 6915, and Wall to Carvajal, May 29, 1749, Estado 4503, all in AGS.

18. On limiting the 1749 expedition to the South Atlantic, see Wall to Carvajal, Apr. 24, 1749, Estado 6915, Carvajal to Wall, May 21, 22, 1749, Wall to Carvajal, May 29, 1749, Estado 4503, all in AGS; Bedford to Keene, Apr. 24, 1749, SP 94/135, 177r–178v (another copy in Add. Mss. 43,423, 78r–79v); earl of Sandwich to Bedford, Apr. 14, 1749, Add. Mss. 43,423, 81r–82r. On Carvajal's dissatisfaction, see Carvajal to Wall, May 21, 1749, Estado 4503, AGS. On questions of free navigation, see Wall to Carvajal, June 16, 1749, Estado 6915, AGS. The Spanish quotation is in Carvajal to Wall, June 27, 1749, Estado 6915, AGS; the English quotation is in Bedford to Keene, June 5, 1749, SP 94/135, 271r–272r (another copy in Add. Mss. 43,423, 105r–106r); Bedford to Keene, Oct. 26, 1749, SP 94/136, 167r–172r.

enemies. He mentioned Anson's argument that, if his ships had reached Cape Horn earlier and had been able to sail around South America with less difficulty, they would easily have been able to conquer Spanish possessions on South America's western shores. Wall went further, observing widespread English belief in the advantages to be gained by attacking Spain in America. He noted that the English often spoke also of assaulting the poorly defended Spanish Philippines. Most important, Wall reiterated his contention that the British government had wanted the *Porcupine* and the *Raven* to establish a base in the South Atlantic from which wartime operations could be launched against Spain. He concluded that it was "indubitable also that they [the English] will send in times of war a squadron into the South Sea." Such concerns left Spanish officials watchful for any signs of further British moves toward the Pacific.[19]

They quickly found some. Rumors circulated in England in late 1749 that the *Porcupine* was being prepared again for a voyage to the South Sea and, indeed, as French observers had noted, for an attempt to discover a Northwest Passage, possibly by sailing from Japan to North America's west coast. Vaulgrenant, the French ambassador in Spain, mentioned these rumors to Carvajal and Ensenada, who in turn passed them to Wall. Carvajal noted that one of Anson's former officers, the same John Campbell who had been slated to lead the 1749 mission, was to command the ships. He enjoined Wall to be vigilant and to make sure that Spain had a good spy in the Admiralty. Wall and Jorge Juan investigated the matter. Wall talked to Bedford, and Bedford assured him there was no British government involvement in the mission, there being no desire to antagonize the Spanish or to drive them toward closer relations with France. Wall allayed Carvajal's fears about this expedition by reporting its cancellation, but Carvajal wished nonetheless that Spain's American colonies would take appropriate precautions so that not even a secret British expedition could offend Spanish interests. Wall promised, British denials notwithstanding, that he would continue to monitor British ships suspected of having Pacific destinations. Ensenada instructed Don Félix de Abreu, the secretary of the Spanish embassy in London, to do the same.[20]

Events in the spring and summer of 1750 seemed for a time to justify this vigi-

19. Wall to Ensenada, Sept. 8, 1749, Estado 42772, AHN: "Es indubitable tambien que embiaran en tiempo de Guerra una Esquadra a la mar del Sud."

20. "Historical Chronicle, September 1749," and "Historical Chronicle, December 1749," in *Gentleman's Magazine*, XIX (1749), 426–428, 568–571, esp. 427, 570–571; Williams, *Great South Sea*, 260–264; Carvajal to Wall, Jan. 25, 1750, Estado 6917, Carvajal to Wall, Dec. 10, 1749, Wall to Carvajal, Dec. 29, 1749, both in Estado 6914, and Wall to Carvajal, Jan. 22, 1750, Estado 6915, all in AGS; Carvajal to Wall, Jan. 25, 1750, Félix de Abreu to Ensenada, June 25, 1750, both in Estado 4263, AHN.

lance. From London, Abreu reported rumors that three English frigates with English officers and crews were preparing, in the Italian port of Livorno, under the patronage of the Holy Roman Empire, for a voyage of Pacific exploration. Abreu suspected that Anson was trying to conceal the voyage from Spanish eyes by preparing it outside of England. He feared, as Carvajal and Wall had the year before, that the French would try to use a British voyage into Spanish waters as a way to foment Anglo-Spanish jealousies ["se agarrarian de esta ocasión para sembrar zelos entre los dos cortes"]. On receiving Abreu's letters, Ensenada suspected the British were attempting to revive their frustrated expedition of the previous year. That he considered the matter significant is evident from the fact that he spoke with King Ferdinand about Abreu's report. Ensenada then reminded Abreu and the Spanish ambassador in France, Francisco Pignatelli, that Ferdinand wanted clear information about such matters. Ensenada directed Abreu and Pignatelli to thoroughly investigate the rumored expedition, observing that it was *"sumamente importante"* to determine its destination, object, and patron. Vague and variable reports made this difficult. The destination of the expedition changed with each dispatch. Sometimes it was the Levant, sometimes Ethiopia, the East Indies, Brazil, China, or Coromandel. Wall asked Bedford and Newcastle about the expedition, and both denied any British governmental involvement. In this case, Wall was inclined to believe them, but Ensenada insisted that Wall be certain. Ultimately, Spanish inquiries confirmed that the ships were not bound for the Spanish Indies, and the matter blew over. Ensenada, Abreu, and Ferdinand's interest in the rumors, however, demonstrates the extent of Spanish unease about the potential British threat to the Pacific.[21]

The summer of 1751 brought a new rumor of a British expedition to the South Sea. In June, Wall in London and Pignatelli in Paris began sending reports of British preparations in the Thames for another Pacific exploration voyage. Pignatelli had received his information from French foreign minister Puysieulx, who had received it from French naval minister Rouillé. Pignatelli transmitted

21. For Abreu's warnings, see Abreu to Carvajal, May 21, 28, 1750, Estado 6914, AGS; Hilton, "Las Indias en la diplomacia española," 471; Abreu to Pignatelli, May 25, 1750, Estado 6490, AHN (Abreu quotation). On Ensenada's reaction, see Ensenada to Pignatelli, June 8 (quotation), July 6, July 13, 1750, Estado 6493, AHN; Ensenada to Abreu, June 8, 15, 22, 1750, Ensenada to Wall, July 13, 1750, all in Estado 4263, AHN. On changing destinations, see Abreu to Carvajal, May 28, 1750, Estado 6914, AGS; Ensenada to Abreu, June 22, 1750, Abreu to Ensenada, July 9, 16, 1750, Estado 4263, AHN; Abreu to Pignatelli, June 15, July 23, 1750, Estado 6490, AHN; Pignatelli to Ensenada and Carvajal, June 1, 1750, Pignatelli to Ensenada, June 22, 29, 1750, Estado 6491, AHN. On inquiries to Wall and Bedford, see Abreu to Carvajal, May 28, 1750, Estado 6914, AGS; Wall to Carvajal and Ensenada, June 3, 1750, Ensenada to Wall, June 29, 1750, Estado 4263, AHN. On the end of the affair, see Carvajal to Abreu, Aug. 17, 1750, Estado 4263, AHN; Hilton, "Las Indias en la diplomacia española," 471.

Rouillé's memoir to both Ensenada and Carvajal because he thought it to be of "*tanta importancia.*" Ensenada transmitted it to King Ferdinand. The public explanation for the voyage in English newspapers was that the leader of the mission, Commodore George Rodney, was on his way to assume Newfoundland's governorship and that en route he would seek an uncharted North Atlantic island. The rumor ran that Rodney actually intended to sail around South America and found a settlement on Juan Fernández. Pignatelli and Rouillé described Rodney as a new executor of one of Anson's 1749 plans to set up a base in the southern oceans. Rouillé's memoir noted that Anson and other Admiralty officials had dined with Rodney on board on May 23, at which time they had given him his final instructions. Rouillé went further and relayed to the Spanish the French concerns that longtime British interest in forming Pacific establishments and gaining control of Pacific entrances at the Americas' northern and southern extremities had become more intense since the dispatch of British expeditions in search of a Northwest Passage and the publication of Anson's *Voyage round the World.* Rouillé's warnings found attentive Spanish listeners. Wall, Pignatelli, and Ensenada all felt that Rodney's support of the 1740s British attempts to find a Northwest Passage from Hudson Bay was worthy of mention, and this gave Rouillé's claim greater credibility.[22]

Because of the "*importancia de la materia,*" Wall initiated inquiries. Ensenada and King Ferdinand were concerned and asked Wall for more reports. Vaulgrenant thought the affair had renewed Carvajal's concerns about the projects discussed in Anson's account of his voyages. After conversations with Newcastle and Bedford, Wall became convinced that the public story was the correct one and that Spain need have no fear about Rodney's going to the Pacific. In September, he reported that Rodney had failed to find the North Atlantic island in question and intended to return to Europe.[23]

22. On Pignatelli's reports, see Wall to Carvajal, June 3, 1751, Estado 6919, AGS; Puysieulx to Pignatelli, June 5, 1751, "Copia de la carta de Mr Rouillé escrita en Versailles al marques de Puyzieulx el cinco de Junio de 1751," Pignatelli to Carvajal, June 8, 1751 (quotation), and Wall to Pignatelli, June 11, 1751, all in Estado 4512, AGS; Rouillé, June 5, 1751, 64r–66r, Puysieulx to Vaulgrenant, June 7, 1751, 70v, both in CP, Espagne, 508; Puysieulx to Pignatelli, June 4, 1751, Estado 6496, AHN; Pignatelli to Carvajal, June 8, 1751, Pignatelli to Ensenada, June 8, 1751, both in Estado 6497, AHN. On Ensenada's transmission, see Ensenada to Pignatelli, June 21, 1751, Estado 6497, AHN. On rumors of Rodney, see Wall to Carvajal, June 3, 1751, Estado 6919, AGS; Wall to Pignatelli, June 11, 1751, "Copia de la carta," Pignatelli to Carvajal, June 8, 1751, all in Estado 4512, AGS; Pignatelli to Wall, June 8, 1751, Estado 6497, AHN; Puysieulx to Pignatelli, June 8, 1751, CP, Espagne, 508, 67r–67v. On Rouillé's warnings and reactions to them, see Wall to Carvajal, June 3, 1751, Estado 6919, AGS; Ensenada to Pignatelli, June 21, 1751, Pignatelli to Ensenada, July 5, 1751, both in Estado 6497, AHN.

23. Wall to Pignatelli, June 11, 1751, Estado 4512, AGS; Wall to Pignatelli, June 11, 1751, Estado 6497, AHN; Ensenada to Wall, June 21, 1751, Estado 4263, AHN; Vaulgrenant to Puysieulx, July 5,

Though Spanish officials were unlikely to forget this succession of Pacific alarms, French diplomats made sure they remembered by repeatedly alluding to them as evidence of Britain's hostile designs on the Spanish Empire. This was particularly true in 1754 and 1755, as events in the Ohio Valley were bringing French and British forces into open conflict. One example appears in a September 1754 memoir sent by Rouillé to Duras for transmission to the Spanish king and court. Articulating to Spanish officials the same misgivings discussed among French statesmen, this memoir attempted to persuade the Spanish of a British plan to dominate the continent of North America and the commerce of Europe. The memoir argued that Anson's ambitions — as revealed in *A Voyage round the World* — to seize a route from the Atlantic to the Pacific, as well as the varied efforts the English had employed in the last war to obtain establishments in Spanish territory, constituted "incontestable proofs of the reality of this design." In March 1755, Duras mentioned again Anson's plans to Ferdinand. In July 1755, in telling Duras how to persuade the Spanish court of British aggressiveness, Rouillé suggested that British ambitions and greed were not truly directed at the French in Canada, Canada's importance being insufficient to justify England's enormous expenditures on naval armaments. He averred that they were instead directed at Spain's possessions, Mexico in particular, and that "this truth, already proved by the published account of Anson's voyages," could also be confirmed by successive English enterprises in the Bay of Honduras, on the Mosquito Coast, and in Florida. Surely, French diplomats could reason, Spanish fear of this manifest British peril must surpass any concerns about the French presence in North America.[24]

DESPITE THIS PLETHORA of rumors and alarms, the British government failed to get a ship into the Pacific between 1749 and 1755. Spanish reactions to these reports and warnings expose, however, pervasive and recurring uneasiness at the highest levels of the Spanish government about British Pacific designs. British writings and actions during and after the War of Jenkins' Ear suggested an apparent intent to open the Pacific to British ships. These vessels would reconnoiter the South Atlantic and South Pacific and establish bases in or near Spanish

1751, CP, Espagne, 508, 111v–112r; Wall to Pignatelli, June 11, 1751, Estado 4512, AGS; Wall to Carvajal, June 17, July 2, 1751, Estado 6919, AGS; Carvajal to Pignatelli, June 28, 1751, Estado 6496, AHN; Wall to Carvajal, Sept. 23, 1751, Estado 6919, AGS; Ensenada to Wall, Oct. 11, 1751, Carvajal to Wall, Oct. 11, 1751, Estado 4263, AHN.

24. Sept. 27, 1754, CP, Espagne, 516, 108r: "preuves incontestables de la réalite de ce dessein" (Louis XV had Duras transmit another copy of this memoir to Ferdinand in May 1755; see May 14, 1755, CP, Espagne, 517, 346r); Duras to Rouillé, Mar. 20, 1755, CP, Espagne, 517, 218r–218v; Rouillé to Duras, July 27, 1755, CP, Espagne, 518, 86v–87v.

waters. In times of war, this new British familiarity with southern latitudes and these new naval stations would facilitate the interruption of Spanish imperial communications and the plundering of Spanish shipping and cities. Most fundamentally, Spain's monopoly of Pacific navigation would be irrevocably broken. Spanish policy existed to prevent such dangers. Diplomatic persuasion had so far sufficed to limit them, but how long could Spanish words restrain Britain's increasing power and expansive ambitions? Spanish anxieties regarding these ominous possibilities must be added, then, to other Spanish concerns about British logging in Central America, encroachments on Florida, contraband trade with the Spanish Empire, and seizures of Spanish ships during the Seven Years' War, as well as about the danger the formidable British fleet and growing population of Britain's American colonies might pose to Spain's American possessions.

France shared these concerns and offered Spain an alliance that would have addressed them. The addition of French soldiers and ships to Spain's balance sheet would have given Spanish diplomats the option of intimidating British aggressors should persuading them prove impossible. But Spain chose to maintain good relations with Britain until 1757 and neutrality until 1761. French statesmen throughout the 1750s kept asking why Spain was rejecting a Franco-Spanish alliance aimed at protecting their respective possessions from this manifestly dangerous British aggression. One answer to this question lies in contemporary Spanish concerns about French exploration of the unexplored North American West.

12

FRENCH BORDERLANDS ENCROACHMENTS
AND SPANISH NEUTRALITY

This book's first chapter mentioned the arrival of French traders Jean Chapuis and Louis Feuilli in New Mexico in 1752, their interrogation by Spanish officers, and their subsequent incarceration in Spain. The matter did not end with the traders in jail, moving instead into French and Spanish diplomatic conversations and deliberations. In November 1754, the Spanish Council of the Indies recommended that Ferdinand VI discuss the wayfarers in "official correspondence" with the French court. In January 1755, Don Jaime Masones de Lima, Spain's ambassador in France, complained to French foreign minister Rouillé about Chapuis and Feuilli. Not only had the traders violated Spanish laws by appearing in New Mexico, but they seemed to have done so with the connivance of French officials. The commander at Michilimackinac, Duplessis Falberte, had given them a passport, and his counterpart at Fort Chartres in Illinois, Jean-Baptiste Benoît de Saint Claire, had given them a license to discover a route to New Mexico and trade with the Spanish there. Masones de Lima inferred that Chapuis and Feuilli's mission had been conceived, prepared, and executed under the auspices of a French governor who could not be ignorant that he was meddling illegally in Spanish affairs. This offended the Spanish king regarding the matter on which he was the "most sensitive" ["le plus sensible"]. It was, moreover, a repetition of similar treaty infractions ["une récidive de ... plusieurs autres semblables infractions de Traités"] of which several other French governors in the region were guilty. It was difficult, according to Masones de Lima, to believe these governors had acted without orders from superiors.[1]

Such conduct had consequences. Writing to Rouillé of Chapuis and Feuilli's

1. "Governor Velez and the French Intrusion of 1752," in Alfred Barnaby Thomas, [ed. and trans.], *The Plains Indians and New Mexico, 1751–1778: A Collection of Documents Illustrative of the History of the Eastern Frontier of New Mexico* (Glendale, Calif., 1940), 82–110, esp. 87; Jaime Masones de Lima to Antoine-Louis Rouillé, Jan. 8, 1755, CP, Espagne, 517, 17r–24r.

expedition, Masones de Lima stated, "The good relations [bonne harmonie] that unite the two crowns of Spain and of France, and the treaties that subsist between them suffer equally from this enterprise which attacks the king, my master, in the sharpest manner [de la maniere la plus vive]." Ferdinand demanded the most rigorous punishments for Saint Claire to deter any other French governor who might contemplate similar enterprises. Masones de Lima reiterated his concerns in March.[2]

These complaints came at a delicate moment in Franco-Spanish relations and derive historical importance from their relation to the larger context of 1750s Anglo-French rivalry. In the years after the War of the Austrian Succession, French and British ministers anticipated renewed strife between their nations and were weighing their respective belligerent capacities. They noted that alliance with Spain might impart a decisive advantage in an uncertain conflict. France possessed fewer warships than Britain and had observed the creditworthy British government's ability to finance the War of the Austrian Succession by borrowing huge sums at low rates of interest. French ministers hoped to balance Britain's naval supremacy with the assistance of Spain's fleet and Britain's financial resources with the acquisition of Spanish silver. British strategists wanted naturally to enhance their empire's naval and financial advantages and to forestall French attempts to compensate for them. Attentive to Spain's importance, French and British diplomats importuned their Spanish counterparts.[3]

In this vying for Spanish favor, French diplomats perceived an advantage. Spain and France were ruled by branches of the same Bourbon family and were respectful of the same Catholic Church. They had fought together against Brit-

2. Masones de Lima, Mar. 25, 1755, CP, Espagne, 517, 238r–239r; Masones de Lima to Rouillé, Jan. 8, 1755, Estado 4511, AGS; Duplesis Falberte, July 30, 1751, Tomás Vélez Cachupín, Aug. 6, 1752, Tomás de Sena, Aug. 7, 1752, Tomás Vélez Cachupín, Aug. 9, 1752, in José Antonio Pichardo, *Pichardo's Treatise on the Limits of Louisiana and Texas: An Argumentative Historical Treatise with Reference to the Verification of the True Limits of the Provinces of Louisiana and Texas . . .* , ed. and trans. Charles Wilson Hackett, trans. Charmion Clair Shelby, 4 vols. (Austin, Tex., 1971), III, 364–369, esp. 364–368; Herbert E. Bolton, "French Intrusions into New Mexico, 1749–1752," in John Francis Bannon, ed., *Bolton and the Spanish Borderlands* (Norman, Okla., 1964), 150–171, esp. 163–168.

3. Lawrence Henry Gipson, "British Diplomacy in the Light of Anglo-Spanish New World Issues, 1750–1757," *AHR*, LI (1946), 627–648, esp. 637–647; Sylvia-Lyn Hilton, "Las Indias en la diplomacia española, 1739–1759" (Ph.D. diss., Universidad Complutense de Madrid, 1979; rpt. Madrid, 1980), 552–554, 654; Richard Lodge, ed., *The Private Correspondence of Sir Benjamin Keene, K.B.* (Cambridge, 1933), xv; Didier Ozanam, "Les origines du troisième Pacte de famille (1761)," *Revue d'histoire diplomatique*, LXXV (1961), 307–340, esp. 309; John Lynch, *Bourbon Spain, 1700–1808* (Oxford, 1989), 161; Richard Pares, *War and Trade in the West Indies, 1739–1763* (Oxford, 1936), 556–557; Max Savelle, *The Diplomatic History of the Canadian Boundary, 1749–1763* (New Haven, Conn., 1940), 56; Jeremy Black, "Anglo-Spanish Naval Relations in the Eighteenth Century," *Mariner's Mirror*, LXXVII (1991), 235–258, esp. 235–236, 252.

ain in the most recent European war, and British expansion in the Americas posed, in French eyes, a manifest danger to Spanish and French imperial interests. As French officials looked with concern on British projects involving Hudson Bay, the Ohio Valley, Central America, and South American approaches to the Pacific, they expected Spanish ministers would do the same.

In such circumstances, French diplomats heeded Spanish complaints about Chapuis and Feuilli but found it difficult to take them as seriously as their Spanish counterparts did. Mid-1750s French foreign office personnel were trying to relate recent events to the larger picture of European imperial competition. On this grand canvas, obscure New Mexico border incidents seemed to tend toward the trifling. Spain's empire stretched from Manila to Madrid, from Baja California to Buenos Aires. When British ships and British statesmen were coveting and encroaching upon the most traditionally lucrative, the most politically and geographically central, and the most diplomatically inviolable possessions of this vast Spanish Empire, was it really plausible that the appearance of a handful of Gallic peddlers at a northern outpost could shape the conduct of empires?

The answer turns out to be yes. Examination not only of the correspondence passing between French and Spanish diplomats but also of the letters and reports circulating among Spanish officials reveals Spanish imperial anxieties underappreciated by French officials. The concerns Spanish diplomats expressed regarding French New Mexico encroachments were genuine. Much of this solicitude sprang from Spanish worries about the North American geographic unknown.

Limited though they were in their understanding of western North American geography, Spanish officials had enjoyed at least one possible consolation: perhaps the same conditions constraining Spanish reconnaissance east of New Mexico would limit the westward movement of French scouts from Louisiana. Spanish geographic horizons might be circumscribed, but perhaps the Spanish colony was safe. The continuing appearance of French parties like that of Chapuis and Feuilli belied this comforting notion, suggesting that subjects of a rival empire had found a way to negotiate the physical and human obstacles impeding European plains expeditions. The isolation from other European colonies that had hindered New Mexico's development no longer promised its security.

In creating a new Franco-Spanish contact point, westering French travelers like Chapuis and Feuilli made the Southwest a more pressing topic of Franco-Spanish discussion. By demonstrating the feasibility of travel between Louisiana and New Mexico and by raising the question of how much farther west French scouts might go, they turned Spanish and French geographic and political in-certitudes into Franco-Spanish geopolitical issues. Spanish officials were forced to decide just how menacing French intentions were and just how dangerous French capabilities, Indian inclinations, and American geography would allow them to be. French diplomats had to assess and attend to Spanish sensibilities so as to prevent French southwestern forays from hamstringing efforts to secure Spanish assistance against Britain.

Such diplomatic exertions proved inadequate, and European uncertainty regarding North American geography marred Franco-Spanish relations. Unsure about what French scouts might find north of New Mexico and fearful that it might include a practicable river route to the Pacific, Spanish officials overestimated the dangers Louisiana presented to New Spain and the Spanish Lake. This diminished the appeal of a French alliance. At the same time, French diplomats' failure to appreciate the extent of and reasons for Spanish southwestern anxieties or to forestall the expeditions inflaming them undermined attempts to soothe Spanish sensitivities and attract Spanish aid. In the end, these missteps and misunderstandings would benefit Britain.

French Diplomats Contemplate Spanish Insecurities

Much of French western reconnaissance's international importance derives from the discrepancy between Spanish and French views of its implications and, indeed, from the difference more generally between Spanish and French understandings of Spanish imperial interests. French diplomatic documents from the Archives des affaires étrangères and the Archives nationales, many of them intended for French foreign office eyes only, show that French diplomats found it difficult to believe Spanish statesmen were genuinely worried about French encroachments on Spanish North American territory. This was in part because French diplomats considered the British threat to the Spanish Empire so dire that they could not see how any French menace could be comparable. It was also because the French sincerely believed, or at least had thoroughly convinced themselves, that Louisiana served as a barrier protecting northern New Spain from British aggression. They inferred that Spanish officials must welcome the French colony's existence. One French foreign office memoir, written some time after the publication of the official account of Anson's voyages, mentioned that, in light of the French North American colonies' protective value for Spain's possessions and of English projects to secure Pacific island bases and usurp Spanish South Sea commerce, "relative to the common enemy, our interests and those of Spain are entirely connected, and it would assuredly be less troublesome for her that we introduce ourselves into her establishments than that the English do"; "Spain, in the case where she cannot prevent the establishments of the English or of foreigners, should always give us the preference."[4]

This memoir at least recognized that Spain might see a French colonial presence as the least of many necessary evils. Other French officials had their attention so concentrated on the magnitude of the British threat to the Spanish Empire that they seemed incapable of acknowledging Spanish perceptions of a French challenge to Spain's American interests. In May 1753, after hearing of a complaint from Spanish foreign minister Carvajal about unsettled Franco-Spanish North American boundary disputes, French minister of foreign affairs François-Dominique Barberie de Saint-Contest (1751–1754) denied to his own ambassador in Spain any knowledge of such disputes on his part or on the part of minister of marine Rouillé. Later in the same year, French diplomats and their counselors discussed rumors that Spain was dispatching to New Mexico troops intended for hostile operations against Louisiana. They found it hard to credit the notion that these Spanish forces were really aimed at French possessions. The duc de Noailles felt with others that the Spanish ought to consider French Louisiana a shield protecting Spain's possessions rather than a sword threatening them. He told French ambassador to Spain the duc de Duras in November 1753 that the matter of the British in Honduras was "so important for the Spanish Monarchy" he was inclined to believe, as did Duras, that this was "the real and hidden object of the armaments" going to New Mexico under the "pretext" of attacking Louisiana. Duras accepted Noailles's reasoning and told Saint-Contest, "I am convinced that we are but a pretext which the minister [Ensenada] uses to conceal his true project concerning the Dutch and the English." "How could one believe" that these rumored military preparations "concern directly our Mississippi colonies?" The Spanish ministry knew "perfectly" the disposition of the French government, its "conspicuous deference," of which it gave "new assurances every day."[5]

4. "Memoire sur l'Amérique," MD France, 2008, 83r–83v, 86v–88v.

5. François-Dominique Barberie de Saint-Contest to duc de Duras, Apr. 13, May 22, 1753, CP, Espagne, 513, 326v–327r, 416r–417v; duc de Noailles to Duras, Nov. 30, 1753, CP, Espagne, 511, 234r–234v ("The Spanish themselves ought to consider our establishments in Louisiana as the most secure barrier and rampart that they can have for guaranteeing themselves against the invasions of a nation [England] so enterprising, so unjust, and of which the cupidity has not the slightest limit. It is essential to make the Spanish aware of this truth" ["Les Espagnols meme doivent considerer nos etablissemens dans la Louisiane, comme une Barriere et un rampart le plus assuré qu'ils puissent avoir pour se garantir des invasions d'une nation aussi entreprenante, aussi injuste, et dont la cupidité n'a point de bornes. Il est essentiel de faire sentir cette verité aux espagnols"]), 233r ("si importante pour la Monarchie d'Espagne"; "c'est l'objet réel et Caché des armements que se sont au nouveau Mexique sous le pretexte d'attaquer la Loüisiane"); Dec. 31, 1753, CP, Espagne, 511,

Two years later, as Britain and France moved closer to open hostilities, French officials remained convinced that Spain would recognize Britain as the menace to and France as the protector of Spain's empire. In 1755, after assuming the post of minister of foreign affairs, Rouillé felt no need to doubt the "tender affection of the King of Spain for the King" of France, and he remained convinced that, in the event of an Anglo-French war, Spain's interests would sooner or later bring it around to France's point of view. In May of the same year, a French memoir sent to the Spanish court to persuade it of English malevolence and French benevolence claimed confidently that Spain need only compare grasping British conduct with French "tranquillity" in the neighborhood of Spain's American possessions to "know the difference between the designs of the one and the other power."[6]

French diplomats were not blind, and, at a time when they were wooing the Spanish court, they certainly had no desire to offend Spanish sensibilities. Some French diplomats acknowledged, some came to recognize, and some occasionally remembered that Spanish officials possessed reason for concern about the French North American presence. In 1750, the comte de La Galissonière — governor of Canada from 1747 to 1749, a member of the Anglo-French boundary com-

---

244r–244v (see also "Differens entre le roi et le roi d'Ange. en Amere . . . ," August 1754, MD, Amérique, 24, 177r–177v); Duras to Noailles, Dec. 21, 1753, CP, Espagne, 512, 185v–187v; Duras to Saint-Contest, Nov. 16, 1753, CP, Espagne, 514, 442v–443r, 451v: "Je suis persuadé que nous ne sommes qu'un pretexte dont le ministére se sert pour dissimuler son veritable projet sur les hollandois ou sur les Anglois"; "Comment pourroit on croire que l'armament designé dans l'extrait des nouvelles puisse regarder directement nos colonies du Mississipi?"; "le ministere d'Espagne qui connoit si parfaitement les dispositions de celui de France, et les deferences marqueés, dont il lui donne touts les jours de nouvelles assurances." Rouillé felt certain that these Spanish preparations, of whatever kind they were, could not "concern Louisiana." "For a long time all has been tranquil between these two nations [France and Spain] in this continent. . . . The successive governors the king has sent there have always complied with the greatest exactitude with the orders that his majesty renews on all occasions for the maintenance of the union that the ties of blood and friendship have so happily established between him and his Catholic Majesty [Ferdinand]" ["Il y a long temps que tout est tranquille entre les deux Nations dans ce Continent. . . . Les Gouverneurs successifs, que Le Roy y a envoyés, se sont constament conformés avec la plus grande exactitude aux ordres que Sa Mté. leur renouvelle dans toutes les occasions pour le maintien de l'union que les liens du sang et de l'amitié ont si heureusement etablie entr'Elle et S.M.C."]. It was in Spain's interest that Louisiana attain a certain stability, for the French colony was "the surest barrier against the designs of the English regarding the possessions of Spain in New Mexico" ["c'est la barriere la plus sûre contre les vües des Anglois sur les possessions d'Espagne au nouveau Mexique"]. See Rouillé, "Observations sur les nouvelles venües d'Espagne," Dec. 21, 1753, CP, Espagne, 514, 511r–511v.

6. For Rouillé's view, see Rouillé to Duras, Apr. 25, 1755, CP, Espagne, 517, 281r–281v. For the May memoir, see Sept. 27, 1754, CP, Espagne, 516, 109r: "Sa Majesté Catolique n'a qu'à comparer cette conduite avec la tranquilité des François du coté des établissements Espagnols pour connoître la différence des vues de l'une et de l'autre Puissance" (Louis XV had Duras transmit another copy of this memoir to Ferdinand in May 1755; see May 14, 1755, CP, Espagne, 517, 347r).

mission of the early 1750s, and the author of a famous 1750 memoir on the strategic importance of the French North American colonies — argued that France should not extend Louisiana to the west for fear of inciting Spanish jealousy. Rouillé warned Louisiana governor Philippe de Rigault de Vaudreuil (1743–1752) not to authorize any New Mexico missions without explicit royal permission. In 1755, minister of marine Jean-Baptiste Machault (1754–1757) transmitted one of Duras's letters about Spanish sensitivities to Louisiana's Governor Kerlérec and ordonnateur Vincent Guillaume Le Sénéchal Auberville. Machault told the French officials to "avoid with the greatest care" giving any "legitimate cause for complaint" to the Spanish governors of the neighboring provinces, enjoined them to take advantage of all possible opportunities to establish and strengthen the union that ought to subsist between the two crowns, warned them particularly about Spanish sensitivity on the question of illicit commerce, and advised them that any trade with Spain be conducted with the "greatest prudence and circumspection." Louis XV would not "authorize or tolerate any enterprise that could occasion disputes with" Spain "about this matter."[7]

Recognizing that the Spanish might be jealous of French possessions was not tantamount, however, to acknowledging that they should be.

Spanish Borderlands Anxieties

Examination of papers passing among Spanish officials demonstrates, on the contrary, that the Spanish protests in question arose from serious and genuine worries about French American expansion, solicitudes exacerbated by Spanish uncertainty regarding western North American geography.

Ever since La Salle's 1681–1687 Mississippi River and Gulf of Mexico explorations and the 1699 founding of Louisiana, Spanish officials had fretted about French incursions into Spain's North American borderlands. Commercial interloping was one concern. New Spain's frontier regions enjoyed no exemption from the policy of reserving American markets for Spanish merchants, and Spanish regulations prohibited trade between colonies like New Mexico and Louisiana. Spanish officials worried that, despite these restrictions, French traders would draw frontier provinces out of the Spanish economic orbit. Spanish goods had to travel the long road from Veracruz and Mexico City, and the costs of transporting items by mule train to the frontier drove volume down

7. Comte de La Galissonière, "Memoire sur les colonies de la France dans l'Amerique septentrionale," December 1750, MD, Amérique, 2, 26v–27r; Rouillé to Philippe de Rigault de Vaudreuil, Oct. 2, 1750, AC, B 91, 405r–405v, Rouillé to Louis Billouart de Kerlérec and Vincent Guillaume Le Sénéchal d'Auberville, Mar. 2, 1755, 245r–245v, both in AN.

and prices up. Taxes and import and export duties levied on goods originating in Spain pushed charges higher still. On the other hand, products competitive with those offered by Spanish traders could be obtained more cheaply in Louisiana, and French merchants could use the Mississippi and its tributaries to move them. Ordinary inhabitants of northern Spanish provinces like New Mexico, plagued by a scarcity of affordable and high-quality manufactured goods, were more likely to embrace than reject foreign traders and their coveted merchandise. Church bells rang in celebration when the Mallets reached Santa Fe in 1739. Hoping the bells would toll for them, too, other French parties followed. In 1749, three Frenchmen appeared at a Taos trade fair; by March 1750, six more had made it to the same place; in 1751, four more, one of whom was second-time visitor Pierre Mallet, arrived at Pecos; in 1752, Chapuis and Feuilli entered this story. As local officials' concerns about the French arrivals mounted, their welcome grew colder, but bad hospitality might not suffice to deter future French expeditions. To mercantilist Spanish officials, the commercial implications were unsettling. If frontier provinces placed their scanty funds into French rather than Spanish hands, draining wealth from rather than circulating wealth within the empire, they would defeat one of the main reasons for having colonies in the first place. This made fastidious Spanish regulators eager to contain French expansion and keep French merchants at a distance.[8]

Worse than the danger of illegal trading was that of military invasion. By the mid-1700s, more than a century had elapsed since the most active period of Spanish New World conquest. Spanish missions, mines, presidios, and towns moved intermittently north in what is now Mexico, but Spain had devoted most of its attention since the early seventeenth century to preserving rather than extending its American empire, and instances of far-ranging eighteenth-century Spanish expansion (in Texas and California, for example) were often by-products of these defensive impulses. Mid-eighteenth-century French inclinations, in contrast, seemed to many Spanish officials to be aggressive and expansionist. Some of these Spanish statesmen had been born while Louis XIV was still alive and still threatening to dominate Europe and the Spanish Empire. Most of them viewed France warily. The clear-eyed Carvajal wrote of France's

8. Vélez Cachupín to Juan Francisco de Güemes y Horcasitas [count of Revilla Gigedo], June 19, 1749, Mar. 8, 1750, Vélez Cachupín to Revilla Gigedo, Feb. 25, 1751, both in Pichardo, *Pichardo's Treatise*, III, ed. and trans. Hackett, trans. Shelby, 308–310, 325–329, 336–340, esp. 309–310, 327, 336–338; Bolton, "French Intrusions into New Mexico," in Bannon, ed., *Bolton and the Spanish Borderlands*, 150–171; Henri Folmer, "Contraband Trade between Louisiana and New Mexico in the Eighteenth Century," *NMHR*, XVI (1941), 249–250, 264; Geoffrey J. Walker, *Spanish Politics and Imperial Trade, 1700–1789* (Bloomington, Ind., 1979), 13–14; David Weber, *The Spanish Frontier in North America* (New Haven, Conn., 1992), 187.

many betrayals of Spanish interests and attempts to seize Spanish territory. The members of a junta advising Ferdinand on American affairs, in a 1752 letter to Ensenada discussing Louisiana-based French territorial incursions, wrote of the expansionist French temperament. New Spain's viceroy Revilla Gigedo (1746– 1755) referred to France as *"esta nacion tan ambulativa."*[9]

New Spain could easily entice restless and rapacious Frenchmen. Northern Mexico's silver mines were a tempting target for all Spain's envious European rivals. Spanish officials knew that New Mexico and Texas, in contrast, lacked much in the way of wealth to attract European invaders, but they could also see that, should such invaders come, the colonies would have trouble resisting them. The same long miles and poor roads hindering commerce between Mexico and New Mexico would make it hard for troops based elsewhere to reach New Spain's northern marchlands in time to confront Indian or European attackers, and the Spanish frontier's difficult terrain and territorial extent made it hard for the small numbers of troops stationed there to repel marauders.[10]

This would be especially true if the invaders were numerous, and Spanish officials were consequently concerned about indications that Louisiana's French population was growing. They had heard rumors of the French government's sending hundreds of families to increase Louisiana's population and companies of regular troops to augment its offensive military capacities. In 1751, New Mexico's Governor Vélez Cachupin relayed dubious French traders' accounts of the Louisiana governor's having 2,500 men available for operations against New Mexico. In spring 1752, official correspondence discussed a report from the governor of Havana of 5,600 French troops bound for Saint-Domingue and possibly destined thereafter for New Spain. Because the European populations of the French and Spanish provinces were small, the addition of a few thousand settlers or soldiers could impart a decisive regional advantage to one empire. The white population of 1763 Louisiana numbered about 4,000. The Spanish population of Texas was about 1,200 in 1760; of New Mexico in 1765, roughly 9,500. Though scarcely a powerhouse relative to American colonies like Virginia, Massachusetts, or New France, Louisiana constituted a possible staging area for military operations against no-less-precarious Spanish settlements in Texas and New Mexico.[11]

9. On Carvajal, see Hilton, "Las Indias en la diplomacia española," 638. For junta and viceroy, see Julián de Arriaga, Sebastián de Eslava, and Francisco Fernández Molinillo to marqués de La Ensenada, May 25, 1752, expediente 1, no. 26 ("Los Franceses por su muchedumbre, y por su genio . . . abriguen la maxima de estenderse"), and Revilla Gigedo to Ensenada, June 22, 1752, expediente 2, both in Estado 3882, AHN.

10. Revilla Gigedo to Ensenada, June 22, 1752, Estado 3882, expediente 2, AHN.

11. On Louisiana's population and reinforcements, see Molinillo and Arriaga to Ensenada, Feb-

The French colony by itself, and the larger empire of which it formed a part, was only one portion of the problem confronting mid-eighteenth-century Spanish officials. The Indians between New Mexico and the Mississippi, especially the Comanches and nations allied with them, figured also among Spanish concerns. Indeed, one reason Mississippi Valley Frenchmen were so alarming to 1750s Spanish officials was because of the apparent connection between their activities and a change in the pattern of relations among the Indian peoples and French subjects west of New Mexico. In earlier chapters about the limits of Spanish and French reconnaissance, we have already seen, from French and Spanish observers, how plains and prairie Indian communities and the animosities among them had made Spanish and French exploration between the Mississippi and the Rockies hazardous. If inter-Indian and Euro-Indian conflicts played such a decisive role in inhibiting western reconnaissance, it stands to reason that, if western Indian nations or a European power could effect a truce among a belt of Indian and European communities between the Great River and New Mexico, then ways across plains and prairies might open.

In the late 1740s and early 1750s, Vélez Cachupín was reporting that this had happened. From his interrogations of French traders arriving in New Mexico, the governor concluded that, at some point during the 1740s, French agents had succeeded in calming disputes between Comanche bands and their Wichita (often called Jumanos or Jumanes by Vélez Cachupín) and Pawnee neighbors. In 1750, he was "persuaded" that "French policy, with the idea of enlarging and extending their colony, has influenced the desire of the Jumanes Panipiques to make peace with the Cumanches." In 1752, he wrote that Chapuis and Feuilli had confirmed "the confederation and friendship of the Comanches with the Jumanos and Pawnees." Such "confederations" filled Vélez Cachupín "with a just fear" of the "treacherous intentions" of the French. It wasn't that efforts to promote harmonious interactions among these French and Indian peoples were uniformly successful, as French references to continued Comanche threats and depredations attested. A truce, moreover, might not last. It might restrain only a part of the nation nominally bound by it. It might be superseded by subsequent Spanish-Indian agreements. Nonetheless, the fact of French efforts to ally with

ruary 1752, expediente 1, no. 25, and Revilla Gigedo to Ensenada, June 22, 1752, expediente 2, both in Estado 3882, AHN; Fernando Sánchez Salvador to Junta de Guerra, May 15, 1751, AGI, Guadalajara 137. On rumors of troops for New Mexico, see Vélez Cachupín to Revilla Gigedo, Feb. 25, 1751, in Pichardo, *Pichardo's Treatise*, ed. and trans. Hackett, trans. Shelby, III, 336–340, esp. 337. On troops to Saint-Domingue, see Arriaga, Eslava, and Molinillo to Ensenada, May 25, 1752, Estado 3882, expediente 1, no. 26, AHN. For a modern estimate of Louisiana's population, see Daniel H. Usner, Jr., *Indians, Settlers, and Slaves in a Frontier Exchange Economy: The Lower Mississippi Valley before 1783* (Chapel Hill, N.C., 1992), 108, 279.

the Comanches and Wichitas, the result in the form of French parties' reaching New Mexico, and the possibility that some form of Comanche-Wichita-Pawnee-French alliance was beginning to take shape and might turn out to be durable were troubling.[12]

It could make Frenchmen and Indians much more dangerous. With a Wichita-Comanche truce established, the French could employ Wichitas as intermediaries in Franco-Comanche commerce. This was especially frightening to New Mexicans because guns and ammunition were among the most alluring items French traders had to offer and were about the last things Spaniards wanted to see in Comanche hands. "With these weapons," Vélez Cachupín warned, the Comanches "will be greatly feared in this province. They can ruin it if they remain here." Wichitas could, moreover, guide French parties to Comanche country and convince Comanche bands to respect French persons and property. This would make French westward travel easier. Indeed, as Vélez Cachupín noted, with "the confidence and good reception which the French find among the Cumanches, through the recommendation of the Jumanes, and the information which they have acquired concerning our settlements of New Mexico from these same Cumanches" (and from the first Mallet party), the French were now able to "send troops as far as the environs of the pueblo of Taos without the knowledge of this government." And if the French could get to Taos, they could eventually extend their arms around New Mexico.[13]

Spanish officials had more than Frenchmen to worry about. The challenge to New Mexico was not just that Wichitas and Comanches could further French

12. Vélez Cachupín to Güemes y Horcasitas, Mar. 8, 1750, in Pichardo, *Pichardo's Treatise*, ed. and trans. Hackett, trans. Shelby, III, 325–329, esp. 326 (see also marqués de Altamira, Jan. 9, 1751, 329–333, esp. 330–331; Vélez Cachupín to Revilla Gigedo, Feb. 25, 1751, 336–340, esp. 338); "Vélez [Cachupín] to Revilla Gigedo, Santa Fe, Sept. 18, 1752," in Thomas, [ed. and trans.], *Plains Indians and New Mexico*, 108–110, esp. 109–110. For another copy of the letter, see *Pichardo's Treatise*, III, 370.

13. On the Comanches' ruining New Mexico, see Vélez Cachupín to Revilla Gigedo, Nov. 27, 1751, 68–76, esp. 75–76, Altamira to [Revilla Gigedo?], Apr. 26, 1752, 76–80, esp. 79, and "Instruction of Don Thomas Véliz Cachupin, 1754," 129–143, esp. 135, all in Thomas, [ed. and trans.], *Plains Indians and New Mexico*; Robert Ryal Miller, ed. and trans., "New Mexico in Mid-Eighteenth Century: A Report Based on Governor Vélez Capuchín's Inspection," *SWHQ*, LXXIX (1975), 166–181, esp. 172–173; Fernando Sánchez Salvador to Junta de Guerra, May 15, 1751, AGI, Guadalajara 137; Elizabeth A. H. John, *Storms Brewed in Other Men's Worlds: The Confrontation of Indians, Spanish, and French in the Southwest, 1540–1795*, 2d ed. (Norman, Okla., 1996), 316–317; Weber, *Spanish Frontier*, 186–188. On easier French westward movement, see Vélez Cachupín to Güemes y Horcasitas, Mar. 8, 1750, in Pichardo, *Pichardo's Treatise*, ed. and trans. Hackett, trans. Shelby, III, 325–329, esp. 325–327 (see also marqués de Altamira, Jan. 9, 1751, 330–331; Vélez Cachupín to Revilla Gigedo, Feb. 25, 1751, 336–340, esp. 338); and see "Vélez [Cachupín] to Revilla Gigedo, Santa Fe, Sept. 18, 1752," in Thomas, [ed. and trans.], *Plains Indians and New Mexico*, 109; Miller, ed. and trans., "New Mexico in Mid-Eighteenth Century," *SWHQ*, LXXIX (1975), 172–173. On the French encompassing New Mexico, see Vélez Cachupín to Revilla Gigedo, Feb. 25, 1751, in *Pichardo's Treatise*, III, 338.

objectives: it was also that Frenchmen could help Indians, most pointedly the Comanches, advance theirs. Over the previous two centuries, Spanish explorers had seen that Plains peoples without horses, firearms, or European allies could check Spanish forays east of New Mexico. By the mid-eighteenth century, Spanish officials were having to ask themselves what Plains peoples with all three could do. The Comanches had taken advantage of the post–Pueblo Revolt dispersal of Spanish horses and had transformed themselves into a mounted Plains people. They had profited from liberal French trading practices to acquire firearms. They appeared now to be working with Spain's frequent rival and potential adversary. This combination of mobility, lethality, and diplomacy was making the Comanches into something far more frightening than the "Escanxaques" who had repelled Oñate or the Pawnees who had killed Villasur. Those Indians fought with Spaniards far from New Mexico. Comanches were taking the fight to the colony itself.[14]

As scholars of trans-Mississippi North America have often remarked, the rise of the Comanches, along with that of the other mounted Plains nations, forms one of the salient developments of eighteenth-century western history. By roughly 1700, Comanche bands with horses were moving down from the Great Basin into eastern Colorado and western Kansas. Spanish documents begin mentioning them in 1706. Mounted Comanche hunters and warriors could move faster and farther than had been possible on foot. They could attack Indian and Spanish communities and escape successful pursuit. They could steal horses from isolated Spanish settlements, trade them with Indian and French groups to the north and east, and use them to acquire animal products and human captives to trade with New Mexicans for agricultural items and metal, pottery, and leatherwork. With horses and, in some cases, firearms, Comanche bands had, by the 1730s and 1740s, driven Apache communities west and south off the plains, and they had joined the array of factors pushing Wichita groups southeast. By the 1740s, the Comanches were dominating commerce in the region between the Mississippi and New Mexico. In the same decade, even as New Mexico-

14. The Comanches represented the most prominent Indian threat, but not the only one. By 1753, the ability of French trader Louis Juchereau de Saint-Denis (son of the celebrated explorer and trader of the same name) to dissuade Nadotes, Tawakonis, and Caddos from a planned attack on Spanish outposts around Los Adaes, near modern Natchitoches, had convinced Mexico City officials that the French exerted considerable control over Spanish frontier Indians. In this case, Saint-Denis had protected Spaniards, but the Spanish could not be certain that Frenchmen might not put their influence to malicious uses in the future. In 1753, Texas governor Jacinto de Barrios y Jauregui warned that the French could destroy Texas settlements by directing local Indians against them. See John, *Storms Brewed*, 345; Herbert Eugene Bolton, *Texas in the Middle Eighteenth Century: Studies in Spanish Colonial History and Administration* (Berkeley, Calif., 1915), 71.

Comanche trade continued, Comanche raids on New Mexico were so severe as to confront Spanish officials with the specter of depopulation in certain areas and perhaps ultimately with the "ruin" of the province as a whole. In the 1740s and 1750s, eastern Comanches were moving to the middle Red River Valley and, along with other "Norteños," were terrifying many of the area's Indian inhabitants as well as the Spanish Texas outposts to which they were appealing for help. We should not exaggerate mid-eighteenth-century Comanche power. When Spanish forces could concentrate in the right place under a resolute leader, they inflicted stinging and costly defeats on Comanche bands. On the global stage of empire, the Comanches didn't cut much of a figure. But on the plains east of New Mexico, an area where European power was more notional than actual, where, in the memorable phrase of one historian, the European empires were "giants wrestling at the very ends of their fingertips," coordinated Comanche rancherías could hold sway.[15]

15. For a forceful synthetic treatment of eighteenth-century Comanche history, see the early chapters in Pekka Hämäläinen, *The Comanche Empire* (New Haven, Conn., 2008). An especially astute treatment of the larger question of northern Spanish expansion and defense, one that places the Comanche efflorescence in a larger context, can be found in Luis Navarro García, *Don José de Gálvez y la comandancia general de las Provincias Internas del Norte de Nueva España* (Seville, 1964). Other strong works on the Comanches include Gary Clayton Anderson, *The Indian Southwest, 1580–1830: Ethnogenesis and Reinvention* (Norman, Okla., 1999), 95–99, 204–209; Morris W. Foster, *Being Comanche: A Social History of an American Indian Community* (Tucson, Ariz., 1991), 32, 36–41; Pekka Hämäläinen, "The Western Comanche Trade Center: Rethinking the Plains Indian Trade System," *Western Historical Quarterly*, XXIX (1998), 485–513; John, *Storms Brewed*, 231–232, 250–257, 306–314; Thomas W. Kavanagh, *Comanche Political History: An Ethnohistorical Perspective, 1706–1875* (Lincoln, Neb., 1996), 28–29, 37, 57–58, 61, 69–70, 73; Thomas, [ed. and trans.], *Plains Indians and New Mexico*, 16–18, 58–59; Ernest Wallace and [E.] Adamson Hoebel, *The Comanches: Lords of the South Plains* (Norman, Okla., 1952), 6–14, 22, 35–44, 209–216, 245–250, 288–289. On elusive Comanche raiders, see Vélez Cachupín to Güemes y Horcasitas, Mar. 8, 1750, in Pichardo, *Pichardo's Treatise*, ed. and trans. Hackett, trans. Shelby, III, 326–328. On New Mexican ruin, see Thomas, [ed. and trans.], *Plains Indians and New Mexico*, 16–18, 58–59; Vélez Cachupín to Revilla Gigedo, Nov. 27, 1751, 68–76, esp. 75, "Instruction of Don Thomas Véliz Cachupín," Aug. 12, 1754, 129–143, esp. 135, both ibid.; John, *Storms Brewed*, 313. The "fingertips" quotation comes from a private letter to me from historian of late antiquity Peter Brown. The Comanches were clearly formidable. Should we follow the title of Pekka Hämäläinen's recent book and designate them as a third "empire" taking shape between those of Spain and France? I won't comment here on the period after 1763, but for the decades before that date, I think the term "empire" would go too far. Comanche bands were a fundamentally different kind of entity than the Inca, Aztec, Russian, Chinese, Spanish, French, and British empires discussed in this book. The Comanches had no capital or clearly identifiable and lasting ruler. They preyed on many victims and worked with many allies but lacked acknowledged subjects. Had a 1750s Comanche chief sought admission to the imperial club, it is hard to imagine the Habsburgs, Ottomans, Mughals, Romanovs, and Ch'ing letting him in. An instructive comparison is Attila and the Huns. When the eastern Roman emissary Priscus met Attila, he found an acknowledged ruler of many nations, the rude beginnings of a capital and palace complex, and hints of a nascent bureaucracy (see J. B. Bury, trans., "Priscus at the Court of Attila," http://www.georgetown

There was something ironic in these developments. In a continent that had lacked horses before the Spanish reintroduced them, during the same period in which the Chinese Empire was breaking for all time the power of Eurasia's steppe nomads, Spanish and Indian communities were suddenly having to cope with North America's version of the Scythians and Huns of distant European memory. Groups like the Comanches were a spectacular anachronism, choosing the path of predatory plains nomadism in North America at just the historical moment when the ultimate military obsolescence of this way of warfare was being demonstrated in Asia. This does not seem to have troubled the eager members of Comanche war bands or consoled the lonely inhabitants of New Mexico ranches. It would take more than a century for North America to catch up with the Old World.

Spanish officials feared these armed and mounted Indians, and they feared westering Frenchmen, and the presence of each magnified fear of the other. With ships capable of reaching Peru and traversing the Pacific, and with troops stationed in Louisiana, Saint-Domingue, and trans-Pyrenean France, the French Empire represented a global threat to Spanish interests. French alliance with an Indian nation like the Comanches accentuated one manifestation of this general challenge because Comanche cooperation could give French forces a decisive advantage over sparse Texas and New Mexico garrisons. Mid-eighteenth-century Comanches and other trans-Mississippi Indians constituted a serious regional problem. The Comanches were not going to burn Madrid or Mexico City—at least, not in a future foreseeable from 1750—but they could terrify Santa Fe or San Antonio. Franco-Indian cooperation could, moreover, enhance the borderlands Indian challenge. French firearms made Indian raids more lethal and intimidating. Trans-Mississippi Indians like the Comanches could not project power outside the Southwest, but they might coordinate their razzias with a French Empire capable of doing so. The force of ten or fifteen thousand Comanches might be added not just to four thousand Louisianians but to "twenty million Frenchmen."[16]

Marginal as New Mexico and Texas were to the Spanish Empire, many miles from viceregal capitals and an ocean away from Madrid, Spain's metropolitan officials took reported borderlands dangers seriously. They directed frontier

---

.edu/faculty/jod/texts/priscus.html [accessed June 29, 2009]). I have not seen evidence that a mid-eighteenth-century Spanish traveler would have found the same among the early Comanches. The Huns of 448 were showing signs of becoming a rival empire; the Comanches of 1748 were not.

16. In the 1830s and 1840s, Comanche forces would launch devastating raids deep into northern Mexico. See especially Brian Delay, "Independent Indians and the U.S.-Mexican War," *AHR*, CXII (2007), 35–68. The Comanche population figure is from Hämäläinen, *Comanche Empire*, 66.

officials to monitor and impede French activities and rewarded them for doing so. In December 1753, Ensenada recommended to Viceroy Revilla Gigedo that Vélez Cachupín retain his current office. The minister of the Indies cited the governor's retaliations against the Comanches, his not allowing "the French . . . to trade . . . illicitly" in New Mexico "in the past year," and his "not permitting the French to establish themselves" in New Mexico "under any pretext." An earlier 1752 royal order had charged Revilla Gigedo himself "to be ever on the alert, and to exercise the most vigilant care, with respect to the operations of the French—in case that they may endeavor to expand into or to advance through the region of Texas, New Mexico and other provinces." With this instruction came news that the king had already sent four hundred muskets and bayonets to the viceroy and that he planned to send six thousand more. Revilla Gigedo was directed to distribute these weapons in the manner most conducive to frontier security. June 1755 royal instructions to the new viceroy of New Spain, the marqués de Las Amarillas, reiterated the need to be "very vigilant concerning the operations of that nation; and in case of the French planning to expand into Texas, New Mexico, or any other part of my dominions to notify them in order that they may restrain themselves." The same letter enjoined monitoring of "the distance of the French" from Spanish mining camps and of opportunities for French interference with Spanish commerce, "holding in mind that this was the principal impelling aim for [France] advancing into that province of Louisiana." Spanish frontier officers reporting to their superiors, imperial officials conferring among themselves, and diplomats rebuking their French counterparts were expressing the same grave concerns about the French menace to Spanish imperial security.[17]

17. For Ensenada's recommendation, see "The Marqués de La Ensenada to Revilla Gigedo, Madrid, Dec. 16, 1753," in Thomas, [ed. and trans.], *Plains Indians and New Mexico,* 145. For royal instructions to Revilla Gigedo, see "The King to the Viceroy, the Count of Revilla Gigedo," June 20, 1751, July 26, 1752, Dec. 16, 1753, in Charles Wilson Hackett, "Policy of the Spanish Crown Regarding French Encroachments from Louisiana, 1721-1762," in George P. Hammond, [ed.], *New Spain and the Anglo-American West: Historical Contributions Presented to Herbert Eugene Bolton,* I, *New Spain* (Lancaster, Pa., 1932), 107-146, esp. 123-124, 132-136 (note that Hackett provides two slightly different translations of the quotation). The letter to Amarillas went on to say, "It is fitting that you endeavor to thwart their plans for expansion in those regions [areas reachable from Louisiana], availing yourself of force or industry as your prudence dictates, never losing from sight such a grave affair." See "The King to the Viceroy, the Marquis of Las Amarillas, Aranjuez, June 30, 1755," in Hackett, "Policy of the Spanish Crown," in Hammond, [ed.], *New Spain and the Anglo-American West,* I, 107-145, esp. 136-139 (original in Julian de Arriaga, *Instrucciones que los vireyes de Nueva España dejaron a sus sucesores . . .* [Mexico City, 1867], 94-103, esp. 96-97).

French Pacific Threats

Spanish protests arose also from another, related issue, one Spanish diplomats omitted from conversations with the functionaries of a potentially hostile power. For the Spanish Empire, one danger was that Frenchmen coming out of the unexplored territory north and west of New Mexico would appear in greater numbers, perhaps as traders to dominate New Mexico's economy, perhaps as soldiers to conquer the province. Another was that hostile Comanches, reinforced by French weapons and French support, would launch attacks from these same regions so devastating as to make the colony unsustainable. To these dark visions, the unexplored West added a third: that French explorers would make their way through the mysterious lands beyond New Mexico and descend a river like the Colorado to the Gulf of California or the Pacific. Spanish officials knew of the French explorer-traders who had appeared in New Mexico, but these might represent only one component of the French western advance. Other French scouts might be bypassing Spanish settlements and moving toward the Pacific without Spaniards' hearing about it.

The clearest and most significant expression of these Spanish fears appears in a 1751 series of memoirs by Don Fernando Sánchez Salvador, a captain of the Sonora and Sinaloa cavalry. Sánchez Salvador argued that France was eagerly seeking the Pacific ["bǎ circulando anciosamente solicita de alcanzar el Mar de el Sur"]. He warned that, in their exploration of the mountains around New Mexico, French scouts might find and descend the Colorado or Gila's upper waters. Recalling a geographic notion discussed in Chapter 2, he contended that the Colorado River divided into two branches. One, as was well known, flowed into the Gulf of California. The other Sánchez Salvador called the Carmelo. He believed it emptied into the Pacific itself on the coast of upper California at about 36°N.

Sánchez Salvador described the first Mallet expedition to Santa Fe as one detachment of the larger French effort to reach the Pacific. The Mallets, in his view, were reconnoitering Spanish conquests while their comrades explored other territory. Sánchez Salvador also noted the more recent French New Mexico arrivals reported by Vélez Cachupín. Fearing the consequences of French exploration, Sánchez Salvador counseled defensive measures. He warned that a French Pacific coast establishment would be "very damaging to the Philippine trade" ["muy perjudicial al comercio de Philipinas"]. He recommended conquest of the Gila and Colorado Indians and construction along their rivers of presidios capable of obstructing French—or English or Dutch, for that matter—navigation. He also advised construction of a presidio at the Carmelo River's mouth, where it could assist returning Manila galleons and sight and report lurking foreign ships. Another fort on the Tres Marías Islands could prevent those islands' becoming a refuge for foreign vessels.[18]

The topic of Sánchez Salvador's memoirs makes them interesting; their reception makes them important. Sánchez Salvador wrote his memoirs in Mexico City, where he sat in council with figures like Viceroy Revilla Gigedo and the marqués de Altamira. The viceroy forwarded Sánchez Salvador's memoirs to Spain, and there, too, they aroused interest. In early 1752, at Ferdinand's order, José de Goyeneche, the *fiscal* (a kind of legal and general counsel) of the Council of the Indies, submitted Sánchez Salvador's memoirs to this governing body of the Spanish Empire. Goyeneche agreed that France sought a Pacific port and looked greedily upon the Colorado as a route to it. Because these French designs were of "so much importance" ["de tanta importancia"] and because their neglect was so potentially damaging, he averred that the viceroy of New Spain should "apply his principal attention" ["aplique su principal atencion"] to them, pursuing without hesitation all the measures necessary to people and fortify the Colorado as Sánchez Salvador had proposed. Goyeneche's warnings reached a receptive audience. The Council of the Indies reiterated in July 1752 claims about the danger of French parties from the east descending the Colorado and Carmelo and dominating the Pacific from Monterrey. King Ferdinand deemed Sánchez Salvador's proposals important enough to merit Ensenada's attention, and Ensenada found them significant enough to earn consideration from Carvajal. In summer 1753 dispatches to Ferdinand, Revilla Gigedo was still emphasizing Sánchez Salvador's proposals. What prevented their implementation was, not lack of imperial commitment, but rather a 1751 Pima Indian revolt. On hearing of this uprising, the Council of the Indies saw that Sonora faced a problem more immediate than canoe-borne French invaders, and it recommended that execution of Sánchez Salvador's ideas be delayed until after Pimería Alta had been pacified.[19]

18. Sánchez Salvador wrote four memoirs dated Mar. 2, 1751, and addressed to the king. A fifth, dated May 15, 1751, is addressed to the Junta de Guerra in Mexico City. For those referred to here, see AGI, Guadalajara 137, 34–67. Original documents summarizing the content of Sánchez Salvador's memoirs and other papers relating to them include "Extracto del expediente que trata de la reduccion a curatos de las missiones de las provincias de Sinaloa, y Sonora . . . ," June 15, 1752, AGI, Guadalajara 137, 299–332; Council of the Indies to the king, July 7, 1752, AGI, Guadalajara 419a, 65–117. For secondary discussions of these memoirs, see Charles Edward Chapman, *The Founding of Spanish California: The Northwestward Expansion of New Spain, 1687–1783* (New York, 1916), 16, 20–23, 30–44; Sylvia L. Hilton, "El límite noroccidental del imperio hispanoamericano, 1513–1784," in Iñigo Abbad y Lasierra, *Descripción de las costas de California*, ed. Hilton (Madrid, 1981), 15–59, esp. 26; Navarro García, *José de Gálvez*, 88–91.

19. For Goyeneche's warnings, see Jan. 16, 1752, AGI, Guadalajara 137, 80–83. On May 15, 1752,

Had Spanish officials known how difficult it was to reach the Pacific through the lands north of the Spanish frontier, French southwestern incursions would have appeared a serious regional problem with, because of Mexico's alluring silver mines, potentially imperial ramifications. As Spanish imperial historian Sylvia L. Hilton has noted, Spanish geographic uncertainty extended the implications of these encroachments, nurturing Spanish fears that a practicable water route to the Pacific might exist in North American lands beyond the horizon of Spanish familiarity. This pushed the implications of French inroads to the Pacific, with consequences potentially not just regional, or even imperial, but global. Geographic uncertainty gave French peddlers in canoes geopolitical significance comparable to that of British sailors in ships of the line.[20]

It also connected them to French activities in other parts of the Americas. Just as Spanish officials pondered simultaneously possible British voyages around Cape Horn and the obstinate British presence on the Mosquito Coast, they evaluated the dangers of French southwestern exploration within a context including contemporary French interlopers in Darien. A scattering of Frenchmen had inhabited the Darien coast since the pirate days of the late seventeenth century. These tended to be, depending on one's point of view, either unsavory types or colorful characters escaping the legal and fiscal impositions of other Caribbean lands. In the 1750s, rogues trying to evade the constraints of provincial officers and societies became a matter of concern for Europe's metropolitan governments. In September 1750, Spanish embassy secretary Abreu wrote from

---

Goyeneche reiterated his concerns. He spoke again of the "French who walk with intrepidity on our coasts" ["los Franceses, que caminan con intrepidez sobre n^ras costas"] and of the "serious damage" ["el grave perjuicio"] to New Spain that would be caused by an "advantageous post that advanced the French nation" ["puesto ventajoso, que adelante la Nacion Francesa"]. Again he thought the viceroy of New Spain should devote his primary attention to carrying out the projects proposed by Sánchez Salvador. See AGI, Guadalajara 137, 288–291. On the Council of the Indies' concern about French descents, see Council of the Indies to the king, July 7, 1752, AGI, Guadalajara 419a, 82v–83v. For Ferdinand and Ensenada's views, see Ensenada to José de Carvajal y Lancaster, Oct. 5, 1751, AGI, Guadalajara 137, 78–79. On Revilla Gigedo's continued emphasis, see Chapman, *Founding of Spanish California*, 62. On the Pima revolt, see Council of the Indies to the king, July 7, 1752, 116r–116v, Ferdinand VI, Goyeneche, and Council of the Indies to Revilla Gigedo, Oct. 4, 1752, 125v, and Council of the Indies, "Sobre el levantamiento de los Indios de la Pimeria Alta," July 19, 1753, 151v–152r, all in AGI, Guadalajara 419a. For a secondary discussion of the Pima revolt, see Russell Charles Ewing, "The Pima Uprising of 1751: A Study of Spanish-Indian Relations on the Frontier of New Spain," in Adele Ogden and Engel Sluiter, eds., *Greater America: Essays in Honor of Herbert Eugene Bolton* (Berkeley, Calif., 1945), 259–280.

20. Hilton mentions the way geographic ignorance heightened Spanish imperial insecurities in "El límite noroccidental del imperio hispanoamericano," in Abbad y Lasierra, *Descripción de las costas de California*, 26: "El desconocimiento de la geografía noroccidental del imperio inducía a las autoridades españolas a temer que los comerciantes y cazadores franceses operando en Luisiana podrían hallar y seguir algún río que les llevase hasta el Pacífico o el golfo de California."

London to Carvajal in Spain and Pignatelli in France that the British secretary of state for the Southern Department, the duke of Bedford, was reporting the dispatch of a French warship to the Gulf of Darien. Abreu's own investigation uncovered rumors that Darien's French settlers had sent a deputation to the governor of Saint-Domingue and were offering to accept the protection of the French government. Abreu took as obvious the undesirability of a French settlement or fort on the Isthmus of Panama and between Cartagena and Portobello. Pignatelli asked Puysieulx and Rouillé about the rumors; they vigorously denied them, and this allayed Pignatelli's concerns.[21]

The matter did not rest. In April 1751, Bedford told Wall of a Dutch report mentioning hundreds of French renegades in Darien. More significant, the report asserted that a French warship with all the necessary equipment to found a settlement was anchored in the Gulf of Darien and that three more ships and two thousand men were expected. Similar accounts of a thirty-two-gun French ship, of a French alliance with Darien's Indians, and of possible French fortifications came from the viceroy of New Granada and the governor of Panama. Ensenada was one of the figures reading these dispatches, and just as he was concerned about the challenge to Spanish imperial security posed by Britons in Honduras and Nicaragua, he worried also about the danger presented by the French in Darien. Ensenada relayed the recent Darien rumors to Pignatelli and directed him to ask Louis XV to prohibit French activities there and punish French subjects engaged in them. Pignatelli talked again to Puysieulx; Puysieulx passed a memoir from Rouillé to Pignatelli. Rouillé conceded that Darien's French and Indian inhabitants had asked French governor Hubert de Brienne, comte de Conflans, of the Leeward Islands to permit a French outpost in the area. Rouillé asserted, however, that royal orders forbade such establishments, and he denied that Conflans had granted the Darien settlers' request. Puysieulx declared that Conflans was ordered not to involve himself with Darien affairs and that French subjects conspiring with Darien's Indians would be punished. French responses again placated Spanish officials.[22]

21. On Darien and its European inhabitants, see Pares, *War and Trade*, 196, 552. For Abreu's report and reactions to it, see Francisco Pignatelli to Carvajal, Oct. 11, 1750, Estado 6487, Pignatelli to Ensenada, Sept. 14, 1750, Estado 6488, Félix de Abreu to Pignatelli, Sept. 7, 1750, Estado 6492, all in AHN; Pignatelli to Carvajal, Sept. 11, 1750, Estado 4509, Abreu to Carvajal, Sept. 10, 1750, Estado 6917, both in AGS.

22. Richard Wall to Carvajal, Apr. 22, 1751, Estado 6919, AGS; Hilton, "Las Indias en la diplomacia española," 631; May 17, 1751, Estado 6497, AHN; Pignatelli to marquis de Puysieulx, May 31, 1751, Puysieulx to Pignatelli, June 10, 1751, both in Estado 6496, AHN; Pignatelli to Carvajal, June 8, 1751, Estado 4512, AGS (copy in Estado 6497, AHN); "N. 2.° Copia de la carta de Mr Rouillé escrita en Versailles al marques de Puycieulx el 4. de Junio de 1751," "N. 3.° Copia de la carta de Mr Rouillé escrita en Versailles al marques de Puyzieulx el cinco de Junio de 1751" (copy in Estado 6496, AHN),

But alarms continued, and Spanish officials remained vigilant. The spring 1752 correspondence discussing the governor of Havana's report that 5,600 French troops might be bound for New Spain had mentioned Darien as another possible destination. Havana's governor was considered a gullible character, but Sebastián de Eslava, Julián de Arriaga, and Francisco Fernández Molinillo (members of the junta counseling Ferdinand on American affairs) nonetheless advised Ensenada that the viceroy of Santa Fe (de Bogotá), the governor of Cartagena, and the president of Panama should thwart French Darien projects. In another memoir, Eslava recommended improving Portobello's fortifications, because Darien's security depended on that city's invulnerability. A July 1752 royal order enjoined Spanish naval vessels to seek, harass, and seize French ships found off Darien. Spanish officials were monitoring Darien as carefully as they were New Mexico and Texas.[23]

In doing so, they discerned multiple instances of apparent or potential French aggression, instances they took to be manifestations of a larger design. Possible French moves toward the Pacific in Darien lent credence to the notion of a French advance on the Californias and vice versa. Like the multiple British Pacific probes alarming French diplomats in the same period, each case of apparent French expansionism made it more plausible to Spanish observers that other French actions were motivated by the same nefarious intent. Spanish officials did not think of French southwestern traders and Central American ruffians as independent actors but rather as executors of some obscure and sanctioned scheme. Viceroy Revilla Gigedo, writing in 1752 of the vastness of France's North American territories, of Frenchmen recently appearing in New Mexico, and of French troops and settlers arriving in Louisiana, concluded that French activities were grounds for suspecting "hidden plans" ["designios ocultos"] menacing Spain's lightly garrisoned frontier provinces. Altamira averred in 1753 that Chapuis and Feuilli's proposed trade with New Mexico was a pretext for "other hidden and more pernicious ends." In 1751, 1754, and 1755, in Madrid and Mexico City, figures such as Altamira, Amarillas, Eslava, Ensenada, Molinillo, and Arriaga felt that they were observing, not a collection of isolated events in Darien, Texas, and New Mexico, but rather a cohesive French effort to dominate Spanish trade, territory, and waters. French troops in Saint-Domingue might

both in AGS; Lucio Mijares Pérez, "Programa político para América del marqués de la Ensenada," *Revista de historia de América*, LXXXI (1976), 82–130, esp. 97–99; Pignatelli to Ensenada, June 14, July 12, 1751, Ensenada to Pignatelli, June 28, 1751, all in Estado 6497, AHN.

23. Arriaga, Eslava, and Molinillo to Ensenada, May 25, July 19, 1752, Estado 3882, expediente 1, nos. 26, 28; Mijares Pérez, "Programa político para América del marqués de la Ensenada," *Revista de historia de América*, LXXXI (1976), 82–130, esp. 96–97, 109.

proceed to Darien or New Spain. French traders from Louisiana and Canada might precede French troops to Texas and New Mexico. From Texas and New Mexico, French soldiers could reach northern Mexico's mines by land. By navigable rivers that might exist north and east of New Mexico, French troops could reach California, Sonora, and the Pacific. Through Darien, French soldiers could cross to the South Sea. Once in the Pacific, the French could trade with or terrorize Acapulco, Peru, and the shipping connecting them to Spain and the Philippines.[24]

These accumulating concerns generated Masones de Lima's early 1755 protest to French foreign minister Rouillé. The views expressed in internal Spanish foreign office and imperial administrative papers indicate that Masones's complaints constituted a sincere articulation of Spanish governmental sentiment, not a duplicitous diplomatic ploy fashioned to fend off French alliance overtures. The series of Spanish documents culminating in Masones's protest make this especially clear. In them, a succession of important Spanish governmental figures uses similar language to speak about the same subjects in a common tone. On November 2, 1754, the fiscal prepared a report on Chapuis and Feuilli. Using language reminiscent of American informants like Revilla Gigedo, Altamira, and Sánchez Salvador, he argued that Chapuis and Feuilli's "principal object" was, not illicit commerce, but rather to "explore and reconnoiter the vast territories of barbarous nations" lying between Spanish and French dominions. He wrote about the "vast ideas" of the French, how Chapuis and Feuilli's incursion was of "such damage" ["tanto perjuicio"] to the Spanish king's dominions, and how, for years, the French had "greedily aspired" to "the riches of his provinces."[25]

Similar language appears in a report from a November 27, 1754, Council of the Indies missive to King Ferdinand. It spoke of "the attempts which on different occasions the French have made to enter our dominions, not only through New Mexico but also through the province of Texas." The council suggested that Chapuis and Feuilli "had come to explore and reconnoiter with others the lands and dominions of your Majesty by availing themselves of the pretext of trade with the Spaniards." It recommended "that official correspondence should be

24. Revilla Gigedo quoted in June 22, 1752, Estado 3882, expediente 1, no. 31, Agustin de Las Cuevas to Revilla Gigedo, Jan. 24, 1753, expediente 1, no. 32, and Revilla Gigedo to Ensenada, June 22, 1752, expediente 2, no. 32, all in AHN; Altamira quoted in Bolton, "French Intrusions into New Mexico," in Bannon, ed., *Bolton and the Spanish Borderlands*, 168 (see 171 also); see also Bolton, *Texas in the Middle Eighteenth Century*, 72–73; Chapman, *Founding of Spanish California*, 43–44; John, *Storms Brewed*, 346; Mijares Pérez, "Programa político para América del marqués de la Ensenada," *Revista de historia de América*, LXXXI (1976), 108–109.

25. Fiscal to Council of the Indies, Nov. 2, 1754, AGI, Guadalajara 329, 325r–328v.

entered into by his Majesty with the Court of France, protesting with the great-est efficacy that the vassals of that crown are contravening the laws of these king-doms and the treaties of peace; that they are trying to disturb the good relations of the two courts by entering the dominions of your Majesty." It demanded pun-ishment of French Fort Chartres commander Saint Claire and greater French governmental control of unruly French subjects. Later descriptions of the coun-cil's report show that it was motivated also by Sánchez Salvador–inspired fears of a French descent of the Colorado.[26]

Coming then in rapid succession were a December 28, 1754, report from Arriaga to Wall reproducing the language of the preceding paragraphs; a Decem-ber 30, 1754, letter from Wall to Masones de Lima following the lead of Arriaga; and, finally, Masones de Lima's January 8, 1755, protest to Rouillé. Access to Spanish correspondence of the period, a privilege French officials lacked, dis-closes Spanish concerns underappreciated by French diplomats. Spanish offi-cials perceived not simply a British menace to the Spanish Empire but simulta-neous threats from France and Britain. French diplomats' failure to grasp this caused them to mishandle their overtures to the Spanish government, weaken-ing French appeals for an anti-British Franco-Spanish alliance.[27]

Effects of Spanish Geographic Ignorance and French Southwestern Threats on Spanish Foreign Policy

Simultaneous examination of Spanish concerns about French and British Pacific threats illuminates the considerations animating Spanish policy between 1750 and 1757. When French diplomats spoke of Anson's South Sea raid and sub-sequent rumored British Pacific missions, they revived Spanish officials' memo-

26. Council of the Indies to the king, Nov. 27, 1754, 331–341 (trans. from "Council of the Indies to His Majesty, Madrid, November 27, 1754," in Thomas, [ed. and trans.], *Plains Indians and New Mexico*, 82–89), and Council of the Indies, "Expediente sobre la aprehension que Dn Jacinto de Barrios . . . ," Oct. 22, 1756, 345–366, both in AGI, Guadalajara 329.

27. Arriaga noted the French official support for Chapuis and Feuilli's mission, and he recalled the Council of the Indies claim that on "various occasions" the French had "tried to introduce themselves into the dominions" of the Spanish king in New Mexico as well as Texas. He indicated that Chapuis and Feuilli came to New Mexico not only to initiate trade but also to "explore and re-connoiter with others the land and dominions" of the king. Speaking for the Council of the Indies, Arriaga recommended a protest to the French court and demanded the punishment of Saint Claire. See Arriaga to Wall, Dec. 28, 1754, Estado 4511, AGS. Wall, writing to Masones de Lima of the incur-sions of Chapuis and Feuilli, noted that this attempt to explore and open illegal trade with the Span-ish dominions, an enterprise "authorized by a French governor or commandant" in the region, con-travened "the treaties and the good harmony" prevailing between France and Spain. He asserted that it was not the "only time French Governors in America" had "committed similar infractions." See Wall to Masones de Lima, Dec. 30, 1754, Estado 4511, AGS (copy in Estado 6512, AHN).

ries of the horde of French ships dominating Peruvian and Chilean maritime commerce during the first two decades of the eighteenth century. When they mentioned British Central American log-cutters' proximity to the Pacific Ocean, they alerted Spanish officials to French Darien settlers' nearness to the South Sea. When they warned of British maritime aggression and of possible British advances on the Pacific, they reminded Spanish ministers of French western incursions and potential French descents of the Colorado.

Spanish geographic ignorance contributed to the belief that France and Britain constituted dangers of comparable magnitude. Uncertainty about western geography transformed a region appearing as a mountainous barrier on modern maps into a riparian highway in Spanish officials' imagination. As a result, French western exploration and trade contributed to a balance of fear in the minds of Spanish ministers. For these imperial officials, it appeared not just that British moves toward the Pacific or French conduct in the Southwest threatened Spanish imperial security but that the activities of both countries did so simultaneously. For each British menace, there arose a French equivalent. This raised the potential cost of assisting France against Britain. If the French Empire should achieve a decisive victory over the British in North America, then France might be able to direct more of its resources toward the conquest of Spanish America's northern borderlands and Pacific littoral. French conduct made France seem a likely Spanish enemy rather than a presumptive Spanish ally, and Spain therefore needed protection against both British and French aggression.

This reinforced the rationale underlying Spanish officials' pursuit of vigilant neutrality and superficial Anglo-Spanish amity. In fact, shrill French warnings about the magnitude of British naval power might have made such relations seem even more desirable. Carvajal recognized that Britain and France coveted Spanish imperial trade and territory. French moves toward the North American Southwest were one reason France belonged on the list of predators. Carvajal believed, however, that naval preeminence made the British Empire the foremost menace to Spain's overseas empire. Where Ensenada held that Spain could brandish French might to overawe British ministers, Carvajal thought the combined forces of France and Spain too weak to confront Britain's fleet. Rather than ally with the weaker maritime power against the stronger and become the target of British naval assaults, Carvajal deemed the maintenance of outwardly friendly relations with Britain more prudent. British statesmen hoping for peacetime gains could be counted on to forego overtly hostile acts against Spain, and British imperialists hoping to prey upon Spain's American possessions could be induced to employ British power to protect the Spanish Empire from Albion's Gallic rival. In short, hostile British naval power could be made temporarily useful. With notable prescience, Carvajal opined also that British merchants and ministers were, for the most part, more interested in Spanish trade than Spanish territory. An ephemeral British hegemony would likely cost Spain some commerce but few possessions. The French Empire, on the other hand, with its expansionistically inclined Louisiana colony looking out on Texas, New Mexico, and northern Mexico, directly threatened Spanish imperial territory. What Carvajal felt Spain needed, and what he hoped amiable Anglo-Spanish relations would give time to build, was a Spanish fleet powerful enough to defend the empire against Britain and France. In Spanish eyes, as French westward exploration and trade increased the risk of a Franco-Spanish alliance, French assertions regarding British Pacific designs raised the cost of antagonizing Britain. Allowing open Anglo-Spanish hostilities could give Britons the opportunity to grasp in war the Spanish Lake they could only eye in peace.[28]

Spanish officials wanted neither Britain nor France to dominate North America or Europe. They wanted each rival to check the other's power. Between 1748 and 1757, the threat of British incursions in the Pacific and French encroachments on the Spanish frontier made it seem less advisable for Spain to confront Britain and less desirable for Spain to support France. A neutrality giving the Spanish Empire time to develop the ability to defend itself seemed the logical policy.

Implications

What havock would a junction of two powerful crowns on ticklish American disputes have made upon poor Great Britain!

— BENJAMIN KEENE to the bishop of Chester, May 4, 1756

Spanish understandings of British and French Pacific threats shaped Spanish policy before and during the Seven Years' War's first phase. Had those Spanish interpretations differed, Spanish choices might have, also, and the momentous imperial struggle between Britain and France might have unfolded otherwise.[29]

28. Joseph de Carvajal y Lancaster, "Pensamientos de D. Joseph de Carvajal y Lancaster," in Antonio Bermejo de la Rica, *La colonia del Sacramento: Su origen, desenvolvimiento y vicisitudes de su historia* (Toledo, Spain, 1920) (see also 42–44); Hilton, "Las Indias en la diplomacia española," 630, 644–646; Jean O. McLachlan, *Trade and Peace with Old Spain, 1667–1750: A Study of the Influence of Commerce on Anglo-Spanish Diplomacy in the First Half of the Eighteenth Century* (Cambridge, 1940), 132–133. Wall concurred with this kind of reasoning, averring that Britain could do the "greatest damage" to Spain's American dominions and that "amistad" with Britain was therefore "most useful" for Spain. See Wall to Carvajal, June 9, 1749, Estado 6914, AGS.

29. For the epigraph, see B[enjamin] K[eene] to the bishop of Chester, May 4, 1756, in Lodge, ed., *Private Correspondence of Sir Benjamin Keene*, 473–475, esp. 474.

In the early 1750s, British victory in a Franco-British struggle appeared by no means inevitable, and British and French ministers shared the belief that Spanish assistance might tip the balance in their favor. Britain's triumphant 1763 circumstances can easily obscure its less favorable position a decade earlier. France enjoyed significant advantages compensating for Britain's naval superiority. Though less numerous, French ships were better built than those of the British navy. And although it had fewer ships than Britain, France possessed more soldiers: in 1756, the French army numbered 200,000 men, whereas only 30,000 British troops were stationed in Britain. Strategists on one side of the channel hoped, and on the other side feared, that an invasion by tens of thousands of Frenchmen might overwhelm English resistance in a way that Bonnie Prince Charlie and his thousands of Scots could not. In North America, the French colonies benefited from a unity of command the squabbling British colonies could only fear or envy—the British Empire was fortunate that France didn't simply take Albany while its famous congress was deliberating. The French Empire, moreover, dominated the interior Saint Lawrence and Mississippi waterways, disposed of Canadian subjects generally better versed in American warfare than the British colonists, and could appeal to a host of Indian allies whose skills might counterbalance Anglo-American numbers.[30]

Its many advantages brought France many victories during the first years of the Seven Years' War. In 1755, France and its Indian allies killed Braddock and destroyed his army; French ships eluded a British blockade and reached Louisbourg and Quebec. In 1756, La Galissonière seized Minorca, General Louis-Joseph de Montcalm took Oswego, and Siraj-ud-Daulah took Calcutta. In 1757, Fort William Henry surrendered, and Lord Loudoun abandoned his plan to attack Louisbourg; Britain's ally Frederick the Great, beset by Austrian, French, and Russian forces, contemplated annihilation. As Lawrence Henry Gipson has described it: "As October 1757 drew to a close, there seemed to be little doubt that the two wars in which France was involved were being won by this nation and its numerous allies." British prospects were so bad that William Pitt, who would devote his considerable energies to despoiling the Spanish Empire a few years later, was trying to secure Spanish assistance by offering to return Gibraltar and quit Honduras.[31]

30. Lawrence Henry Gipson, *The British Empire before the American Revolution*, VI, *The Great War for the Empire: The Years of Defeat, 1754–1757* (New York, 1946), 400–401.

31. For the quotation, see Gipson, *British Empire before the American Revolution*, VII, *The Great War for the Empire: The Victorious Years, 1758–1760* (New York, 1949), 125. For Pitt and Gibraltar, see Hilton, "Las Indias en la diplomacia española," 573; Pares, *War and Trade*, 561–562; Savelle, *Diplomatic History of the Canadian Boundary*, 88; Pitt to Benjamin Keene, Aug. 23, 1757, Keene to

He did not get this Spanish assistance, but, before 1761, neither did France, and this lingering Spanish neutrality contributed to Britain's early survival and ultimate success. Because Spain and its Neapolitan and Sardinian possessions stayed out of the war, Britain could more easily maintain a fleet east of Gibraltar, continue its trade with the Levant, and acquire information from British Mediterranean consuls. More fundamentally, Spanish neutrality meant that Britain confronted fewer, worse-funded enemies. In 1754, Spain possessed roughly forty-five ships of the line and nineteen frigates, with thirty more large vessels under construction. France counted some forty-five ships of the line and thirty frigates in 1755; Britain, about ninety and seventy. In addition to ships, the Spanish government had, exceptionally, money. As Spanish officials had hoped would be the case, while Spain stayed out of war, funds accumulated in the treasury. When Ferdinand VI died in 1759, the Spanish government enjoyed a surplus of six million pesos. Fortunately for Britain, during the uncertain early years of the war, its fleet had only to fight French ships, its treasury to outspend one Bourbon power. French and British statesmen acted on the belief that Spanish intervention could affect the war's outcome, but even with the benefit of hindsight, we cannot say with any certainty that the addition of sixty-four ships to France's initially victorious forces would have enabled it to win the war in the years before Pitt and his lieutenants were able to destroy France's navy and mobilize the British Empire's human and financial resources. The entry of Spanish forces in 1762, after years of French defeats and the demise of France's fleet, brought only new venues for British victories. We can only suggest that earlier Spanish support for France could have produced a substantially but unpredictably different outcome.[32]

Pitt, Sept. 26, 1757, in [William Stanhope Taylor and John Henry Pringle], eds., *Correspondence of William Pitt, Earl of Chatham,* 4 vols. (London, 1838–1840), I, 247–256, 263–277; Basil Williams, *The Life of William Pitt, Earl of Chatham,* 2 vols. (London, 1913), I, 339–340. Pitt was asking for Minorca as compensation for Gibraltar.

32. Vera Lee Brown, "The Spanish Court and Its Diplomatic Outlook after the Treaty of Paris, 1763," *Smith College Studies in History,* XV (1929–1930), 7–38, esp. 30; Julian S. Corbett, *England in the Seven Years' War: A Study in Combined Strategy,* 2 vols. (London, 1907), I, 21, II, 373; Gipson, *British Empire before the American Revolution,* VI, 401–402; Hilton, "Las Indias en la diplomacia española," 653–655; Lynch, *Bourbon Spain,* 157, 161, 167, 174, 178, 347; Pierre Muret, *La prépondérance anglaise (1715–1763)* (Paris, 1937), 465; Vicente Palacio Atard, *Las embajadas de Abreu y Fuentes en Londres, 1754–1761* (Valladolid, Spain, 1950), 17; Richard Middleton, *The Bells of Victory: The Pitt-Newcastle Ministry and the Conduct of the Seven Years' War, 1757–1762* (Cambridge, 1985), 31, 217; François Rousseau, *Règne de Charles III d'Espagne (1759–1788),* 2 vols. (Paris, 1907), I, 21–22; Williams, *Life of William Pitt,* I, 302; J. Leitch Wright, Jr., *Anglo-Spanish Rivalry in North America* (Athens, Ga., 1971), 101–102. Note that estimates of naval strength vary depending on source, month, and judgment of ship classification or condition.

This possibility raises questions about venerable interpretations of the British triumph over France in North America. In 1939, looking back at the Anglo-French struggle for the continent, Gipson saw a contest between "two contrasting and rival systems: the free competitive enterprise of English colonials . . . as against state planning that continually circumscribed the activities of those of France." Almost a century earlier, Francis Parkman had seen a battle between "Liberty and Absolutism, New England and New France. . . . Here [in New France] was a bold attempt to crush under the exactions of a grasping hierarchy, to stifle under the curbs and trappings of a feudal monarchy, a people compassed by influences of the wildest freedom." Both authors saw France's powerful and centralized government ultimately hindering the development of the French North American empire, vitiating its ability to compete with a less rigid, less regulated British Empire in which individual initiative flourished and colonies consequently prospered.[33]

In the 1750s, French diplomats sought the assistance of Spanish ships, soldiers, and silver to counterbalance the reputed advantages of English liberty. The reality of French individual initiative and the reputation of French governmental power frustrated this design. With the international situation so delicate and the imperial stakes so high, the last thing French statesmen wanted was French colonial officials and traders antagonizing the Spanish government by essaying illicit trade with New Mexico and Texas or French vagabonds alarming Spanish officials by settling in Darien. For all its power, the French imperial government could not contain all its nominal subjects. The force of central authority diminished somewhere between Versailles and Santa Fe. Spanish officials, however, convinced of Paris's long reach, believed the actions of French men in North America must reflect in some fashion the intentions of French ministers in Europe. The Gallic frontiersman's independence was as elusive or irrelevant to Spanish imperial understanding as it was confounding to French diplomacy. Had French absolutism limited the actions of France's trans-Mississippi subjects as effectually as its reputation suggested it could, France might not have alienated Spanish officials in the 1750s, and an earlier Bourbon alliance might have given English liberty a stiffer challenge.

33. Lawrence Henry Gipson, *The British Empire before the American Revolution*, IV, *Zones of International Friction: North America, South of the Great Lakes Region, 1748–1754* (New York, 1939), 6; Francis Parkman, *France and England in North America*, I, part 1, *Pioneers of France in the New World* (New York, 1983), I, 14.

PART V.

THE ELUSIVE WEST AND THE OUTCOME OF THE SEVEN YEARS' WAR

13

FRENCH GEOGRAPHIC CONCEPTIONS AND
THE 1762 WESTERN LOUISIANA CESSION

On November 3, 1762, in the waning days of the Seven Years' War, a beleaguered France ceded the trans-Mississippi remnants of the colony of Louisiana to Spain. This cession has always been something of an enigma; its necessity is not immediately obvious. Spain had not occupied western Louisiana, Britain had not conquered it, and neither was demanding it. Trans-Mississippi Louisiana remained, so far as European diplomacy was concerned, under French dominion, the last piece of a remarkable continental venture.

From 1524 to 1762, French scouts, missionaries, traders, officials, and settlers had tried to explore North America's territories, harvest its resources, cultivate its soils, and variously convert, contain, exploit, inveigle, ally with, and rule over a wide selection of its native peoples. Few in number, the French in America had been bold in conception and grand in achievement, traversing a good bit and claiming the better part of the continental interior. When, after losing Canada and yielding eastern Louisiana to Britain, France ceded its trans-Mississippi territories to Spain, it became — save for a brief Napoleonic interlude — an empire to be remembered in Quebec rather than reckoned with along the Mississippi. Even with Canada and eastern Louisiana gone, retaining trans-Mississippi Louisiana would have meant keeping much, and giving it away amounted to sacrificing a great deal. France had spent sixty-three years, thousands of lives, and tens of millions of livres on the colony. Louisiana's performance had disappointed, but its dimensions and potential were enormous. The 1762 Louisiana cession has always raised the question of why France relinquished so easily the last mainland piece of a North American empire it had been striving for two centuries to build.[1]

---

1. William R. Shepherd, in the 1904 *Political Science Quarterly*, argued that France displayed "the utmost indifference as to the fate of its American colony," seeing Louisiana as not merely "destitute of intrinsic value" but as an actual liability. He identified France's desire to maintain Spanish

The figure most responsible for the 1762 decision to cede Louisiana was France's leading minister, Étienne-François, duc de Choiseul. Choiseul served as minister of foreign affairs from 1758 to 1761, retained personal control of Spanish diplomacy after that date, and dominated French foreign policy until 1770. He was vigorous, able, and, as the strong servant of a weak king, played a critical role in shaping the policy of his government. Choiseul walked into a crisis when he stepped into office. The Seven Years' War had begun with a series of French victories, but Frederick the Great's tenacity, Williams Pitt's direction, and Britain's seemingly inexhaustible financial and naval resources were, as war dragged on, overwhelming France. Choiseul needed to find a way, if not to gain French victory, at least to minimize the costs of defeat. He was well suited to the task. A protégé of Louis XV's influential mistress Madame de Pompadour, Choiseul, unlike many court favorites, evinced talents well beyond the art of pleasing the powerful. He possessed a quick and supple mind and could not only grasp the

---

"subservience" as the key motive for the cession. Shepherd relied almost exclusively on Spanish archival documents, records insufficient by themselves for an investigation of French thinking about Louisiana. He did not explain why, if France cared so little about Louisiana, it had established and maintained a colony there at all. Arthur S. Aiton used French documents for a 1931 *American Historical Review* essay and concluded that the argument that Louisiana "was regarded as worse than valueless by both France and Spain" was "not substantiated by the record." It was only the desperate military and financial situation at the end of the Seven Years' War that drove the French government to part with the colony. Aiton averred that French minister the duc de Choiseul offered New Orleans and trans-Mississippi Louisiana to Spain as compensation for Spain's surrender of Florida to Britain, hoping this would induce Spain to sign an otherwise unfavorable and humiliating peace treaty. Although Shepherd and Aiton's articles provide solid answers to the questions of why France had to yield something to Spain and why that cession took place on November 3, 1762, their explanations as to why trans-Mississippi Louisiana was the item given are less persuasive. They took into account too little of the colony's territory and too few years of its existence. A complete understanding of the cession requires that we go farther west in space and farther back in time than they did. See William R. Shepherd, "The Cession of Louisiana to Spain," *PSQ*, XIX (1904), 439–458, esp. 439, 454, 457; Arthur S. Aiton, "The Diplomacy of the Louisiana Cession," *AHR*, XXXVI (1931), 701–720, esp. 719–720; and see E. Wilson Lyon, *Louisiana in French Diplomacy, 1759–1804* (Norman, Okla., 1974); Zenab Esmat Rashed, *The Peace of Paris, 1763* (Liverpool, U.K., 1951).

For the most insightful and best documented discussion of the Louisiana cession — and, indeed, the most impressive treatment of mid-eighteenth-century French colonial policy as a whole — see Pierre Henri Boulle's remarkable 1968 dissertation on "The French Colonies and the Reform of Their Administration during and Following the Seven Years' War" (Ph.D. diss., University of California, Berkeley), esp. 565, 574–577. Boulle saw the origins of the Louisiana cession in Choiseul's desire to make the French Empire more compact and defensible. I agree with Boulle's idea of a French strategic withdrawal from North America and his emphasis on elimination of colonial burdens. In this chapter, I will supplement his argument by exploring the changing French geographic notions making 1762 Louisiana seem an unnecessary or even an undesirable possession. I differ from Boulle in placing greater weight on the risks Choiseul was willing to run in pursuit of territories he deemed potentially valuable, such as Guiana and the Falkland Islands, and in highlighting the importance for French policy of trans-Mississippi Louisiana's diminishing allure.

magnitude of France's wartime difficulties but also devise and revise a variety of strategies, like ceding western Louisiana to Spain, to address them.[2]

One way to understand why Choiseul and France gave up trans-Mississippi Louisiana and to appreciate the role French conceptions of the uncharted American West played in the decision is to examine the records preserved in the Archives des affaires étrangères. Choiseul played his cards close to the vest, and his private papers remain unavailable, but foreign office letters and memoirs passing between him and French representatives in Spain and England disclose considerations shaping wartime French decisions. Comparing foreign office documents from the late 1740s to the early 1760s with those from earlier decades reveals a conceptual transformation. From the first seventeenth-century French forays west of the Great Lakes and Mississippi River to the 1762 loss of Louisiana, metropolitan French officials remained—despite the best efforts and considerable achievements of French explorers, missionaries, and traders—unsure of the true character of most of western North America. A high degree of geographic uncertainty persisted; approaches to this nescience and the regions it involved changed. Between 1748 and 1762, French foreign ministry officials became increasingly skeptical about the potential value of undiscovered western North America. The interaction of four factors contributed to this growing incredulity. After the 1748 Treaty of Aix-la-Chapelle, intensifying Anglo-French imperial rivalry forced the French foreign office to rigorously reassess the value of France's colonies. At the same time, French explorers remained unsuccessful in their search for western features deemed essential for the region's utility, and scholarly debate was undermining the credibility of French geographers who had posited these features' existence. These developments coincided with a general movement in mid-eighteenth-century French intellectual life toward a more critical and empirically demanding attitude regarding a range of received notions about the human and natural world. The combination of these factors sapped earlier French notions about the importance of the lands beyond the Mississippi.[3]

2. See Guy Chaussinand-Nogaret, *Choiseul (1719–1785): Naissance de la gauche* ([Paris], 1998); Jacques Levron, *Choiseul, un sceptique au pouvoir* ([Paris], 1976); Roger H. Soltau, *The Duke of Choiseul: The Lothian Essay, 1908* (Oxford, 1909).

3. On the unforthcoming Choiseul and his papers, see Boulle, "French Colonies and the Reform of Their Administration," v–vi; Boulle, "Some Eighteenth-Century French Views on Louisiana," in John Francis McDermott, ed., *Frenchmen and French Ways in the Mississippi Valley* (Urbana, Ill., 1969), 15–27, esp. 27; H. M. Scott, "Religion and Realpolitik: The Duc de Choiseul, the Bourbon Family Compact, and the Attack on the Society of Jesus, 1758–1775," *IHR*, XXV (2003), 37–62, esp. 45–46.

PLATE 3. Louis-Michel van Loo, *Portrait of Etienne-François, Duke of Choiseul*. N.d.
Oil on canvas. The Bridgeman Art Library

Changing Assessments of Louisiana, 1748–1762

The war now kindled in America, has incited us to survey and delineate the immense
wastes of the western continent by stronger motives than mere science or curiosity could
ever have supplied.
— SAMUEL JOHNSON, "Review of Lewis Evans," 1756

From about 1748 to 1762, a combination of circumstances altered the way French
diplomats thought about the unknown American West and lowered the standing of trans-Mississippi Louisiana. Sharpening Anglo-French rivalry after the
War of the Austrian Succession was one of them. Descrying the storm clouds
of a momentous Anglo-French struggle, French officials had to start deciding
which of their imperial possessions they could shelter under the inadequate um-
brella of French financial and military resources. They had to find ways to rank
the importance of their colonies and other imperial objectives, and this forced a
more exacting assessment of individual colonies' worth than had been necessary
in the comparatively tranquil decades after the Treaty of Utrecht. Such concerns
became more acute as the Seven Years' War unfolded and a succession of losses
forced French officials to consider which territories might have to be yielded to
other powers at the negotiating table.[4]

Indeed, it is tempting to assert that, after debacles like the loss of Canada,
the French position was so dire that French officials simply gave up on or lost
interest in the Mississippi colony now so completely exposed to British arms. It
is true that the seizure of Quebec in 1759 and the surrender of Canada in 1760
shook the French government's commitment to retaining Louisiana. After news
of these events had reached France, foreign office documents discussing the
colony increasingly acknowledged the crumbling of French North American
power and pondered the implications of this disintegration for the empire's remaining continental territories. Many authors—including, on occasion, Choiseul—had argued and continued to contend that the threat of Canadian raids
had pinned British forces in the Northeast and that Louisiana would be too isolated and underpopulated to defend itself when France's Canadian defeats allowed British troops to move south. Louisiana's advocates might aver, as they
had earlier in the century, that Franco-Indian alliances would compensate for
Anglo-American numerical and naval superiority, but this was a much harder
case to make when a century and a half's painstaking cultivation of Franco-

4. For epigraph, see Samuel Johnson, "Review of Lewis Evans, Analysis of a General Map of the
Middle British Colonies in America, 1756," in Donald J. Greene, ed., *The Yale Edition of the Works of
Samuel Johnson, X, Political Writings* (New Haven, Conn., 1977), 200.

Indian relations had failed to provide Canada with adequate protection against British invaders. Louisiana's survivability and desirability had fallen far enough in French estimation by summer 1761 for Choiseul to propose exchanging it for a Spanish loan or an early Spanish entrance into the war. For reasons discussed in the next chapter, Spain refused the offer.[5]

The colony's status rose and fell with events outside and the mood within French bureaus and palaces, however, and it would be too simple to say that Quebec and Montreal's demise sealed the fate of New Orleans and western Louisiana. Even with Canada lost, and notwithstanding Choiseul's summer 1761 dangling of Louisiana before Spain, many and powerful French officials urged the colony's retention. Some denied its defense required Canada. Some saw the Saint Lawrence colony as an onerous burden contrasting with an inexhaustibly rich Louisiana. Some felt transfer of Canada's population to Louisiana would invigorate and entrench the Mississippi colony. Some thought the current war's peace settlement or the next conflict's fortunes might restore Canada to France. The French ambassador in Spain, Pierre-Paul, marquis d'Ossun, although accepting that Canada's fall heightened Louisiana's vulnerability, recommended succoring Louisiana rather than ceding it. Choiseul himself exhibited recurring attachment to the colony. In summer 1760, he rebuffed Charles III (1759–1788) of Spain's interest in acquiring the province. In December 1761, with the knowledge that the formation in August 1761 of a Franco-Spanish alliance meant that Spanish help was on the way, and in a mood of optimism regarding French naval recovery, Choiseul declared it "certain that this colony merits a closer attention than has been accorded it up to the present. . . . When circumstances permit, I shall neglect none of the advantages that a colony so useful can produce. . . . We intend to send aid to Louisiana." Louisiana remained, in the eyes of many of those writing for the French foreign office, a large and fertile country enjoying an agreeable climate and great agricultural potential. Erratic but vigorous French inclinations to keep it remained evident even with Canada and its *"arpentes de neiges"* gone. The question of why France ceded its trans-Mississippi territories despite such leanings remains in need of an answer. To find one, it is necessary to

5. On Louisiana's indefensibility without Canada, see duc de Choiseul to marquis d'Aubeterre, June 19, 1759, 64r–65r, Choiseul to marquis d'Ossun, Oct. 29, 1759, 344r–346r, both in CP, Espagne, 515; "Interêts de la France par rapport a l'Amerique ou vües qu'on doit avoir en negotiant la paix pour ce qui concerne nos colonies," Apr. 15, 1761, 92v–93r, letter to Choiseul, ". . . si la possession du Canada est bien importante pour la France . . . ," May 2, 1761, 143r–143v, both in MD, Amérique, 25; "Apuntaciones originales con su copia, hechas por el duque de Choiseul para hablar en el consejo sobre la contra-memoria de la Inglaterra," July or August 1761, Estado 4544, AGS; Charles-Edwards O'Neill, "L'autodétermination: L'idée révolutionnaire de la Louisiane en 1768," *Revue d'histoire diplomatique*, CVIII (1994), 3–25, esp. 3.

consider how post-1748 statesmen conceptualized western Louisiana and North America.[6]

One consequence of the interwar and wartime stresses forcing French diplomats to reflect upon the relative and long-term value of their colonies was a more rigorous way of talking about western American territories. From 1712 to 1747, French foreign ministry memoirs had speculated openly and confidently about attributes of the undiscovered American West and had used these conjectural characteristics as arguments for Louisiana's importance. From the late 1740s, authors increasingly shunned such speculative reasoning.

A good example can be seen in the influential 1750 / 1751 memoir on the French North American colonies by Canadian governor and Anglo-French boundary commission member the comte de La Galissonière. La Galissonière declared he would forego the conjectures earlier authors had indulged in: "The uncertain future products both of Canada and Louisiana will not be adduced as arguments, although the expectation of them is based on an immense country, a great people, fertile lands, forests, quarries, and mines already discovered." In

6. On the possible return of Canada, see Ossun to Choiseul, Nov. 29, 1761, CP, Espagne, 534, 218r–223v; "Reflexions sur quelques points du memoire historique de la negociation de la France avec l'Angleterre pendant l'année 1761: Et moyens de fixer les desseins ambitieux des Anglois en 1762," July 9, 1762, CP, Espagne, 536, 413v–414v, 422r; letter to Bussy, "Mémoire sur les limites à donner à la Loüisiane du côté des colonies anglois et du côte du Canada, en cas cession de ce dernier pays," Aug. 10, 1761, CP, Angleterre, 444, 150r–159r; "Observations topographiques sur l'intérêt de l'Espagne dans la présente négociation relativement à ses possessions dans l'Amérique septentrionale," August 1762, CP, Angleterre, 446, 311v, 314v–318r; "Memoire pour la transmigration proposée du Canada à la Loue," April 1761, 49r–59r, "Sur la Loüisiane," April 1761, 69r–73v, "Moyen de peupler la Louisiane: Encouragements à donner aux habitants du Canada, pour passer au Missisipy," June 1761, 74r–78r, all in CP, États-Unis, supplément 6; "Extraits et observations," 1761, MD, Amérique, 25, 202r–208v; "Observations sur les preliminaires de la paix," Sept. 2, 1762, MD, Angleterre, 48, 20r–28r.

Might France have ceded trans-Mississippi Louisiana to Spain with the hope of recovering the colony in the next war? Choiseul and Ossun's 1762 correspondence reveals no such intent, nor do their 1763–1770 letters discussing Louisiana. For more on this matter, see Pierre H. Boulle, "French Reactions to the Louisiana Revolution of 1768," in John Francis McDermott, ed., The French in the Mississippi Valley (Urbana, Ill., 1965), 143–158, esp. 156–157; Allan Christelow, "French Interest in the Spanish Empire during the Ministry of the Duc de Choiseul, 1759–1771," HAHR, XXI (1941), 515–537, esp. 530; Christelow, "Proposals for a French Company for Spanish Louisiana, 1763-1764," MVHR, XXVII (1941), 603–611, esp. 604–605; Mildred Stahl Fletcher, "Louisiana as a Factor in French Diplomacy from 1763 to 1800," MVHR, XVII (1930), 367–376, esp. 375; O'Neill, "L'auto-détermination," Revue d'histoire diplomatique, CVIII (1994), 4.

For Ossun's suggestion of succor, see Ossun to Choiseul, Oct. 23, 1760, CP, Espagne, 530, 118r; Ossun to Choiseul, Nov. 29, 1761, CP, Espagne, 534, 218r–223v. For Choiseul's rebuff of Charles, see Choiseul to Ossun, July 15, 1760, CP, Espagne, 529, 74r. Choiseul quotation a translation from Choiseul to Ossun, Dec. 15, 1761, CP, Espagne, 534, 304v–305r, quoted in Aiton, "Diplomacy of the Louisiana Cession," AHR, XXXVI (1931), 711–712; see also "Extrait d'une lettre écrite par M. le duc de Choiseul à M. le mis d'Ossun le 25. Decembre 1761," Estado 4547, AGS.

the 1710s and 1720s, authors had stood one step shy of entering the income from undiscovered silver lodes in their account books. In La Galissonière's memoir, "nothing is said of the mines that it is claimed have been discovered in this district [the Illinois country]. Independently of the fact that we are not sufficiently informed, we should not think of them until the district has been sufficiently developed in men, wheat, and cattle." Rather than surmise, La Galissonière preferred to await additional empirical evidence. His attempt to omit speculation did not prevent him from wanting to retain Louisiana; he considered both it and Canada strategically important. Nor, in fact, did it prevent him from speculating. He waxed poetic about the profits to be drawn from the buffalo of the Illinois country and the plains beyond. But his declared intent to avoid conjecturing about the value of the imperfectly known North American West and the significant instances in which he followed this intent mark a departure from earlier habits of thought.[7]

Other 1750s and early 1760s writings about Louisiana display similar attitudes. An anonymous circa 1752 report requested and received by French ambassador to Spain the duc de Duras furnishes a good example. In its discussion of Louisiana, the paper observed that partisans of the colony had praised and were still praising it to the skies, making of it a second Potosí. All Louisiana's inhabitants were guilty of the most "extraordinary exaggerations," and "all their exaggerations" were not "founded on anything." The author foresaw, nonetheless, a bright future for the colony because of its economic potential and immense, underpopulated territories. Unlike earlier measured voices, however, this author based his optimistic predictions primarily on products such as silk, meat, timber, and beaver pelts coming from known Louisiana regions rather than on the projected fruits of far-off lands. He emphasized Louisiana trade with proximate Texas rather than distant New Mexico. Like La Galissonière, the author talked of Louisiana's value but tried to exclude fanciful speculations about unknown regions. Many other foreign office authors lauded Louisiana's value and advocated its retention. But they supported their positive views by citing products of the colony's more familiar eastern regions rather than by offering conjectures about the unexplored West.[8]

7. "Sur les colonies de la France, dans l'Amerique-septentrionale," appears in 1751, 110r–144v, and Dec. 2, 1750, 21r–38r, both in MD, Amérique, 24. Translation from "Memoir of La Galissonière," 1751, in Theodore Calvin Pease, ed., *Anglo-French Boundary Disputes in the West, 1749–1763* (Springfield, Ill., 1936), 5–22, esp. 9, 18–20.

8. Quotations from "Description de l'Amerique meridional tant des côtes de la mer du Nord que d'une partie de celle du Sud . . . ," MD, Amérique, 2, 70v, 106r–107v, 111r, 113v–126r. The estimated date comes from a past-tense reference in the memo to 1751, the duc de Duras's 1752–1755 term as ambassador to Spain, and new ambassadors' habit of soliciting memoirs early in their

Quickening Anglo-French imperial competition also changed foreign office approaches to New Mexico. With war looming, securing a Spanish alliance against Britain became more important than acquiring Spanish silver from New Mexico. Consequently, French diplomats generally showed more concern about contraband trade offending Spanish officials and less interest in western trade routes making Louisiana profitable. This change of priorities led to the recommendations against westward Louisiana expansion and in favor of restrictions of Louisiana trade expeditions discussed in the previous chapter. It contributed also to the 1753 suggestion of Louis XV's advisor Noailles to Duras that Louisiana's wealth should rest on fertile soil rather than contraband commerce and to Duras's 1754 observation that Spanish officials were "furiously jealous" of French North American possessions and frequently querulous about French territorial usurpations. Where earlier reports emphasized commercial opportunities in a supposedly wealthy and accessible New Mexico despite Spanish prohibitions, most post-1748 memoirs and instructions saw those Spanish interdictions precluding traffic. Where earlier memoirs had highlighted the value of western Indians as partners in French operations against Spanish territories, later foreign office efforts were frequently devoted to trying to allay Spanish fears of Franco-Indian partnership. New Mexico's imaginary wealth remained enticing. It was, for the moment, however, at least theoretically out of reach, since diplomatic considerations and ministerial cautions militated against French traders and raiders pursuing it.[9]

tenure to inform themselves about their new hosts. The unnamed author had previously corresponded with Maurepas. Rather than buffalo, his imaginative indulgence was thousands of western Indians waiting to join hundreds of French American hunters in an invasion of Mexico—an atypical and less-than-immediately useful vision at a time when French diplomats were increasingly trying to foster cooperative Franco-Spanish relations. For the other memoirs, see Ossun to Choiseul, Nov. 29, 1761, CP, Espagne, 534, 218r–223v; M. de Marolles to Choiseul, "Reflexions sur quelques points du memoire historique," July 9, 1762, CP, Espagne, 536, 413v–414v, 420r–420v, 422r; "Memoire pour la transmigration proposée du Canada à la Loue," April 1761, 49r–59r, "Sur la Loüisiane," April 1761, 69r–73v, "Moyen de peupler la Louisiane: Encouragements à donner aux habitants du Canada, pour passer au Mississipy," June 1761, 74r–78r, all in CP, États-Unis, supplément 6; letter to Bussy, "Mémoire sur les limites à donner à la Loüisiane," Aug. 10, 1761, CP, Angleterre, 444, 150r–159r; "Observations topographiques," August 1762, CP, Angleterre, 446, 311v, 314v–318r; "Extraits et observations," 1761, CP, Amérique, 25, 202r–208v; "Observations sur les preliminaires de la paix," Sept. 2, 1762, CP, Angleterre, 48, 20r–28r.

9. Antoine-Louis Rouillé to Philippe de Rigault de Vaudreuil, Oct. 2, 1750, AC, B 91, 405r–405v; Rouillé to Louis Billouart de Kerlérec and Vincent Guillaume Le Sénéchal d'Auberville, Mar. 2, 1755, AC, B 101, 245r–245v; comte de La Galissonière, "Memoire sur les colonies de la France dans l'Amerique septentrionale," December 1750, MD, Amérique, 2, 26v–27r; duc de Noailles to duc de Duras, Nov. 30, 1753, CP, Espagne, 511, 233v–234r ("Mon Sisteme particulier n'est pas D'etablir la richesse, ni la puissance de cette Colonie sur un commerce de contrebande avec l'Espagne, mais sur la culture et la production Des terres que sont trés fertiles"); Duras to Noailles, January

In part because of this official and diplomatic New Mexican inaccessibility, 1748–1762 foreign ministry documents displayed less enthusiasm about Louisiana's proximity to the Spanish colony. Earlier memoirs stressed Louisiana's positive worth as a route to a fabulously rich New Mexico and gave secondary consideration to Louisiana's negative value as a barrier between the Spanish colony and British invaders. Post-1748 memoirs and correspondence generally reversed the emphasis. In a 1753 memoir, minister of marine Rouillé wrote of Louisiana as "the most sure barrier against the designs of the English on the possessions of Spain in New Mexico" ["la barriere la plus sûre contre les vües des Anglois sur les possessions d'Espagne au nouveau Mexique"]. He did not elaborate on the New Mexican assets this barrier would protect, nor did he suggest France could benefit from them itself rather than merely prevent Britain from doing so. Other authors described Louisiana as an obstacle in the same fashion. In fact, on numerous occasions in this later period, authors such as Noailles, La Galissonière, minister of foreign affairs Puysieulx, and Rouillé spoke about Louisiana as the shield of the Spanish possessions in old rather than New Mexico, subtly shifting the direction of French policy away from protection of the undiscovered West and toward the safety of a better-known Spanish colony. Louisiana remained important to France because of its position near New Mexico and Mexico, but this significance now arose more pressingly from the fearful logic of diplomacy and less immediately from the seductive lure of silver.[10]

Along with increasingly acute Anglo-French imperial rivalry, the limited reach of French North American reconnaissance was vitiating foreign office perceptions of the potential value of the unexplored West. French scouts' failure to reach the continent's western slope or to locate a great river descending it to the

1754, CP, Espagne, 511, 198r. See also "Mémoire pour servir d'instruction au sieur Maréchal duc de Noailles . . . ," Mar. 30, 1746, 265–269, "Mémoire du maréchal de Noailles pour demander les ordres du rois sur la commission sont S. M. l'a chargé auprès du roi d'Espagne," Mar. 27, 1746, 269–275, both in A. Morel-Fatio and H. Léonardon, eds., *Recueil des instructions données aux ambassadeurs et ministres de France depuis les traités de Westphalie jusqu'à la Révolution française*, XII, *Espagne*, III, *1722–1795* (Paris, 1899), 265–269, 269–275.

10. For Rouillé, see his "Observations sur les nouvelles venües d'Espagne au sujet des preparatifs qui s'y sont faits pour l'Amerique," Dec. 21, 1753, CP, Espagne, 514, 511v. For other authors' view of Louisiana as obstacle, see "Differens entre le roi et le roi d'Ange. en Amere. Afrique . . . et sur l'Interruption de la common. et celle de la negoton.," August 1754, MD, Amérique, 24, 177r–177v; Noailles to Duras, Dec. 31, 1753, CP, Espagne, 511, 244r; Abbé François-Joachim de Pierre de Bernis to Aubeterre, Aug. 29, 1758, CP, Espagne, 523, 439r; "Observations topographiques," August 1762, MD, Amérique, 24, 317r–318r. For the shift to old Mexico, see marquis de Puysieulx to Vaulgrenant, June 10, 1749, CP, Espagne, 503, 46r; Noailles to Duras, June 3, 1753, CP, Espagne, 511, 141r; Rouillé to Duras, Feb. 25, 1755, CP, Espagne, 517, 153v–154r; Ossun to Choiseul, Nov. 29, 1761, CP, Espagne, 534, 219r; Galissonière, "Memoire sur les colonies de la France dans l'Amerique septentrionale," MD, Amérique, 2, 26v, 33v.

Pacific fed growing doubts about the existence of the Sea and River of the West, at least in the form envisioned earlier during the seventeenth and eighteenth centuries. The utility of the River of the West had always depended on its being relatively easy to get to, on the far side of a surmountable height of land, for instance. But as French scouts like the La Vérendryes and Saint-Pierre pushed farther west, their reports and other French geographic writings were mentioning "lofty mountains" ["montagnes fort hautes"] rather than gentle slopes. Spirited believers in North America's porousness would press their case into the nineteenth century—one would become president of the United States and launch an expedition to test his theories. Mid-eighteenth-century French officials could not yet know that the Rockies would surpass anything visible from Monticello. But they could begin to wonder.[11]

While the course of French reconnaissance was casting doubt on imagined western geographic features, exploration and scholarship were simultaneously undermining the credibility of French geographers advocating their existence. In 1753, British geographer John Green launched a series of attacks on French scholars Joseph-Nicholas Delisle and Phillippe Buache, in particular on the way they had used the spurious account of Admiral Bartholomew Fonte's 1640 voyage in support of their ideas about a Northwest Passage. A document that appears to have entered the French foreign ministry papers in 1755 contained Green's contentions. He accused Buache and Delisle of citing Fonte as evidence even though they recognized the account's unreliability. He charged them with employing the Fonte voyage to substantiate the notion of a Western Sea, even though the Fonte document said nothing about such a body of water. He arraigned Buache and Delisle with deliberately misplacing at latitude 63° north on their maps lakes and rivers located ten degrees farther south in the Fonte document. Their intention, he claimed, was to render the existence of these watercourses compatible with reported Russian discoveries. Green also accused another French geographer, J. N. Bellin (*ingénieur de la marine et du Dépôt des cartes et plans de la marine*), of using unverified longitudinal figures, of placing

11. Quotations from "Abridged Memorandum Respecting the Map Which Represents the Establishments and Discoveries Made by the Sieur de La Vérendrye and His Sons," [1749], and "Report, April, 1750," in Lawrence J. Burpee, ed., *Journals and Letters of Pierre Gaultier de Varennes de La Vérendrye and His Sons: With Correspondence between the Governors of Canada and the French Court, Touching the Search for the Western Sea,* trans. [W. D. LeSueur] (Toronto, 1927), 483–488, 489–492, esp. 487, 490–492. See also "1757: Memoir of Bougainville," in Reuben Gold Thwaites, ed., State Historical Society of Wisconsin, *Collections,* XVIII, *The French Regime in Wisconsin,* 3 vols. (Madison, Wis., 1908), III, 167–195, esp. 189; "Brief Report or Journal of the Expedition of Jacques Legardeur de Saint-Pierre . . . ," in Joseph L. Peyser, ed. and trans., *Jacques Legardeur de Saint-Pierre: Officer, Gentleman, Entrepreneur* (East Lansing, Mich., 1996), 180–191, esp. 189.

too much faith in antiquated geographic ideas, and of ignoring recent evidence from ships' logs contradicting his theories. Bellin, though himself a target of Green's censure, seconded Green's criticism of those who took the Fonte account seriously. Cartographers and geographers in England, Spain, Russia, and France followed these arguments and joined in the animadversions on Buache and Delisle's work. Green's attacks pertained to French views of regions northwest of Louisiana, but his disparagement of eminent French figures implicitly impugned the French geographic community as a whole.[12]

Green's criticisms regarding French maps of northwestern North America and the cavils of other figures concerning charts of other parts of the continent reverberated not only within the republic of letters but also in ministries of state. The presence, among the foreign office papers, of a French version of Green's polemic provides one indication of this, as ministry personnel generally collected, translated, composed, and transmitted memoirs because they perceived their utility for those directing and conducting French diplomacy. Attentiveness to geographic disagreements was also to be expected from the nature of foreign-office work. When pondering North American policy, French diplomats often consulted geographers and their maps of the continent. Noailles, Duras, and the secretary of the French embassy in Spain, the abbé de Frischmann, spoke of referring to J. B. Bourguignon d'Anville's and J. N. Bellin's respective 1746 and 1755 maps of North America and Bellin's 1744 and 1750 charts of Louisiana. Choiseul claimed to have solicited geographers' advice when formulating his ideas about Louisiana's boundaries. Only a careless diplomat would ignore the geographic part of geopolitics, and cartographic contentions should therefore have weighed on policy decisions.[13]

More telling than documentary presence and logical expectations are the contemporary cartographic dissatisfactions French diplomats themselves were

12. John Green, "Remarques à l'occasion de la nouvelle carte marine de l'Amerique septentrionale et meridionale," "Remarques à l'occasion de la nouvelle carte marine de l'Amerique," MD, Amérique, 22, 62r–68r. Estimate of 1755 date of entry based on January and May 1755 dates on bracketing documents of similiar paper and size. The previous document also discusses problems with French cartography: duc de Mirepoix to Rouillé, "Differences entre les cartes francses. et angses. de l'Am. septale," Jan. 6, 1755, 60r. Accounts of Green and his criticisms appear in Lucie Lagarde, "Le passage du Nord-Ouest et la mer de l'Ouest dans la cartographie française du 18e siècle, contribution à l'etude de l'oeuvre des Delisle et Buache," *Imago mundi*, XLI (1989), 19–43, esp. 30–38; and Glyn[dwr] Williams, *Voyages of Delusion: The Northwest Passage in the Age of Reason* (London, 2002), 249–258.

13. Duras to Noailles, Dec. 21, 1753, CP, Espagne, 512, 188r; Abbé de Frischmann to Rouillé, Feb. 10, 1756, CP, Espagne, 519, 96r–96v; Noailles to Duras, June 3, 1753, 36r, "Examen du projet de faire passer les habitans du Canada à la Louisianne," Feb. 8, 1759, 39v (Bellin drew the maps in Charlevoix's frequently mentioned 1744 description of New France), both in CP, États-Unis, supplément 6; for Choiseul, see marquis de Grimaldi to Richard Wall, Sept. 13, 1761, Estado 4545, AGS.

MAP 37. Jean-Baptiste Bourguignon d'Anville, *Amérique septentrionale.* 1746.
Special Collections, University of Virginia Library

uttering and acknowledging. Noailles alluded in November 1753 to the incompleteness and imperfections of French North American maps. Ambassador to England Charles Pierre Gaston François de Lévis, duc de Mirepoix (1749–1755), declared in a January 1755 dispatch to Rouillé that British secretary of state Thomas Robinson (1754–1755) "told me our French maps and their English maps differed entirely as to the location and course of the Ohio or Beautiful River; that the error amounted to more than three hundred leagues, and that our French maps even differed from each other." (This document appeared in the foreign ministry collections directly before Green's previously cited attack on Delisle, Buache, and Bellin.) An August 1761 memoir to the French minister plenipotentiary in London, François de Bussy, observed that the Appalachians "are ill represented on ordinary maps" and that mistaken ideas about them had been gained from "inexact maps which show these mountains other than they are." Some of these cartographic complaints can be imputed to the usual diplomatic sparring and the habitual frustrations of Europe-bound officials trying to comprehend a large and distant world. Nonetheless, at a time when recognizable geographic accuracy was critical, the foreign office was seeing the validity of French cartographic images of many parts of North America come increasingly into question. When the experts quarreled, which maps and cartographers should French diplomats trust? When even the Appalachians defied accurate

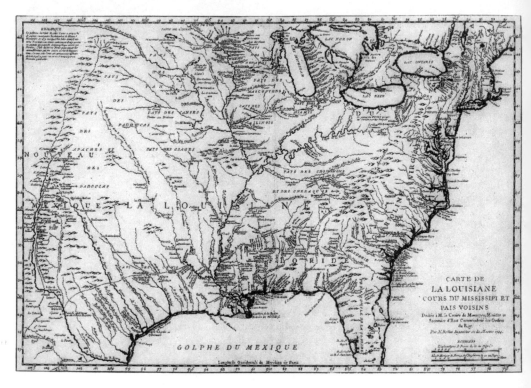

MAP 38. Jacques-Nicolas Bellin, *Carte de la Louisiane, cours du Mississippi et pais voisins*. 1744.
Library and Archives Canada, NMC 6528

representation, how could French statesmen credit old conjectures about the trans-Mississippi West's South Sea routes and ore-bearing mountains?[14]

Growing geographical skepticism can also be seen, paradoxically, in what foreign office documents increasingly failed to discuss. Early-eighteenth-century documents spoke frequently about a Sea of the West and a River of the West leading to it and the South Sea. Foreign ministry reports and correspondence from 1748 to 1762 omit discussion of these subjects. La Galissonière and Rouillé assessed or oversaw French efforts to find the Sea and River of the West, but they left these western geographic features out of their foreign office geopolitical ruminations. Noailles not only failed to mention the fabled waterways but

14. Noailles to Duras, Nov. 30, 1753, CP, Espagne, 511, 234v–235r; Mirepoix to Rouillé, Jan. 16, 1755, MD, Amérique, 22, 60r (trans. from Pease, ed., *Anglo-French Boundary Disputes*, 87; Pease used a CP, Angleterre, 438 copy of the document). See also "Mémoire en forme d'éxtrait tiré des ouvrages Anglois qui paroissent actuellement sur les colonies de l'Amérique septentrionale . . . ," "Joint a la lettre de M. Rouillé a M. le duc de Mirepoix du 24 May 1755," CP, Angleterre, 439, 116v. For Bussy, see Pease, ed., *Anglo-French Boundary Disputes*, 354–355, trans. of "Mémoire sur les limites à donner à la Loüisianne," Aug. 10, 1761, CP, Angleterre, 444, 156v–157v.

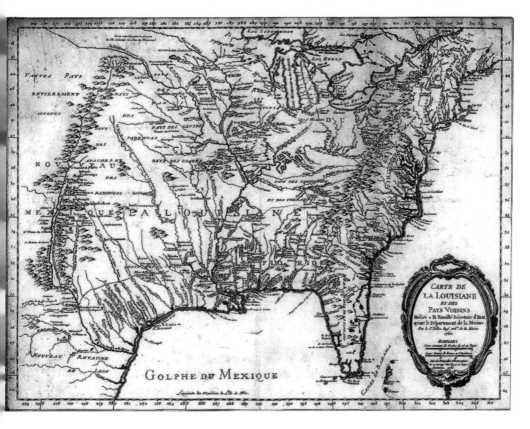

MAP 39. Jacques-Nicolas Bellin, *Carte de la Louisiane et des pays voisins*. 1750.
Library and Archives Canada, NMC 54995

also told Duras that they "should not try at all to establish ourselves on any river that falls into Spanish possessions" ["Je pense que nous ne devons point chercher à nous etablir sur aucune Riviere qui tombe Dans les possessions Espagnols"]. Given Spain's pretended monopoly of South Sea navigation, this statement could easily be interpreted as excluding the French Empire from any North American watercourse on the far side of the Continental Divide.[15]

15. Maurepas to Galissonière, Mar. 1, 6, 1748, AC, B 87, 217r–217v, 219v; Rouillé to marquis de La Jonquière, May 4, 1749, AC, B 89, 266r–266v; Rouillé to La Jonquière, Apr. 15, 1750, AC, B 91, 247r–247v; Rouillé to Michel-Ange Duquesne de Menneville, June 1, 1754, AC, B 99, 208r–208v; "Extrait des registres de l'Académie royale des sciences du 6. Septembre 1752," Service hydrographique, 3 JJ 67, 7, Archives nationales, Paris; "Lettre de Delisle à la Galissoniere sur la decouverte d'un passage dans la mer du Sud," July 14, 1750, letter 1, Service hydrographique, 3 JJ 68, section 22; La Jonquière and François Bigot to Rouillé, Oct. 20, 1750, C 11 A, 95, 91r, La Jonquière to Rouillé, Feb. 27, Oct. 8, 1750, C 11 A, 95, 130r, 276v, Duquesne to Rouillé, Oct. 26, 31, 1753, C 11 A, 99, 106r–106v, 120r, and Jean Jacques Macarty Mactigue to Rouillé, May 20, 1753, C 13 A, 37, 190r–190v,

The silence of foreign ministry figures regarding Sea and River of the West Pacific connections contrasts with their overt interest in the South Sea and approaches to it. As discussed in Chapters 10 and 11, Rouillé, Puysieulx, and Duras had repeatedly spoken of their concerns about British South Sea encroachments. French officials had pondered British attempts to find a Northwest Passage from Hudson Bay. If taken seriously, the possible existence of routes to the Pacific through lands west of Louisiana would have been highly relevant to French diplomats interested in South Sea issues. Omission of the Sea and River of the West from post-1748 foreign office documents reinforces the conclusion that these particular imagined western North American waterways had become a matter of secondary interest in ministry geopolitical contemplations. They remained important topics in mid-eighteenth-century France, but they were no longer finding their way into the international assessments shaping French foreign policy. Previously central aspects of the unexplored West had become marginal concerns, scholarly issues rather than strategic considerations.

This silence continued into the Seven Years' War's final phase and was once again especially striking in contexts where the supposed existence of North American routes to the Pacific would have been especially pertinent. Although discussion of a Hudson Bay Northwest Passage stopped appearing in foreign ministry documents after the mid-1750s, discussion of British maritime designs continued. Ossun argued that Britain had formed under Cromwell a design to achieve commercial dominance and political despotism and that one could regard it as *"certain"* that England intended to conquer western America—or at least to place it in a state of "servile dependence"—by controlling the Gulf of Mexico's entrance and exit and by arranging "a secure and short communication with the Pacific" through Panama. Discussion of Louisiana's value was a regular feature of Ossun's correspondence, and if he had thought that Louisiana afforded access to the Pacific, one would expect him to mention it here in his discussion of communication with the Pacific through Central America.[16]

A similar example appears in a July 1762 memoir *ingenieur du roi* M. de Marolles sent to Choiseul. Marolles elaborated on a long-term British project to attain universal monarchy through domination of Europe's commerce with the Indies. Control of "capital points" ["points capitaux"] would serve as a means

---

all in Documents des colonies, Archives nationales, Paris. For the Noailles quotation, see Noailles to Duras, Nov. 30, 1753, CP, Espagne, 511, 233v–234r.

16. "Ainsi l'on peut regarder comme certain que l'Angleterre se propose de faire la Conquêtte de toute l'Amerique Occidentalle, ou au moins de la mettre dans une dépendance servile en se rendant Maitresse de l'Entrée et de la Sortie du Golphe du Mexique, et en se ménageant une communication seure et courte avec la mer du Sud par celuy de Panama" (Ossun to Choiseul, Nov. 29, 1761, CP, Espagne, 534, 218r–223v).

to this end. These capital points were sites enabling their possessor to dominate trade and territory by opening or obstructing navigation. Gibraltar and Port Mahon provide good examples. Marolles envisaged renewed British attacks on American capital points such as Cartagena, Portobello, and Panama. Conquest of them would allow Britain to fall upon Spain's Chilean, Peruvian, and Mexican Pacific territories. Marolles also discussed Canada and Louisiana. Canada's constellation of portages and riparian and lacustrine settlements dominated northern river traffic and the interior route to Louisiana, and Marolles designated the northern French colony a capital point. He left Louisiana off the list. If Marolles had thought Louisiana offered a practicable route to the Pacific, the colony would have constituted territory perhaps as strategically important as Panama and more geopolitically significant than Canada. He wrote nothing of the possibility. As a decision about Louisiana's fate drew near, key foreign office territorial evaluations were excluding its Pacific possibilities.[17]

French diplomatic conduct seems to bear this out. In October 1762, Choiseul and Louis XV, frustrated by Spanish reluctance to accept British terms for a peace settlement, appear to have offered trans-Mississippi Louisiana to Britain. It is hard to imagine Choiseul doing this if he had thought western Louisiana afforded easy Pacific access. Seven Years' War defeats had convinced him that he must focus, after the conflict, on making French naval power competitive with Britain's. In keeping with this theme, he was arguing in July 1762 that North American territorial losses would harm France neither immediately nor in the future, provided France wasted no time in rebuilding its navy and preparing its defenses for the next Anglo-French conflict. Choiseul saw this revived and enhanced French naval power as necessary to counter what he had spoken of in the past as a long-term British design to dominate Europe through wealth and power obtained from British supremacy at sea and in the Americas ["Le Projet des Anglois depuis le commencement du Siecle . . . est de maintenir sur le Continent une balance de Puissance, et d'emporter totalement cette balance sur Mer, et en particulier en Amerique. Les Anglois savent que ce sont les Colonies en Amerique, l'Empire de la Mer et du Commerce qui doivent donner la Loi en Europe"].[18]

17. Marolles to Choiseul, "Reflexions sur quelques points du memoire historique," July 9, 1762, CP, Espagne, 536, 413v–414v, 420r–420v, 422r. For a similar example discussing Louisiana's value and the danger of Britain's reaching the Pacific through Central America, while leaving out the possibility of attaining the South Sea from Louisiana, see "Observations topographiques," August 1762, CP, Angleterre, 446, 311v, 314v–318r.

18. For Choiseul's postwar focus on French naval power, see Boulle, "French Colonies and the Reform of Their Administration," 583; Jean-Étienne Martin-Allanic, Bougainville navigateur et les découvertes de son temps, 2 vols. (Paris, 1964), I, 6, 70–71. For his 1762 assessment of territorial losses,

After the war, Choiseul's efforts to foster French Pacific commerce and forestall a British South Sea presence would demonstrate the place of the far side of the world in his maritime ambitions. Sometime between November 3, 1762, and February 10, 1763, Choiseul met with French officer Louis-Antoine de Bougainville and received his proposal to found a Falkland Islands settlement to assist French vessels. Distant islands like the Falklands would be difficult to defend against the British navy, and, knowing of Britain's 1749 interest in them, Choiseul could see that France might someday have to. He nonetheless endorsed Bougainville's plan. Spanish protests forced the withdrawal of Bougainville's Falklands colony, but Choiseul remained interested in the southern seas. In 1766, he received reports that Britain had formed its own establishment either in the Falklands or on the other side of Cape Horn. He advised the Spanish government to destroy the outpost if it lay in the Pacific, and he promised French support in whatever crisis ensued. Choiseul warned the British government that a Falklands settlement could lead to war. In the same year, Choiseul had Ossun suggest the Spanish government cede the Philippines to France, and he expressed interest in a joint Franco-Spanish Philippines trading company. Choiseul repeatedly displayed an inclination to get France into the Pacific and keep Britain out of it. Given these goals, offering western Louisiana to Britain would have made no sense if Choiseul had believed it easy to pole upstream from New Orleans, portage the Continental Divide, and drift with the current to the Sea of the West and California.[19]

---

see copy of dispatch from Choiseul to Ossun, July 22, 1762, Estado 4547, AGS. For the quotation, see Choiseul to Conseil du Roi, July 14, 1759, MD, Angleterre, 54, 119r–119v.

19. On Choiseul and Bougainville, see Martin-Allanic, *Bougainville navigateur*, I, 6–8, 70–72, 83–85, 101; Julius Goebel, *The Struggle for the Falkland Islands: A Study in Legal and Diplomatic History* (1927; rpt. New Haven, Conn., 1982), 193–202, 225–228, 224; Chaussinand-Nogaret, *Choiseul*, 198. Bougainville wanted also to search for rumored southern oceanic bodies of land. On Choiseul's pugnacious rhetoric concerning the Falklands, see Choiseul to Grimaldi, Oct. 2, 1766, 152r–155r, Choiseul to Ossun, Nov. 25, 1766, 345r–346r, both in CP, Espagne, 547; Choiseul to François-Michel Durand de Distroff, Sept. 15, 1766, CP, Angleterre, 471, 136r–144r. In John Fraser Ramsey, *Anglo-French Relations, 1763–1770: A Study of Choiseul's Foreign Policy*, Publications in History, XVII (Berkeley, Calif., 1939), 143–263, esp. 175–176, Ramsey offers a different interpretation of these events. He argues that Choiseul knew the British post lay in the Falklands and that his counsel to the Spanish government to attack a Pacific establishment therefore ran no risk of drawing France into hostilities Choiseul hoped, in reality, to avoid. Ramsey based this interpretation on a September 2, 1766, letter from French chargé d'affaires in London François-Michel Durand de Distroff reporting the dispatch of a British squadron to the Falklands. It is not clear, however, that this information precluded the possibility of a British Pacific presence. More to the point, in a September 15 letter to Durand, Choiseul avowed his uncertainty regarding the precise position of the British southern island outpost and thus whether it lay in the Atlantic or Pacific (see also Ramsey, 181). At least in his own mind, Choiseul's counsel to the Spaniards could have entailed hostilities. At any

Choiseul does not seem to have believed that unexplored western Louisiana would provide a water route to the Pacific. When meeting with Choiseul after November 3, 1762, Bougainville had spoken not only of the Falklands but also of establishing a French outpost north of California and reinforcing Louisiana from it ["former un établissement au Nord de la Californie et les vues du ministre se dirigeront à établir une communication entre cet établissment et la Louisiane"]. (The Louisiana cession was kept secret from the public for several years, and Bougainville does not seem to have known that, after November 3, no French Louisiana remained to reinforce; Bougainville was keeping his proposal secret, and Choiseul does not appear to have been aware before the Louisiana cession of this particular account of American geography.) Choiseul withheld approval of the California-to-Louisiana component of Bougainville's plan, and his discussions with Bougainville reveal no indication that Choiseul believed in the existence of a Sea or River of the West useful to French navigation. Postwar correspondence between Choiseul and the French ambassadors in Britain and Spain sustains this view. When British commodore John Byron sailed into the Pacific in 1764 on a voyage the British government described as a search for a Northwest Passage, Choiseul exhibited interest in British South Sea activities but no apparent fear that a passage existed for Byron to locate.[20]

Foreign ministry documents from late in the Seven Years' War also disclose a departure from previous attitudes concerning Louisiana's vast extent. Earlier

rate, Choiseul's bellicose advice to the Spanish government affirmed the principle that a British Pacific presence justified an immediate military response.

On Choiseul and the Philippines, see Ossun to Choiseul, Sept. 8, 1766, CP, Espagne, 547, 38v–39r; Choiseul to Ossun, June 9, 1767, CP, Espagne, 549, 37r. The Spanish government rejected both of his suggestions.

20. On Choiseul and Bougainville, see Martin-Allanic, *Bougainville navigateur*, I, 6–8, 10–11, 70–72, 83–85, 101. The case of French naval officer and North American traveler Jean-Bernard Bossu resembles Bougainville's. Bougainville served in America in the 1750s, Bossu in the 1750s and early 1760s. Bougainville contemplated a water connection between Louisiana and California, Bossu a land connection between Asia and America. Both seem to have stood outside the circle of pre–Louisiana cession foreign office deliberations. Bossu also provides an interesting point of comparison with La Galissonière's somewhat ambivalent approach to conjecture. Bossu might avow that "only discovery" of an Asian-American land connection could "prove its existence," that "conjecture proved nothing," but he could not resist discussing reports that led him to think a land bridge's reality "likely." His treatment of migrating Asian elephants, roaming Welsh princes, and northwest American–northeast Asian trade places him among those still attached to, even if slightly embarrassed by, American wonders. See Jean-Bernard Bossu, *Travels in the Interior of North America, 1751–1762*, ed. and trans. Seymour Feiler (Norman, Okla., 1962), 104–105, 209–215. For Choiseul's postwar letters, see correspondence between Choiseul and Ossun, June–August 1766, CP, Espagne, 546, 134r–134v, 183r–184r, 307r–307v, 324r–329v, 355r–357r; correspondence between Choiseul, comte de Guerchy, and Durand, June–July 1766, CP, Angleterre 470, 123r–123v, 157r–159r, 163r–165v, 176r–177r, 183r–184v, 203r–203v, 212r–212v; Sept. 15, 1766, CP, Angleterre, 471, 136r–144r.

reports had generally described Louisiana's great size as a positive attribute, both in and of itself and because it gave France access to more of the resources these unknown lands might hold. Some later memoirs continued to laud Louisiana for these reasons. Between 1759 and 1762, however, documents indicative of the most powerful French figures' opinion increasingly depicted size as a neutral, or even a negative, colonial feature. An August 10, 1761, memoir sent to guide French minister plenipotentiary Bussy in his negotiations with Britain provides a clear expression of this sentiment:

> For the object of the belligerents in America in the present war is in no wise mere extent of territory but rather the fruits which may be enjoyed from the lands over which they contend.
>
> . . . If it [that which renders Canada and Louisiana valuable to France] is taken into proper account, the extent of the territory ceded, however great, becomes a matter of indifference. Possibly it may even be beneficial. Advantages concentrated in a small area seem preferable to the same advantages dispersed over a larger area.
>
> . . . The produce of the land and not its extent induces a desire for its possession to the eyes of the wise.[21]

Others shared this author's sentiment. Bougainville had argued in February 1759 that extent of land or a large population failed by themselves to establish a state as a great power. An anonymous spring 1761 memoir made a similar argument in stating, "It is not the extent of lands that makes the power of a kingdom, it is their fruitfulness, and the number of their peoples" ["Ce n'est pas l'etendüe des terres qui fait la puissance d'un Royaume, c'est leur fecondité, et le nombre des peuples."]. Choiseul declared, "I do not think as they formerly used here that it is necessary to have many colonies. . . . It pertains to the perfection of the constitution of the kingdom to have enough American possessions to supply its needs in that sort [commodities not produced in France, such as coffee, sugar, and indigo]. But it should have no more than suffices for its needs." Because the practically infinite western reach of Louisiana's territory formed one of the colony's most salient characteristics, dismissing the importance of immensity depreciated the distant territories that had once figured so prominently in Louisiana assessments, reducing the colony's appeal. Other reasons

---

21. Translation of "Mémoire sur les limites à donner à la Loüisianne," CP, Angleterre, 444, 150r–150v, 158v, from "Memoir on the Boundaries of Louisiana," Aug. 10, 1761, in Pease, ed., *Anglo-French Boundary Disputes*, 345–358, esp. 345–346, 358.

remained for retaining Louisiana, but the colony counted less when weighed against other possessions and considerations.[22]

In forming their opinions about Louisiana and the undiscovered territories west of it, French statesmen might additionally have felt the influence not only of specific geographic disputes and particular geopolitical conceptions but also of a more general evolution of European intellectual attitudes. Salon conversations, lecture hall revelations, pointed essays, and learned tomes might have subtly shaped the deliberations occurring within and the policy recommendations emanating from the foreign office. One way to assess this possibility is to consider the changes in eighteenth-century habits of mind identified by intellectual historians and compare them to tendencies visible in foreign ministry documents.

Scholarly views of the role of theoretical and empirical reasoning provide a good starting point. Intellectual historians such as Ernst Cassirer, Peter Gay, and Daniel Mornet have described changing patterns of thought among educated eighteenth-century French men and women, one aspect of which was the growth of Isaac Newton and John Locke's influence at the expense of René Descartes's. From roughly the mid-seventeenth century, Descartes's approaches and ideas had enjoyed tremendous prestige in France, subtly directing many individuals' outlook. Like Euclidean geometry, Cartesian reasoning began with accepted principles from which it moved to logical deductions. It tried to erect a philosophical system with internal rigor sufficient to escape the uncertainties and errors arising inevitably from imperfect human perceptions. Reasoned conclusions from certain axioms would yield true statements about the external world.

Newton, Locke, and their followers emphasized a different approach. They stressed acquiring empirical data through observation and collection, then applying inductive reasoning to this material to discern underlying principles. From the analysis of empirical facts would come the construction of theories. Institutions such as the Académie des sciences and publicists such as Voltaire disseminated versions of these Newtonian and Lockean ideas. Eighteenth-century thinkers and later historians surveying their activities have seen the

22. For Bougainville's view, see his "Premiere mémoire où l'on traite la question, s'il convient on non d'abandonner le Canada?" Feb. 8, 1759, MD, Amérique, 24, 260v: "Il ne suffit pas pour la puissance des Etats d'avoir une grand etendüe de terres, même d'avoir un grand nombre d'habitans; il faut de plus avoir dequoi les mettre en oeuvre, et dequoi défrayer leur mouvement. Un Etat sans commerce et sans richesses est une espece de terre morte." For the 1761 memoir, see CP, États-Unis, supplément 6, 66v–67r (dating from position in a group of 1761 memoirs). For the Choiseul quotation, see Choiseul to bailli de Solar, May 28, 1762 (trans. from Pease, ed., *Anglo-French Boundary Disputes*, 431–432).

years around the middle of the eighteenth century as the period in which the Lockean and Newtonian empirical emphasis became a recognizably widespread and powerful feature of elite French thought. Voltaire averred that the empirically minded philosophes came to dominate public opinion around the middle of the century. Denis Diderot opined in 1754 that he was observing an intellectual turning point and a scientific revolution. Jean le Rond d'Alembert wrote in 1759 of the change in ideas and method becoming evident at the century's midpoint. Étienne Bonnot de Condillac argued in 1749 that philosophers should concentrate on collecting phenomena and should avoid the "mania for systems" that had misled earlier thinkers. George-Louis Leclerc de Buffon's observation, experience, and description-emphasizing *Natural History* appeared in the same year. On the basis of examples like these, Mornet contends that Descartes's influence on French thought had diminished by 1740 and that a preference for empirically rather than theoretically derived knowledge had become commonplace by about 1750. Gay mentions 1738 to 1756. Cassirer speaks of the middle of the century as the moment of triumph for the position that received notions required empirical confirmation, and empirical investigation formed the necessary foundation of general ideas. If these are valid generalizations, and if larger intellectual trends shaped the thought of midcentury foreign ministry personnel, then projected but unverified western features like the silver lodes logically lying in mountains extending north from Mexico and the great river presumably draining territories west of the Continental Divide should have lost at least part of their appeal.[23]

The existence of alluring western features had rested, however, not simply on deductions but also on the testimony of European geographers and travelers, western Indians, and westering Frenchmen. Such reports had often been imprecise, incomplete, and indirect—and perhaps less than entirely honest— but in many cases they had at least derived from people who had actually seen some part of North America, unlike numerous figures sketching and pondering it in comfortable Parisian bureaus. In assessing ideas about western geography, more was involved than simply weighing empirical versus theoretical reasoning. Urgent also were the questions of what constituted credible, reliable, and sufficient information. Who was to be believed? What made knowledge? Under-

23. Ernst Cassirer, *The Philosophy of the Enlightenment*, trans. Fritz C. A. Koelln and James P. Pettegrove (Princeton, N.J., 1951), vii, 3, 6–9, 12, 22, 28, 46–47, 52, 54–55, 74, 77–81; Peter Gay, *The Enlightenment: An Interpretation*, II, *The Science of Freedom* (1966; rpt. New York, 1977), 83, 136, 138–139, 151, 165; Daniel Mornet, *La pensée française au XVIIIe siècle*, 5th ed. (Paris, 1938), 87–98; Roy Porter, *The Enlightenment*, 2d ed. (Hampshire, U.K., 2001), 2. For detailed consideration of the example of Buffon, see Jacques Roger, *Buffon: A Life in Natural History*, ed. L. Pearce Williams, trans. Sarah Lucille Bonnefoi (Ithaca, N.Y., 1997), 65, 68, 73–76, 82, 91–92, 99.

lying these questions was the complexity of ostensibly simple concepts like observation, data collection, and empirical information. The eighteenth century might have lacked the bewildering insights of quantum physics, but it did have Jonathan Swift, David Hume, and Voltaire to remind it to look critically at what it was being told. Earlier chapters mentioned responses to these kinds of epistemological challenges in early-eighteenth-century French cartographic and geodetic endeavors in France, Russia, China, and South America as well as in Cook, Vancouver, and Lewis and Clark's later explorations. These responses included meticulous record keeping; generation, collection, and comparison of multiple observations; employment of mechanical measuring devices; relation of data to comprehensive classification schemes; critique and verification of venerable and developing propositions. With the exception of Bering's efforts on the fringes of Alaska, this spirit and these techniques, however, had not reached mid-eighteenth-century western North America.

This is one reason why Lorraine Daston and Katharine Park's discussion of eighteenth-century epistemological issues in *Wonders and the Order of Nature, 1150–1750* is so apposite. Daston and Park discuss the long evolution of European attitudes about "wonder and wonders," "prodigies," and "marvels." They write of a growing reluctance among elite eighteenth-century Europeans to credit accounts of wonders and marvels, in part because accepting such things came to be seen as a mark of commonness. Even while treating a topic distant from the torturous operation of eighteenth-century diplomacy, their book conveys the tenor of changing foreign office treatment of the undiscovered West. They argue that marvels' declining esteem in eighteenth-century "European high culture" had "less to do with some triumph of rationality . . . than with a profound mutation in the self-definition of intellectuals. For them wonder and wonders became simply vulgar, the very antithesis of what it meant to be an *homme de lumières,* or for that matter a member of any elite." "Leading Enlightenment intellectuals," Daston and Park inform us, "did not so much debunk marvels as ignore them. On metaphysical, aesthetic, and political grounds, they excluded wonders from the realm of the possible, the seemly, and the safe." A kind of raised-nose skepticism was coming to characterize the cultivated.[24]

French diplomatic personnel were no strangers to Enlightenment fashions. Less lastingly famous and intellectually accomplished for the most part than

24. Lorraine Daston and Katharine Park, *Wonders and the Order of Nature, 1150–1750* (New York, 1998), 18–19, 329–331, 343–350, 361. For an article making a similar point regarding eighteenth-century French perceptions of Pacific peoples, see Tom Ryan, "'Le Président des Terres Australes': Charles de Brosses and the French Enlightenment Beginnings of Oceanic Anthropology," *Journal of Pacific History,* XXXVII (2002), 157–186.

the celebrated philosophes, mid-eighteenth-century French diplomats inhabited nonetheless the same Enlightenment world Cassirer, Mornet, Gay, Park, and Daston evoke. They attended savant-graced salons and read the Enlightenment works scholars peruse still. Some foreign-office functionaries hoped not simply to study the works of intellectuals but to compose them. Scholars have remarked that *l'esprit des lois* sat upon many French functionaries' desks, and who could say the pen advising Choiseul today might not supplant Montesquieu tomorrow? Men of letters and men of state interacted and overlapped. Choiseul exchanged enough letters with Voltaire to fill a volume. Jean-Jacques Rousseau served in the mid-1740s as French ambassadorial secretary in Venice and asserted Choiseul's later interest in offering him another position. Channels existed through which intellectual developments taking place outside the foreign office could influence thinking within it.[25]

More to the point, foreign ministry documents exhibit the kinds of changes one would expect to find if those exterior events were making themselves felt. A disdain for speculative reasoning and a tendency to omit mention of unproved ideas becomes evident in the foreign office documents around 1748, a development compatible with, and indeed supportive of, the generalizations of historians such as Cassirer, Gay, and Mornet. Foreign ministry documentary trends fit even more closely the observations of Daston and Park. Imagined features of the undiscovered West had something of the marvelous about them. In early-eighteenth-century foreign office documentary descriptions, the region's supposedly fertile lands sounded Edenic. Projected western mines might be as wondrous as Peru's Potosí had been. A River and Sea of the West could dissolve the difficulties of land carriage and unite the Atlantic world with the kind of Indian civilizations the Spanish had found or the Asian outposts Columbus had sought. And many of the figures from whom reports of such western wonders were drawn—traders, wood runners, commercial agents, Indians—were, by the standards of polished mid-eighteenth-century Parisian society, vulgar, or even *sauvage*. No surprise if those hoping to meet evolving mid-eighteenth-century standards of intellectual deportment hesitated to embarrass themselves by treating buckskin testimonials too seriously. At any rate, after 1748, the old marvels were dropping out of written evidence of foreign office deliberations. Such a de-

25. Boulle, "French Colonies and the Reform of Their Administration," 338; Pierre Calmettes, *Choiseul et Voltaire d'après les lettres inédites du duc de Choiseul a Voltaire* (Paris, 1902); E. Daubigny, *Choiseul et la France d'outre-mer après le traité de Paris: Étude sur la politique coloniale au XVIIIe siècle, avec un appendice sur les origines de la question de Terre-Neuve* (Paris, 1892), 24–25; Jean-Jacques Rousseau, *The Confessions of Jean-Jacques Rousseau*, trans. J. M. Cohen (London, 1953), 277–303, 470, 472, 511.

velopment cannot be ascribed with certainty to the trend Daston and Park have identified, but it is at least in accordance with it.

Ceding Louisiana

In September 1762, Choiseul heard that supposedly impregnable Havana had fallen to British forces. The British government set Florida or Puerto Rico as the price of its return to Spain. In October, the French representative in London was warning of British plans to launch a seaborne invasion of Mexico in the next campaign. He advised the impossibility of making peace after the British Parliament reconvened on November 10, and a reinvigorated war party cast aside all talk of ending hostilities when continued conflict promised more victories. British conquest of Mexico would bring control of one of the world economy's key silver sources, and with it the possibility that Britain would use this bullion to purchase allies sufficient to dominate European affairs. French funds were exhausted, the French navy destroyed. The Spanish government, stung by Havana's capture and loath to give up Florida, was claiming it preferred to continue a disastrous war rather than consent to a dishonorable peace. Choiseul felt that he must offer something adequate to induce Spain to sign a settlement before it and France lost still more of their possessions.[26]

Western Louisiana was one option, but not the only one. In the Peace of Paris, France managed to hold on to the islands of Guadeloupe and Martinique, possessions that might have captured Spanish attention. Charles III had indicated his empire's interest in the fate of various Caribbean islands and had declared Spain's claims to them more valid than those of Britain or France. After the Seven Years' War, Charles and his ministers would devote considerable effort to refashioning Cuba on the model of the French and British Caribbean plantation colonies. The sugar and other tropical commodities pouring out of Guadeloupe and Martinique made them precious to Choiseul and other French thinkers, but even the manifest profitability of many Caribbean islands could not render automatic the appeal of insular over continental territory. The Spanish portion of Saint-Domingue had not overshadowed Louisiana in 1720. Because of remarkable production increases, Saint-Domingue, Martinique, and Guadeloupe counted for more in 1762 than they had four decades before. Nonetheless, Seven Years' War victor Britain, in a position to take what it wanted from prostrate France, famously chose Canada over Guadeloupe. Moreover, with Saint-

26. Correspondence between Nivernois and Choiseul-Praslin, September–October 1762, CP, Angleterre, 447, 143r–380v; Aiton, "Diplomacy of the Louisiana Cession," *AHR*, XXXVI (1931), 716–720; Rashed, *Peace of Paris*, 169–187.

Domingue's plantation output leaping, the French Empire could have sacrificed a smaller island and still have outproduced its Caribbean rivals. Louisiana could have supplied provisions for the slaves on whom Saint-Domingue's economy brutally relied. Still, if boundless western Louisiana were going to compete with circumscribed but unmistakably valuable possessions like Caribbean islands, it was going to have to promise something really fabulous, and the fabulous had receded from foreign ministry calculations.[27]

Choiseul offered western Louisiana to Spain. In French diplomatic eyes, the colony and larger continent beyond it differed from what they had been decades before. It was not so much that French officials' knowledge of the American West had grown; parts of the western plains had become more familiar, but the mountain and Pacific West remained beyond the horizons of French geographic familiarity. It was more that attitudes about uncertain regions had changed. Foreign office writings had once filled the unknown with the products of speculation and surmise. They had presumed something of value lay in the West's unknown territories. Louisiana's fertile lands had extended along navigable rivers leading to the Pacific. The colony would open trade with China and a silver-rich New Mexico. Its limitless extent had promised boundless mineral, commercial, and agricultural possibilities.

27. On Charles's interest in Caribbean islands, see his instruction to Grimaldi, Mar. 26, 1760, rpt. in Vicente Palacio Atard, *El tercer Pacto de familia* (Madrid, 1945), 301 (see also 60–64). On the relative value of Caribbean and other territories, see Boulle, "French Colonies and the Reform of Their Administration," 566–568; Jean-Pierre Poussou, Philippe Bonnichon, and Xavier Huetz de Lemps, *Espaces coloniaux et espaces maritimes au XVIIIe siècle: Les deux Amériques et la* [sic] *Pacifique* (Paris, 1998), 113–118, 127–137; Paul Butel, *L'économie française au XVIIIe siècle* (Paris, 1993), 114–120, 151; Clarence Walworth Alvord, *The Mississippi Valley in British Politics: A Study of the Trade, Land Speculation, and Experiments in Imperialism Culminating in the American Revolution*, 2 vols. (Cleveland, Ohio, 1917), I, 49–78; William L. Grant, "Canada versus Guadeloupe: An Episode of the Seven Years' War," *AHR*, XVII (1912), 735–743; Philip Lawson, "'The Irishman's Prize': Views of Canada from the British Press, 1760–1774," *Historical Journal*, XXVIII (1985), 575–596; L. B. Namier, *England in the Age of the American Revolution* (London, 1930), 317–327; Richard Pares, *War and Trade in the West Indies, 1739–1763* (Oxford, 1936), 216–219, 223–226; Jack M. Sosin, *Whitehall and the Wilderness: The Middle West in British Colonial Policy, 1760–1775* (Lincoln, Neb., 1961), 9–10.

The idea that Guiana might provision France's island plantations helped inspire Choiseul's support for a disastrous 1763–1764 effort to strengthen the French South American colony. From Paris, Choiseul failed to recognize the hardships inherent in settling a region perhaps best known to Americans as the later site of a fearsome French penal colony. Tropical diseases decimated the unprepared 1760s French colonists. See Boulle, "French Colonies," 625–630; Daubigny, *Choiseul et la France d'outre-mer*, 31–45; Jean Tarrade, "De l'apogée économique à l'effondrement du domaine colonial (1763–1830)," in Jean Meyer, Jean Tarrade, and Annie Rey-Goldzeiguer, *Histoire de la France coloniale*, 3 vols., I, *La conquête* (Paris, 1991), 275–440, esp. 296–301; François Regourd, "Kourou 1763: Succès d'une enquête, échec d'un project colonial," in Charlotte de Castelnau-L'Estoile and François Regourd, eds., *Connaissances et pouvoirs: Les espaces impériaux (XVIe–XVIIIe siècles), France, Espagne, Portugal* (Pessac, France, 2005), 233–254.

At the end of the Seven Years' War, French diplomats treated the unknown as expendable. The Louisiana of 1762 was no longer limitless. In discussions of the colony's strategic importance, the possibility that it offered maritime access to the Pacific no longer appeared. Its rivers led to growing mountains more than the long-coveted South Sea. Louisiana's potential wealth rested on better development of already-known eastern lands rather than discovery of new western ones. The colony was oriented as much toward Mexico as New Mexico, and it did more to keep Spanish silver away from the British than to put it in the hands of the French. When, in fall 1762, Britain was threatening a naval invasion of Mexico, even Louisiana's role as a land barrier became irrelevant.

Over the course of the first two-thirds of the eighteenth century, a mental transformation preceded and shaped a North American territorial dispensation. From 1712 to 1747, optimism regarding the undiscovered West had made Louisiana's future value uncertain but potentially infinite. From 1748 to 1762, foreign office skepticism about and inattention to the region made Louisiana's value surer, but smaller. The demands of intensifying Anglo-French imperial competition, the frustrations of French North American exploration, attacks on French cartographers' credibility, and an inclination to question received notions nourished these geographical doubts. A bounded Louisiana, its value confined to its known products, was not worthless, but it was worth less: less than what it had been, less than other French colonies to the south. At the end of the Seven Years' War, when French and Spanish military losses, financial exhaustion, and need for peace made territorial sacrifices necessary, heightened geographical skepticism allowed French diplomats to assume the expendability of the Great West and consequently of the French colony pointing to it.

Implications

I have been told by Englishmen, and not only by such as were born in America but also by those who came from Europe, that the English colonies in North America, in the space of thirty or fifty years, would be able to form a state by themselves entirely independent of Old England. But as the whole country which lies along the seashore is unguarded, and on the land side is harassed by the French, these dangerous neighbors in times of war are sufficient to prevent the connection of the colonies with their mother country from being quite broken off.

—PETER KALM, 1748

The Louisiana cession terminated the French continental North American empire. Contemporary observers and later scholars have argued that the end of this French imperial presence was one factor emboldening the thirteen colonies to revolt against Britain. Removal of France from the trans-Mississippi West decreased the chance that the Catholic, absolutist, and Indian-friendly power British colonists had long feared would use their rebellion as an opportunity to retake Canada or eastern Louisiana. The burning New England villages lighting Anglo-French wars had long reminded the British colonies of the danger of being alone in the New World with France. After 1762, fear of France no longer inhibited the varied forces impelling the colonies toward separation. As western Indians lost one potential imperial ally, eastern Anglo-Americans lost one potential imperial enemy. In the period after the Revolution, with the exception of a few years when Jefferson dreaded the appearance of Napoleon's legions in New Orleans, the young American republic confronted potent but disunited Indians and a grand but brittle Spanish empire in the West rather than a dynamic French dominion in the Mississippi Valley. The 1762 Louisiana cession helped clear the way for American independence; both the cession and the Louisiana Purchase helped to open the West for American expansion. The continental destiny of the English colonies and their successor republic was less manifest before France's defeat in the Seven Years' War.[28]

28. Adolph B. Benson, ed., *Peter Kalm's Travels in North America: The America of 1750; The English Version of 1770*, 2 vols. (New York, 1964), 139–140; Fred Anderson, *Crucible of War: The Seven Years' War and the Fate of Empire in British North America, 1754–1766* (New York, 2001), xvi; George Louis Beer, *British Colonial Policy, 1754–1765* (New York, 1907), 1, 160–161, 171–173; Jack P. Greene, "The Seven Years' War and the American Revolution: The Causal Relationship Reconsidered," *Journal of Imperial and Commonwealth History*, VIII (1980), 85–105; Lawrence Henry Gipson, *The British Empire before the American Revolution, VI, The Great War for the Empire: The Years of Defeat, 1754–1757* (New York, 1946), 10–11; James H. Hutson, "The Partition Treaty and the Declaration of American Independence," *JAH*, LVIII (1972), 877–896. For a clever essay questioning the connection between French eviction and American revolution, see John M. Murrin, "The French and Indian War, the American Revolution, and the Counterfactual Hypothesis: Reflections on Lawrence Henry Gipson and John Shy," *Reviews in American History*, I (1973), 307–318. I agree with Murrin that tensions between British North American colonies and empire preceded France's imperial departure and were therefore not caused by it; but I feel the colonists were more willing to allow grievances to flower into rebellion without France's glaring at Philadelphia from New Orleans.

14

SPAIN'S ACCEPTANCE OF
TRANS-MISSISSIPPI LOUISIANA

In the closing years of the Seven Years' War, the unusually forceful and capable Charles III replaced Ferdinand VI as king of Spain. During the same period, an increasingly formidable and menacing Britain challenged Spanish imperial security. Few rulers look with indifference upon a growing threat to their cherished dominions, and Charles found the New World progress of British arms especially alarming because his plans for Spanish national revival depended on more efficient exploitation of Spain's colonial resources. Losing control or possession of significant portions of the Spanish Empire could handicap the new ruler's nascent reform efforts. Charles, later immortalized as the huntsman king of a Goya portrait, ran the danger of commencing his reign as prey for the goutish William Pitt.

Despite sharing a family name with the king of France and a dislike of British cannons with French admirals, Charles was neither reflexively anti-British nor automatically pro-French. Spanish interests drove him, and, like his royal predecessor, he saw that both Britain and France posed dangers to them. He was particularly attentive to developments in North America, where the British, French, and Spanish empires shared long frontiers and nourished incompatible aspirations. Charles had no illusions about returning to sixteenth-century Spanish North American preeminence, but he hoped to forestall British or French continental dominance. Befitting a man of regular habits and sober judgment, he envisaged a balance among the three European empires in North America. This would enable Spain to maneuver its stronger rivals against each other, to ally with the more distantly threatening of them against the more immediately dangerous, and to derive imperial security from the inability of either enemy to concentrate on Spanish territory. For such an equilibrium to exist, of course, Spain, Britain, and France must all retain North American territories.[1]

1. In 1742, the threat of a British naval bombardment of Naples had forced Charles, at that time king of the Two Sicilies (1734–1759), to withdraw his troops from operations against Britain's ally

PLATE 4. Francisco José de Goya y Lucientes, *King Charles III as a Huntsman.* 1786–1788. Oil on canvas. The Bridgeman Art Library

Such considerations raise questions about Spain's 1762 acceptance of trans-Mississippi Louisiana from France. Though Spanish officials had long decried the French colony's contribution to illicit trade with Spanish possessions, its provision of arms and allies to western Indians, and its location within invasion range of northern Mexico, they recognized its compensating strategic utility. While Paris ruled New Orleans, avaricious, expansionist, and multiplying British colonists had to fight through French territory before they could reach Mexican silver mines, and scheming British imperialists had to reckon with two North American imperial antagonists. After Spain accepted trans-Mississippi Louisiana, Spanish subjects on the western bank of the Mississippi gazed upon British lands on the opposite shore. French aid against Britain could be requested but not assumed, and though Spanish officials might hope the great river would separate the rival empires' subjects, it was more likely that the currents of the Mississippi and its tributaries would carry them into conflict. Britain's recent victories over France and Spain suggested that Spain alone could not withstand British might. By receiving western Louisiana, balance-minded Spain had ratified a North American imperial disequilibrium.[2]

Austria. On Charles's personality and inclinations, see Vera Lee Brown, "The Spanish Court and Its Diplomatic Outlook after the Treaty of Paris, 1763," *Smith College Studies in History*, XV (1929–1930), 7–38, esp. 13–18; Allan Christelow, "Economic Background of the Anglo-Spanish War of 1762," *JMH*, XVIII (1946), 22–36, esp. 25; Antonio Domínguez Ortiz, *Carlos III y la España de la Ilustración* (Madrid, 1990), 18, 28, 32, 37, 47–51, 56–58; John Lynch, *Bourbon Spain, 1700–1808* (Oxford, 1989), 247–250; Zenab Esmat Rashed, *The Peace of Paris, 1763* (Liverpool, U.K., 1951), 37. For a strong treatment of Spanish governmental reforms under Charles III, see Stanley J. Stein and Barbara Stein, *Apogee of Empire: Spain and New Spain in the Age of Charles III, 1759–1789* (Baltimore, 2003).

2. Scholarly literature regarding Spain's 1762 acceptance of western Louisiana is cogent but incomplete. Historians have produced excellent treatments of aspects of Spanish North American policy but not a complete explanation of its twists and turns. Scholars have noted the anomalousness of Spanish acceptance of western Louisiana: the increased likelihood of Anglo-Spanish conflict, the challenges of governing immense new territories, the risks involved in France's departure. See Herbert E. Bolton, "Defensive Spanish Expansion and the Significance of the Borderlands," John Francis Bannon, ed., *Bolton and the Spanish Borderlands* (Norman, Okla., 1964), 32–64, esp. 45–46; Claude de Bonnault, "Le Canada et la conclusion du Pacte de famille de 1761," *Revue d'histoire de l'Amérique française*, VII (1953), 341–355; D. A. Brading, "Bourbon Spain and Its American Empire," in Leslie Bethell, ed., *The Cambridge History of Latin America*, I, *Colonial Latin America* (Cambridge, 1984), 399; Brown, "Spanish Court and Its Diplomatic Outlook," *Smith College Studies in History*, XV (1929–1930), 15–16; Julius Goebel, *The Struggle for the Falkland Islands: A Study in Legal and Diplomatic History* (1927; rpt. New Haven, Conn., 1982), 217; Charles Wilson Hackett, "Policy of the Spanish Crown Regarding French Encroachments from Louisiana, 1721–1762," in George P. Hammond, [ed.], *New Spain and the Anglo-American West: Historical Contributions Presented to Herbert Eugene Bolton*, I, *New Spain* (Lancaster, Pa., 1932), 107–145, esp. 109; Elizabeth A. H. John, *Storms Brewed in Other Men's Worlds: The Confrontation of Indians, Spanish, and French in the Southwest, 1540–1795*, 2d ed. (Norman, Okla., 1996), 377; Richard Pares, *War and Trade in the West Indies, 1739–1763* (Oxford, 1936), 565; Francis P. Renaut, *Le Pacte de famille et l'Amerique: La politique coloniale franco-espagnole de 1760 à 1792* (Paris, 1922), 46, 50–51; William R. Shepherd, "The

One possible explanation for Spain's acceptance of trans-Mississippi Louisiana is that Spanish officials had become so alarmed by the French encroachments on New Mexico and the Pacific discussed in Chapter 12 that they deemed preventing French westward expansion more important than maintaining a North American balance of power. In the latter years of the Seven Years' War, Spanish officials remained aware of these menacing French activities, but Charles III and his ministers appear nonetheless to have viewed the risk of continued French frontier incursions as a price worth paying for the French Louisiana counterweight to Britain's Atlantic seaboard colonies.[3]

<hr />

Cession of Louisiana to Spain," *PSQ*, XIX (1904), 439–458, esp. 455; J. Leitch Wright, Jr., *Anglo-Spanish Rivalry in North America* (Athens, Ga., 1971), 439–458, esp. 107, 112.

On Charles's idea of a North American balance of power, Vicente Palacio Atard has written the fundamental article, and other scholars have developed or referred to the idea. See Allan Christelow, "French Interest in the Spanish Empire during the Ministry of the Duc de Choiseul, 1759–1771," *HAHR*, XXI (1941), 515–537, esp. 522; Lawrence Henry Gipson, *The British Empire before the American Revolution*, VIII, *The Great War for the Empire: The Culmination, 1760–1763* (New York, 1954), 246; Lynch, *Bourbon Spain*, 317–318; Didier Ozanam, "Les origines du troisième Pacte de famille (1761)," *Revue d'histoire diplomatique*, LXXV (1961), 307–340, esp. 315–318; Vicente Palacio Atard, *El tercer Pacto de familia* (Madrid, 1945), 48, 50, 261, 290; Palacio Atard, "El equilibrio de América en la diplomacia del siglo XVIII," *Estudios americanos*, I (1949), 461–479; Palacio Atard, *Las embajadas de Abreu y Fuentes en Londres, 1754–1761* (Valladolid, Spain, 1950), 3, 5, 38–39; Pares, *War and Trade*, 13, 128; Rashed, *Peace of Paris*, 42, 47; Max Savelle, "The American Balance of Power and European Diplomacy, 1713–78," in Richard B. Morris, ed., *The Era of the American Revolution* (New York, 1965), 140–169, esp. 168; Savelle, *The Origins of American Diplomacy: The International History of Angloamerica, 1492–1763* (New York, 1967), 448–451, 456, 489; Shepherd, "Cession of Louisiana to Spain," *PSQ*, XIX (1904), 440.

Neither scholars interested in the 1762 transfer of Louisiana to Spain nor those interested in the strategic assumptions guiding Spanish policy have explained the relation between this idea of a North American imperial equilibrium and the Spanish acceptance of Louisiana that rendered such a balance impossible.

3. Scholars have generally disregarded possible connections between Spanish concerns about Pacific intrusions and Spanish acceptance of western Louisiana. For the most part, historians who have discussed the Pacific have not talked about the Louisiana cession, and those treating the cession have not extended their interest to include the South Sea. Those who have mentioned both subjects, Julius Goebel, Sylvia-Lyn Hilton, and David J. Weber, have done so in works that were not primarily concerned with the causal relationship between them. See Charles Edward Chapman, *The Founding of Spanish California: The Northwestward Expansion of New Spain, 1687–1783* (New York, 1916), 55; Warren L. Cook, *Flood Tide of Empire: Spain and the Pacific Northwest, 1543–1819* (New Haven, Conn., 1973), 47; Donald Cutter and Iris Engstrand, *Quest for Empire: Spanish Settlement in the Southwest* (Golden, Colo., 1996), 170; Goebel, *Struggle for the Falkland Islands*, 217; Iñigo Abbad y Lasierra, *Descripción de las costas de California*, ed. Sylvia L. Hilton (Madrid, 1981), 26–36; Hugo O'Donnell, *España en el descubrimiento, conquista y defensa del Mar del Sur* (Madrid, 1992); Jean-Pierre Poussou, Philippe Bonnichon, and Xavier Huetz de Lemps, *Espaces coloniaux et espaces maritimes au XVIIIe siècle: Les deux Amériques et la* [sic] *Pacifique* ([Paris], 1998), 336; David J. Weber, "Bourbons and Bárbaros: Center and Periphery in the Reshaping of Spanish Indian Policy," in Christine Daniels and Michael V. Kennedy, eds., *Negotiated Empires: Centers and Peripheries in the Americas, 1500–1820* (New York, 2002), 79–103, esp. 80–81. Knowing that the American West affords no easy water route from the Mississippi Basin to the Pacific coast, modern historians have

A better explanation emphasizes Spanish concerns about growing British imperial power. Spanish diplomats in the early 1760s remained acutely aware of the British Pacific forays of earlier decades. The 1757 publication of a new history of California by Padre Andrés Marcos Burriel made Spanish officials even more cognizant of the magnitude of British ambitions than they had been earlier in the decade. Then, in 1761, reports of Russian exploration on North America's west coast caused Charles to suspect that Britain and Russia were cooperating to find a route from the North Pacific to Spain's South Sea possessions. What appears to have alarmed Charles most was the possibility that a war-weary French government would cede trans-Mississippi Louisiana to Britain in the hope that territorial sacrifice would induce Britain to end hostilities. Britain would then have access to all the rumored southwestern Pacific routes France was thought to have been seeking, as well as ownership of territory ominously close to northern Mexican silver mines. Charles had to prevent this. His acceptance of western Louisiana eliminated the basis for a tri-imperial North American balance of power, but it forestalled a possible French cession to Britain that would have produced the same result while simultaneously extending British dominion west of the Mississippi and toward the Pacific.[4]

Spanish Concerns about the North American Balance of Power

Spain alone motionless in the midst of this conflict of the universe.
—LOUIS-ANTOINE DE BOUGAINVILLE, October 2, 1758

From its inception, the Seven Years' War had constituted a serious matter for the Spanish Empire. But so long as Spain remained neutral and no clear victor emerged, Spanish officials could observe events with some degree of detachment. Spanish funds were accumulating, the Spanish navy was growing, Spanish

been able to easily and unconsciously neglect any search for a relationship between issues involving Louisiana and matters pertaining to the distant ocean.

4. Scholars such as Bolton, Shepherd, Renaut, Wright, and Arthur S. Aiton have made the correct historical argument that it was Charles's fear of French cession of trans-Mississippi Louisiana to Britain that drove him to accept the region. Understanding of Spanish policy has remained incomplete, however, because earlier scholars have omitted identification and foregone analysis of important sources of Charles's concern about British westward expansion. In addition, previous work has left unexplained the puzzling vicissitudes of Spanish policy that led Charles to express an interest in acquiring Louisiana from France in the summer of 1760, Spanish officials to decline a French offer of the province in summer 1761, and finally, Charles to accept trans-Mississippi Louisiana in November 1762. Attention to Spanish officials' geographic notions in the years before 1762 and to the connections between these ideas and contemporary Spanish concerns about the increasingly ominous disparity of power in North America makes possible a fuller and more coherent explanation of Spain's Louisiana acquisition.

administration was undergoing gradual reform. Spain's American possessions, unlike those of its northern neighbors, were not on fire. In the months before succeeding the aged and ill Ferdinand VI, Charles was primarily concerned with maintaining the safety of Spain's colonies. When Charles moved from the Neapolitan to the Spanish throne in August 1759, the Seven Years' War had begun turning in Britain's favor, but its outcome remained uncertain. Charles still hoped that neither Britain nor France would gain a North American position so dominant as to allow a concentration of hostile forces against New Spain's frontier provinces, and he wanted to continue the neutrality that was strengthening Spain. After the British capture of Quebec in September 1759, however, complete British conquest of France's North American possessions became a possibility, and Charles viewed the career of British forces with mounting concern.[5]

Beginning in autumn 1759, references to a fragile American equilibrium become abundant in Spanish diplomatic correspondence. Indeed, although it is not always easy to discern the motives driving the close-lipped Charles's decisions, he expressed himself openly and frequently on the issue of the British threat to the North American balance of power. In November 1759, Charles directed Spanish representative in London Félix de Abreu to inform Britain's George II that Charles could "not look with indifference" at the way in which the British conquest of Quebec, and thus Britain's near mastery of Canada, offended "the equilibrium in that New World, which was established by the Peace of Utrecht." Charles's instructions indicated further that, because of his many American possessions, the Spanish king was the monarch most interested in this balance. March 1760 royal instructions to the marqués de Grimaldi—Spain's minister at the Hague, a Genoese abbé-turned-diplomat—repeated Charles's concern about the precariousness of the North American balance of power.[6]

5. Louis-Antoine de Bougainville, *Adventure in the Wilderness: The American Journals of Louis Antoine de Bougainville, 1756–1760*, ed. and trans. Edward P. Hamilton (Norman, Okla., 1964), 283; Christelow, "Economic Background of the Anglo-Spanish War," *JMH*, XVIII (1946), 23, 25; Palacio Atard, *El tercer Pacto de familia*, 28–30, 48–50, 66; Pares, *War and Trade*, 566.

6. Brown, "Spanish Court and Its Diplomatic Outlook," *Smith College Studies in History*, XV (1929–1930), 13–14; Christelow, "Economic Background of the Anglo-Spanish War," *JMH*, XVIII (1946), 22–23; Gipson, *British Empire before the American Revolution*, VIII, 246; Lynch, *Bourbon Spain*, 249; Rashed, *Peace of Paris*, 37; Renaut, *Le Pacte de famille et l'Amerique*, 24; "Carta de Esquilache a D. Félix Abréu, fechada en Zaragoza el 14 de Noviembre de 1759 . . . ," in Palacio Atard, *El tercer Pacto de familia*, 295–296, esp. 295: "No puede S. M. mirar con indiferencia lo mucho que ofenden estas conquistas al equilibrio en aquel nuebo Mundo, que se estableció por la Paz de Utrecht, siendo el Príncipe más interesado en él conforme a la gran parte que le toca"; see also "Instrucción a Grimaldi para su embajada en la Haya, fechada en el Buen Retiro a 26 de Marzo de 1760," ibid., 298. Grimaldi was reputedly a "Man of Parts," though one British observer was never "able to discover any Talents in Him, except a peculiar Gift of Noise and Impudence, and . . . the

Charles's diplomats seem to have shared his sentiments. Foreign minister Wall (in his correspondence with Abreu), Joaquín Pignatelli de Aragón y Moncayo, conde de Fuentes (who replaced Abreu in London in May 1760), and Masones de Lima (the Spanish ambassador in Paris) reiterated Charles's solicitudes about the implications should France or Britain gain too great an advantage in North America. Fuentes warned the Spanish court repeatedly about the dangers of English aggression. In January 1761, speaking in the context of rumors about a British expedition directed at French Louisiana, he asked, "If the British take possession of Martinique and attack the Mississippi, who will maintain the equilibrium in America?" Spain could never let Britain establish itself in Louisiana, for "today it will be at the expense of France, tomorrow at that of Spain." In June 1761, Fuentes fretted that France's loss of Canada would lead also to that of Louisiana, and this would damage irreparably the American balance of power while simultaneously bringing the British still closer to still more Spanish possessions. In August 1761, discussing Pitt's demands for English ownership of the Ohio to its junction with the Mississippi, Fuentes warned that this would place the British in a position to easily conquer French Louisiana and that such a conquest would facilitate future British surprise assaults on nearby Mexico. He therefore saw future British attacks on French North American possessions as indirect blows to Spanish interests. The boundaries of French Louisiana currently under discussion were thus "the most important object for the security of the king's dominions."[7]

Despite these anxieties about British North American conquests, Spain held back from formal alliance with France until summer 1761 and from a declaration of war on Britain before January 1762. Initially, in 1759 and early 1760, Charles offered his services ("buenos oficios") as a mediator, hoping that Spanish participation would secure an Anglo-French peace leaving neither Britain nor France predominant in North America.[8]

---

confident assurance with which He delivers His opinion upon all Subjects." See earl of Bristol to William Pitt, Mar. 5, 1759, PRO 30/47/12.

7. For Wall's concerns, see Jaime Masones de Lima to Richard Wall, Dec. 23, 1759, Estado 6945, AGS; Wall to Félix de Abreu, Nov. 14, Dec. 13, 1759, Wall to Masones de Lima, Mar. 21, 1760, conde de Fuentes to Wall, Dec. 14, 1759, and Abreu to Wall, Jan. 1, 1760, all in Estado 4098, AHN. On Fuentes's warnings, see Palacio Atard, El tercer Pacto de familia, 96, 104–105, 117; Palacio Atard, Las embajadas de Abreu y Fuentes, 64, 68. For the first Fuentes quotation, see Fuentes to Wall, Jan. 30, 1761, Estado 6948, AGS (another copy in Estado, 4282-1, AHN). For Fuentes's concern about the loss of Canada, see Fuentes to Wall, June 30, 1761, Estado 6949, AGS (another copy in Estado 4282-2, AHN). For the final Fuentes quotation, see Fuentes to Wall, Aug. 28, 1761, Estado 6950, AGS: "el objeto mas importante para la seguridad de los Dominios del Rey."

8. Gipson, British Empire before the American Revolution, VIII, 246–247; Ozanam, "Les origines du troisième Pacte de famille," Revue d'histoire diplomatique, LXXV (1961), 318–319; Palacio Atard,

Continuing Spanish Fears of French North American Aggression

Part of the reason Charles and his ministers hesitated to offer France more than mediation was their recollection of early 1750s concerns about French activities in New Mexico, Texas, and Darien. Continuity of personnel characterized Spain's government after Charles's accession, and many of the officials who had expressed such concerns in the early 1750s retained important positions in the early 1760s. Wall and Arriaga remained minister of foreign affairs and secretary of state for the Indies, respectively, and letters concerning Spanish policy toward North America continued to flow to and from them. As the discussion of territorial cessions and boundaries became critical in the closing years of the war, Charles and Wall consulted Arriaga on pertinent aspects of North American geography, and Wall corresponded with Spanish ambassadors such as the marqués de Grimaldi (who replaced Masones de Lima as Spanish ambassador to France at the beginning of 1761) about the North American territories Britain wished to obtain and France might have to sacrifice.[9]

Between 1755 and 1762, French subjects and officials had continued to give Spanish ministers reason for worry about the dangers of French North American expansion. In March 1756, New Spain's viceroy the marqués de Las Amarillas (1755–1760) reported that the French commander at Natchitoches had proposed establishing a jail on the Franco-Spanish border to facilitate the mutual return of deserters. This proposal led Wall, Arriaga, and the Council of the Indies to reexamine a 1752 letter from Ensenada to Viceroy Revilla Gigedo concerning an earlier French version of the idea. As the letter mentioned, Ferdinand had rejected the French suggestion, fearing the Frenchmen Spain would return as deserters might actually be spies investigating Spain's northern provinces. Ferdinand further enjoined his American officials to guard against possible French attempts to encroach on Spanish provinces such as Texas and New Mexico. Reporting on this issue in 1756, the fiscal reiterated the long-standing Spanish dis-

---

El tercer Pacto de familia, 48–52; Palacio Atard, Las embajadas de Abreu y Fuentes, 39–42; Pares, War and Trade, 568–569; Rashed, Peace of Paris, 45–51; François Rousseau, Règne de Charles III d'Espagne (1759–1788), 2 vols. (Paris, 1907), I, 32–34, 46–47; Savelle, Origins of American Diplomacy, 448.

9. See for example, Wall to Julián de Arriaga, July 6, 1762, Wall to marquis de Grimaldi, July 12, 1762, both in Estado 4550, AGS; Grimaldi to Wall, July 22, 1762, Wall to Grimaldi, Sept. 5, Oct. 23, 1762, all in Estado 4551, AGS (also in Estado 4176–1, AHN); Renaut, Le Pacte de famille et l'Amerique, 22; Rousseau, Règne de Charles III, I, 16. Although some have argued that Wall and Arriaga's influence declined after Spain's entry into the Seven Years' War or that Arriaga was a weak figure generally (Louis Blart, Les rapports de la France et de l'Espagne après le Pacte de famille [Paris, 1915], 9; Lynch, Bourbon Spain, 251; Rousseau, Règne de Charles III, I, 18–19), the documents above indicate that Wall and Arriaga continued to play an active role in the formation of Spanish North American policy.

trust of French merchants trying to infiltrate Spanish markets and French spies trying to examine Spanish territory. In 1760, Charles agreed to the mutual restitution of deserting soldiers, but not of anyone who might be an illicit trader, explorer, or agent.[10]

Amarillas's March 1756 letter also mentioned the October 1754 arrest on the lower Trinity River in Texas of a party of French traders led by Joseph Blancpain. As Herbert E. Bolton demonstrated in his 1913 article "Spanish Activities on the Lower Trinity River, 1746–1771," much of what Spanish officials feared and disliked about French interlopers could be found in Blancpain. He was trading in Spanish territory. His commerce included selling guns to Indians. He carried a license from the governor of Louisiana authorizing him to trade and encouraging him to befriend Indian nations and bring their leaders to New Orleans. Blancpain might presage a French settlement or engineer a Franco-Indian alliance, and he appeared to be acting as a representative of the French government. Louisiana's Governor Kerlérec protested Blancpain's arrest, averring that he had been on French land, and he proposed to Spanish governor Jacinto de Barrios y Jauregui of Texas a joint commission to ascertain which empire properly owned the area. The last thing Spanish imperial officials wanted, however, was to have questions raised about their ownership of Texas. Instead, Governor Barrios and Viceroy Amarillas favored establishing a Spanish presidio and village in the area to thwart future French activities.[11]

Spanish diplomats also continued to discuss other French traders who had been involved in similarly worrisome activities around New Mexico. In May 1757, the marquis d'Aubeterre, France's ambassador to Spain from 1757 to 1760, requested the release of Chapuis and Feuilli, reminding Spanish officials of the illicit commerce and western exploration leading to their imprisonment in the first place.[12]

10. Marqués de La Ensenada to conde de Revilla Gigedo, July 26, 1752, Wall to Arriaga, Aug. 28, Sept. 18, 1756, Arriaga to Wall, Sept. 18, 1756, all in Estado 4534, AGS; marqués de Las Amarillas to Arriaga, Mar. 14, 1756, 370–377, and Council of the Indies, "Expediente sobre la apprehension que Dn Jacinto de Barrios . . . ," Oct. 22, 1756, 345–366, both in AGI, Guadalajara 329; "The King to the Viceroy of New Spain, Aranjuez, May 4, 1760," in Hackett, "Policy of the Spanish Crown," in Hammond, [ed.], *New Spain and the Anglo-American West*, I, 141–145, esp. 144.

11. Wall to Arriaga, Aug. 28, Sept. 18, 1756, Arriaga to Wall, Sept. 18, 1756, all in Estado 4534, AGS; Amarillas to Arriaga, Mar. 14, 1756, 370–377, Arriaga to duque de Alva, Sept. 15, 1756, 367–368, Council of the Indies, "Expediente sobre la apprehension que Dn Jacinto de Barrios," 345–366, all in AGI, Guadalajara 329; Herbert E. Bolton, "Spanish Activities on the Lower Trinity River, 1746–1771," *SWHQ*, XVI (1913), 339–377, esp. 348–354, 366–367; Bolton, *Texas in the Middle Eighteenth Century: Studies in Spanish Colonial History and Administration* (Berkeley, Calif., 1915), 66, 73–75.

12. Perhaps because he thought several years in a Spanish dungeon was sufficient punishment or because he did not want to further irritate his French royal cousin, Ferdinand ordered their release.

Spanish officials fretted not only about the economically troublesome activities of French smugglers but also about the militarily menacing implications of French arms sales to, and French military alliance with, hostile Texas Indians. In 1757, at the request of Apaches seeking protection from Comanche raiders, Spain established a mission on the San Sabá River in Texas (northwest of San Antonio, near modern Menard). In 1758, a diverse group of Indians including Comanches, Tonkawas, and Hasinais destroyed the mission. The next year, the commander of San Sabá's presidio, Colonel Diego Ortiz Parrilla, led a retaliatory expedition north. Parrilla found his enemies in a Wichita village. A "stockade and moat" protected the town, "the entire front" of which "was crowned with Indians armed with muskets," and "the French flag" flew "in the center of the enclosure, among the military instruments of drum and fife." In the ensuing battle, steady fire from inside the village attested to a "large store of ammunition." Parrilla's Indian scouts claimed that fourteen Europeans within the village were directing its defense. To Parrilla and those with him, all this constituted evidence that Frenchmen must have assisted in the Wichita town's defense and incited the 1758 attack on San Sabá. In spring 1760, Parrilla shared his views with the newly appointed viceroy of New Spain, Joaquin de Montserrat, the marqués de Cruillas.[13]

Spanish officials continued to link such rumored and actual French Texas and New Mexico borderlands activities to a possible French advance on the Pacific through unexplored southwestern territories. A summary of Sánchez Salvador's memoirs concerning the Colorado was considered relevant to a discussion of Blancpain's appearance on the lower Trinity and was included in the document treating his activities. In July 1758, the viceroy of New Spain, Arriaga, the king, the fiscal, and the Council of the Indies exchanged documents concerning the proposed Spanish conquest of the Gila and Colorado rivers and the troublesome Indian unrest in Sonora, Sinaloa, and Nueva Vizcaya. The viceroy was instructed to remain vigilant that no foreign nation establish itself in the region of the Colorado and Gila rivers because of the "damage and fatal consequences that the slightest omission or carelessness in this matter could bring" ["por el perjuicio, y fattales consequenzias, que puede traher la mas lebe omission ò

See Aubeterre to Wall, May 25, 1757, Arriaga to Wall, June 3, 1757, Wall to Aubeterre, June 19, 1757, all in Estado 4534, AGS; Arriaga to José de Goyeneche, June 4, 1757, AGI, Guadalajara 329, 550–551.

13. "Diary of Juan Angel de Oyarzún," in Robert S. Weddle, [ed.], *After the Massacre: The Violent Legacy of the San Sabá Mission; With the Original Diary of the 1759 Red River Campaign*, trans. Carol Lipscomb (Lubbock, Tex., 2007), 103–138, esp. 124–128; Bolton, *Texas in the Middle Eighteenth Century*, 87–90; John, *Storms Brewed*, 351–352; David J. Weber, *The Spanish Frontier in North America* (New Haven, Conn., 1992), 188–191.

descuido eneste asuntto"]. On September 27, 1759, after Charles had become king, another royal order to the viceroy repeated these concerns.[14]

Nor was Sánchez Salvador the last author of memoirs warning of possible foreign use of southwestern rivers to reach the South Sea. Around 1760 or 1761, former New Spanish military officer Pedro de Labaquera wrote four reports for the Spanish king. Like Sánchez Salvador, Labaquera discussed the danger of foreign conquest of Baja California's Cape San Lucas. He mentioned a Dutch ship that had reached Baja California from Batavia in 1746 and the English pirates who had taken refuge there in 1686. He also referred to the Colorado River and the unknown lands ["paises incognitos, cuia extension se ignora"] through which its northern course ran. Labaquera presumed the river's origin lay near French territory ["su origen, o curso, no diste demasiado de los establecimientos Franceses"], and he echoed Sánchez Salvador in warning that this French nation, "by nature ambitious to see the world" ["naturalmente ambiciosa de ver mundo"], might form establishments on the Colorado, reach the Pacific, and prevent New Spain's future northward expansion ["limiten la extencion de estos Dominios à aquel paraje"]. Labaquera's letters reached someone in Spain who felt them significant enough to include among Spain's official imperial documents. The kinds of concerns Sánchez Salvador raised continued to be discussed ten years later.[15]

Thus Spanish concern about French North American expansion remained alive after the beginning of the Seven Years' War and despite the continuing succession of French warnings that Britain posed the real danger to Spanish

14. For the summary of Sánchez Salvador, see Council of the Indies, "Expediente sobre la apprehension que Dn Jacinto de Barrios," Oct. 22, 1756, AGI, Guadalajara 329, 345–366. For the viceroy's instructions, see Arriaga to duque de Alva, July 15, 1758, 535, Fiscal, "Expediente sobre la conquista de los ríos Colorado, y Gila, y sobre las providencias tomadas en Mexico para contener los insultos de los Indios barbaros de la Nueva Vizcaya," July 31, 1758, 963–969, both in AGI, Guadalajara 137. For Charles's renewed warnings, see Chapman, *Founding of Spanish California*, 63.

15. In 1745, the Dutch governor of Batavia outfitted six ships, of which two were English privateers, to take Chinese merchandise across the Pacific for sale in New Spain. In the end, in 1746, only two Dutch ships, the *Hersteller* and the *Hervating*, actually sailed. The *Hervating* reached the tip of Baja California in December 1746 and received a friendly welcome from the isolated and undersupplied Jesuits at San José de Cabo; the *Hersteller* reached Acapulco in March 1747. As this Dutch mission contravened Spanish commercial regulations and flouted the European agreement to leave Pacific navigation to Spain, New Spain's viceroy Revilla Gigedo reiterated orders to the Baja California Jesuits to avoid contact with and withhold assistance from foreigners. He also informed the Spanish government; the Spanish ambassador to the Dutch Republic protested the sending of the ships and requested the chastisement of Batavia's governor. See "Extracto del expediente que trata de la reduccion a curatos de las misiones de las provincias de Sinaloa, y Sonora . . . ," June 15, 1752, AGI, Guadalajara 137, 311; Peter Gerhard, "A Dutch Trade Mission to New Spain, 1746–1747," *PHR*, XXIII (1954), 221–226; Pedro de Labaquera to Charles III, c. 1760 or 1761, AGI, Guadalajara 511, 39–46; Chapman, *Founding of Spanish California*, 63–65.

imperial interests. With the wide array of French threats in mind, the May 1760 royal instructions to Viceroy Cruillas of New Spain instructed him to take "all possible precautions" so that "the entrance [of the French in the area of the lower Trinity] may be blocked, doing likewise at all the places and sites through which the French, established in Canada, are introducing themselves into New Mexico." Charles wanted the viceroy to secure a Comanche alliance that would hinder French westward movement, and he wanted Franco-Spanish communication and trade in the Natchitoches region eliminated.[16]

Meanwhile, Spanish diplomats, even as they discussed a Franco-Spanish alliance against Britain, continued to voice their fears of French aggression to their French interlocutors. In July 1761, French ambassador Ossun reported that Wall had reminded him of Spain's long-standing concerns about French penetration of Mexico from Louisiana. Wall said that Spain was willing to recognize the legitimacy of French claims to Louisiana ["on reconnût la légitime propriété de la France sur cette Colonie"], but he felt that France should agree to a Louisiana boundary settlement allaying Spanish fears of French expansion toward Mexico ["il pensoit que la France . . . devroit se prêter au réglement des limites de la Louisiane, et qu'il conviendroit de les fixer de maniére à dissiper les craintes que l'Espagne avoit toujours conservées"]. Into 1761, Spain retained a healthy fear of the French empire Britain was vanquishing.[17]

Spanish Fears of British and Russian Pacific Encroachments

But just as Spanish officials remembered the dangers of French southwestern expansion, they also recalled the threat of British Pacific incursions. In part, this continued fear of British South Sea projects reflected general Spanish late–Seven Years' War concerns about British maritime domination. Such anxieties are evident in a March 31, 1761, Franco-Spanish maritime alliance proposal Wall sent to Grimaldi. Its preamble spoke of "the system that the British court has formed and follows to reign despotically in all the seas and to seize indifferently the dominions and principal ports of the East and West Indies." It appeared, the preamble continued, that England hoped no one would be able to navigate without its permission or conduct more than the passive commerce England

16. Translated quotation from "King to Viceroy of New Spain," in Hackett, "Policy of the Spanish Crown," in Hammond, [ed.], New Spain and the Anglo-American West, I, 141–145; marqués de Cruillas to Goyeneche, July 10, 1761, AGI, Guadalajara 330.

17. Marquis d'Ossun to duc de Choiseul, July 16, 1761, CP, Espagne, 533, 122r–128r (rpt. in Theodore Calvin Pease, ed., Anglo-French Boundary Disputes in the West, 1749–1763 [Springfield, Ill., 1936], 329–331); Arthur S. Aiton, "The Diplomacy of the Louisiana Cession," AHR, XXXVI (1931), 701–720, esp. 704.

allowed. In an April 12, 1761, discussion with Louis XV and Choiseul, Grimaldi declared that Spain's interest in an alliance arose in response to this British design. The formal Spanish proposal to France in May contained the same language. The June 22, 1761, version of the project for an alliance spoke of the "ambitious projects of the British Court and the despotism it attempts to arrogate in all the seas." The English nation had demonstrated clearly that it wanted to be the "absolute mistress of navigation." Spanish officials were evaluating specific British navigational threats, around Cape Horn or in Hudson Bay, for example, in light of this grand British design.[18]

In fact, the same documents disclosing Spanish fears of France's finding a southwestern river route to the Pacific also revealed Spanish concerns that England might find a comparable riparian or oceanic passage farther north. In 1745, cavalry captain and former alcalde of San Juan de Sonora, Don Gabriel de Prudhom Butron y Mujica, baron de Heyder, had written of British desires, articulated by Daniel Coxe in his *Description of the English Province of Carolana*, to open a route between the English colonies and the Pacific by way of the Colorado River. Labaquera spoke generally of the danger of foreign ships' appearing on the coasts of Baja California or other parts of Pacific New Spain, and he alluded specifically to the English renewal of "the idea of discovering a Northwest Passage to the South Sea" ["renovando los Ingleses la idea de descubrir paso por el Norueste al Mar del Sur"]. He worried that they might discover a connection between the Colorado River system and a branch of the ocean farther north ["comunicacion por el Norte, con el mar exterior"].[19]

18. "Primitivo proyecto español, remitido por Wall a Grimaldi el 31 de marzo de 1761, para una alianza marítima con Francia," rpt. in Palacio Atard, *El tercer Pacto de familia*, 304: "Es tan visible el sisthema que ha formado y sigue la Corte Británica de reinar con despotismo en todos los mares y apoderarse indiferentemente de los Dominios y principales Puertos de las Indias Orientales y Occidentales que sirven de escala a la navegación, o son como Llabe para entrar y comunicarse cada qual con los propios, aspirando al parecer la nación Inglesa a que nadie navegue sin necesitar su salvo-conducto, y que nadie haga otro comercio que el pasibo que le permita." For the formal proposal, see May 1761, Estado 4542, AGS; see also CP, Espagne, 532, 281r, 289r; conversation quoted in Palacio Atard, *El tercer Pacto de familia*, 128. For the final quotations, see "Proyecto de convención formado por Wall y remitido a Grimaldi el 22 du junio de 1761," ibid., 330: ". . . los ambiciosos Proyectos de la Corte Británica y el Despotismo que intenta arrogarse en todos los Mares. La Nación Inglesa muestra claramente en sus procederes, con especialidad de diez años a esta parte, que quiere hacerse Dueña absoluta de la Navegación, y no dejar a las demás sino un comercio pasivo y dependiente." Copies also in Estado 4542, AGS, and CP, Espagne, 532, 334r.

19. Daniel Coxe, *A Description of the English Province of Carolana, by the Spaniards Call'd Florida, and by the French La Louisiane* (1722; rpt. Gainesville, Fla., 1976), 62–63; "[Statement of Don Gabriel de Prudhom Butron y Mujica, 1746]," in Donald Rowland, "The Sonora Frontier of New Spain, 1735-1745," in Hammond, [ed.], *New Spain and the Anglo-American West*, I, 147–164, esp. 162–164; Labaquera to Charles III, c. 1760 or 1761, AGI, Guadalajara 511, 40v, 42v–43r; Chapman, *Founding of Spanish California*, 65.

More significant than the works of either Prudhom or Labaquera was Padre Andrés Burriel's 1757 *Noticia de la California*. Burriel tried to answer the question of why territory as apparently unforgiving and unpromising as California merited Spanish attention. He argued that Spain's western American coastal possessions could not enjoy security while California remained outside Spanish control ["no pueden tener seguridad las Costas Americanas sobre el Mar del Sùr, mientras no estuviere sujeta à Dios, y al Rey Catholico la California"]. California had often given shelter to corsairs and pirates, who had "captured many Spanish ships, disrupted all the trade of the South Sea, and disturbed the tranquillity of those remote provinces" ["apresado muchos Navios Españoles, turbado todo el Comercio de el Mar de el Sùr, y alterado la quietud de aquellas remotas Provincias"]. In light of these depredations, Burriel worried about the consequences should a European power establish coastal California colonies or military outposts. Burriel recalled Anson's claim that, if he had taken Valdivia, he could have shaken Peru. The padre feared that a comparable risk would arise for Mexico ["semejantemente serìa muy gran de eltemor, y el riesgo del Imperio Mexicano"] if a foreign power achieved mastery over California. In connection with this danger, Burriel claimed that Russian explorers in 1741 had reached lat 55°36′N, a little more than twelve degrees north of Cape Blanco, for Burriel the northern limit of Spanish California ["ultimo Termino, conocido hasta ahora de nuestra California"].[20]

Burriel viewed skeptically many accounts of northwestern American geography and of the long-sought Northwest Passage. In appendix VII of the *Noticia*, for example, he noted that a thorough investigation of Spanish archives failed to sustain reports of the 1640 Fonte voyage from Peru to the Arctic Ocean that Delisle and Buache had cited and John Green had questioned. Burriel recognized, however, that discrediting particular reports was something quite different from disproving the existence of alleged features of North American geography. With regard to what actually lay north of Cape Blanco and south of the limits of Russian exploration, he conceded on the last page of his history, "*Ignoro. Nescio. Yo no lo sé.*" This ignorance left him fearful that a nation such as England might discover in North America some kind of route to the Pacific imperiling Spanish territories on the Americas' western shores.[21]

Along these lines, Burriel spoke of the "stubborn and much talked-about English attempts to find a passage to the South Sea through the North of America,

20. Miguèl Venegas and [Andrés Marcos Burriel], *Noticia de la California, y desu conquista temporal, y espiritual hasta el tiempo presente sacada de la historia manuscrita, formada en Mexico año de 1739 por el Padre Miguèl Venegas . . .* , 3 vols. (Madrid, 1757), III, 4–5, 12–16, esp. 12.
21. Ibid., 436.

and Hudson and Baffin Bays" ["las ruidosas, y porfiadas tentativas de los Ingleses, para hallar un Passage al Mar del Sùr por el Norte de America, y Bahìas de Hudson, y Baffins"] and of the "high thoughts and vast ideas" ["los altos pensamientos, y vastas ideas"] entertained by some Englishmen concerning "this desired and controversial passage" ["este deseado, y controvertido Passage"]. Should such a passage be found, California and other Spanish domains would be "very near lands possessed by England" ["muy vecinos à Pàises posseìdos por la Inglaterra"]. The English might descend on New Mexico, the Gila and Colorado river regions, and California. What, then, would become of the Spanish frontier? Burriel mentioned English notions of crossing the Pacific from the west and forming colonies on the west coast of the Americas. Lands such as Jamaica and Georgia, he recalled, had once been part of the Spanish dominions but were now English. The same could happen to the coast north of California ["puedan tambien serlo las Costas al Norte de la California"].[22]

Notable because of its warning about the potential dangers uncharted lands north of New Spain might harbor for the Spanish Empire, Burriel's work was also important because of its audience. The appearance of the *Noticia* in the 1750s reflected official interest in his topic. The Council of the Indies had directed and licensed its publication. Burriel intended the work for Ensenada and Carvajal. In 1754, Burriel had corresponded with Wall and with Padre Pedro Altamirano, the procurator general of the Indies, about the Fonte account of a Northwest Passage. Sentiments echoing Burriel's appeared in the works of prominent Spanish imperial theorists like Pedro Rodríguez Campomanes's 1762 *Reflexiones sobre el comercio español a Indias.*[23]

Burriel's lengthy and well-documented work gave Spanish imperial officials reason for worry, and members of the Spanish court appear to have done so. In the March 1761 royal instructions to Spain's ambassador in Russia, Pedro Luján

22. Ibid., 13, 16, 238–239, esp. 239. Chapman offers a nice introduction to Burriel's work and to the sources from which much of it was drawn in *Founding of Spanish California*, 56–61. For a recent book on one Anglo-American who tried to reach the Pacific by crossing Eurasia, see Edward G. Gray, *The Making of John Ledyard: Empire and Ambition in the Life of an Early American Traveler* (New Haven, Conn., 2007).

23. Joseph Garcia de Leon y Pizarra and Fernando Joseph Mangino, "Extractos de providencias para el descubrimto del mar del Sur, y Californias, desde la conquista de Indias, y para la exclusion impuesta à todas las naciones extrangeras de navegar aquellos mares," 1790, Estado 2848, carpeta 6, AHN; Fidel Fita, "Noticia de la California, obra anónima del P. Andrés Marcos Burriel, emprendida en 1750, impresa en 1757 y traducida después en varias lenguas de Europa. . . . ," *Boletín de la real academia de la historia*, LII (1908), 396–438, esp. 396–397, 401–403, 407, 409, 416; G[lyndwr] Williams, "An Eighteenth-Century Spanish Investigation into the Apocryphal Voyage of Admiral Fonte," *PHR*, XXX (1961), 319–327, esp. 323; Pedro Rodríguez Campomanes, *Reflexiones sobre el comercio español a Indias (1762)*, ed. Vicente Llombart Rosa (Madrid, 1988), 25, 124, 129.

Jiménez de Góngora y Silva, the marqués de Almodóvar, Charles III enjoined him to be particularly attentive to the conduct and schemes of the English representatives in Saint Petersburg because of the lack of sincerity and good faith the British court was exhibiting on matters involving Spain. Charles also spoke specifically about Russian and British exploration. He charged Almodóvar to investigate how far the Russians had gone in their attempts to reach California ["indagar á qué términos han llegado los descubrimientos de los rusos en las tentativas de su navegacion á la California"], because the Russians had had greater success than other nations and because "the studied silence of that court and of that of London in this matter becomes suspicious" ["se hace sospechoso el estudiado silencio de esa córte y la de Lóndres en este asunto"]. The matter could be "of great significance for both" ["de grande entidad para ambas"], and it was of great importance that the Spanish court be "solidly instructed of the ideas of the above courts, in order to impede their progress" ["sólidamente instruidos de las ideas de las expresadas córtes, para impedir sus progresos"]. From at least 1761, Spanish concern about Russian North Pacific activities was official as well as scholarly.[24]

Almodóvar was not as alarmed about these Russian activities as his superiors were. He transmitted an October 1761 report to Wall downplaying the threat Russian North Pacific exploration posed to Spanish possessions. Almodóvar conceded that in the distant future the Russian Empire's eastern provinces might be capable of great advances, but at the moment, the danger to Spain's American possessions from Russian exploration was "so remote that it scarcely merits consideration" ["me parece tan remoto que apenas merece consideracion"]. Almodóvar's twenty-two-page report evinced a familiarity with the course of eighteenth-century Russian eastward expansion derived from conversations with officials and scholars in Saint Petersburg. He described the 1728 Bering expedition and the 1741–1742 Bering-Chirikov voyages. His estimates of the southward extent of Russian exploration were comparable to those of Burriel; he noted that the most southern point reached by the Russians was about 56° north, some thirteen degrees north of Cape Blanco. But Almodóvar found this Russian southward movement less ominous than did Burriel. He observed that it was more than three thousand leagues from Saint Petersburg to Kamchatka, assistance for expeditions from the Russian provinces closest to the Pacific was little and late, and Russian expeditions directed toward the Pacific thus took a

24. "Instrucción de los que vos . . . marques de Almodóvar, . . . habeis de observar en desempeño del cargo de mi Ministro plenipotenciario cerca de la Emperatriz de la Rusia," Mar. 9, 1761, Estado 6618, AGS (partially reprinted in Abbad y Lasierra, *Descripción de las costas de California*, ed. Hilton, 197–198).

great deal of time to prepare. He suggested, in an echo of old Russian claims, that "these [Russian] voyages are more useful for the advancement of geography than for the enlargement of empire" ["Estos viages mas pueden servir para el adelantamiento de la Geographia, que para el aumento del Imperio"]. It might be easier to approach Asia from America than America from Asia, and consequently the Russians might have more to fear from the Spaniards than vice versa ["Yo no sé si puede decirse con verdad que tanto pueden temer los Españoles a los Rusos en las costas de la America, como los Rusos a los Españoles en las de Asia; y que tal vez es mas facil que los Americanos vengan a hacer conquistas en las costas de Siberia; que el que los Rusos vayan à hacerlas à nuestra America"].[25]

Almodóvar also scoffed at danger to the North Pacific from the English. He denied that they were in any way involved in the Russian expeditions there. All their attempts to reach the South Sea from the northeast had been "up to now useless" ["hasta aora inutiles"]; even if ships could reach the Pacific from the Arctic, the recent Russian voyages had demonstrated this would be "useless for commerce" ["inutil para el comercio"] because of the dangers of navigation in the rigorous northern climate. Besides, it was possible that a Northwest or Northeast Passage only existed "on geographic globes or in the heads of some geographers" ["El camino mas corto, que tanto se busca, y se desea por el Nord-Este, y Nord-Oeste de la Europa para pasar a buscar en el Mar pacifico las Costas de la America es muy posible, que no exista sino en las Glovos Geographicos o en las cavezas de algunos Geographos"].[26]

How Wall and Charles reacted to Almodóvar's report is unclear. Where Sánchez Salvador's reports generated numerous official documents expressing concerns about a possible foreign descent of the Colorado, documents revealing official attitudes about Almodóvar's views are not yet available. Even without such documents, a few points are evident. Although Almodóvar did not admit his ignorance as eloquently as did Burriel, he offered no information about the Pacific coast south of 56°N. Almodóvar's report could neither establish nor disprove the existence of a water route to the Pacific emerging on this stretch of coastline. Although he was skeptical about the existence or value of a Northwest Passage, Almodóvar acknowledged that the British had eagerly sought one. His report omitted discussion of the rumored river routes to the Pacific that might lie farther south in North America—routes whose possible existence had so alarmed Sánchez Salvador, Labaquera, Burriel, and the Council of the Indies. The arctic impediments to navigation that the Russians had experienced coming

25. Almodóvar to Wall, Oct. 7, 1761, Estado 86B, N. 100, 7–28, AGI (a slightly different version reprinted in Abbad y Lasierra, *Descripción de las costas de California*, ed. Hilton, 198–205, esp. 204).
26. Ibid.

from Siberia and that the British were likely to experience coming from Hudson Bay or Canada need not apply to these southern rivers or to segments of the Pacific coast south of 56°N. Almodóvar's report gave good reasons for Spanish officials to refrain from panic about Russian North Pacific expeditions, but it was solace insufficient to assuage Spanish fears that a British presence in the trans-Mississippi West might facilitate Albion's attempts to encroach upon the South Sea.

Spain's 1762 Acceptance of Western Louisiana

And to this acquisition [Canada], had we, during the late war, either by conquest or treaty, added the fertile and extensive country of Louisiana, we should have been possessed of perhaps the most valuable territory upon the face of the globe, attended with more real advantages than the so-much-boasted mines of Mexico and Peru.
— *Journals of Major Robert Rogers*, 1765

Si los Ingleses ocupasen la Luisiana y aliasen con los Indios no reducidos, serían funestas las conseqüencias para el dominio español en la América Septentrional.
— PEDRO RODRÍGUEZ CAMPOMANES, 1762

With Spanish concerns about British and French activities in the unexplored West and the increasing precariousness of a North American imperial equilibrium in mind, it is possible, despite the frustrating silences of the Spanish and French documents, to reach an interpretation of the twists and turns of Spanish Louisiana policy consistent with available Spanish words and observable Spanish actions.[27]

Responding to continued worries about French North American expansion and to new opportunities created by French defeats in Canada, Spanish officials evinced in the summer of 1760 a desire to acquire Louisiana. In July of that year, Ossun reported that Charles had voiced an interest in some kind of territorial exchange placing Louisiana in Spanish hands (upon hearing of this from Ossun, Choiseul hastily rejected the idea). The scanty references to Charles's suggestion in the available documents make clear neither how serious he was about this nor exactly what he meant by "Louisiana." Ossun does not say whether Charles had in mind the entire Mississippi Basin, the lands west of the Mississippi, or the French lands along the Gulf Coast. At this stage in the war, Spain's taking possession of all or part of Louisiana could still have left a North American balance of

27. Robert Rogers, *Journals of Major Robert Rogers: Containing an Account of the Several Excursions He Made under the Generals Who Commanded upon the Continent of North America, during the Late War* (Ann Arbor, Mich., 1966), 196; Campomanes, *Reflexiones sobre el comercio español*, 32.

power in place. France had lost the city of Quebec but had not yet surrendered Canada. It was not yet clear that Britain would want to obtain Canada in the war's peace settlement nor that Britain would be in a position to force France to accede to its demands. In October 1760, after the September 7 French surrender of Canada, Prince Albertini Giovanni Battista San Severino (the principal Neapolitan representative in London, and, because Charles had ruled Naples [1734–1759] before taking the Spanish throne, another of Charles's agents there) was still predicting that Britain would return at least part of Canada to France. If France remained in control of Canada, Spain could take Louisiana while retaining the basis for a North American imperial equilibrium. The same would be true if Spain obtained only a part of Louisiana. British forces would still have to defend the English colonies against the threat of French aggression. Spain could still seek the assistance of whichever of its two North American rivals seemed the most useful and the least menacing. Moreover, Spain had not yet committed itself to joining the war on France's side; so Charles could reasonably hope to acquire Louisiana without also incurring the immediate obligation to defend it against British attacks.[28]

When, a year later, Spain enjoyed an opportunity to obtain what it had requested, changed circumstances resulted in a different Spanish position. By 1761, Charles had grown so concerned about the global progress of British arms that he had come to support the idea of allying with France. In spring and summer 1761, Spanish and French representatives conducted negotiations leading to the August 15, 1761, Bourbon Family Compact (*Pacte de famille*). Under its terms, Spain agreed to enter the war on May 1, 1762. During the summer 1761 negotiations leading to the compact, French officials offered Louisiana "with widest boundaries" to Spain in exchange for either a loan of 3.6 million piasters or an early Spanish declaration of war. Curiously, in light of Charles's announced interest in obtaining Louisiana a year earlier, Spanish negotiators rejected the offer.[29]

Available Spanish and French diplomatic documents throw little light on the

28. On Charles's 1760 interest in acquiring Louisiana, see Ossun to Choiseul, July 4, 1760, 32r, Choiseul to Ossun, July 15, 1760, 74r, both in CP, Espagne, 529; Pease, ed., *Anglo-French Boundary Disputes*, 276–280; Aiton, "Diplomacy of the Louisiana Cession," *AHR*, XXXVI (1931), 702–703. On predictions of Canada's return, see Bonnault, "Le Canada et la conclusion du Pacte de famille," *Revue d'histoire de l'Amérique française*, VII (1953), 351.

29. Choiseul to Ossun, July 31 or Aug. 1, 1761, 210r–212r, Ossun to Choiseul, Aug. 10, 17, 1761, 240r–245v, 320r–325r, all in CP, Espagne, 533; Grimaldi to Wall, Aug. 2, 1761, "Apuntaciones originales con su copia hechas por el duque de Choiseul para hablar en el consejo sobre la contramemoria de la Inglaterra" (partially reprinted in "Choiseul's Speech to the Council," Aug. 1, 1761, in Pease, ed., *Anglo-French Boundary Disputes*, 336–341), both in Estado 4544, AGS; Aiton, "Diplomacy of the Louisiana Cession," *AHR*, XXXVI (1931), 704–712.

reasons for Spain's 1761 refusal to receive Louisiana, but Spanish military limitations were likely a major consideration. It was evident that accepting Louisiana might very well increase Spanish defense burdens. By summer 1761, Grimaldi, having observed the attitudes of his French hosts, had already become uneasy about the French commitment to guarding Louisiana; and although Choiseul had offered to provide French forces to defend the colony even if Spain took formal possession of it, Spanish officials could reasonably suspect that wartime pressures might result in French troops' abandoning a possession French diplomats had relinquished.[30]

Spanish officials could not take the burden of defending Louisiana lightly. Historians have noted that, after coming to the Spanish throne, Charles grew increasingly concerned about the state of Spain's armed forces and that this was one factor making him reluctant to intervene militarily in the Anglo-French struggle. The buildup and reforms of the 1750s had not yet produced a military establishment capable of holding its own against Britain's, and Charles did not want to rush unprepared forces into harm's way. Nor, when ready Spanish units were few, did Spanish officials want to assume new defensive responsibilities. Despite keen Spanish desire to recover Minorca, Wall refused a summer 1761 French offer of it because he did not yet want the onus of garrisoning the island. Similar reasoning might have influenced the Spanish decision to leave Louisiana in French hands. More generally, events such as the 1680 Pueblo Revolt, the 1751 Pima rebellion, and post-1706 Comanche raids had already revealed New Spain's frontier vulnerabilities. Taking on territories stretching from the Rockies to the Appalachians and lying within striking distance of the British forces that had just conquered New France could scarcely make borderland defense easier — and might divert Spanish forces needed elsewhere in the empire.[31]

Perhaps more significant, receiving all of Louisiana in summer 1761 was potentially far more disruptive to a North American imperial equilibrium than taking all or part of the colony would have been a year earlier. Britain now ruled Canada, and Pitt was demanding Britain keep the French colony as part of the war's settlement. If France yielded Canada to Britain and Louisiana to Spain, the French North American mainland presence necessary for a tripartite North American balance of power would no longer exist. Spain might suddenly have to

30. Grimaldi to Wall, July 14, 1761, Estado 4543, Grimaldi to Wall, Aug. 2, 1761, Estado 4544, both in AGS; Aiton, "Diplomacy of the Louisiana Cession," *AHR*, XXXVI (1931), 711.

31. Aiton, "Diplomacy of the Louisiana Cession," *AHR*, XXXVI (1931), 711; Gipson, *British Empire before the American Revolution*, VIII, 248, 250; Ozanam, "Les origines du troisième Pacte de famille," *Revue d'histoire diplomatique*, LXXV (1961), 318–319; Pares, *War and Trade*, 569; Rashed, *Peace of Paris*, 59, 139; Rousseau, *Règne de Charles III*, I, 50; Savelle, *Origins of American Diplomacy*, 460; Palacio Atard, *El tercer Pacto de familia*, 209.

defend a huge expanse of vulnerable territory with inadequate forces and without much in the way of French help. Despite the last decade's concerns about the dangers of Louisiana's westward expansion, in the summer of 1761, Spanish officials were willing to tolerate continued French possession of the colony.

Instead of taking Louisiana, Spanish diplomats tried, between summer 1761 and November 1762, as the progress of British arms continued, to secure a North American border settlement that would not only maintain a French continental presence and prevent French encroachments on New Spain but also deny British access to the Mississippi and Gulf of Mexico. The Spanish challenge was to keep Louisiana in France's hands and out of Britain's. Seeing the growing desperation of French officials, Grimaldi and Wall expressed concern in summer 1761 about French willingness to yield part of Louisiana to Britain in exchange for British concessions elsewhere. Initially, Spanish fears had centered on the possibility that Britain would gain territory on the Gulf of Mexico, making it more difficult for Spain to exclude British ships from its waters. Indeed, Spanish statesmen came to favor a French alliance in part because it would make monitoring French diplomats easier and forestall this kind of French cession. Ossun observed in July 1761 that one of Spain's essential interests in negotiating a French alliance was preventing his government from treating Louisiana like Canada. Grimaldi and Wall agreed that keeping Louisiana out of British possession justified by itself Spain's negotiations with France. They remonstrated repeatedly against French inclinations to yield Gulf Coast or Mississippi River territory to Britain. One tactic Charles and Grimaldi employed was denying to Ossun and Choiseul the validity of French and English claims to Louisiana and Georgia and asserting that France could not, therefore, alienate territory to Britain without Spanish permission. This argument was unlikely to persuade anyone who wasn't Spanish, or at least working for Spain, and Grimaldi and Charles quickly tried the opposite approach. In August 1762, they mentioned again Spain's willingness to acknowledge France's Louisiana claims as part of a three-power boundary settlement maintaining Britain's distance from the Gulf and Mississippi River. These were valiant efforts to preserve a particular vision of North America, but the reality was that the pace of British victories and French collapse meant that Spain was going to have to adapt to circumstances rather than influence them.[32]

32. On Spanish concerns about France's yielding portions of Louisiana to Britain, see Grimaldi to Wall, Feb. 13, 1761, Estado 4548, Grimaldi to Wall, July 14, 1761 (rpt. in Pease, ed., *Anglo-French Boundary Disputes*, 319–320), Wall to Grimaldi, July 31, 1761, both in Estado 4543, and Grimaldi to Wall, Aug. 2, 1761, Estado 4544, all in AGS. On Spanish fears of Britain's gaining territory on the Gulf of Mexico, see, for example, "Copia de capitulo del marqs de Grimaldi," June 28, 1762, Estado

In October 1762, France again offered trans-Mississippi Louisiana to Spain, and this time Spain accepted the offer. In contrast with the summers of 1760 and 1761, Spain could do little else. Grimaldi and Wall had worried that France might offer portions of Louisiana to Britain; they now observed that France had already done so with the eastern part of the colony (on September 20). Frustrated, moreover, by Spanish reluctance to accept British peace terms, Choiseul was also making it plain to his Spanish counterparts that France was at liberty to give Britain western Louisiana as well ["Représentés avec force à Madrid, Monsieur, la liberté où est le Roi de céder ou même de faire évacuer ces possessions"].[33]

Having brandished the stick of ceding western Louisiana to Britain, France then held out the carrot of offering it to Spain. On October 9, recognizing Spanish discontent over the loss of eastern Louisiana and its gulf coastline to Britain and still seeking some way to induce Spain to consent to a humiliating settlement with Britain, Louis XV offered the trans-Mississippi remnant of French Louisiana to Spain, suggesting that Spain might, in turn, exchange it for a captured Spanish possession such as Havana. Sometime in the same month, Louis also offered western Louisiana directly to Britain as a substitute for the Spanish colony of Florida. British negotiators refused, preferring to acquire territory along which eastbound ships sailed, from which Spaniards and Indians had attacked Georgia and South Carolina, to which South Carolina and Georgia slaves

---

4550, AGS. For Ossun's July 1761 observation, see Ossun to Choiseul, July 27, 1761, CP, Espagne, 533, 148r–148v, 158v. For Wall and Grimaldi's emphasis on keeping Louisiana out of British hands, see Grimaldi to Wall, July 14, 1761, Wall to Grimaldi, July 31, 1761, both in Estado 4543, Choiseul to Ossun, July 22, 1762, Estado 4547, "Copia de capitulo del marqs de Grimaldi," June 28, 1762, Estado 4550, and Grimaldi to Wall, July 22, 1762, Estado 4551, all in AGS; Wall to Grimaldi, Sept. 29, 1762, Estado 4176, AHN; Savelle, *Origins of American Diplomacy*, 498–502. For discussions of the validity of non-Spanish claims to Louisiana territory, see Grimaldi to Choiseul, Aug. 13, Sept. 15, 1762, Estado 4551, AGS; Wall to Grimaldi, Aug. 2, 1762, Estado 4176-2, AHN; Ossun to Choiseul, Aug. 2, 1762, CP, Espagne, 537, 4r–12v (partially reprinted in Pease, ed., *Anglo-French Boundary Disputes*, 496–499); Shepherd, "Cession of Louisiana to Spain," *PSQ*, XIX (1904), 442–446.

33. On the French offer of eastern Louisiana to Britain and Spanish reactions thereto, see Grimaldi to Wall, July 14, 1761, Estado 4543, Choiseul to Ossun, July 22, 1762, Estado 4547, "Copia de capitulo del marqs de Grimaldi," June 28, 1762, Estado 4550, and Grimaldi to Wall, July 22, 1762, Estado 4551, all in AGS; Wall to Grimaldi, Sept. 29, 1762, Estado 4176-1, AHN; Ossun to Choiseul, Aug. 2, 1762, 4r–12v, "Copie d'une dépêche de Mr. le duc de Choiseul à Mr. le mis. d'Ossun à Versailles le 20 7.bre. 1762," 160r–183v (partially rpt. in Pease, ed., *Anglo-French Boundary Disputes*, 525–534), both in CP, Espagne, 537; Aiton, "Diplomacy of the Louisiana Cession," *AHR*, XXXVI (1931), 715–717. New Orleans, though east of the river, was not included in this cession.

On Choiseul's threat to cede western Louisiana to Britain, see "Copie d'une dépêche de Mr. Le duc de Choiseul," 163v–164r, Louis XV to Charles III, Nov. 3, 1762, 292v–293r, both in CP, Espagne, 537; Wall to Grimaldi, Oct. 23, 1762, Estado 4551 (also in Estado 4176-1, AHN), Louis XV to Charles III, Nov. 3, 1762, and Charles III to Louis XV, Estado 4552, all in AGS; Aiton, "Diplomacy of the Louisiana Cession," *AHR*, XXXVI (1931), 717; Shepherd, "Cession of Louisiana to Spain," *PSQ*, XIX (1904), 448.

had fled, and in which land-hungry Anglo-Americans could settle. Charles also rejected the idea of keeping Florida Spanish at the expense of making western Louisiana British. Wall observed that British acquisition of either Florida or Louisiana damaged Spanish interests but that Charles dreaded Britain's gaining Louisiana more than he feared losing Florida ["mas teme S.M. que la Luisiana quede en poder de ellos que de perder la Florida"]. Spain had considered Florida important for centuries because galleons sailed through the Straits of Florida on their voyage to Spain. That Charles considered trans-Mississippi Louisiana more significant attests to its perceived value. On November 3, Charles acceded to the French offer of western Louisiana.[34]

Toward the end of the Seven Years' War, Spain's leaders and diplomats pursued two objectives: protecting northern New Spain and the Pacific from French aggression and maintaining a French North American presence capable of containing British westward expansion. As long as France retained Canada, Spain could achieve both objectives by acquiring all or large parts of Louisiana. French Canada would continue to attract the lion's share of British attention while Spanish Louisiana could shield New Spain. Charles's July 1760 interest in Louisiana flowed logically from these objectives. When it became increasingly clear after September 1760 that Britain would retain Canada after the war, it became evident also that Spanish acquisition of Louisiana would remove France from North America and preclude a North American balance of power: hence Spain's rejection of France's summer 1761 offer of Louisiana. So Spanish officials pursued into the fall of 1762 a border settlement that would simultaneously maintain and contain a French North American presence. French willingness to

34. On the initial French offer of western Louisiana to Spain, see Choiseul to Ossun, Oct. 9, 1762, 215r–219v, "Copie de la lettre du roi au roi d'Espagne," "Joint à la depêche de la cour du 9 Oct. 1762," 221r–222v, Ossun to Choiseul, Oct. 22, 1762, 271v–272r, all in CP, Espagne, 537. On Anglo-French discussions of Louisiana's transfer, see Wall to Grimaldi, Oct. 23, 1762, Estado 4551, Louis XV to Charles III, Nov. 3, 1762, Estado 4552, both in AGS; Wall to Grimaldi, Oct. 23, 1762, Charles III to Louis XV, Oct. 23, 1762, both in Estado 4176–1, AHN; Louis XV to Charles III, Nov. 3, 1762, 292v–293r, Ossun to Choiseul, Nov. 12, 1762, 318v–319r, both in CP, Espagne, 537; Aiton, "Diplomacy of the Louisiana Cession," AHR, XXXVI (1931), 717; Shepherd, "Cession of Louisiana to Spain," PSQ, XIX (1904), 448. On Charles's fear of Britain's gaining Louisiana, see Wall to Grimaldi, Oct. 23, 1762, Estado 4551, AGS; Charles III to Louis XV, Oct. 23, 1762, Estado 4176–1, AHN. On Florida's traditional importance, see Pontchartrain, "Memoire concernant les colonies, le commerce, et la navigation, pour Messrs. les plenipotentiares du roy," Jan. 2, 1712, MD, France, 1425, 85v; "Memoir on English Aggression," October 1750, in Pease, ed., Anglo-French Boundary Disputes, 3; Campomanes, Reflexiones sobre el comercio español, 28; Pares, War and Trade, 23–24; Savelle, "American Balance of Power and European Diplomacy," in Morris, ed., Era of the American Revolution, 151; Wright, Anglo-Spanish Rivalry in North America, 5, 110. On Charles's acceptance of Louisiana, see Wall to Grimaldi, Oct. 23, 1762, Estado 4551, Wall to Grimaldi, Nov. 13, 1762, Estado 4552 (two letters), all in AGS; Ossun to Choiseul, Oct. 22, Nov. 12, 1762, CP, Espagne, 537, 267v, 317r.

sacrifice all or part of Louisiana to Britain undermined this solution. If France abandoned North America and ceded Louisiana to Britain at the same time, Spain would lose any hope of a North American imperial equilibrium and would simultaneously confront the British westward expansion Spanish policy had endeavored to prevent.

In the available Spanish documents, neither Charles nor Wall nor Grimaldi discloses his reasons for fearing British acquisition of trans-Mississippi Louisiana. Scholars who have considered the question of why Spain accepted the French colony emphasize a desire to keep Britain as far as possible from Mexico and its silver mines. This was undoubtedly an important factor. A complete account of Spanish policy requires inclusion of another concern shaping Spanish decisions: namely, that British acquisition of trans-Mississippi Louisiana would forward British designs on the Spanish Pacific. Britain, with its search for a Northwest Passage, Anson's Pacific raid, Central American logging camps, and planned and rumored 1749, 1750, and 1751 South Sea voyages, had declared its Pacific ambitions. Spanish officials had repeatedly expressed concern about these aspirations. They had also voiced their worries that the undiscovered American Southwest's great rivers might provide another route to the Spanish Lake, and French explorers might find and exploit it. But in autumn 1762, the threat of the triumphant British Empire overshadowed the dangers of French American expansion. British acquisition of trans-Mississippi Louisiana might, if the geographic rumors were true, give Britain the long-sought passage allowing easy extension of British naval power into the Pacific. By November 1762, British Seven Years' War victories had eliminated the possibility of a North American imperial equilibrium. British dominion over the American West and advance toward the Pacific would tip the global balance of empire still further in Britain's favor. Charles and Spain could not risk this. Spain had to get western Louisiana out of the shaking hands of France and keep it away from the grasping hands of Britain. Accepting France's offer of trans-Mississippi Louisiana made the best of a bad situation.[35]

Spanish concerns about the Pacific need to be kept in perspective. In November 1762, Spanish officials possessed two independently sufficient reasons for forestalling British possession of trans-Mississippi territory. Even if they had known the American West contained no practicable passage to the Pacific, the British menace to Mexico's silver mines would still have provided ample cause

35. On Spanish concern about British encroachments on silver mines, see, for example, Aiton, "Diplomacy of the Louisiana Cession," *AHR*, XXXVI (1931), 718; John, *Storms Brewed,* 377; Shepherd, "Cession of Louisiana to Spain," *PSQ*, XIX (1904), 449–451. For one expression of fear of British Pacific expansion, see Campomanes, *Reflexiones sobre el comercio español,* 27, 31–32.

for the Spanish Empire to oppose British westward expansion. Because the value of those mines to Spain and the British lust for bullion was beyond doubt, this was likely the most important force driving Spanish policy. But if, through the operation of some kind of alchemy gone wrong, those Mexican silver mines had magically lost all their value in the fall of 1762, Spanish fear of a British advance on the Pacific through the unexplored West would have furnished Charles with reason enough to take western Louisiana.

PROBLEMS OF SPANISH borderlands governance and defense remained intractable after Charles's acceptance of Louisiana. In one sense, the change of imperial colors on maps of North America was rather abstract. French cession of trans-Mississippi Louisiana to Spain did not mean that France had actually controlled the territories in question nor that Spain suddenly would. The cession was simply an agreement that France would not henceforth try to obtain such eminence. Western Indians who had never troubled themselves too much about French notions of sovereignty, and who had not been a party to the agreement rather presumptuously giving their lands away, were not overly bothered by the transfer of French sovereignty to Spain. "The Scratch of a Pen" changed many aspects of life in North America while having little effect on many others. French and Spanish officials had not seen in the trans-Mississippi West an indigenous power they deemed worthy of a seat at the metropolitan negotiating table, and they tried to do to western Indians what other European states would soon do to Poland. Nations like the Comanches would contest on American ground what had been decided in European capitals. Much of the history of the Spanish-Indian borderlands in later decades would revolve around Spanish officials' trying and failing to exert authority over western peoples and territories. One can only speculate about how the experiences of western Indians would have differed had the European imperial power perhaps least immediately inimical to their interests retained an imperial presence in New Orleans and the lands beyond the Mississippi.[36]

Ultimately, revolutions would swamp the efforts of Charles, his ministers, and their successors to preserve and enhance the Spanish Empire. The French Revolution would give Napoleon his chance to rule France, overawe Spain, and retake western Louisiana. A failed intervention in the Haitian Revolution would render it expedient for him to sell Louisiana to a newly independent Anglo-

36. For a judicious appraisal of the 1762–1763 peace settlement's consequences, or lack of consequences, in North America, see Colin G. Calloway, *The Scratch of a Pen: 1763 and the Transformation of North America* (Oxford, 2006). For a bold argument denying its importance for parts of the West, see Pekka Hämäläinen, *The Comanche Empire* (New Haven, Conn., 2008), 68–69.

American republic as covetous of Florida and more dangerous to Mexico than the British Empire had been. The Spanish Empire's own revolutions would deprive it of its remaining mainland North American possessions, and an independent Mexico would find itself, as Spanish and French diplomats had long feared, despoiled of its northern territories by the descendants of the British colonists of the Atlantic seaboard. Despite considerable acumen, adeptness, and application, Charles and his agents Wall and Grimaldi could delay, but they could not prevent, the loss of Spain's North American possessions.

15

OLD VISIONS AND NEW OPPORTUNITIES

Britain and the Spanish Empire at the End of the Seven Years' War

Florida, he [Pitt] said, was no compensation for the Havannah; the Havannah was an important conquest.... From the moment the Havannah was taken, all the Spanish treasures and riches in America, lay at our mercy. Spain had purchased the security of all these, and the restoration of Cuba also, with the cession of Florida only. It was no equivalent....

...They ["the friends of the ministry"] expatiated on the great variety of climates which that country [North America] contained, and the vast resources which would thence arise to commerce. That the value of our conquests thereby ought not to be estimated by the present produce, but by their probable increase. Neither ought the value of any country to be solely tried on its commercial advantages; that extent of territory and a number of subjects are matters of as much consideration to a state attentive to the sources of real grandeur, as the mere advantages of traffic; that such ideas are rather suitable to a limited and petty commonwealth ... than to a great, powerful, and warlike nation.

— "MOTION ON THE PRELIMINARIES OF PEACE," December 9, 1762

One character making occasional appearances in earlier chapters has been Henry Ellis (1721–1806). Like Arthur Dobbs, Ellis was one of those second-tier figures of eighteenth-century British imperial history who frequently influenced or exemplified important historical developments. In a sparkling 1970 essay, John Shy used Ellis to help illuminate the "spectrum of imperial possibilities" following the Seven Years' War. In this book, Ellis has spoken confidently of the great islands of necessity existing in the North Pacific. He has participated in the 1746–1747 British expedition in search of a Northwest Passage from Hudson Bay, and his account of the voyage, and the continuing openness to the existence of the passage it exhibited, seems to have been one of the key documents that alarmed late 1740s and early 1750s French officials with the scale of British ambitions. Ellis went on to serve as governor of Georgia (1757–1760) and then principal advisor on colonial affairs to the secretary of state for the Southern Department, Charles Wyndham, earl of Egremont (1761–1763). This last position was a significant one, because, as Ellis's biographers have noted, Egremont was an inexperienced man in a powerful office, open to and in need of advice from a vigorous character.[1]

This was an ominous situation for the Spanish Empire. Ellis's earlier writings had evinced the kind of bold and competitive spirit so worrisome to Britain's adversaries. In the preface to his *Voyage to Hudson's-Bay,* Ellis had wanted to use a Northwest Passage to open relations with American peoples who had "had continual Wars with, the Spaniards." He had wished "to revive all that Ardour and Dilligence which was so conspicuous" in the age of Francis Drake and John Hawkins when "we first opened a Passage to the *East* and *West-Indies; . . .* when almost every Port in *England* fitted out Vessels to share in that Commerce by which the *Spaniards* and *Portuguese* had been so suddenly and so surprisingly enriched." He had argued that finding a Northwest Passage, by "making such an addition to our Trade, as may afford new Funds for discharging old Debts," would assist British finances, a crucial consideration for a nation adding up the bills of the War of the Austrian Succession in 1748. At the end of the Seven Years' War, Britain's debts were higher and a staggering Spanish Empire far more vulnerable than had been the case in 1748. Ellis was another in a series of eighteenth-century Britons hoping to use unrestrained looting of the Spanish Empire to finance British grandeur. In 1761 and 1762, he was in a position to forward such a design, and he did so, but he did not push it as far as might have been expected. Though Ellis was recognizably an older version of the adventurous young man who had run so far away to sea that he had ended up freezing in Hudson Bay, in certain key respects, Ellis's counsel to Egremont bore more resemblance to the ideas of the Louisiana-ceding duc de Choiseul than to the demands of the world-conquering William Pitt.[2]

Reasons for the late–Seven Years' War territorial choices of the British Empire should not have had much in common with those driving France's 1762 western Louisiana cession. The empires' wartime experiences ran in different directions. The French Empire went from early victories to inexorable defeats.

1. "Commons Debate: Motion on the Preliminaries of Peace," Dec. 9, 1762, in R. C. Simmons and P. D. G. Thomas, eds., *Proceedings and Debates of the British Parliaments Respecting North America, 1754–1783,* 6 vols. (Millwood, N.Y., 1982), I, 416–424, esp. 420–423. On Ellis's influence, see Edward J. Cashin, *Governor Henry Ellis and the Transformation of British North America* (Athens, Ga., 1994), 153, 163, 169–170, 265; John Shy, *A People Numerous and Armed: Reflections on the Military Struggle for American Independence,* rev. ed. (Ann Arbor, Mich., 1990), 45–50.

2. Henry Ellis, *A Voyage to Hudson's-Bay, by the "Dobbs Galley" and "California," in the Years 1746 and 1747, for Discovering a North West Passage . . .* (Dublin, 1749), iii–x.

Its dire end-of-the-war circumstances forced desperate maneuvers and painful sacrifices upon its leaders. Britain passed from initial debacles to remarkable triumphs. In the early 1760s, its statesmen could seemingly take their pick of Spanish and French imperial territories. A logical consequence would be a British official elation contrasting with French diplomatic resignation. Hints of this are visible in the second part of the commons debate above: its praise of imperial magnitude stands in opposition to the small-is-beautiful emphasis of many early 1760s French imperial theorists.

It is odd, then, when viewing eighteenth-century British governmental documents regarding the Spanish Empire, to find attitudinal developments resembling those in French official papers concerning the North American West. French foreign office optimism regarding trans-Mississippi territories between 1711 and 1747 gave way, from 1748 to 1762, to a growing skepticism about their value. A comparable pattern appears in documents discussing the value to the British Empire of commercial and territorial expansion in Spanish America. Overbearing optimism at the end of the War of the Spanish Succession and around the War of Jenkins' Ear yields increasingly, in the early 1760s, to a sober and cautious restraint.

What takes this Anglo-French comparison beyond the realm of diverting but immaterial coincidence is the opportunity it provides to extend previous chapters' arguments and elucidate a puzzling feature of British imperial policy. When early 1760s British territorial acquisitions are viewed in light of the eighteenth-century Anglo-imperial conduct so preoccupying previous chapters' French and Spanish statesmen, British ministers appear to have acted even more anomalously at the end of the Great War for Empire than did their French and Spanish counterparts. During the century's first six decades, the British Empire's subjects and statesmen had tried repeatedly to penetrate Spanish imperial markets and pry away Spanish imperial wealth. They had founded a South Sea Company in 1711, requested Spanish American trading sites and commercial privileges at the Utrecht negotiations, besieged Spanish American ports during the War of Jenkins' Ear, and essayed Cape Horn and Hudson Bay routes to the Spanish Lake in the 1740s. In late 1762, after a succession of victories over France and Spain, the British government enjoyed the choice of continuing hostilities in the expectation of future conquests or insisting upon extensive and remunerative cessions as the price of peace. If ever Britons wanted to despoil the Spanish Empire of maritime jewels like Havana, Veracruz, Portobello, Darien, Buenos Aires, the Falklands, Juan Fernández, Panama City, or Manila and to achieve diplomatically sanctioned and territorially sustained commercial dominance of

Spanish America, the end of the Seven Years' War would seem to have been the time. Instead, British ministers settled for Florida, site of the most famous swamp in North America.

The choice raised questions for critics then and historians later. William Pitt and his many supporters outside Parliament bemoaned the wasted opportunity to gut the Spanish rival and enrich the British Empire. Florida disappointed them. Unlike Peru and New Spain, it possessed no silver mines. Unlike Darien, it offered no South Sea access. Unlike coastal Mexico, Panama, Colombia, and Argentina, it possessed neither important ports at which slaves and manufactures could be sold nor populous hinterlands from which mineral and vegetable wealth could be immediately extracted. Florida "compared," parliamentarian William Beckford averred, "for barrenness to Bagshot Heath." After all Britain's Seven Years' War sacrifices, after two centuries of fearing, reviling, and coveting Spain's empire, was this all? Unremunerative wilderness and unruly Indians? It is hard to imagine the founders of the South Sea Company and the brandishers of Jenkins' ear making the same choices.[3]

One way to better understand this apparent imperial change of direction is by evaluating Britain's Florida acquisition in the light of France's western Louisiana cession. Juxtaposition of similarly evolving British and French notions of American opportunities points to a development spilling across the usual imperial boundaries. The appeal of comparable notions about the possibilities of American empire diminished before and during the Seven Years' War. A marvelous Indies of Pacific passages, South Sea markets, cooperative Amerindians, collaborating creoles, agricultural paradises, and precious minerals lost ground to a prosaic but profitable Anglo-French New World of colonial cultivation and consumption. Imperial maturation was instructing a colonial reevaluation. Advantages from the Spanish Empire and the mysterious West had proved arduous to acquire and difficult to retain. Those from many of the less-glamorous British and French Caribbean and British Atlantic seaboard colonies were turning out to be both substantial and reliable. For Britain, moreover, benefits from the longest sought and most splendid Indies of them all, the Indian subcontinent, were rather surprisingly turning out to be suddenly and lucratively available. Centers of imperial gravity were shifting and with them the orientation of European policy.[4]

3. Beckford quoted in "Commons Debate: The Address," Nov. 25, 1762, in Simmons and Thomas, eds., *Proceedings and Debates*, I, 402–403, esp. 402.

4. On the British side of post–Seven Years' War imperial reevaluations, see P. J. Marshall, *The Making and Unmaking of Empires: Britain, India, and America, c. 1750–1783* (Oxford, 2005), 82.

Early 1760s British Approaches to the Spanish Empire

When we had spent a hundred millions, should we throw away the fruit, rather than spend twelve more? Let a man so narrow-minded stand behind a counter, and not govern a kingdom. . . .

. . . The successes which had attended the British arms in all parts of the world, and the immense advantages gained in our trade, . . . would more than compensate the great expence we had been at . . . a consideration that had been overlooked by those who were complaining of the heavy burthen of the war.

—WILLIAM PITT's November 13, 1761, and May 12, 1762, parliamentary speeches

In December 1762, the ministry of John Stuart, earl of Bute, submitted preliminary articles of peace to Parliament. Britain was to receive Florida from and return Havana to Spain. There were 319 votes to approve the articles and only 64 to reject them. The better part of the deciding voices in British government favored a peace leaving the Spanish Empire essentially intact, where decades before, public and Parliament had forced a war to loot it.[5]

The peace terms had much to commend them. George III, his royal favorite and leading minister Bute, and many others thought obtaining Florida from Spain and Canada, eastern Louisiana, and the Newfoundland fisheries from France sufficient reward for a victorious war. Bute and many other British grandees wanted to gain security for the Atlantic-American colonies that had incited and endured the late unpleasantness. Canada, eastern Louisiana, and Florida placed miles between England's Atlantic settlements and the hostile trans-Mississippi empire with which they would share the continent, while simultaneously offering plenty of room for future Anglo-colonial expansion. At the same time, British ministers and diplomats hoped acquisition of such heretofore unprofitable and uncoveted continental territories would strike European nations as too innocuous to justify formation of a coalition against British hegemony. More basically, as the Anglo-French conflict had dragged on and British expenses had mounted, statesmen with an eye for balance sheets had come to see — despite Pitt's harangues — the imperative of ending a "ruinous war" before Britain sank "from a dream of ambition to a state of bankruptcy." Political exigencies also played a part: Bute and his allies loathed Pitt and favored a quick peace that would deprive him of any opportunity to return to power.[6]

---

5. "Commons Debate: The Address," Nov. 13, 1761, and "Commons Debate: Committee of Supply," May 12, 1762, in Simmons and Thomas, eds., *Proceedings and Debates,* I, 363–363, 399–400, esp. 363, 399.

6. Quotations are from Richard Rigby and Charles Townshend, in "Commons Debate: Motion for Muster Rolls of Hessian Troops," Jan. 25, 1762, ibid., 385–386, esp. 386. For a recent treatment of

It is fairly clear which arguments and considerations convinced 1762-1763 British leaders, less evident why such reasoning was so overwhelmingly persuasive. Arguments like those of Pitt's and Bute's camps had appeared before, but during the wars of the Spanish Succession and Jenkins' Ear, the more aggressive approaches to the Spanish Empire had borne fruit in chartered companies, diplomatic demands, and declarations of war. Britons of the 1710s and 1730s might not have captured much Spanish American territory, but they were certainly captivated by it, and it is unlikely that imperial enthusiasts in earlier decades would have walked away from an opportunity like 1762. The hold of the Spanish Empire upon the British imagination had loosened enough for figures like Bute and his followers to deem it possible and wise to do so, for the counter's counsel to outweigh Pitt's exhortations.

It was not that the old schemes were forgotten or the old conquering spirit had vanished. August warriors and imperialists like Anson and Pitt were, after all, still alive and, in Anson's case, in a position to dispatch expeditions. Venerable ideas made fresh appearances in government deliberations. Examples can be seen in documents collected by Egremont to assist British strategists and diplomats. One paper of unknown authorship, "Hints relative to the Preliminaries for a Peace with France and Spain," recommended that Britain assert claims to Darien and Campeche to counter Spanish efforts to end British access to Central American logwood. Another document, which appears to date from before the War of Jenkins' Ear, recommended assaulting Portobello, Darien, and Panama City, hoped for an "open Passage to the South Seas" there, and discussed attacks on Buenos Aires, Veracruz, and Mexico City. Ellis favored seizure of both Havana and Buenos Aires. In British hands, he argued, these cities would hinder movement of Spanish goods through the Straits of Florida and from the South American interior, cutting off Peru and New Spain's inhabitants from Spain itself. These documents fit nicely with those of 1711 or 1739.[7]

Anson's 1740s accomplishments having extended definitions of the possible,

---

1762-1763 British policy discussions, see Fred Anderson, *Crucible of War: The Seven Years' War and the Fate of Empire in British North America, 1754-1766* (New York, 2001), 487-517.

7. On Anson, see Alan Frost, *The Global Reach of Empire: Britain's Maritime Expansion in the Indian and Pacific Oceans, 1764-1815* (Carlton, Victoria, Australia, 2003), 33. On claims in Central America, see "Hints relative to the Preliminaries for a Peace with France and Spain," n.d., PRO 30/47/7. For War of Jenkins' Ear projects, see "Projects of Attacks on New Spain," n.d., PRO 30/47/14, 192r-202r. I infer this document dates from before the War of Jenkins' Ear because its discussion of the feasibility of attacking Spanish possessions such as Cartagena makes no mention of British experiences during that war. It is perhaps telling that ministers had to go back decades to find enough comprehensive and enthusiastic conquest proposals. For Ellis's views, see Henry Ellis to earl of Egremont, Jan. 16, 1762, PRO 30/47/14, 244v-245r.

the Pacific projects aired in the early 1760s were, in some respects, bolder than those of earlier decades. Supplementing his Atlantic plans, Ellis suggested to Egremont in January 1762 that Britain consider dispatching a squadron to the South Sea as it had sent Anson during the War of Jenkins' Ear. Should Spain, suffering, Ellis hoped, from the loss of Havana and Buenos Aires, try to restore trade by rounding Cape Horn or threading the Straits of Magellan, "an English Squadron sent to the South Seas would thoroughly embarrass and constrain her to exhaust her Revenues in Fortifying her Towns in that part of the Globe; as well as compel her to send thither a considerable Share of her Naval Force, which would expose her to all manner of Insults and losses, in other parts of the World."[8]

Bolder still were plans to take Manila. In what seems like an echo of earlier decades, a proposal for the attack on the city spoke of its "Known Wealth and Opulency," of the Acapulco galleon's treasures lying there ready for plunder, of Manila's possessions facilitating British East Asian trade, and of the town's vulnerability to British attack. In a similar vein, a "Rough Sketch of an Expedition to M[anil]a" argued that seizing the city would interrupt "all trade or intercourse betwixt the E. Indies and the Spanish American provinces in the South Seas." The sketch contended further, betraying a rather optimistic notion of Pacific distance, that from "a British settlement . . . in some one of the neighboring Philippine Islands," "the Spanish provinces in the South Seas, both of South and North America" could be "with great success . . . insulted and plundered." More striking, of course, than the articulation of these plans was their execution: though a small Buenos Aires expedition failed entirely, other British forces not only managed to take unassailable Havana on one side of the globe but remote Manila on the other.[9]

8. Ellis to Egremont, Jan. 16, 1762, PRO 30/47/14, 244v–245r. Another possible indication of British interest in the Pacific comes from Henry Hutchinson, a figure active in the promotion of Anson's Pacific expedition. Hutchinson claimed around October 1762 to have "had the Honour of Several friendly Conversations with Lord Granville particularly Relating to the Spanish West Indies, and the South Sea to our Mutual Satisfaction." See Add. Mss. 47,014c, 79r–79v. For excellent accounts of Hutchinson and his activities, see Glyndwr Williams's *Great South Sea: English Voyages and Encounters, 1570–1750* (New Haven, Conn., 1997), and *Prize of All the Oceans: The Triumph and Tragedy of Anson's Voyage round the World* (London, 1999). For an account using some different documents and identifying a more aggressive end-of-the-Seven Years' War British attitude toward the Spanish Empire, see Alan Frost, "Shaking off the Spanish Yoke: British Schemes to Revolutionise Spanish America, 1739–1807," in Margarette Lincoln, ed., *Science and Exploration in the Pacific: European Voyages to the Southern Oceans in the Eighteenth Century* (Woodbridge, Suffolk, U.K., 1998), 19–37.

9. For the proposal to attack Manila, see "Reasons and Consideration upon the Enterprize against the Philippine Ilands," n.d., PRO 30/47/20, third packet of documents, 1r–2v. The "Rough

What distinguishes the early 1760s corpus of documents from those of earlier periods is, not the absence of aggressive schemes, but rather the presence around them of cautious notes indicative of the attitudes that ultimately marginalized Pitt and resulted in a more restrained approach to the Spanish Empire. Where decades before, prudent tones in one paragraph tended to give way to a rash recommendation in the next, in the early 1760s, discretion restrained diplomacy. Viewed as a whole, discussions of the Spanish Empire evinced a more sober sense of the possibilities offered by British conquest than had been the case earlier in the century.

Debates over the Manila expedition provide a good example. In the venture's planning stages, significant discussion centered, not on the taking of Manila, but on the returning of it. The Secret Committee of the East India Company expressed the fear that "Manilha being an object of infinite importance to the Spanish nation, the Company can hardly flatter themselves with holding it when Peace takes place.... Before we are thus established according to human reason, the Company must deliver it back again." The committee went on to request compensation in this case for their expenditures. This represents a sharp departure from Bladen's 1739 contention that seizing Darien would be worth the effort and expense even if the territory were only held a few years. Taking East India Company concerns as well founded, Egremont referred in his correspondence with the Secret Committee to the eventuality in which "this Conquest should be restored by a Treaty of Peace, before the Company shall have received advantages therefrom, adequate to their expences in this expedition." He indicated that in such a case the king would indeed ask Parliament to recompense the company. The Secret Committee was wise to worry: the British government returned Manila to Spain in 1764.[10]

In the contrasting case of Florida, Britain kept for two decades what it had obtained at the Peace of Paris, but the acquisition generated less enthusiasm

---

Sketch of an Expedition to M[anil]a, Mentioned to Lord A[nson] on the 8th, 11th, and 12th Inst. January 1762" is reprinted in Nicholas P. Cushner, ed., *Documents Illustrating the British Conquest of Manila, 1762–1763* (London, 1971), 12–15, esp. 12. On the Buenos Aires expedition, see Lawrence Henry Gipson, *The British Empire before the American Revolution*, VIII, *The Great War for the Empire: The Culmination, 1760–1763* (New York, 1954), 268–269.

10. On the East India Company's concern that Manila would be returned to Spain, see Secret Committee of the East India Company to Egremont, Jan. 14, 1762, in Cushner, ed., *Documents Illustrating the British Conquest of Manila*, 15–17, esp. 16–17 (another copy in PRO 30/47/20, third packet of documents, 5v). On Egremont's views, see Egremont to Secret Committee, Jan. 23, 1762, ibid., 25 (another copy in PRO 30/47/20, third packet of documents, 8r). On the postwar restoration of Manila, see Egremont to Richard Aldworth-Neville, Whitehall, Apr. 21, 1763, 277r, Neville to marqués de Grimaldi, Paris, Apr. 25, 1763, 296r–297v, Egremont to duke of Bedford, Whitehall, May 20, 1763, 25r–25v, all in SP 78/256.

than had the Spanish territories coveted earlier in the century. Some Florida discussions painted the new British colony in rosy hues reminiscent of earlier decades' discussions of Argentina and Guatemala. One document spoke of "a very large quantity of excellent Lands . . . from the equality and warmth of the Climate extremely well suited to the Culture of Silk, Indigo, Cotton, and probably Sugar and Coffee." On the whole, though, expectations for Florida's future value were modest and mixed. Some authors questioned the desirability of obtaining it at all. "Hints relative to the Preliminaries for a Peace with France and Spain" recommended leaving "to the Spaniards the Peninsula of Florida, and the Coasts on the Bay of Apelache," asserting that the region would "be found much more useful in their hands, than if taken into our own." Though he recommended taking Florida, Henry Ellis acknowledged that it had been a money-losing rather than a revenue-generating colony for Spain and that acquiring it "would not in my opinion be of any immediate value to us, nor in any degree hurt Spain, to which it has long been a very heavy Burden, without affording her the least benefit."[11]

For those who did covet Florida, its appeal had more to do with its relation to other British and Spanish territories than with its value in its own right. In keeping with the emphasis on colonial security marking early 1760s British deliberations, one benefit highlighted was that possession of Florida would protect other British colonies. The author of "Heads of Advantages etc. Gained by Treaty, and State of America" argued that acquisition of Florida would give "full security to our Colonies in that Quarter" and would deprive escaped slaves of a Spanish refuge: while the Spanish "continued in Florida our Frontiers would never be settled, as our slaves were seduced to, and towd. Liberty and protection at St. Augustine." A second gain was Florida's use as a base for contraband trade with other Spanish colonies. Ellis's "Advantages Which England Gains by the Present Treaty with France and Spain," composed for parliamentary debates regarding the peace preliminaries, noted that possession of eastern and western Florida and ports like Pensacola and Saint Augustine placed the British "in a much better condition for availing ourselves of the inclinations of the Spaniards to carry on a clandestine trade with us, in time of Peace; a Trade always profitable to us, and which we may now enjoy without risque, as the Spaniards themselves will be the instruments of it." Such advantages were probably better than nothing. They might even cover the cost of administering an impoverished, under-

11. For a rosy view of Florida, see "Heads of Advantages etc. Gained by Treaty, and State of America, Prussian Treaty etc.," n.d., PRO 30/47/7. For more modest views, see "Hints relative to the Preliminaries for a Peace with France and Spain," n.d. (but written sometime after Spain's entry into the war), and Ellis to Egremont, Jan. 16, 1762, 242r, both in PRO 30/47/7.

populated, and peripheral colony. But Florida was a step down from earlier aspirations. Missing from praise of it was the talk of mines, oceanic passages, useful Indian alliances, and a foothold within core Spanish territories. Pensacola was no Potosí.[12]

Modest expectations were evident also in discussion of the relation of Spanish territories to British finances. British officials expressed, as they had in previous conflicts, an awareness of the high costs of colonial war and of the suggestion that Spanish conquests might defray them. In his January 1762 letter to Egremont, Ellis referred, as he had in his *Voyage to Hudson's Bay,* to the dire state of British finances: "We are at length, unhappily forced into a War with Spain, after one so long and so burdensome with France, that there is some danger of this Nation's being undone, unless it is brought to a speedy conclusion.... Public Credit seems already at its utmost stretch; the difficulty of raising the Supplies for the current Year, is notoriously great; and for the next, must, in the nature of things, become abundantly greater, as our Commerce will diminish as the War extends, and the nation be more exhausted." He also mentioned Spain's "rich and extensive Provinces in South America" and discussed the possibility that conquests there might alleviate British financial difficulties. But even as Ellis recognized British financial precariousness and Spanish American resources, he also opined that "nothing that we can acquire by our Arms will in my opinion, be capable of supporting us, or of compensating for the enormous Expence of even one Campaign." Belief in Spanish American wealth remained, but the confidence in Britain's profiting from it, so evident in earlier years, had diminished.[13]

Alongside more moderate financial hopes, late–Seven Years' War British governmental documents evinced less of the truculent spirit and lust for dominion that had marked the writings of figures from earlier decades like Trelawny and still filled the parliamentary testimony of imperialists like Pitt. Where earlier writers had spoken of giving and prescribing laws to the Spanish Empire,

12. "Heads of Advantages etc.," and Ellis, "Advantages Which England Gains by the Present Treaty with France and Spain," late fall 1762, 249r, both in PRO 30/47/7. Ellis also mentioned that Florida would be a useful base for future attacks on Spanish shipping and territory. These comments have the vaguely defensive tone of someone explaining why a treaty that failed to acquire the most valuable Spanish possessions might not be as bad as it seemed. Dating and authorial identification from Cashin, *Governor Henry Ellis,* 160, 266. The "Heads of Advantages etc." argued that British possession of Florida's harbors would create "a facility to carry on an illicit commerce with the Spanish Colonys." Useful discussions can also be found in Allan Christelow, "Contraband Trade between Jamaica and the Spanish Main, and the Free Port Act of 1766," *HAHR,* XXII (1942), 309–343, esp. 314; Richard Pares, *War and Trade in the West Indies, 1739–1763* (Oxford, 1936), 598–601; J. Leitch Wright, Jr., *Anglo-Spanish Rivalry in North America* (Athens, Ga., 1971), 109–110, 114–118.

13. Ellis to Egremont, Jan. 16, 1762, PRO 30/47/14, 240r–241r.

those from the early 1760s recommended caution, sobriety, and restraint. Some warned against the zeal for imperial gain currently animating the populace. The unsigned "Hints relative to the Preliminaries for a Peace with France and Spain" remarked, "Should a favourable opportunity happen for restoring the Public tranquillity; his Majestys Ministers will undoubtedly recollect that the extravagant expectations of the giddy multitude, dazzled by the variety, and greatness of our Success, is one thing, and the true Interests of this Nation another. And, that whilst it may be requisite in point of prudence, to shew some attention to the former, their principal regard, will be due to the latter." The report of parliamentary debate quoted at the beginning of the chapter displays an instance of this political strategy of lauding grandeur while nurturing serenity. The passage's rhetorical tribute to territorial extent and numerous subjects follows a longer section highlighting "the security of our colonies upon the continent" as "the original object of the war" and arguing, "It was not only our best, but our only policy, to guard against all possibility of the return of" dangers to them. The "most capital advantage" Britain could obtain, "worth purchasing by almost any concessions," was removing France from a position where it could menace the Anglo-American mainland colonies.[14]

British imperial emphasis had shifted from acquisition of new Spanish American revenue sources to protection of economic assets the British Empire already possessed. This was most evident in declarations of the importance of shielding Britain's Atlantic seaboard colonies from external threats. The quotations above had a great deal of company. Although the economic value of Britain's mainland North American colonies might still have lagged behind that of imperial jewels like Saint-Domingue or Mexico, their exploding population and growing demand for British manufactures was increasing their importance in the eyes of mid-eighteenth-century British officials. Conserving these developing colonies was a more prominent consideration than it had been even a few decades before.[15]

14. "Hints relative to the Preliminaries for a Peace with France and Spain," n.d., PRO 30/47/7; "Motion on the Preliminaries of Peace," Dec. 9, 1762, in Simmons and Thomas, eds., *Proceedings and Debates*, I, 420–423.

15. On the desire to secure Britain's Atlantic seaboard colonies, see Paul Mapp, "British Culture and the Changing Character of the Mid-Eighteenth-Century British Empire," in Warren Hofstra, ed., *Cultures in Conflict: The Seven Years' War in North America* (Lanham, Md., 2007), 23–59, esp. 43–50; "The Spectrum of Imperial Possibilities: Henry Ellis and Thomas Pownall, 1763–1775," in Shy, *A People Numerous and Armed*, 49. On the growing economic value of the mainland colonies, see Clarence Walworth Alvord, *The Mississippi Valley in British Politics: A Study of the Trade, Land Speculation, and Experiments in Imperialism Culminating in the American Revolution*, 2 vols. (Cleveland, Ohio, 1917), I, 52; Daniel A. Baugh, "Maritime Strength and Atlantic Commerce: The Uses of 'a Grand Marine Empire,'" in Lawrence Stone, ed., *An Imperial State at War: Britain from 1689 to 1815*

The other instance of conservative emphasis, more striking because the assets to be protected had been so recently acquired, concerned the East India Company and Bengal. The growth of British power in India was one of the most important of the many important results of the Seven Years' War. During the conflict, Robert Clive's desire to replace Bengal's nawab with a more pliable figure and the East India Company's recognition of the need for an Indian source of income to pay its soldiers led the British enterprise to modify its traditional policy of avoiding direct rule over large areas of the subcontinent. It assumed control of taxation in significant portions of Bengal. Despite initial hesitation about becoming entangled in Indian governance, East India Company officials and the British ministers with whom they conferred quickly realized the value of what the fortunes of war had brought them. Among the earl of Egremont's papers are a number of letters and reports pertaining to the place of India in the late–Seven Years' War negotiations and policy considerations. They reek of revenue. In a July 1761 letter to William Pitt, company chairman Laurence Sulivan alluded to the "Great Benefits" arising from "possession of Countries either by cession or usurpation, whose revenues must maintain Armies and draw riches to Europe." In Bengal, "the Territories granted the Company are Provinces abounding in manufactures and tillage, whose revenues are great and encreasing." In another memoir in Egremont's papers, Sulivan observed, "The English are at present possessed in Bengal of Territories etc. which produce Seven Hundred thousand Pounds Sterling p Annum, and on the coasts of Choromandel, they may . . . possess Territories etc whose Annual Income amounts to near half that Sum."[16]

East India Company directors, cognizant of the value of the privileges they

---

(London, 1994), 185–223, esp. 196, 211; George Louis Beer, *British Colonial Policy, 1754–1765* (New York, 1907), 134–139; Linda Colley, *Britons: Forging the Nation, 1707–1837* (New Haven, Conn., 1992), 56, 62–64, 99; Patrick K. O'Brien, "Inseparable Connections: Trade, Economy, Fiscal State, and the Expansion of Empire, 1688–1815," and Jacob M. Price, "The Imperial Economy, 1700–1776," in W[illia]m Roger Louis, ed., *The Oxford History of the British Empire, II, The Eighteenth Century,* ed. P. J. Marshall and Alaine Low (Oxford, 1998), 53–77, 78–104, esp. 53–54, 56, 58, 70–72, 80, 82, 87, 98, 100–101, 103; Pares, *War and Trade,* 217–218; Kathleen Wilson, *The Sense of the People: Politics, Culture, and Imperialism in England, 1715–1785* (Cambridge, 1995), 56, 158–160.

16. On the East India Company's assuming control of taxation, see "Mr Sullivans Paper," c. 1761–1762, PRO 30/47/20, second packet of documents, 12r–17v; P. J. Marshall, "The British in Asia: Trade to Dominion, 1700–1765," in Louis, ed., *Oxford History of the British Empire,* II, ed. Marshall and Low, 498–503. On the East India Company's role in Seven Years' War peace negotiations, see Lucy Sutherland, "The East India Company and the Peace of Paris," in Sutherland, *Politics and Finance in the Eighteenth Century,* ed. Aubrey Newman (London, 1984), 165–176; Marshall, *Making and Unmaking of Empires,* 156. For Sulivan's views, see Laurence Sulivan to William Pitt, East India House, July 27, 1761, second packet of documents, 1r–1v, "Mr Sullivans Paper," 13r, both in PRO 30/47/20; see also "Hints relative to the Preliminaries for a Peace with France and Spain," in PRO 30/47/7, which mentions and criticizes the East India Company's acquisition of revenue-collecting privileges in India.

had gained, cast a skeptical eye on projects that might jeopardize them. This was particularly true of the Manila expedition. Beyond the objections mentioned above about wasting money and effort to acquire a city that would soon be lost, the East India Company also worried about diverting troops and money from a more valuable property. In a January 1762 memoir to Egremont concerning East India Company participation in the Manila expedition, the company's Secret Committee acknowledged the Manila plan's importance and promised assistance in executing it, "so far as their abilities and the safety of their trade and settlements may allow." The paper went on to state, however, that "the Committee are under the necessity of acquainting Your Lordship that their Affairs in Bengal and Madrass are in a very Critical Situation. The Security of those important acquisitions gained at an immense expence to the Company will require their utmost care and attention. They are surrounded with enemies and false friends and a misfortune to either of these may prove their ruin, for the revenues and large investments from thence and China are the only resource to support them and benefit Government." The company was thinking more about the security of its Indian acquisitions than the desirability of new ventures.[17]

More generally, governmental figures such as leading minister Bute, plenipotentiary in France the duke of Bedford, and ambassador to Spain George William Hervey, earl of Bristol (1758–1761), perceived that pushing British conquests and demands too far could endanger recent gains by antagonizing other nations. They feared that attaining the kind of American imperial dominance fire-eaters like Trelawny and Pitt had craved would turn other European powers against Britain. Bute worried that overly aggressive British demands in the peace negotiations would cause Britain to be "thought to aim at *universal dominion.*" Bristol warned repeatedly of Spanish concern about Britain's Seven Years' War victories. He wrote to Pitt in August 1761, "The great and uncommon success, which has for so many Years attended Our just Cause, now begins to excite the jealousy of this Court," and to Egremont in November, "I have long observed the Jealousy of Spain at the British Conquests." Bedford warned that too complete a British victory over its imperial rivals would prompt fears that Britain sought a monopoly of naval power and would lead to formation of a hostile coalition of maritime nations. The stress was on the danger rather than the desirability of dominant imperial power. British statesmen feared being seen, as Chapters 10 to 14 have shown they were seen, as aspirants to hegemonic power.[18]

17. Secret Committee to Egremont, East Indian House, Jan. 14, 1762, in Cushner, ed., *Documents Illustrating the British Conquest of Manila*, 15–16 (also in PRO 30/47/20, third packet of documents, 4r–4v).

18. Earl of Bute to duke of Bedford, July 12, 1761, in John [Russell] Bedford, *Correspondence of*

What does this restraint signify? By the 1750s and early 1760s, significant parts of both British and French officialdom were becoming increasingly disenchanted with earlier visions of imperial opportunity. There is no reason why a somewhat parallel trajectory in the development of two empires' ideas could not arise for entirely separate reasons, but there is some basis for considering commonalities here, not least of all because Spanish America was, in many respects, Britain's undiscovered West.

For both the French and British empires, experience had disappointed one set of expectations. For France, western Louisiana's explored territories had failed to yield a Northwest Passage, silver mines, an agricultural paradise, or a spectacularly profitable trade with Mexico or New Mexico, and French officials had grown skeptical about the value of territories that remained unknown to them. For Britain, gains at the expense of the Spanish Empire had proved hard to get and difficult to keep. Whereas cities like Manila and Havana had fallen to British forces, Buenos Aires in the current war and Cartagena in the previous one had frustrated British efforts. The combination of tropical diseases and Spanish weapons had meant that even a successful military campaign like that against Havana had come at a frightful cost. Spanish American populations had largely rejected British suggestions that they would be better off politically independent of Spain and economically and politically dependent on Britain. Spain's soldiers, its guarda costas, and its officials had resisted British expansion vigorously, if not always successfully. The profits from British licit and illicit trade with Spanish America had rarely lived up to promoters' extravagant promises. British gains had brought, and threatened again to bring, balancing responses from rival powers, making success more problematic in some respects than failure. Experience had taught many British statesmen to be wary of hoping for too much from Spanish America.[19]

In contrast, other areas of colonial endeavor had proved surprisingly valuable. For both France and Britain, parts of their own colonial empires were paying enormous dividends by the mid-eighteenth century. Whereas Louisiana and

---

John, Fourth Duke of Bedford: Selected from the Originals at Woburn Abbey, 3 vols. (London, 1842–1846), III, 29–36, esp. 33; earl of Bristol to Pitt, Segovia, Aug 10, 1761, 29r–30r, Bristol to Egremont, Escorial, Nov. 2, 1761, 200r–209r, and Bristol to Egremont, Madrid, Dec. 6, 1761, 338v–339r, all in SP 94/164; Bedford to Newcastle, May 9, 1761, in Theodore Calvin Pease, ed., Anglo-French Boundary Disputes in the West, 1749–1763 (Springfield, Ill., 1936), 294–296; Anderson, Crucible of War, 484.

19. J. H. Parry provides a nice overview of these issues in Trade and Dominion: The European Overseas Empires in the Eighteenth Century (New York, 1971). See also Geoffrey J. Walker, Spanish Politics and Imperial Trade, 1700–1789 (Bloomington, Ind., 1979), and Jean O. McLachlan, Trade and Peace with Old Spain, 1667–1750: A Study of the Influence of Commerce on Anglo-Spanish Diplomacy in the First Half of the Eighteenth Century (Cambridge, 1940).

Canada had generally drained money from French coffers, the French sugar islands of Martinique, Guadeloupe, and Saint-Domingue were yielding staggering profits. Indeed, so rapid was the growth in these islands' profitability that, by the 1770s or 1780s, Saint-Domingue alone might have been generating more wealth than the entire Spanish American empire. Although Britain's sugar colonies never rivaled prerevolutionary Saint-Domingue, the growing importance of the Anglo-American mainland colonies was becoming evident to imperial officials. None of this meant that British and French officials had stopped coveting Spanish imperial trade and silver, but it did mean that both empires could draw wealth from the Americas without conquering Spanish territory. Someone like Ellis, with recent experience running a threatened North American frontier colony, could still cast an appreciative eye on Spanish America and the South Sea while putting his principal emphasis, not on the acquisition of Spanish imperial gems, but rather, as Shy noted, on the security of what Britain already had in North America.[20]

With recent gains in India, moreover, Britain was now, somewhat unwittingly, in a position to begin enjoying the East Indies treasures Spain's Genoese explorer had been looking for three centuries before. In one sense, the extra-European portions of the Seven Years' War had been, not a set of separate conflicts in North America, the Caribbean, Africa, India, and the Philippines, but a single struggle for the eastern and western Indies. With a seemingly secure and paying position in the Caribbean and the North American continent, and with a promising toehold in India itself, the British Empire could afford to leave Spain's Indies to Spain.

Ellis, after Egremont's death in August 1763, found he could afford four decades of pleasant, learned, and peripatetic retirement, living off the income from his colonial offices and family estates. He could contemplate from his winter refuges in southern Europe the painful loss of thirteen of the colonies he had hoped to secure for the British Empire and the renewed search under Cook and others for the Northwest Passage he had tried to find. Ellis died in Naples in 1806, a long way from a passage to the West.

20. On the value of the French sugar islands, see D. A. Brading, "Bourbon Spain and Its American Empire," in Leslie Bethell, ed., *The Cambridge History of Latin America*, I, *Colonial Latin America* (Cambridge, 1984), 389–439, esp. 426; James E. McClellan, *Colonialism and Science: Saint Domingue in the Old Regime* (Baltimore, 1992), 2.

CONCLUSION

This study has argued that perceptions of western American geography influenced the course of imperial diplomacy, that ideas about the undiscovered West contributed to the origins, unfolding, and outcome of the mid-eighteenth century's Great War for Empire. Unease about the implications of British Hudson Bay exploration helped draw France into war with Britain. Spanish concerns about French westward exploration reinforced Spain's neutral tendencies, keeping the Iberian empire out of the Seven Years' War until its entry was too late to forestall French defeat or British victory. Increasing French skepticism about the value of the unexplored West and lingering Spanish disquiet about the danger it might hold shaped a diplomatic settlement removing France from the continent and pushing Spain east to the Mississippi.

With respect to interimperial relations, the North American Far West gained significance from the tendency of European statesmen to understand it in terms of familiar territories like Peru and Mexico and idealized realms like China and Japan. The Spanish colonies, with their mineral resources and large, productive, and tractable indigenous and creole populations were indisputably important. The possibility that the North American West would resemble or lead to these places elevated the region's significance in imperial strategists' eyes. The same could be said of the undiscovered West's relation to East Asia. The configuration of northwestern America and northeastern Asia remained unclear to early-eighteenth-century Europeans, and the prospect that advanced Asian peoples might frequent or inhabit North America connected western lands to Indies visions. The Pacific tied these Spanish and Asian dreams together. The contents of the South Sea, and the boundary between North American lands and North Pacific waters was still uncertain for early-eighteenth-century Europeans. The Pacific surely led to Spanish and East Asian riches. It might hold new Japans and untouched spice islands. European imagination often filled strange lands and waters with extrapolations from other real or imagined areas.

The more subtle way to explicate the undiscovered West's significance is to think of it in terms not just of unknown space, but also of unrealized time. French, Spanish, and British geopolitical thinkers could see that western North America represented a significant part of the imperial and mercantile future.

Whether the region would hold profit or peril, bitter disappointment or splendid destiny, was as yet uncertain. What was clear was that wise statesmen ought to plan for the possibilities. Enlightened imperial thinkers observed that previous centuries' geographic discoveries had revolutionized international affairs and that current statesmen were still grappling with the implications of Columbus's voyage. There was every reason to believe that future statecraft would continue to react to reconnaissance, and there was some basis for hoping that current statesmen might anticipate and inflect coming events. Eighteenth-century trends were evident enough and mid-eighteenth-century European faith in the utility of systematic thinking and the efficacy of government action sufficiently strong to suggest that forethought might pay. The West might afford new and better routes to Spanish Empire and Lake, offer new resources to cultivate or unearth, and hold new peoples to contain or convert, trade or ally with. A sage statesmen might think ahead and find a way to put these bounties at his empire's disposal.

This western futurity brings the story of the region's relation to the Seven Years' War in line with other aspects of that contest. A striking feature of the war as a whole is the extent to which its component conflicts arose not simply in reaction to earlier events but in anticipation of future developments. Austria's Count Kaunitz engineered a revolutionary anti-Prussian alliance not only to avenge Frederick the Great's seizure of Silesia but also because of the realization that a rising and unscrupulous Prussia might endanger Austria in the future. Frederick launched his preemptive 1756 invasion of Saxony both as a reaction to the Austro-Russo-Franco alliance forming against him and in hopes of forestalling the disaster this coalition portended. French India governor Joseph-François Dupleix's attempts to gain control of indigenous revenue sources both responded to past French financial shortfalls and sketched the kind of Euro-Asian empire soon to be engineered by Robert Clive and the British East India Company. Washington and his French adversaries west of the mountains had been sent to head off each other's imperial expansion. In all these cases, statesmen were discerning the direction of events, looking ahead, and trying to adapt policy to prevailing circumstances — or execute plans to modify them.

It is also noteworthy how far awry such schemes often went. Dupleix and his successors' efforts to outmaneuver the British East India Company ended in total French defeat and momentous British victory in India. Frederick's efforts to stave off Prussian destruction came within a hairsbreadth of leading to his kingdom's annihilation and his own suicide. Kaunitz's efforts to crush Prussia and recover the Habsburgs' lost province not only left Prussia and Frederick shakily standing and in possession of Silesia, but also created a legend of Prusso-

German indomitability that would nourish future German statesmen far more dangerous to Austria than Frederick. When Washington and Braddock went west to counter France, the Virginian found defeat and humiliation, the general defeat and death, the British Empire a nearly disastrous war.

In a comparable fashion, geographic uncertainty contributed to the tendency of mid-eighteenth-century international relations to frustrate the designs of European statesmen. French officials hoped to avoid war with Britain and check the growth of British imperial power. Their exaggerated concerns about apparent British Pacific advances helped draw France into a conflict that elevated the British Empire to the leading position among European states. French diplomats hoped to avoid antagonizing their Spanish counterparts with untimely and ill-considered activities in the mysterious West and to secure prompt Spanish assistance against the common British foe. The Spanish government found more to fear from French conduct in lands beyond the Spanish frontier than the diplomats of Paris imagined and saw in neutrality the course best calculated to protect Spanish overseas territories. This temporary avoidance of military involvement made Spain's eventual intervention in the Anglo-French conflict more costly and less effective. Judicious French statesmen wisely ceded the American lands championed by their credulous countrymen; those territories turned out to be more valuable than the wildest dreamers had imagined. The track record of European American policies informed by geographic expectations was disappointing.

This pattern of failure calls attention to telling features of mid-eighteenth-century European and American international relations. One of these was the gap between statesmen's ability to discern the probable course of events and their capacity to direct it. Related to this was the manner in which the quest for or enjoyment of immediate security entailed eventual loss. International actors were often like characters in a tragic play, running into what they were running from, or at least toward what they would have liked to avoid.

For far western Indians, this tragic quality of international relations had to do with the ambivalent implications of isolation and division. The Far West's peoples were in many respects insulated from the great mid-eighteenth-century European imperial struggle. Although European diplomats might have felt powerful enough to ignore what the 1763 inhabitants of today's Oregon might think of the Peace of Paris, 1763 Oregonians were probably remote enough not to have cared much about, or even to have heard the terms of, the settlement. Europeans were still a long way away from much of the West. But the kinds of maritime and land expeditions discussed in this book were removing the protection from European expansion remoteness had provided, and distance from European doings might have left the farthest western peoples less ready than they might have been for the Euro-American onslaught that did come. Moreover, although the apparent disunity of the American West's native peoples probably rendered them less inviting and vulnerable to swift European geographic comprehension and political conquest than the more centralized Aztec and Inca empires, want of political cohesion might have precluded the kind of development, concentration, and employment of resources that hindered European and American imperialism in at least some countries on the other side of the Pacific.

With regard to mid-eighteenth-century European imperial relations, seemingly inescapable difficulties arose from the European state system's near inability to accommodate changing configurations of power. Early-eighteenth-century Europeans could not be sure what the Far West contained, but it might hold something capable of contributing to imperial wealth and state power, something like the Northwest Passage captivating Rogers's imagination. In a competitive, territorially insecure, diplomatically unstable, and conflict-ridden international system, one state's gains threatened others' security. This relation of imperial might and western potential both stimulated interest in the undiscovered West and impeded understanding of the region. Spanish scouts and missionaries reconnoitered vulnerable regions; the Spanish government tried to conceal its geographic expertise from outsiders. French explorers and merchants trudged west toward Spanish markets even as Paris urged colonial officials to avoid offending Spanish sensibilities. British ships probed Hudson Bay's western shores for a Northwest Passage while French colonies, traders, and allies tried to obstruct British land exploration. An accumulation of hindrances complicated European statesmen's assessments of western geography and clouded their judgments of the region's significance.[1]

A related dilemma afflicted European imperial policy at the end of the Seven Years' War. Amid the wreckage of that conflict, French, British, and Spanish statesmen were all wrestling with the confounding relationship between military costs and colonial value. Varying approaches to the undiscovered West would not resolve the problem. As European wars grew more expensive, colonial revenues became increasingly essential. But as colonies became more valuable to possess, the envy of rival powers made them more expensive, difficult, and hazardous to defend. Imperial officials were trying to assemble colonial territories that were valuable but not vulnerable, lucrative but not provocative, secure but still dependent. They were sifting their policies, trying to reject failed

---

1. For an insightful discussion of the dynamics of the mid-eighteenth-century European state system, see Walter L. Dorn, *Competition for Empire, 1740–1763* (New York, 1963).

and troublesome colonial efforts and launch more effectual initiatives. Long-standing land power France abandoned a North American continental empire in favor of island possessions and dubious Falklands and South American ventures. The traditionally maritime and ethnocentric British Empire acquired rights over huge dominions and large alien populations on two distant continents while eschewing the prize coastal Spanish American possessions it had long sought. The overextended and underpopulated Spanish Empire obtained thousands of square miles of unsubjugated, thinly peopled, and administratively expensive North American territory. These disparate imperial approaches hinted, and common imperial disappointments would soon demonstrate, that it was not really a question of finding some novel and easy path to imperial success. It is not clear that there was any way these empires could control the developments they had to confront and, at any rate, all would soon find themselves losing much of what they had endeavored so long to gain and retain. Neither acquiring little known trans-Mississippi territory in the case of Spain, nor sacrificing it in the case of France, nor standing apart from and foregoing the equivalent of it in the case of Britain would prevent these empires' losing key possessions in the decades to come. Sometimes discerning deeper realities, sometimes blind to them, sometimes paddling frantically against the current, the frail barks tossed in the running waters.[2]

2. J. H. Elliott presents a suggestive discussion of common post–Seven Years' War imperial quandaries in *Do the Americas Have a Common History? An Address; Presented on the Occasion of the Celebration of the 150th Anniversary of the Founding of the John Carter Brown Library, 13 November 1996* (Providence, R.I., 1998), 37. He develops the idea in *Empires of the Atlantic World: Britain and Spain in America, 1492–1830* (New Haven, Conn., 2006), 292–324.

INDEX

Black (ethnic or racial classification), 219, 221n. 23
Blackfeet, 201–202, 204, 219, 237, 247–248, 248n. 17, 251, 270n. 10
Black Hills, 167, 203, 235–236
Black Legend, 84
Black River, 295
Bladen, Martin, 266, 274, 279–280, 420
Blancpain, Joseph, 395–396
Board of Trade, 266–267, 274
Bolingbroke, Henry St. John, first Viscount, 139
Bolivia, 115
Boothia Peninsula, 225
Boston, 186
Botany Bay, 18
Bougainville, Louis-Antoine de, 18, 52, 252, 376–378, 391
Bourbon Family Compact, 405
Bourgmont, Étienne Véniard, sieur de, 200, 229
Bow Indians, 236, 240
Bowrey, Thomas, 133
Braddock, General Edward, 267, 284, 354, 431
Brazil, 29, 154, 180, 290n. 10, 326
Breton language, 228
Bristol, George William Hervey, earl of, 317, 425
Britain and the British Empire: French desire to avoid war with, 2, 284; and contest for North America, 14–15, 21–22, 255–256, 284, 387, 433; Seven Years' War territorial gains of, 15, 165, 413–427; historical importance of, 16; and Pacific exploration and settlement, 18, 26, 377; interest of, in Pacific trade and expansion, 18, 26, 108, 132–140, 261–264, 272–275, 283–329, 375, 419–420, 425; and historiographical neglect, 19; value of sources from, 20–24; and degree of interest and involvement in West, 22, 40, 179, 183, 266, 268–270; as threat to Spanish Empire, 22, 25, 104, 132–138, 261–310, 387–427; growing power and ambitions of, 22, 305–310, 315, 329, 431; and interest in Spanish American silver, 25, 112–113, 120, 132–135, 167, 271, 275, 277, 286, 296, 301, 306, 416, 427; and origins of Seven Years' War, 26, 284, 283–310, 333, 429, 431; and territorial reevaluation at end of Seven Years' War, 26, 413–427, 433; and China trade, 112–113; and War of the Span-

ish Succession, 132–138; and Utrecht settlement, 132–142, 165, 271, 283, 286, 303–304; as rival of France, 148, 162–163, 283–356; and Spanish America as undiscovered West, 264; maritime character and prestige of, 264, 291–292, 302; and War of Jenkins' Ear, 270–282; and efforts to secure Spanish alliance or neutrality, 312–329, 331–333; and French cession of western Louisiana, 359, 375–376, 378, 383, 385–386; interest in New Mexico of, 368; and Spanish acceptance of western Louisiana, 387–393, 397–411. See also Byron, John; Cook, James; England; Vancouver, George
Brobdingnag, 7, 14
Brosses, Charles de, 107, 292–293
Buache, Phillippe, 186–188, 293, 369–371, 400
Bucareli y Ursúa, Antonio María de, 82
Buenos Aires, 134, 276–277, 280, 282, 287–290, 290n. 10, 332, 415, 418–419, 426
Buffalo, 48n. 15, 73n. 5, 79n. 11, 80, 95, 216, 234n. 2, 244, 248, 366
Buffon, George-Louis Leclerc de, 380
Burriel, Padré Andrés Marcos, 391, 400–403
Bussy, François de, 289–292, 297, 371, 378
Bute, John Stuart, earl of, 417–418, 425
Byron, George Gordon, 288n. 8
Byron, John, 288n. 8, 377

Cabeza de Vaca, Alvar Núñez, 41, 57
Cabrillo, Juan Rodríguez, and Cabrillo-Ferrelo expedition, 41, 55–56, 79
Caddoan Indians, 72, 202, 234, 341n. 14
Cadiz, 30, 41, 120, 129, 137, 139, 270, 314
Cagnoni, Carlos, 290n. 10
Cajuenches, 74
Calcutta, 354
California: as island, 7, 33; and perceived proximity to New Mexico, 32; Spanish voyages along, 32, 44, 54, 72, 109; Spanish exploration of, 52, 87; and connections to other regions, 62, 81, 239, 249, 345; ethnic and linguistic diversity in, 72; state of, 82; and British Empire, 104, 107, 135, 277, 300, 320–321, 401; Russian advances toward, 183, 400; and China, 193; and peoples to north, 218; French advances toward, 332, 345–350, 376–377; eighteenth-century Spanish expansion into, 337; Spanish assessment of, 400–404. See also Baja California; California coast

California coast, 44, 54, 56, 69, 79–82

Callao, 115, 141, 277

Cameahwait, 251

Campbell, John (author), 104, 106

Campbell, John (naval officer), 325

Campeche, 296, 418

Campomanes, Pedro Rodríguez, 401, 404

Canada: historiographical approaches in, 16n. 8; and access to western lands, 21, 30, 65, 148, 150, 164, 172, 240, 249–250, 268, 308, 350, 375, 398, 404; limited value and fearsome climate of, 154, 179, 328, 364–365, 378, 383, 427; and regions to west, 215, 218, 244, 311; as theater of Seven Years' War, 284; French loss and British conquest of, 310, 359, 363–364, 386, 392–393, 404–407, 409, 417; and origins of Seven Years' War, 335. *See also* France and the French Empire

Canary Islands, 102

Candide (Voltaire), 195

Cantemir, Antiochus, 290n. 10

Canton, 111–112

Cape Blanco, 32, 41, 44–45, 55, 74, 400, 402

Cape Disappointment, 53

Cape Horn: as land's end, 20; and difficulty of rounding, 123, 125–126, 148, 162, 288n. 8, 325; British interest in, 271, 277, 286, 289, 293–294, 298, 307, 310n. 35, 314, 322, 347, 376, 399, 415, 419

Cape Mendocino, 44

Cape of Good Hope, 101, 105

Cape San Antonio, 290

Cape San Lucas, 53, 54, 397

Cape Verde Islands, 178

Captives. *See* Slavery and slave trade (Indian)

Caribbean Sea and islands, 19, 65, 123, 147–148, 178, 271, 276, 284, 287, 309, 347, 383–384, 416, 427. *See also individual islands*

Carlisle, Pa., 267

Carmelo River, 345–347

Carolina, 111, 269, 276. *See also* North Carolina; South Carolina

Cartagena, 125, 271, 274, 281, 316, 348–349, 375, 426

Cartesian reasoning, 379–380

Cartier, Jacques, 168, 195

Cartography: and representations of geographic uncertainty, 6–14, 15n. 6, 31–40; evaluating influence of, 15n. 6, 20–21; western Indian, 71, 77–78, 167–169, 194, 225–227, 254–255; Aztec, 92; French, 170–193, 369–

372; Russian, 181–185, 199; British, 266–268, 371

Carvajal y Lancaster, José de, 316–319, 321–327, 334, 337, 346, 348, 352–353, 401

Carver, Jonathan, 219, 254

Casa Grande ruins, 62

Casas Grandes, 60

Caserniers, 213

Cassini, César-François, 176–177

Cassini, Jean-Dominique, 174, 178, 189, 190

Castañeda de Najera, Pedro de, 4, 50, 53, 72

Castelldosríus, Manuel de Oms olim y Sentmenat, marqués de, 131

Castro, Clemente Guillén de, 320

Cathcart, General Charles, 272, 274–275, 280

Cavendish, Thomas, 54, 116

Cayenne, 178

Caynigua Indians, 76

Central America: Spanish exploration of, 25, 52, 87; Spanish settlements in, 104; and attempted Scottish Darien colony, 132, 272, 274; Indians of, 275, 278–280, 297n. 19; British activities in, 286, 295–297, 304, 307, 309, 313–314, 316, 329, 332, 352, 374, 410, 418; French activities in, 347–349, 352, 365, 394. *See also individual countries*

Chaco Boreal, 52

Chaco Canyon, 96

Chamillart, Michel, 129

Champlain, Samuel de, 168, 195

Chamuscado-Rodríguez expedition, 73n. 5

Chancas, 66–68

Chapuis, Jean, 29–30, 235, 330–332, 337, 339, 349–350, 395

Charles, archduke of Austria, 130

Charles II (king of Spain), 122, 130, 138

Charles III (king of Spain), 364, 383, 387–412

Charles V (Holy Roman Emperor), 4, 58, 72

Charles XII (king of Sweden), 183

Charleston (Charles Town), 265

Charlevoix, Father Pierre-François-Xavier de, 196, 203–204, 218, 227, 243, 370n. 13

Chemeguavas, 86

Chemevet Indians, 81

Chesterfield Inlet, 299

Chicomoztoc. *See* Aztlan

Chihuahua, 60

Chilcotins, 231

Chile: as Spanish and Inca imperial frontier, 64, 69, 90, 95, 106; Spanish exploration of, 92, 95; Spanish settlements in, 104; and Juan

Fernández, 107; French trade with, 123–143, 148, 162, 322, 352; British interest in, 132–135, 278, 287, 293, 297–298, 301, 323

China: French and Jesuit surveying and mapping of, 24–25, 171–172, 187–194, 198, 228–234, 238, 255–256, 266, 268, 381; Spanish notions concerning, 32; European trade with, 101, 111–116, 119, 198, 272n. 11, 397n. 15, 425; as goal of European exploration, 104, 111–113, 161–162, 168, 262, 326, 384, 429; and climate of thought, as comparable to Louisiana, 154n. 6. *See also individual cities*

Chinchay-suyu, 69

Chinese Empire, 114, 171–172, 180, 189–191, 198, 228–234, 238, 343

Chipewyan Indians, 241n. 10, 269

Choctaws, 266

Choiseul, Étienne-François, duc de, 360n. 1, 360–364, 370, 374–378, 382–384, 398, 404, 406–408, 414

Churchill (Factory, Prince of Wales's Fort), 201, 224n. 25, 241, 302

Cíbola, 58, 80

Cicuyc (Pecos), 50

Cieza de León, Pedro, 67, 69, 88, 90, 92n. 26

Clark, William, 2, 5, 101, 117, 204, 210–211, 222–223, 233, 242–243, 247, 250, 251, 254, 311, 381

Clatsops, 117

Clive, Robert, 424, 430

Cochimí-Yuman language family, 72

Cocking, Matthew, 237n. 5

Cocomaricopas, 73, 80n. 13, 86

Colbert, Jean-Baptiste *(contrôleur général des finances)*, 173–174

Colbert, Jean-Baptiste, marquis de Torcy. *See* Torcy, Jean-Baptiste Colbert, marquis de

Colhuacan. *See* Aztlan

Colla-suyu, 69

Colombia, 67, 95, 416

Colorado, 41, 49, 81, 341

Colorado Basin, 81, 87

Colorado Delta, 62

Colorado Desert, 45

Colorado River: extraregional significance of, 20; Spanish exploration of, 33–34, 41, 45, 50, 60–61, 69, 70, 72–73, 79; Indian violence along, 50–51, 85–86; and connections to other regions, 80–81, 84; and Lake Cahuilla, 82–83; salinity of, 210; Spanish fears regarding, 345–347, 351–352, 396–399, 401, 403

Colorado River valley, 91

Columbia Plateau, 249

Columbia River, 55, 222–223, 242, 248n. 17, 249

Columbus, Christopher, 4, 53, 101–102, 200, 233, 382, 427, 429

Comanches: as obstacles to exploration, 49, 235, 236n. 4, 254; Indian reports of, 81, 236n. 4, 244n. 13; geographic horizons of, 97; language of, 202; and horses, 251; as danger to Spanish frontier, 339–345, 396, 406; and imperial status, 342n. 15; Spanish interest in alliance with, 398; and Peace of Paris, 411

Compagnie des Indes, 150, 156, 162

Compagnie royale de la Mer pacifique, 126, 128, 131, 147

Company Land, 33, 107n. 9, 111

Complete Collection of Voyages and Travels, A (Harris and Campbell), 104

Concepción (Chile), 127, 136

Condillac, Étienne Bonnot de, 380

Conflans, Hubert Brienne, comte de, 348

Consag, Fernando, 33n. 6

Conseil de la marine, 150, 162

Conseil des finances, 150

Continental Divide, 76n. 9, 243, 253, 373, 376, 380

Contraband trade, 65n. 37, 119, 271, 296, 313, 316, 329, 367, 389, 395–396, 421, 426

Cook, James, 18, 32, 52, 55, 113, 195, 212, 242n. 11, 248–249, 381, 427

Coolidge, Ariz., 49

Cooper, James Fenimore, 147, 154

Copala, Lake of, 59–60

Copper, 58, 61, 110, 154, 155, 242n. 11, 248–249, 253, 261

Coppermine River, 249

Copper River, 249

Coquimbo, 133

Corodeguachi Presidio, 48

Coromandel, 326, 424

Coronado, Francisco Vásquez, de, 4, 40–41, 45–46, 50, 53, 57–59, 72, 79–80, 95, 265

Coronelli, Vincenzo, 63

Cortés, Hernán, 4, 53, 55, 87, 89–90, 92–94, 101

Cossacks, 179, 181, 218

Costa, Antonio da, 289

Costa, Juan da, 289–292

Costansó, Miguel, 69, 73n. 4, 79, 86, 89

Council of the Indies, Spanish, 320, 330, 346, 350, 394, 396, 401, 403

Course of Empire, The (DeVoto), 6–7

Empirical reasoning, 379–380
Enchanted City of the Caesars, 52, 66
Encuche, 77
England, 55, 104, 134, 140, 161, 173, 180, 271–272, 287, 291–292, 298, 302, 305–307, 318, 321, 323, 325–326, 328, 361, 370, 374, 385, 398, 400–401, 414, 417. *See also* Britain and the British Empire
Enlightenment, 41, 112, 381–382
Ensenada, Zenón de Somodevilla y Bengoechea, marqués de La, 318–319, 321, 324–327, 334, 338, 344, 346, 348–349, 352, 394, 401
Escalante, Father Silvestre de Vélez de, 49, 74, 81, 94, 96, 220, 254
Escanxaques, 47, 77, 341
Escobar, Father Francisco de, 61, 83
Eslava, Sebastián de, 349
Espejo, Antonio de, and Espejo expedition, 60–61, 71
Espinosa, Fray Isidro Félix, 76n. 9, 234n. 2
Esprit des lois, L' (Montesquieu), 382
Essay on Tea, An (Hanway), 112
Esteban the Moor, 57–58, 80
Euclidean geometry, 379
Eurasia, 102, 171, 179–180, 182, 189, 255, 343
Europe: cartographers of, 6, 171, 181, 191–193, 229; empires of, 21, 180, 182; mental distance of, from America, 23, 53, 123, 371, 385, 427; engagement of, with western America, 29; and limited knowledge of nearby regions, 84, 173; financing wars of, 101, 124, 163; economic relation of, to Americas, 114, 118–120, 281, 283, 307, 316; western, 119, 180; military relation of, to Americas, 277; Americas and balance of power in, 281, 283, 303, 307, 315, 328, 353; Louis XIV's attempt to dominate, 337; and Darien French, 347; and India, 424
Evans, Lewis, 267
Exploration: Spanish, of Pacific coast, 22, 41; Spanish, in Southwest, 22, 33–34, 45–46; Spanish, Central American and Mexican, 24, 25, 41, 43, 52–53, 57, 59–60, 87–98; Spanish, of Peru, 24, 53, 56; Spanish, South American, 25, 41, 43, 51–52, 55, 87–98; French, western, 25, 62–64, 98, 101, 164–172, 194–257; Spanish, western, 29–98; Spanish, of Colorado River, 33–34, 41, 50, 60–61; Spanish, of Texas, 41; Anglo-American, 101; British, of North American West, 101, 264–270; Russian, of North Pacific, 107n. 9,

108–109, 172, 179, 182–187, 193, 238, 245, 292, 370, 391, 400–404; and credibility of French explorers, 194–199
Eyak-Athabaskan language family, 72, 202

Falberte, Duplessis, 330
Falkland Islands, 127, 293–294, 322n. 13, 360n. 1, 376–377, 415, 433
Far East, 172, 189. *See also* East Asia, East Indies
Far West. *See* West, North American
Ferdinand VI (king of Spain), 318, 326–328, 330–331, 335n. 6, 338, 346, 349–350, 355, 387, 392, 394, 395n. 12, 396
Fernández Molinillo, Francisco, 349
Feuilli, Louis, 29–30, 235, 330–332, 337, 339, 349–350, 395
Fidler, Peter, 248n. 17
Finley, John, 267
Fisheries, North Atlantic, 307, 417
Fleury, Cardinal André-Hercule de, 142
Florentine Codex, 95
Florida, 3, 64, 264, 316, 328–329, 360n. 1, 382, 408–409, 412–413, 415–417, 420–422; Straits of, 418
Font, Fray Pedro, 51, 85, 231
Fonte, Bartholomew, 186, 369–370, 400–401
Forks of the Ohio, 1, 3
Forks of the Saskatchewan, 167
Fort Bourbon, 240
Fort Chartres, 330, 351
Fort Chipewyan, 241
Fort Duquesne, 267
Fort Le Pas, 268
Fort Necessity, 1, 284
Fort Oswego, 265, 354
Fort William Henry, 354
Foss, Theodore, 188–189
Fou sang, 193
France and the French Empire: and Louisiana cession, 3, 20, 147, 149, 359–386, 414, 416, 426; and contest for North America, 14–15, 21, 268, 387; and loss of North American territories, 15, 147, 149, 359–412, 415, 433; and perceived insignificance of North American territories, 16, 162; and Pacific exploration, 18, 164–165; interest in Pacific trade of, 18, 123–143, 148; and historiographical neglect of Pacific interests, 19; value of sources from, 20–24; and ideas about western

Guiana, 360n. 1, 384n. 27, 433
Guignes, Joseph de, 193
Gulf Coast, 404
Gulf of California, 33n. 6, 35, 55, 61, 82, 345
Gulf of Darien, 272, 348
Gulf of Guayaquil, 56
Gulf of Honduras, 275, 316. See also Bay of
    Honduras
Gulf of Mexico, 178, 276, 308, 336, 374, 407
Gulf of San Miguel, 272
Gulliver, Lemuel, 7
Gulliver's Travels (Swift), 14, 195

Habsburgs, 430
Haitian Revolution, 411
Hamilton, Dr. Alexander, 113
Hamilton, Alexander (statesman), 132
Hanway, Jonas, 112, 120
Harley, John Brian, 15n. 6
Harrington, Lord. See Stanhope, William, Lord
    Harrington
Harris, John, 104
Hasinais. See Assinais
Havana, 275–276, 279–281, 316, 338, 349, 383,
    408, 413, 415–419, 426
Hawaiian Islands, 103
Hawkins, John, 414
Hearne, Samuel, 240–241, 245
Henday, Anthony, 204, 236–238, 264, 268,
    270n. 10
Hennepin, Louis, 197
Henry, Alexander, Sr., 241n. 10, 248n. 17
Herodotus, 24, 84
Hervey, George William, earl of Bristol. See
    Bristol, George William Hervey, earl of
Heyder, baron de, 399–400
Hezeta, Bruno de, 32, 55
Hidalgo, Fray Francisco, 31, 74, 76, 234n. 2
Hidatsas, 76, 202, 242, 247
Histoire des navigations aux terres australes
    (Brosses), 107, 292
Historiography: of Seven Years' War, 1–3,
    285n. 3, 286n. 4, 313n. 2; of cartography,
    14n. 6; of early North American West, 15–17;
    of early modern Pacific, 17–19; regarding
    French cession of Louisiana, 359n. 1; re-
    garding Spanish acceptance of Louisiana,
    389n. 2, 390n. 3, 391n. 4
History of the Incas (Sarmiento de Gamboa),
    66
History of the Indies of New Spain (Durán), 59

Hobbes, Thomas, 257
Hogiopa Indians, 70
Hokkaido, 108
Holy Roman Empire, 326
Honduras, 308, 318, 334, 348, 354
Hopis, 86, 222n. 23
Horses, 216–217, 221, 235, 239, 244, 247–252,
    254, 341
Howard, John, 265–266
Huayna Capac, 56, 93, 95
Hubert, Marc Antoine (ordonnateur), 150, 160,
    162
Hudson, Henry, 262
Hudson Bay: and causes of Seven Years' War,
    3, 264, 283–410; European uncertainty con-
    cerning regions beyond, 5, 164, 212, 229,
    236, 262, 269; and Northwest Passage, 5,
    19, 22, 32, 101, 106, 125, 136, 148, 162, 187,
    261–264, 272, 283–410, 314, 327, 374, 401,
    404, 413–415, 432; British posts around, 22,
    265, 268–270; British exploration of, 22, 26,
    107, 282, 292, 297–298, 332, 399, 429; and
    Utrecht settlement, 136, 302; as boundary of
    Louisiana, 150; French moves toward, from
    south and east, 167, 225, 236; cartographic
    representations of, 225; Indian hostilities in
    regions west of, 236–237, 239–240; Indian
    knowledge of regions west of, 243, 249, 253;
    historical importance of, 262, 303–304
Hudson River Valley, 265
Hudson's Bay Company, 201, 204, 219,
    224n. 25, 225–226, 240, 245, 248n. 17, 261–
    262, 265, 268–270, 286n. 4, 299
Humboldt, Alexander von, 116
Hume, David, 381
Huns, 343
Hutchinson, Henry, 419n. 8
Hydrographic Office (of British Admiralty),
    174

Ibarra, Francisco de, 60
Iberville, Pierre Lemoyne d', 147, 265
Idaho, 41
Illinois country, 29–30, 65, 217, 235, 265, 330,
    366
Illinois Indians, 229
Illinois River, 150
Imperial rivalry, 5, 9, 14–16, 21–22, 255, 293, 303,
    305–310, 331–333, 353–356, 363, 365, 387–393,
    429–433
Inca Empire, 66, 71, 87–98, 186, 432

Inca road system, 69, 95–96, 98
Incas, 6, 43, 52, 55, 66–67, 88–98
Inchin, 298
India, 104, 284, 307, 416, 424–425, 427, 430
Indian Ocean, 105–106, 307
Indians, South American, 133, 278, 324–325. *See also individual tribes*
Indians, western: and familiarity with western American geography, 5, 70–71, 172, 229, 239; as sources of geographic ideas and information, 24–25, 58, 61, 69–98, 161, 166–167, 171, 172, 194–196, 199, 233–257, 269–270; difficulty of communications with and among, 25, 69–98, 194–257; disunity of, hostilities among, and limited movement of, 25, 84–86, 233–257, 431–432; wayfinding skills of, 46, 223–224; as guides, 46, 49–50, 167, 200–202, 233–234, 238, 240–242, 253, 268–269; as obstacles to Spanish exploration, 46–53; linguistic diversity of, 72–74, 194, 201–202, 223, 228, 255; as actual or prospective French allies, 147–148, 160, 268, 363–364, 366n. 8, 367, 389, 395–396; as French trading partners, 148, 151, 154, 160, 389, 395–396; as obstacles to French exploration, 155, 200–201, 233–238, 255–258; and trade with British, 268–269; as threats to Spanish colonies, 339–344, 396; and tragic qualities of Seven Years' War, 431–432. *See also* Slavery and slave trade (Indian); *individual tribes*
Indies, 101, 104, 119, 122, 261n. 1, 262, 314, 416, 427, 429. *See also individual islands*
Inside Passage, 61–62, 244n. 13
Inuits, 262
Iowas, 202, 243–244
Iron, 212, 216–217, 219–220, 221n. 23, 242n. 11, 248–249, 253
Iroquoia, 265
Iroquois, 168
Isham, James, 201, 226, 245n. 14

Jaguallapais, 81
Jalcheduns, 82, 86
Jalliquamais, 74, 86
Jamaica, 135, 271, 274, 281n. 23, 295, 401
Jamajabs, 73n. 5, 81, 86
Japan, 56, 76, 107–112, 148, 161–162, 183, 219, 325, 429. *See also individual cities*
Japanese fishermen, 218, 249
Jefferson, Thomas, 4, 101, 369, 386

Jemez pueblo, 70
Jenkins' Ear, War of, 142, 270–283, 287–288, 313, 316, 320, 328, 415, 418–419
Jérémie, Nicolas, 212, 227, 240–241
Jesso. *See* Yedso
Jesuits: in Spanish borderlands, 32–33, 49, 73, 320, 397n. 15; in France, China, and hinterlands of French America, 172, 187–194, 198, 202, 228–230, 232–234, 238
Jesús María, Fray Francisco Casanas, 234n. 2
Johnson, Samuel, 29, 312, 363
Journey from Prince of Wales's Fort in Hudson's Bay to the Northern Ocean in the Years 1769, 1770, 1771, and 1772 (Hearne), 240
Juan, Jorge, 119, 321, 325
Juan de Fuca, Strait of, 32
Juan Fernández, Island of, 106–107, 137, 288n. 8, 289, 293–294, 298, 327, 415
Jumanos (Jumanes), 50, 339–340. *See also* Wichitas
Jumonville's Glen, 1
Juneau, 164

Kalm, Peter, 385
Kamchatka, 112, 182–183, 218, 402. *See also* Great Northern Expedition
Kangxi emperor, 189–190, 228, 232, 233
Kansas, 41, 74–75, 200, 341
Kansas Indians, 76n. 9, 202
Kaunitz-Rittberg, Wenzel Anton von, 430
Keene, Benjamin, 317, 319, 323, 353
Keller, Ignatius, 49
Kelsey, Henry, 236–237, 268
Kentucky, 267
Kerlérec, Louis Billouart de, 252, 336, 395
Kino, Father Francisco Eusebio, 7, 32–33, 48, 62, 65n. 37, 69–70, 73–74, 80–81, 85–88, 94, 108, 265
Kinongé8ilini Indians, 244, 245n. 14
Kiowa-Tanoan language family, 72
Kiowas, 236n. 4
Kirilov, Ivan, 183
Knife River, 247
Knight, James (associated with sugar-planting interest), 274n. 13, 281n. 23
Knight, James (Hudson's Bay Company), 225, 262
Knowles, Commodore Charles, 288
Kootenay Indians, 247, 248n. 17
Korea, 107n. 9, 191
Kuril Islands, 108–109

250, 333, 337–338, 380; as base for northern exploration and expansion, 44, 53, 63, 65, 68, 261n. 1, 397; intelligibility of indigenous geographic information in, 87–88, 172; as distinct from regions to north, 96; Spanish settlements in, 104; as model for other empires, 180; Russian moves toward, 183; and Spanish commercial policy, 336; vulnerability of, at end of Seven Years' War, 392, 406, 415–416; and United States, 412

Mexico City, 4, 29–30, 58, 65, 77–78, 116, 276, 336, 343, 346, 349, 418

Miami River, 265

Michilimackinac, 2–3, 330

Middleton, Captain Christopher, 262, 301–302

Miguel (Indian), 76–78, 84, 95

Mills, Utah, 220

Mines, 92, 147, 154–157, 224, 275, 279–280, 338, 347, 350, 365–366, 382, 389, 391, 410–411, 416, 422, 426. See also Gold; Silver

Minetares, 248

Minnesota River, 197n. 5

Minorca, 354, 355n. 31, 406

Mirepoix, Charles Pierre Gaston François de Lévis, duc de, 303, 305, 371

Mission française, 189–191, 198

Mississippi Delta, 147

Mississippi River, 2, 15, 20, 31, 65, 75, 148–150, 154, 157–159, 164, 167, 199, 224, 229, 232, 239, 243, 252, 264–265, 270, 308, 334, 336–337, 339, 341, 354, 359, 361, 364, 389, 391, 393, 404, 407, 411, 429

Mississippi Valley, 15–16, 147–148, 155, 339, 386, 404

Missouri River: as possible Northwest Passage, 4, 31, 161; and agricultural town-dwelling Indians, 75–76, 212; ambiguous identification of, 75–76, 167, 197n. 5, 205; sources of, and Continental Divide, 97, 150, 162, 203, 225, 252, 253; as route to New Mexico, 155, 160, 216; as key east-west route, 163, 222, 247–248, 248n. 17, 251n. 22; salinity of, 205, 210; and Pawnees, 252

Missouris, 202, 203, 229

Mitchell, John, 266

Moaches. See Utes

Mohave Indians, 82–83

Mohawk River Valley, 265

Moll, Herman, 6–7, 10–11

Moluccas, 103

Moncacht-apé, 222–223, 245–246, 248

Monongahela River, 284

Montaigne, Michel de, 216

Montcalm, General Louis-Joseph de, 354

Monterey, 81, 85

Monterey Bay, 44, 346

Monterrey, Gaspar de Zúñiga Acevedo y Fonseca, conde de, 79n. 11

Montesquieu, Charles-Louis de, 114, 382

Montezuma, 89–90, 92–93, 95

Montezuma I, 59

Monticello, 369

Montreal, 168, 198, 252, 364

Moor, William, 262

Moqui pueblos, 49–50, 81

Moquis, 50–51, 86

Morgan, Henry, 274n. 13

Mornet, Daniel, 379–380

Moscow, 179

Moskvitin, Ivan, 179

Mosquito Coast, 295–296, 316, 318, 328, 347

Mountains, 147–148, 154, 170, 203–205, 369. See also individual ranges

Mountain West, 5, 43

Moyobamba, 67

Moziño, José Mariano, 242n. 11

Müller, Gerhard Friedrich, 108, 183–185

Muret, Pierre, 283

Muscovy, 173, 179

Nadotes, 341n. 14

Nagasaki, 111–112

Nahuatl, 87

Nairne, Thomas, 267

Naples, 405, 427

Napoleon, 173, 386, 411

Narborough, John, 297, 304

Nassonite Indians, 234, 240

Natchitoches, 341n. 14, 394, 398

Nation du Serpent, 150. See also Gens de Serpent; Snakes (ethnic classification)

Natural History (Buffon), 380

Nebraska, 41, 47, 70

Neches River, 234n. 2

Needles, Calif., 81

Nelson River, 167

Nevada, 81

New Britain, 106–107

Newcastle, Thomas Pelham Holmes, duke of, 276, 279–280, 324, 326–327

New England, 55, 186, 308, 356, 386

Newfoundland, 327, 417

New France, 15, 64, 126, 148, 166, 170, 196, 252, 264–265, 338, 356, 406

New Granada, 348

New Guinea, 104, 106

New Hebrides, 104

New Mexico: and course of Seven Years' War, 3; European uncertainty concerning regions beyond, 5, 111, 161–162, 167, 212, 217–218, 227; and French traders and encroachments, 20, 29–30, 47, 155, 198, 217–218, 235, 330–333, 336–337, 339–340, 344–345, 350, 367–368, 390, 394–396, 398, 426; as northern limit of Spanish Empire, 22, 29, 34, 41, 64; French interest in, 22, 155–162, 385; British interest in, 22, 270; Spanish uncertainty concerning regions beyond, 31–34, 70, 74, 79, 345–347, 350; Spanish exploration in and beyond, 45, 49–50, 57, 60, 64, 73n. 5, 77, 87, 94, 264, 332; vulnerability of, 48, 157, 160, 282, 338–345, 349–350, 353, 356; relative poverty and isolation of, 58, 64–65, 157, 332, 338; connections of, to other parts of West, 61, 69, 70, 81, 86, 239, 245, 249; map of, 63; and Indian captives, 69–70, 84; as boundary of Louisiana, 150; French belief in wealth of, 154–159, 212, 367–368, 384–385; perceived British threats to, 163; possible reports of, 216–217, 219–220, 227, 252n. 23; diverse population of, 221; Spanish dispatch of troops to, 334

New Orleans, 198, 252, 265, 360n. 1, 364, 376, 386, 389, 395, 408n. 33, 411

New Spain. *See* Mexico

Newton, Isaac, 178, 379

New World, 34, 40, 53, 63, 168, 180, 275, 281–282, 314, 337, 386–387, 392, 416

New York, 113

New York City, 95

New Zealand, 106

Nez Perces, 247, 251

Nicaragua, 348

Nine Years' War, 134

Niverville, Joseph Boucher de, 204, 244

Niza, Fray Marcos de, 58

Noailles, Adrien Maurice, duc de, 150, 294, 296, 306, 308n. 33, 334, 367–368, 371–372

Nolin, J. B., 6–7, 9

Nootka Sound, 18, 212, 242n. 11

Nootka Sound Indians, 249

North America: eastern, 2, 3, 16; lingering

geographic uncertainty concerning, 4, 6–7, 10–11, 78, 171–172, 194, 262; as site and object of European imperial competition, 5–6, 22, 182, 285, 320, 328, 352–356, 427; and historical importance of different regions, 16–17, 365; relation of, to Asia, 32–33, 111, 171, 217–218; maps of, 36–39, 152–153, 170, 185, 247, 370–371; disappointments of, 53, 62, 66; trading routes and sites of, 61, 96, 244–245, 249–251; fabulous expectations for, 61, 111, 147–163, 217–218, 413; and Northwest Passage, 101, 162, 403; as oceanic boundary, 107; and British colonies, 120, 385; and Russian Empire, 184–187; unreliable sources for geography of, 186–187, 198–199; interconnectedness of, 250; and horse revolution on the plains, 341; and French colonial venture, 359; mental distance of, from Europe, 380; and Charles III's ideas, 387–393; and British gains at end of Seven Years' War, 416–427. *See also* West, North American

North Atlantic, 299, 327. *See also* Atlantic Ocean; South Atlantic

North Carolina, 310. *See also* Carolina

Northeast (North American), 264

North Pacific: British presence in and exploration of, 18, 32, 52, 55, 113, 195, 212, 242n. 11, 248–249, 299, 381, 427; French exploration of, 25, 164–165, 179, 186, 194, 233, 238; Spanish exploration of and notions concerning, 32, 52, 186; European ideas about, 107–110, 413; Russian presence in and exploration of, 108, 179–187, 233, 238, 292, 391, 400–404. *See also* Pacific Ocean

North Platte River, 41

North Pole, 106, 304

Northwest, 187, 246

Northwest Coast Indians, 61–62, 76n. 9, 217–219, 220n. 22, 227, 249, 253

Northwest coast of North America, 6–7, 32, 52, 61–62, 217–218, 248–249. *See also* Pacific coast of North America

North West Company, 219, 241n. 10

Northwest Passage: and Robert Rogers, 1–2, 432; as goal of western exploration, 4, 101, 161–162, 166–170, 261–264, 285–286, 292–311, 372, 377, 400–404; European and Euro-American speculations concerning, 4, 369; and Hudson Bay, 5, 6, 19, 22, 101, 106, 125, 136, 148, 162, 261–263, 285, 292–311, 314, 327, 374, 401–404, 413–414, 427, 432; and importance of the West, 17, 20, 25, 110, 148; and desired or feared access to Spanish possessions, 22, 54, 399–404, 410; Spanish uncertainty concerning, 32, 40, 42, 44, 400; Spanish failure to find, 53–54; and trade with Asia, 111–112; French and British search for, from Pacific, 164–165, 299, 325, 377; French search for, from east, 166–170, 261, 426; unreality of, 168, 170; and causes of Seven Years' War, 285–310. *See also* Passage to India

Norton, Moses, 225

Norton, Richard, 302

Noticia de la California (Burriel), 391, 400

Nova Scotia, 107n. 9

Nuclear America, 63

Nueva Vizcaya, 396

Ochagach. *See* Auchagah

Odyssey, The (Homer), 83

Ohio Company, 267

Ohio River, 1, 3, 265, 307–308, 371, 393

Ohio Valley, 1–3, 19, 284, 285n. 3, 307–309, 328, 332

Oklahoma, 203, 234

Omans, 243

Oñate, Juan de, and Oñate expeditions, 41, 47, 54, 57–61, 71, 76–79, 83–84, 341

Ophir, 104, 111

Opie, John, 290n. 10

Opium, 113

Oregon, 32, 41, 164, 431

Orellana, Francisco de, 67

Orinoco Basin, 65

Orléans, Philippe II, duc de, 142, 149, 149n. 2, 162, 192

Orry, Philibert, 175

Ortelius, Abraham, 105

Osages, 229, 234

Ossun, Pierre-Paul, marquis d', 364, 374, 376, 398, 404, 407

Otos, 47, 229

Otters, sea. *See* Sea otters

Otters (Indian nation), 222

Ottoman Empire, 290n. 10

Ouachipouennes, 213

Ozarks, 203

Pacific Basin, 18, 128, 307

Pacific coast of North America: and persistent European uncertainty, 5; and international rivalry, 15; British exploration of, 18, 32, 52, 104, 264; Spanish exploration of, 22, 41, 43–45, 52, 55–56, 63, 97; Spanish fear of European maritime encroachments on, 54, 320, 325, 401; distant reports of, 61–62, 248n. 17; and European visions of Northwest Indians, 110, 161; French exploration of, 164–165; Spanish fear of Russian encroachments on, 391, 400–404. *See also* Cook, James; Northwest coast of North America; Vancouver, George

Pacific coast of South America, 89, 97, 101, 104, 115

Pacific Northwest, 61–62, 111, 249

Pacific Ocean: and search for Northwest Passage, 2–6, 20, 25, 40, 54, 101, 123, 125, 149, 151, 155, 161–162, 164–165, 167, 186–188, 261–264, 282, 285–311, 416; as goal of Anglo-American land explorers, 4, 241–242, 250, 311; and Sea of the West, 5–6, 8, 31, 149, 161–162, 205, 282, 372–374, 376; imperfect notions of eastern edge of, 6–7, 31–32, 34, 74–75; historical significance of, 17–19; historiographical neglect of, 17–18, 315; European exploration of, 18, 52, 98; Spanish settlements in and around, 18, 25; exclusion of non-Spanish shipping from, 18, 25, 40, 101–104, 135–143, 148, 164–165, 283–329, 373; British encroachments and designs on, 19, 264, 271, 282, 283–310, 314–329, 332, 374, 391, 398–404, 410–411, 413–416, 418–420, 427; and imperial rivalry, 21, 132–143, 285–329, 432; French movement through North America toward, 63, 171, 194, 196, 205, 210–212, 225, 333, 345–347, 372–373, 384, 389, 396–399; as western edge of North America, 72, 81–82, 150, 205, 239, 241; distant reports of, 74–75, 111, 242, 246; appeal of, 101–121, 416, 429; and Spanish silver, 101, 117, 122–143, 416; obscurity of, for Europeans, 102–111, 167, 429; navigational challenges of, 102–103; Russian presence on and exploration of, 108, 179–186, 400–404; piracy in, 123–125, 132; French crossings of, 164, 343; and Anson expedition, 271, 277–278, 280, 287–289, 304, 314, 320–321, 351, 410, 419; and Central America, 272, 274–275, 279, 282, 286; continued French interest in, 374–377. *See also* North Pacific; Spanish Lake

Padoucas. *See* Apaches

Paita, 271, 314

Paiutes, 83, 220, 223, 236n. 4

Pako, 166
Palacio Atard, Vicente, 319
Panama, 101, 104, 115, 132, 272, 279–281, 298, 348–349, 374–375, 416
Panama, Isthmus of, 41, 53, 123, 124, 125, 274, 279–280, 296, 297n. 19, 348
Panama Canal, 162
Panama City, 115, 271–272, 274, 415, 418
Panana Indians, 235, 239
Panni Indians. See Pawnees
Pánuco River, 3
Paraguay, 64
Paraná River, 41, 55
Paris, 5, 124, 167, 194, 229, 288, 304, 314, 326, 356, 389, 393, 431, 432
Paris, Peace of (1763), 5, 103; and French territorial losses, 310–311, 359–411; British gains in, 383–385, 408–427; and Spanish territorial gains, 387–412; and western Indians, 411, 431
Parkman, Francis, 2, 15, 356
Parliament, 383, 416–417, 420
Parrilla, Colonel Diego Ortiz, 396
Passage to India, 101, 167. See also Northwest Passage
Pasto, 69
Patagonia, 52, 288–292, 297, 299
Paullu, 92
Pawnees, 47, 70, 75–76, 202, 235, 243, 252, 252n. 23, 339–341
Pecos, 29, 50, 70, 91, 235, 337
Pecos River, 210
Pecquet, Antoine (the elder), 137–138
Pelham, Henry, 323
Pelones, 75
Pennsylvania, 267, 269
Pensacola, 250, 421–422
Peñalosa, Diego de, 156
Pepys Island, 294
Pérez, Juan, 248
Peru: silver and wealth of, 5, 22, 63–64, 117–119, 124, 274n. 13, 277, 296, 382, 404, 416, 429; and Pacific, 18; Spanish exploration of, 24, 53, 56, 172; French trade with and interest in, 25, 123–143, 148, 162, 164–165, 322, 343, 350, 352; vague British notions of, 29; and access to other parts of South America, 65–67; indigenous geographic information concerning, 87–98, 172; Spanish settlements in, 104; eighteenth-century scientific expeditions in, 118n. 26, 178; allure of, 118–119, 124, 148; British interest in, 132–135, 274n. 13,

275, 277–278, 281, 281n. 23, 301, 304–306, 316, 400, 416, 418; as model for other empires, 180, 186
Peter the Great, 180–184, 189, 228–229
Petit Jour, 166
Philip II (king of Spain), 23–24
Philip V (king of Spain), 122, 130–131, 135
Philippines, 18, 55, 101, 104, 109, 116–117, 271, 277, 281, 325, 345, 350, 376, 419, 427
Pickawillany, 265
Piegan Indians. See Blackfeet
Pigafetta, Antonio, 102
Pignatelli, Francisco, 326–327, 347
Pikaraminioüach nation, 253
Pimas, 48, 50, 65n. 37, 73, 80, 84, 86–88, 346, 406
Pimería Alta, 346
Pindar, Thomas, 134
Pioneers, The (Cooper), 147
Pirates and piracy, 18, 102, 123–125, 132, 272, 274n. 13, 279, 315, 320, 347, 397, 400
Pitt, William (the elder), 324, 354–355, 360, 387, 393, 406, 413–421, 424–425
Pizarro, Francisco, 55–56, 67, 89, 101
Pizarro, Gonzalo, 67
Pizarro de Carvajal, Diego, 67
Plains, 43, 46, 50, 52, 57–58, 65, 72, 73n. 5, 77–78, 85, 221, 248–249, 251, 332, 339, 366; western, 5, 63; southern, 30, 41, 46, 63, 72, 87, 91, 234–235, 238–239, 244; northern, 236n. 4, 239
Plascotez de Chiens, 245
Platte River, 47, 75
Pochteca, 95
Poland, 120, 411
Pompadour, Jeanne-Antoinette Poisson, Madame de, 360
Pond, Peter, 241n. 10, 251n. 22
Pontchartrain, Jérôme Phélypeaux de Maurepas, comte de, 124–126, 128–129, 136, 147, 162
Porcelain, 112–113, 116
Porcupine, 321, 324–325
Port des Français, 164
Port Mahon, 375
Portobello, 115, 271–272, 274, 281, 316, 348–349, 375, 415, 418
Portolá expedition, 71, 79
Portugal, 103, 188, 290n. 10
Posada, Fray Alonso de, 32, 70
Potosí, 115, 118–119, 134, 148, 179, 212, 366, 382, 422

Prior, Matthew, 135, 137–138
Priscus, 342n. 15
Prussia, 430
Ptolemy, 105
Pueblo Indians, 48n. 15, 73n. 5, 220–221
Pueblo Revolt, 48, 49, 65n. 37, 341, 406
Puerto Rico, 383
Pufendorf, Samuel, 119
Puget Sound, 61–62, 73, 244n. 13, 250, 261n. 1
Punta Baja, 53
Puysieulx, Louis-Philogène Brûlart, marquis
    de, 296–297, 300, 303n. 28, 305–306, 326,
    348, 368, 374

Quadruple Alliance, War of the, 139
Quang-tong, 192
Quebec, 309, 354, 359, 363–364, 392, 405
Quechua, 87–88
Quibo, 298
Quicoma, 85
Quiquima Indians, 70, 73, 80
Quiros, Pedro de, 104
Quito, 56, 67, 69, 89, 92, 118n. 26, 132
Quivira, 50, 57–58, 70, 161
Quivira Indians, 47

Raguet, Abbé, 150
Raven, 321, 324–325
Red River (of the South), 97, 160, 234, 238, 342
Reflexiones sobre el comercio español a Indias
    (Campomanes), 401
Regency Council, French, 149, 151, 157
Regis, Father Jean-Baptiste, 191
Repulse Bay (Canada), 299
Revilla Gigedo, Juan Francisco de Güemes
    y Horcasitas, conde de, 34, 338, 344, 346,
    349–350, 394, 397n. 15
Revolutionary War. See American Revolution
Rica de Oro, 109
Rica de Plata, 109
Rio de Janeiro, 290n. 10
Rio Grande, 41, 150, 156, 158, 160, 210
Rio Grande Basin, 81
Rio Grande de São Pedro region, 290n. 10
Rio Grande Valley, 31, 41, 58, 80, 91
Río Salado. See Pecos River
Rivera, Juan María, 83n. 17, 222n. 23
River of the West, 5, 149, 161–162, 166, 253, 261,
    369, 372–374, 380, 382
River Plate, 288n. 8, 290n. 10, 309
Roads, 69, 95–96, 98

Robins, Benjamin, 294n. 14
Robinson, Thomas, 371
Rocky Mountains, 31, 46, 65n. 37, 68, 74, 163,
    167, 171, 201, 203–205, 236n. 4, 238–249, 251,
    268, 369, 406. See also Mountains
Rodero, Gaspar, 32
Rodney, Commodore George, 327
Rogers, Robert, 1–3, 404, 432
Rogers, Woodes, 106, 116, 131, 133n. 15, 137n. 22
Roggeveen, Jacob, 106
Rogue River, 41
Rome, 189
Rouillé, Antoine-Louis, 296–297, 297n. 20,
    300, 303n. 28, 307–309, 326–328, 330, 334–
    336, 348, 350–351, 368, 371–372, 374
Rousseau, Jean-Jacques, 382
Royal Society, 107n. 9
Ruíz, Bartolomé, 56
Russia and the Russian Empire: need for
    distinct treatment of, 21n. 14; French and
    Russian surveying and mapping of, 24–25,
    171–172, 178–185, 189–190, 193, 198–199, 228–
    233, 238, 255–256, 266, 268, 381; and Pacific
    exploration, 107n. 9, 172, 193, 245, 292, 370,
    391, 400; economic relation of, to western
    Europe, 120; rapid expansion of, 178–185,
    189; and contest for North America, 255–
    256, 391; Spanish concerns about Pacific ex-
    pansion of, 400–404
Ryswick, Treaty of, 125n. 5

Sabeata, Juan de, 50
Sacagawea, 251
Sacramento (Portuguese colony), 290n. 10
Saint Augustine, Fla., 264, 421
Saint Claire, Jean-Baptiste Benoît de, 330–331,
    351
Saint-Contest, François-Dominique Barberie
    de, 334
Saint-Denis, Louis Juchereau de (father), 199,
    220n. 2
Saint-Denis, Louis Juchereau de (son),
    341n. 14
Saint-Domingue, 147, 162, 178, 338, 343, 348–
    349, 383–384, 423, 427
Saint John River, 308
Saint Lawrence Island, 182
Saint Lawrence River, 15, 65, 148, 265, 354, 364
Saint Lucia, 308–309
Saint Malo, 122–124, 127, 129–130, 132
Saint Petersburg, 178, 186, 192–194, 238, 402

Virginia, 267, 269, 338
Vizcaíno, Sebastián, and Vizcaíno expedition, 41, 44–45, 55–56, 69, 72
Voltaire, François-Marie Arouet, 292, 379–382
Voyage round the World, A (Anson), 322, 327–328, 333
Voyage to Hudson's-Bay, by the Dobbs Galley and California, in the Years 1746 and 1747, for Discovering a North West Passage (Ellis), 298, 300–301, 414, 422
Voyage to South America (Ulloa and Juan), 123
Vries, Maerten Gerritsz, 108

Waddington, Richard, 283
Wafer, Lionel, 133
Wager, Charles, 274–275, 277, 290–291
Walker, Geoffrey, 131
Wall, Richard, 317–318, 321–327, 348, 353n. 28, 393–394, 394n. 9, 398, 401–403, 407–410, 412
Walpole, Robert, 271, 277, 280, 289–291
Walter, Richard, 293n. 14
Washington, George, 1, 3, 19, 284, 430–431
Weiser, Conrad, 267
Wentworth, General Thomas, 271, 274–275
West, North American: and Northwest Passage, 2, 20, 25, 32, 54, 162, 187, 410–411; European geographic uncertainty concerning, 3–4, 9, 15n. 6, 23–24, 187, 381, 384; and Sea of the West, 6, 196; historical significance of, 16–18, 429; essential diversity of, 19; evidence for, 20–21, 24, 41–42, 197–201; role of, in international affairs, 21, 283–412; British interest and activities in, 22, 264–270, 374, 404, 410; and the Seven Years' War, 23, 283–311, 330–412, 429–433; difficulty of comprehending, 24–25, 41–98, 170–172, 178, 194–247, 381; perceived value of, 25, 65, 156, 196, 301–304, 359–412; Spanish exploration of, 25, 29–98; French ideas about, 25, 147–163, 187, 196, 218, 272, 275, 281–282, 361–386; French exploration of, 25, 166–172, 194–257, 368–369, 415; Spanish geographic ignorance concerning, 29–42, 72, 87, 90, 98–99, 410; relation of, to Asia, 32–33, 171–172, 186–187, 192–193, 429; disappointments of, 53–62, 416; projected features of, 53–64, 148–170, 196–197, 218, 261–262, 429; river systems of, 55, 162, 410; relation of, to other regions,

63, 69, 73; western Indian understanding of, 72–86, 93, 160, 225–257; lack of French Pacific access to, 102, 164–165; Spanish American silver and, 123, 131–132, 143, 154–160, 366–368; French ignorance of, 148, 187, 361, 384; projected mineral wealth of, 154–155; exploration of, from Russia and China, 171–172, 186–187, 192–193; La Harpe map of, 206–207; limited range of Indian journeys in, 239; and Spanish America's relation to British Empire, 264, 426; and feared French encroachment, 345–347, 351–352, 396–398, 403; and American Revolution, 386; and Anglo-American expansion, 386, 410; lack of recognized Indian power in, 411; and tragedy of Seven Years' War, 431–432. *See also* Fur trade
West Africa, 307
West coast. *See* Pacific coast of North America
Western Europe, 20, 112
Western Sea. *See* Sea of the West
West Indies, 134, 274–277, 279, 398, 414, 427
Wheels of Commerce, The (Braudel), 120
White (ethnic or racial classification), 75–76, 213–219, 222n. 23, 235, 239, 243–245, 252n. 23, 253
Whiteness. *See* Aztlan
Wichitas, 339–341, 396. *See also* Jumanos
William III (king of England), 122
Williams River, 33
Winchester, Va., 267
Winnebagos, 219
Winnipeg River, 167
Wisconsin, 251n. 22
Wood, William, 276–277, 280
Wyoming, 199, 236

Yavipais, 86
Yedso, 33, 107n. 9, 108, 111
York (Factory, Fort), 201, 224n. 25, 245, 269n. 9, 270n. 10
Yucatán, 295
Yumas, 73, 73n. 5, 80n. 13, 82, 85–86
Yuta Indians. *See* Utes

Zacatecas, 64
Zárate Salmerón, Gerónimo de, 61n.33
Zuñi pueblo, 60, 62

CPSIA information can be obtained at www.ICGtesting.com
Printed in the USA
LVOW11s0021050815

448813LV00004BA/362/P